INTERDISCIPLINARY MATHEMATICAL SCIENCES*

Published

Vol. 4: Mathematical Theory of Adaptive Control
Vladimir G. Sragovich

Vol. 5: The Hilbert–Huang Transform and Its Applications
eds. Norden E. Huang & Samuel S. P. Shen

Vol. 6: Meshfree Approximation Methods with MATLAB
Gregory E. Fasshauer

Vol. 7: Variational Methods for Strongly Indefinite Problems
Yanheng Ding

Vol. 8: Recent Development in Stochastic Dynamics and Stochastic Analysis
eds. Jinqiao Duan, Shunlong Luo & Caishi Wang

Vol. 9: Perspectives in Mathematical Sciences
eds. Yisong Yang, Xinchu Fu & Jinqiao Duan

Vol. 10: Ordinal and Relational Clustering (with CD-ROM)
Melvin F. Janowitz

Vol. 11: Advances in Interdisciplinary Applied Discrete Mathematics
eds. Hemanshu Kaul & Henry Martyn Mulder

Vol. 12: New Trends in Stochastic Analysis and Related Topics: A Volume in Honour of Professor K D Elworthy
eds. Huaizhong Zhao & Aubrey Truman

Vol. 13: Stochastic Analysis and Applications to Finance: Essays in Honour of Jia-an Yan
eds. Tusheng Zhang & Xunyu Zhou

Vol. 14 Recent Developments in Computational Finance: Foundations, Algorithms and Applications
eds. Thomas Gerstner & Peter Kloeden

*For the complete list of titles in this series, please go to
http://www.worldscientific.com/series/ims

Recent Developments in Computational Finance

Foundations, Algorithms and Applications

Interdisciplinary Mathematical Sciences – Vol. 14

Recent Developments in Computational Finance

Foundations, Algorithms and Applications

Editors

Thomas Gerstner
Goethe University Frankfurt am Main, Germany

Peter Kloeden
Goethe University Frankfurt am Main, Germany

World Scientific

NEW JERSEY • LONDON • SINGAPORE • BEIJING • SHANGHAI • HONG KONG • TAIPEI • CHENNAI

Published by

World Scientific Publishing Co. Pte. Ltd.
5 Toh Tuck Link, Singapore 596224
USA office: 27 Warren Street, Suite 401-402, Hackensack, NJ 07601
UK office: 57 Shelton Street, Covent Garden, London WC2H 9HE

British Library Cataloguing-in-Publication Data
A catalogue record for this book is available from the British Library.

Interdisciplinary Mathematical Sciences — Vol. 14
RECENT DEVELOPMENTS IN COMPUTATIONAL FINANCE
Foundations, Algorithms and Applications

ISBN 978-981-4436-42-7

Printed in Singapore by B & Jo Enterprise Pte Ltd

Preface

Mathematical finance has revolutionized the financial world in the past forty years. A major reason for this success has been the parallel development of efficient computational methods as well as more sophisticated mathematical models. These novel computational tools are the foundation of a new field of research called Computational Finance, whose main task is to calculate as accurately and efficiently as possible the risks that financial instruments generate. This requires an interdisciplinary approach involving a variety of methods from financial mathematics, stochastics, statistics, numerics and scientific computing.

Major impacts on the development of the field were the publication of the monograph of Kloeden and Platen on stochastic numerics in 1992 and, later, the monograph of Glasserman on Monte Carlo methods in 2004. These books, as well as many others, provide the foundations of this rapidly developing subject.

The new computational tools have led to even more sophisticated mathematical models, which in turn require computational methods that work under requirements not handled in the existing textbooks. For example, the theory of stochastic numerics has now been extended to handle non-standard assumptions on the coefficients of the stochastic differential equations. Another significant new development is the multi-level Monte Carlo method of Michael Giles, while others include the use of inverse problem methods, wavelets and backward stochastic differential equations.

This volume consists of a series of cutting-edge surveys of recent developments in the field of computational finance written by leading international experts. Several of the contributions in this volume are based on talks presented at the *International Workshop on Numerical Algorithms in Computational Finance* that was held from July 20-22 in 2011 at the House of Finance of the Goethe University in Frankfurt am Main. These surveys make the subject accessible to a wide readership in academia and the financial world. They may also be of interest to practitioners in many areas in engineering, technology and science beyond finance. Besides reviews of existing results many new, previously unpublished, results are also presented.

The book consists of 13 chapters divided into the three parts: Foundations, Algorithms and Applications.

The first part **Foundations** is devoted to survey and review articles. It begins with the article *Multilevel Monte Carlo methods for applications in finance* by Mike Giles and Lukasz Szpruch which presents a survey of recent progress regarding the multilevel Monte Carlo method. The chapter *Convergence of numerical methods for SDEs in finance* by Peter Kloeden and Andreas Neuenkirch deals with nonstandard assumptions on the coefficients of an SDEs and the effect on the convergence of numerical discretization schemes. In *Inverse problems in finance,* Johann Baumeister gives an overview on inverse problems in finance, which are in general ill-posed and thus require special regularization. The article *Asymptotic and non asymptotic approximations for option valuation* by Roman Bompis and Emmanuel Gobet reviews approximation methods for the derivation of closed-form solutions for option pricing problems.

The second part **Algorithms** covers the algorithmic and numerical aspects of the computational methods. The article *Discretization of backward stochastic Volterra integral equations* by Christian Bender and Stanislav Pokalyuk deals with the approximation of backward SDEs, while the next chapter *Semi-Lagrangian schemes for parabolic equations* by Kristian Debrabant and Espen Robstad Jakobsen covers numerical schemes for nonlinear second order parabolic PDEs. In *Derivative-free weak approximation methods for stochastic differential equations* Kristian Debraband and Andreas Rößler consider stochastic Runge-Kutta methods for the weak approximation of SDEs and in the chapter *Wavelet solution of degenerate Kolmogoroff forward equations* Oleg Reichmann and Christoph Schwab review wavelet Galerkin discretizations for Kolmogoroff forward pricing equations. Finally, in *Randomized multilevel quasi-Monte Carlo path simulation,* Thomas Gerstner and Marco Noll combine the multilevel Monte Carlo method with quasi-random number generation.

The third part **Applications** then deals with specific financial problems. The article *Drift-free simulation methods for pricing cross-market derivatives with LMM* by José Luis Fernández Pérez, María Rodríguez Nogueiras, Marta Pou Bueno and Carlos Vázquez considers a simulation approach for LIBOR rates, which avoids drift-dependent paths. In *Application of simplest random walk algorithms for pricing barrier options* by Maria Krivko and Michael V. Tretyakov proposes a special discretization for barrier options close to the barrier. In the article *Coupling local currency Libor models to FX Libor models*, John Schoenmakers focuses on the coupling of single currency LIBOR models into a joint LIBOR model. The final chaper *Dimension-wise decompositions and their efficient parallelization* by Philipp Schröder, Peter Mlynczak and Gabriel Wittum tackles the pricing of high-dimensional basket options on parallel computers.

We would like to thank the referees for their valuable comments on the submissions. We also thank Marco Noll for his tireless work to bring the manuscript into publishable format. Finally, and most of all, we would like to thank the authors for their informative contributions.

Thomas Gerstner and Peter Kloeden — Frankfurt am Main, 2012

Contents

PART 1
Foundations

Chapter 1

Multilevel Monte Carlo methods for applications in finance

Mike Giles and Lukasz Szpruch
Oxford-Man Institute of Quantitative Finance and Mathematical Institute, University of Oxford

Abstract Since Giles introduced the multilevel Monte Carlo path simulation method [18], there has been rapid development of the technique for a variety of applications in computational finance. This paper surveys the progress so far, highlights the key features in achieving a high rate of multilevel variance convergence, and suggests directions for future research.

1. Introduction

In 2001, Heinrich [28], developed a multilevel Monte Carlo method for parametric integration, in which one is interested in estimating the value of $\mathbb{E}[f(x,\lambda)]$ where x is a finite-dimensional random variable and λ is a parameter. In the simplest case in which λ is a real variable in the range $[0,1]$, having estimated the value of $\mathbb{E}[f(x,0)]$ and $\mathbb{E}[f(x,1)]$, one can use $\frac{1}{2}(f(x,0)+f(x,1))$ as a control variate when estimating the value of $\mathbb{E}[f(x,\frac{1}{2})]$, since the variance of $f(x,\frac{1}{2})-\frac{1}{2}(f(x,0)+f(x,1))$ will usually be less than the variance of $f(x,\frac{1}{2})$. This approach can then be applied recursively for other intermediate values of λ, yielding large savings if $f(x,\lambda)$ is sufficiently smooth with respect to λ.

Giles' multilevel Monte Carlo path simulation [18] is both similar and different. There is no parametric integration, and the random variable is infinite-dimensional, corresponding to a Brownian path in the original paper. However, the control variate viewpoint is very similar. A coarse path simulations is used as a control variate for a more refined fine path simulation, but since the exact expectation for the coarse path is not known, this is in turn estimated recursively using even coarser path simulation as control variates. The coarsest path in the multilevel hierarchy may have only one timestep for the entire interval of interest.

A similar two-level strategy was developed slightly earlier by Kebaier [31], and a similar multi-level approach was under development at the same time by Speight [42; 43].

In this review article, we start by introducing the central ideas in multilevel

Monte Carlo simulation, and the key theorem from [18] which gives the greatly improved computational cost if a number of conditions are satisfied. The challenge then is to construct numerical methods which satisfy these conditions, and we consider this for a range of computational finance applications.

2. Multilevel Monte Carlo

2.1. *Monte Carlo*

Monte Carlo simulation has become an essential tool in the pricing of derivatives security and in risk management. In the abstract setting, our goal is to numerically approximate the expected value $\mathbb{E}[Y]$, where $Y = P(X)$ is a functional of a random variable X. In most financial applications we are not able to sample X directly and hence, in order to perform Monte Carlo simulations we approximate X with $X_{\Delta t}$ such that $\mathbb{E}[P(X_{\Delta t})] \to \mathbb{E}[P(X)]$, when $\Delta t \to 0$. Using $X_{\Delta t}$ to compute N approximation samples produces the standard Monte Carlo estimate

$$\hat{Y} = \frac{1}{N}\sum_{i=1}^{N} P(X^i_{\Delta t}),$$

where $X^i_{\Delta t}$ is the numerical approximation to X on the ith sample path and N is the number of independent simulations of X. By standard Monte Carlo results $\hat{Y} \to \mathbb{E}[Y]$, when $\Delta t \to 0$ and $N \to \infty$. In practice we perform Monte Carlo simulation with given $\Delta t > 0$ and finite N producing an error to the approximation of $\mathbb{E}[Y]$. Here we are interested in the mean square error that is

$$MSE \equiv \mathbb{E}\left[(\hat{Y} - \mathbb{E}[Y])^2\right]$$

Our goal in the design of the Monte Carlo algorithm is to estimate Y with accuracy root-mean-square error ε ($MSE \leq \varepsilon^2$), as efficiently as possible. That is to minimize the computational complexity required to achieve the desired mean square error. For standard Monte Carlo simulations the mean square error can be expressed as

$$\begin{aligned}\mathbb{E}\left[(\hat{Y} - \mathbb{E}[Y])^2\right] =& \mathbb{E}\left[(\hat{Y} - \mathbb{E}[\hat{Y}] + \mathbb{E}[\hat{Y}] - \mathbb{E}[Y])^2\right] \\ =& \underbrace{\mathbb{E}\left[(\hat{Y} - \mathbb{E}[\hat{Y}])^2\right]}_{\text{Monte Carlo variance}} + \underbrace{\left(\mathbb{E}[\hat{Y}] - \mathbb{E}[Y])^2\right)}_{\text{bias of the approximation}} .\end{aligned}$$

The Monte Carlo variance is proportional to $\frac{1}{N}$

$$\mathbb{V}[\hat{Y}] = \frac{1}{N^2}\mathbb{V}\left[\sum_{i=1}^{N} P(X^i_{\Delta t})\right] = \frac{1}{N}\mathbb{V}[P(X_{\Delta t})].$$

For both Euler-Maruyama and Milstein approximation $|\mathbb{E}[\hat{Y}] - \mathbb{E}[Y]| = \mathcal{O}(\Delta t)$, typically. Hence the mean square error for standard Monte Carlo is given by

$$\mathbb{E}\left[(\hat{Y} - \mathbb{E}[Y])^2\right] = \mathcal{O}(\frac{1}{N}) + \mathcal{O}(\Delta t^2).$$

To ensure the root-mean-square error is proportional to ε, we must have $MSE = \mathcal{O}(\varepsilon^2)$ and therefore $1/N = \mathcal{O}(\varepsilon^2)$ and $\Delta t^2 = \mathcal{O}(\varepsilon^2)$, which means $N = \mathcal{O}(\varepsilon^{-2})$ and $\Delta t = \mathcal{O}(\varepsilon)$. The computational cost of standard Monte Carlo is proportional to the number of paths N multiplied by the cost of generating a path, that is the number of timesteps in each sample path. Therefore, the cost is $C = \mathcal{O}(\varepsilon^{-3})$. In the next section we will show that using MLMC we can reduce the complexity of achieving root mean square error ε to $\mathcal{O}(\varepsilon^{-2})$.

2.2. *Multilevel Monte Carlo Theorem*

In its most general form, multilevel Monte Carlo (MLMC) simulation uses a number of levels of resolution, $\ell = 0, 1, \ldots, L$, with $\ell = 0$ being the coarsest, and $\ell = L$ being the finest. In the context of a SDE simulation, level 0 may have just one timestep for the whole time interval $[0, T]$, whereas level L might have 2^L uniform timesteps $\Delta t_L = 2^{-L} T$.

If P denotes the payoff (or other output functional of interest), and P_ℓ denotes its approximation on level l, then the expected value $\mathbb{E}[P_L]$ on the finest level is equal to the expected value $\mathbb{E}[P_0]$ on the coarsest level plus a sum of corrections which give the difference in expectation between simulations on successive levels,

$$\mathbb{E}[P_L] = \mathbb{E}[P_0] + \sum_{\ell=1}^{L} \mathbb{E}[P_\ell - P_{\ell-1}]. \tag{1}$$

The idea behind MLMC is to independently estimate each of the expectations on the right-hand side of (1) in a way which minimises the overall variance for a given computational cost. Let Y_0 be an estimator for $\mathbb{E}[P_0]$ using N_0 samples, and let Y_ℓ, $\ell > 0$, be an estimator for $\mathbb{E}[P_\ell - P_{\ell-1}]$ using N_ℓ samples. The simplest estimator is a mean of N_ℓ independent samples, which for $\ell > 0$ is

$$Y_\ell = N_\ell^{-1} \sum_{i=1}^{N_\ell} (P_\ell^i - P_{\ell-1}^i). \tag{2}$$

The key point here is that $P_\ell^i - P_{\ell-1}^i$ should come from two discrete approximations for the same underlying stochastic sample (see [39]), so that on finer levels of resolution the difference is small (due to strong convergence) and so its variance is also small. Hence very few samples will be required on finer levels to accurately estimate the expected value.

The combined MLMC estimator $\hat{Y}$ is

$$\hat{Y} = \sum_{\ell=0}^{L} Y_\ell.$$

We can observe that

$$\mathbb{E}[Y_\ell] = N_\ell^{-1} \sum_{i=1}^{N_\ell} \mathbb{E}[P_\ell^i - P_{\ell-1}^i] = \mathbb{E}[P_\ell^i - P_{\ell-1}^i],$$

and

$$\mathbb{E}[\hat{Y}] = \sum_{\ell=0}^{L} \mathbb{E}[Y_\ell] = \mathbb{E}[P_0] + \sum_{\ell=1}^{L} \mathbb{E}[P_\ell - P_{\ell-1}] = \mathbb{E}[P_L].$$

Although we are using different levels with different discretisation errors to estimate $\mathbb{E}[P]$, the final accuracy depends on the accuracy of the finest level L.

Here we recall the Theorem from [18] (which is a slight generalisation of the original theorem in [18]) which gives the complexity of MLMC estimation.

Theorem 1. *Let P denote a functional of the solution of a stochastic differential equation, and let P_ℓ denote the corresponding level ℓ numerical approximation. If there exist independent estimators Y_ℓ based on N_ℓ Monte Carlo samples, and positive constants $\alpha, \beta, \gamma, c_1, c_2, c_3$ such that $\alpha \geq \frac{1}{2} \min(\beta, \gamma)$ and*

i) $|\mathbb{E}[P_\ell - P]| \leq c_1 2^{-\alpha \ell}$

ii) $\mathbb{E}[Y_\ell] = \begin{cases} \mathbb{E}[P_0], & \ell = 0 \\ \mathbb{E}[P_\ell - P_{\ell-1}], & \ell > 0 \end{cases}$

iii) $\mathbb{V}[Y_\ell] \leq c_2 N_\ell^{-1} 2^{-\beta \ell}$

iv) $C_\ell \leq c_3 N_\ell 2^{\gamma \ell}$, *where C_ℓ is the computational complexity of Y_ℓ*

then there exists a positive constant c_4 such that for any $\epsilon < e^{-1}$ there are values L and N_ℓ for which the multilevel estimator

$$Y = \sum_{\ell=0}^{L} Y_\ell,$$

has a mean-square-error with bound

$$MSE \equiv \mathbb{E}\left[(Y - \mathbb{E}[P])^2\right] < \epsilon^2$$

with a computational complexity C with bound

$$C \leq \begin{cases} c_4 \epsilon^{-2}, & \beta > \gamma, \\ c_4 \epsilon^{-2} (\log \epsilon)^2, & \beta = \gamma, \\ c_4 \epsilon^{-2-(\gamma-\beta)/\alpha}, & 0 < \beta < \gamma. \end{cases}$$

2.3. *Improved MLMC*

In the previous section we showed that the key step in MLMC analysis is the estimation of variance $\mathbb{V}[P_\ell^i - P_{\ell-1}^i]$. As it will become more clear in the next section, this is related to the strong convergence results on approximations of SDEs, which differentiates MLMC from standard MC, where we only require a weak error bound for approximations of SDEs.

We will demonstrate that in fact the classical strong convergence may not be necessary for a good MLMC variance. In (2) we have used the same estimator for the payoff P_ℓ on every level ℓ, and therefore (1) is a trivial identity due to the telescoping summation. However, in [17] Giles demonstrated that it can be better to use different estimators for the finer and coarser of the two levels being considered, P_ℓ^f when level ℓ is the finer level, and P_ℓ^c when level ℓ is the coarser level. In this case, we require that

$$\mathbb{E}[P_\ell^f] = \mathbb{E}[P_\ell^c] \quad \text{for } \ell = 1, \ldots, L, \tag{3}$$

so that

$$E[P_L^f] = \mathbb{E}[P_0^f] + \sum_{\ell=1}^{L} \mathbb{E}[P_\ell^f - P_{\ell-1}^c].$$

The MLMC Theorem is still applicable to this modified estimator. The advantage is that it gives the flexibility to construct approximations for which $P_\ell^f - P_{\ell-1}^c$ is much smaller than the original $P_\ell - P_{\ell-1}$, giving a larger value for β, the rate of variance convergence in condition *iii)* in the theorem. In the next sections we demonstrate how suitable choices of P_ℓ^f and P_ℓ^c can dramatically increase the convergence of the variance of the MLMC estimator.

The good choice of estimators, as we shall see, often follows from analysis of the problem under consideration from the distributional point of view. We will demonstrate that methods that had been used previously to improve the weak order of convergence can also improve the order of convergence of the MLMC variance.

2.4. *SDEs*

First, we consider a general class of d-dimensional SDEs driven by Brownian motion. These are the primary object of studies in mathematical finance. In subsequent sections we demonstrate extensions of MLMC beyond the Brownian setting.

Let $(\Omega, \mathcal{F}, \{\mathcal{F}_t\}_{t\geq 0}, \mathbb{P})$ be a complete probability space with a filtration $\{\mathcal{F}_t\}_{t\geq 0}$ satisfying the usual conditions, and let $w(t)$ be a m-dimensional Brownian motion defined on the probability space. We consider the numerical approximation of SDEs of the form

$$\mathrm{d}x(t) = f(x(t))\,\mathrm{d}t + g(x(t))\,\mathrm{d}w(t), \tag{4}$$

where $x(t) \in \mathbb{R}^d$ for each $t \geq 0$, $f \in C^2(\mathbb{R}^d, \mathbb{R}^d)$, $g \in C^2(\mathbb{R}^d, \mathbb{R}^{d\times m})$, and for simplicity we assume a fixed initial value $x_0 \in \mathbb{R}^d$. The most prominent example of SDEs in finance is a geometric Brownian motion

$$\mathrm{d}x(t) = \alpha x(t)\,\mathrm{d}t + \beta x(t)\,\mathrm{d}w(t),$$

where $\alpha, \beta > 0$. Although, we can solve this equation explicitly it is still worthwhile to approximate its solution numerically in order to judge the performance of the numerical procedure we wish to apply to more complex problems. Another interesting example is the famous Heston stochastic volatility model

$$\begin{cases} \mathrm{d}s(t) = rs(t)\,\mathrm{d}t + s(t)\sqrt{v(t)}\,\mathrm{d}w_1(t) \\ \mathrm{d}v(t) = \kappa(\theta - v(t))\,\mathrm{d}t + \sigma\sqrt{v(t)}\,\mathrm{d}w_2(t) \\ \mathrm{d}w_1\,\mathrm{d}w_2 = \rho\,\mathrm{d}\,t, \end{cases} \tag{5}$$

where $r, \kappa, \theta, \sigma > 0$. In this case we do not know the explicit form of the solution and therefore numerical integration is essential in order to price certain financial derivatives using the Monte Carlo method. At this point we would like to point out that the Heston model (5) does not satisfy standard conditions required for numerical approximations to converge. Nevertheless, in this paper we always assume that coefficients of SDEs (4) are sufficiently smooth. We refer to [32; 35; 44] for an overview of the methods that can be applied when the global Lipschitz condition does not hold. We also refer the reader to [33] for an application of MLMC to the SDEs with additive fractional noise.

2.5. *Euler and Milstein discretizations*

The simplest approximation of SDEs (4) is an Euler-Maruyama (EM) scheme. Given any step size Δt_ℓ, we define the partition $\mathcal{P}_{\Delta t_\ell} := \{n\Delta t_\ell : n = 0, 1, 2, ..., 2^\ell\}$ of the time interval $[0, T]$, $2^\ell \Delta t = T > 0$. The EM approximation $X_n^\ell \approx x(n\,\Delta t_\ell)$ has the form [34]

$$X_{n+1}^\ell = X_n^\ell + f(X_n^\ell)\,\Delta t_\ell + g(X_n^\ell)\,\Delta w_{n+1}^\ell, \tag{6}$$

where $\Delta w_{n+1}^\ell = w((n+1)\Delta t_\ell) - w(n\Delta t_\ell)$ and $X_0 = x_0$. Equation (6) is written in a vector form and its ith component reads as

$$X_{i,n+1}^\ell = X_{i,n}^\ell + f_i(X_n^\ell)\,\Delta t_\ell + \sum_{j=1}^{m} g_{ij}(X_n^\ell)\,\Delta w_{j,n+1}^\ell.$$

In the classical Monte Carlo setting we are mainly interested in the weak approximation of SDEs (4). Given a smooth payoff $P : \mathbb{R}^d \to \mathbb{R}$ we say that $X_{2^\ell}^\ell$ converges to $x(T)$ in a weak sense with order α if

$$|\mathbb{E}[P(x(T))] - \mathbb{E}[P(X_T^\ell)]| = \mathcal{O}(\Delta t_\ell^\alpha).$$

Rate α is required in condition (i) of Theorem 1. However, for MLMC condition (iii) of Theorem 1 is crucial.We have

$$\mathbb{V}_\ell \equiv \mathrm{Var}\,(P_\ell - P_{\ell-1}) \le \mathbb{E}\left[(P_\ell - P_{\ell-1})^2\right],$$

and

$$\mathbb{E}\left[(P_\ell - P_{\ell-1})^2\right] \le 2\,\mathbb{E}\left[(P_\ell - P)^2\right] + 2\,\mathbb{E}\left[(P - P_{\ell-1})^2\right].$$

For Lipschitz continuous payoffs, $(P(x) - P(y))^2 \le L\,\|x - y\|^2$, we then have

$$\mathbb{E}\left[(P_\ell - P)^2\right] \le L\,\mathbb{E}\left[\left\|x(T) - X_T^\ell\right\|^2\right].$$

It is clear now, that in order to estimate the variance of the MLMC we need to examine strong convergence property. The classical strong convergence on the finite time interval $[0,T]$ is defined as

$$\left(\mathbb{E}\left[\|x(T)-X_T^\ell\|^p\right]\right)^{1/p}=\mathcal{O}(\Delta t_\ell^\xi),\quad \text{for}\quad p\geq 2.$$

For the EM scheme $\xi=0.5$. In order to deal with path dependent options we often require measure the error in the supremum norm:

$$\left(\mathbb{E}\big[\sup_{0\leq n\leq 2^\ell}\|x(n\Delta t_\ell)-X_n^\ell\|^p\big]\right)^{1/p}=\mathcal{O}(\Delta t^\xi)\quad\text{for}\quad p\geq 2.$$

Even in the case of globally Lipschitz continuous payoff P, the EM does not achieve $\beta=2\xi>1$ which is optimal in Theorem (1). In order to improve the convergence of the MLMC variance the Milstein approximation $X_n\approx x(n\,\Delta t_\ell)$ is considered, with i^{th} component of the form [34]

$$\begin{aligned}X_{i,n+1}^\ell=&X_{i,n}^\ell+f_i(X_n^\ell)\,\Delta t_\ell+\sum_{j=1}^m g_{ij}(X_n^\ell)\,\Delta w_{j,n+1}^\ell\\&+\sum_{j,k=1}^m h_{ijk}(X_n^\ell)\left(\Delta w_{j,n}^\ell\Delta w_{k,n}^\ell-\Omega_{jk}\,\Delta t_\ell-A_{jk,n}^\ell\right)\end{aligned}\tag{7}$$

where Ω is the correlation matrix for the driving Brownian paths, and $A_{jk,n}^\ell$ is the Lévy area defined as

$$A_{jk,n}^\ell=\int_{n\Delta t_\ell}^{(n+1)\Delta t_\ell}\left((w_j(t)-w_j(n\Delta t_\ell))\,dw_k(t)-(w_k(t)-w_k(n\Delta t_\ell))\,dw_j(t)\right).$$

The rate of strong convergence ξ for the Milstein scheme is double the value we have for the EM scheme and therefore the MLMC variance for Lipschitz payoffs converges twice as fast. However, this gain does not come without a price. There is no efficient method to simulate Lévy areas, apart from dimension 2 [14; 41; 45]. In some applications, the diffusion coefficient $g(x)$ satisfies a commutativity property which gives

$$h_{ijk}(x)=h_{ikj}(x)\quad\text{for all}\quad i,j,k.$$

In that case, because the Lévy areas are anti-symmetric (i.e. $A_{jk,n}^l=-A_{kj,n}^l$), it follows that $h_{ijk}(X_n^\ell)\,A_{jk,n}^l+h_{ikj}(X_n^\ell)\,A_{kj,n}^l=0$ and therefore the terms involving the Lévy areas cancel and so it is not necessary to simulate them. However, this only happens in special cases. Clark & Cameron [9] proved for a particular SDE that it is impossible to achieve a better order of strong convergence than the Euler-Maruyama discretisation when using just the discrete increments of the underlying Brownian motion. The analysis was extended by Müller-Gronbach [38] to general SDEs. As a consequence if we use the standard MLMC method with the Milstein scheme without simulating the Lévy areas the complexity will remain the same as for Euler-Maruyama. Nevertheless, Giles and Szpruch showed in [22] that by constructing a suitable antithetic estimator one can neglect the Lévy areas and still obtain a multilevel correction estimator with a variance which decays at the same rate as the scalar Milstein estimator.

2.6. *MLMC algorithm*

Here we explain how to implement the Monte Carlo algorithm. Let us recall that the MLMC estimator Y is given by

$$\hat{Y} = \sum_{\ell=0}^{L} Y_\ell.$$

We aim to minimize the computational cost necessary to achieve desirable accuracy ε. As for standard Monte Carlo we have

$$\mathbb{E}\left[(Y - \mathbb{E}[P(X)])^2\right] = \underbrace{\mathbb{E}\left[(Y - \mathbb{E}[\hat{Y}])^2\right]}_{\text{Monte Carlo variance}} + \underbrace{\left(\mathbb{E}[P_L] - \mathbb{E}[P(X)])^2\right)}_{\text{bias of the approximation}}.$$

The variance is given by

$$\mathbb{V}[Y] = \sum_{\ell=0}^{L} \mathbb{V}[Y_\ell] = \sum_{\ell=0}^{L} \frac{1}{N_\ell} V_\ell,$$

where $V_\ell = \mathbb{V}[P_\ell - P_{\ell-1}]$. To minimize the variance of Y for fixed computational cost $C = \sum_{\ell=0}^{L} N_\ell \Delta t_\ell^{-1}$, we can treat N_ℓ as continuous variable and use the Lagrange function to find the minimum of

$$L = \sum_{\ell=0}^{L} \frac{1}{N_\ell} V_\ell + \lambda \left(\sum_{\ell=0}^{L} N_\ell \Delta t_\ell^{-1} - C \right).$$

First order conditions shows that $N_\ell = \lambda^{-\frac{1}{2}} \sqrt{V_\ell \Delta t_\ell}$, therefore

$$\mathbb{V}[Y] = \sum_{\ell=0}^{L} \frac{V_\ell}{N_\ell} = \sum_{\ell=0}^{L} \frac{\sqrt{\lambda}}{\sqrt{V_\ell \Delta t_\ell}} V_\ell.$$

Since we want $\mathbb{V}[Y] \leq \frac{\varepsilon^2}{2}$ we can show that

$$\lambda^{-\frac{1}{2}} \geq 2\varepsilon^{-2} \sum_{\ell=0}^{L} \sqrt{V_\ell / \Delta t_\ell},$$

thus the optimal number of samples for level ℓ is

$$N_\ell = \left\lceil 2\varepsilon^{-2} \sqrt{V_\ell \Delta t_\ell} \sum_{\ell=0}^{L} \sqrt{V_\ell / \Delta t_\ell} \right\rceil. \tag{8}$$

Assuming $\mathcal{O}(\Delta t_\ell)$ weak convergence, the bias of the overall method is equal $c\Delta t_L = cT\,2^{-L}$. If we want the bias to be proportional to $\frac{\varepsilon}{\sqrt{2}}$ we set

$$L_{\max} = \frac{\log\left(\varepsilon/(cT\sqrt{2})\right)^{-1}}{\log 2}.$$

From here we can calculate the overall complexity. We can now outline the algorithm

(1) Begin with L=0;
(2) Calculate the initial estimate of V_L using 100 samples.
(3) Determine optimal N_ℓ using (8).
(4) Generate additional samples as needed for new N_ℓ.
(5) if $L < L_{\max}$ set $L := L + 1$ and go to 2.

Most numerical tests suggests that $L_{\max}$ is not optimal and we can substantially improve MLMC by determining optimal L by looking at bias. For more details see [18].

3. Pricing with MLMC

A key application of MLMC is to compute the expected payoff of financial options. We have demonstrated that for globally Lipschitz European payoffs, convergence of the MLMC variance is determined by the strong rate of convergence of the corresponding numerical scheme. However, in many financial applications payoffs are not smooth or are path-dependent. The aim of this section is to overview results on mean square convergence rates for Euler–Maruyama and Milstein approximations with more complex payoffs. In the case of EM, the majority of payoffs encountered in practice have been analyzed in Giles et al. [20]. Extension of this analysis to the Milstein scheme is far from obvious. This is due to the fact that Milstein scheme gives an improved rate of convergence on the grid points, but this is insufficient for path dependent options. In many applications the behavior of the numerical approximation between grid points is crucial. The analysis of Milstein scheme for complex payoffs was carried out in [11]. To understand this problem better, we recall a few facts from the theory of strong convergence of numerical approximations. We can define a piecewise linear interpolation of a numerical approximation within the time interval $[n\Delta t_\ell, (n+1)\Delta t_\ell)$ as

$$X^\ell(t) = X_n^\ell + \lambda_\ell (X_{n+1}^\ell - X_n^\ell), \quad \text{for} \quad t \in [n\Delta t_\ell, (n+1)\Delta t_\ell) \tag{9}$$

where $\lambda_\ell \equiv (t - n\Delta t_\ell)/\Delta t_\ell$. Müller-Gronbach [37] has show that for the Milstein scheme (9) we have

$$\mathbb{E}\big[\sup_{0\le t\le T} \left\|x(t) - X^\ell(t)\right\|^p\big] = \mathcal{O}(|\ \Delta t_\ell \log(\Delta t_\ell)\ |^{p/2}), \quad p \ge 2, \tag{10}$$

that is the same as for the EM scheme. In order to maintain the strong order of convergence we use Brownian Bridge interpolation rather than basic piecewise linear interpolation:

$$\tilde{X}^\ell(t) = X_n^\ell + \lambda_\ell\,(X_{n+1}^\ell - X_n^\ell) + g(X_n^\ell)\,\left(w(t) - w(n\Delta t_\ell) - \lambda\,\Delta w_{n+1}^l\right), \tag{11}$$

for $t \in [n\Delta t_\ell, (n+1)\Delta t_\ell)$. For the Milstein scheme interpolated with Brownian bridges we have [37]

$$\mathbb{E}\big[\sup_{0\le t\le T}\left\|x(t) - \tilde{X}^\ell(t)\right\|^p\big] = \mathcal{O}(|\,\Delta t_\ell \log(\Delta t_\ell)\,|^p).$$

Clearly $\tilde{X}^\ell(t)$ is not implementable, since in order to construct it, the knowledge of the whole trajectory $(w(t))_{0\le t\le T}$ is required. However, we will demonstrate that combining $\tilde{X}^\ell(t)$ with conditional Monte Carlo techniques can dramatically improve the convergence of the variance of the MLMC estimator. This is due to the fact that for suitable MLMC estimators only distributional knowledge of certain functionals of $(w(t))_{0\le t\le T}$ will be required.

3.1. *Euler-Maruyama scheme*

In this section we demonstrate how to approximate the most common payoffs using the EM scheme (6).

The Asian option we consider has the payoff

$$P = \left(T^{-1}\int_0^T x(t)\,dt - K\right)^+.$$

Using the piecewise linear interpolation (9) one can obtain the following approximation

$$P_l \equiv T^{-1}\int_0^T X^\ell(t)\,dt = T^{-1}\sum_{n=0}^{2^\ell-1} \tfrac{1}{2}\,\Delta t_\ell\,(X_n^\ell + X_{n+1}^\ell),$$

Lookback options have payoffs of the form

$$P = x(T) - \inf_{0\le t\le T} x(t).$$

A numerical approximation to this payoff is

$$P_\ell \equiv X_T^\ell - \inf_{0\le n\le 2^\ell} X_n^\ell.$$

For both of these payoffs it can be proved that $V_\ell = \mathcal{O}(\Delta t_\ell)$ [20].

We now consider a digital option, which pays one unit if the asset at the final time exceeds the fixed strike price K, and pays zero otherwise. Thus, the discontinuous payoff function has the form

$$P = \mathbf{1}_{\{x(T)>K\}},$$

with the corresponding EM value

$$P_\ell \equiv \mathbf{1}_{\{X_T^\ell>K\}}.$$

Assuming boundedness of the density of the solution to (4) in the neighborhood of the strike K, it has been proved in [20] that $V_\ell = o(\Delta t_\ell^{1/2-\delta})$, for any $\delta > 0$. This result has been tightened by Avikainen [3] who proved that $V_\ell = \mathcal{O}(\Delta t_\ell^{1/2} \log \Delta t_\ell)$.

An up-and-out call gives a European payoff if the asset never exceeds the barrier, B, otherwise it pays zero. So, for the exact solution we have

$$P = (x(T) - K)^+ \mathbf{1}_{\{\sup_{0\le t\le T} x(t) \le B\}},$$

and for the EM approximation

$$P_\ell \equiv (X_T^\ell - K)^+ \mathbf{1}_{\{\sup_{0\le n\le 2^\ell} X_n^\ell \le B\}}.$$

A down-and-in call knocks in when the minimum asset price dips below the barrier B, so that

$$P = (x(T) - K)^+ \mathbf{1}_{\{\inf_{0\le t\le T} x(t) \le B\}},$$

and, accordingly,

$$P_l \equiv (X_T^\ell - K)^+ \mathbf{1}_{\{\inf_{0\le n\le 2^\ell} X_n^\ell \le B\}}.$$

For both of these barrier options we have $\mathbb{V}_\ell = o(\Delta t_\ell^{1/2-\delta})$, for any $\delta > 0$, assuming that $\inf_{0\le t\le T} x(t)$ and $\sup_{0\le t\le T} x(t)$ have bounded density in the neighborhood of B [20].

Table 1. Orders of convergence for V_ℓ as observed numerically and proved analytically for both Euler discretisations; δ can be any strictly positive constant.

	Euler	
option	numerical	analysis
Lipschitz	$\mathcal{O}(\Delta t_\ell)$	$\mathcal{O}(\Delta t_\ell)$
Asian	$\mathcal{O}(\Delta t_\ell)$	$\mathcal{O}(\Delta t_\ell)$
lookback	$\mathcal{O}(\Delta t_\ell)$	$\mathcal{O}(\Delta t_\ell)$
barrier	$\mathcal{O}(\Delta t_\ell^{1/2})$	$o(\Delta t_\ell^{1/2-\delta})$
digital	$\mathcal{O}(\Delta t_\ell^{1/2})$	$\mathcal{O}(\Delta t_\ell^{1/2} \log \Delta t_\ell)$

As summarized in Table 1, numerical results taken form [17] suggest that all of these results are near-optimal.

3.2. *Milstein scheme*

In the scalar case of SDEs (4) (that is with $d = m = 1$) the Milstein scheme has the form

$$X_{n+1}^\ell = X_n^\ell + f(X_n^{2^l})\Delta t_\ell + g(X_n^\ell)\Delta w_{n+1}^\ell + g^{'}(X_n^\ell)g(X_n^\ell)((\Delta w_{n+1}^\ell)^2 - \Delta t_\ell), \quad (12)$$

where $g' \equiv \partial g/\partial x$. The analysis of Lipschitz European payoffs and Asian options with Milstein scheme is analogous to EM scheme and it has been proved in [11] that in both these cases $V_\ell = \mathcal{O}(\Delta t_\ell^2)$.

3.2.1. *Lookback options*

For clarity of the exposition we will express the fine time-step approximation in terms of the coarse time-step, that is $\mathcal{P}'_{\Delta t_\ell} := \{n\Delta t_{\ell-1} : n = 0, \frac{1}{2}, 1, 1+\frac{1}{2}, 2, ..., 2^{\ell-1}\}$. The partition for the coarse approximation is given by $\mathcal{P}_{\Delta t_{\ell-1}} := \{n\Delta t_{\ell-1} : n = 0, 1, 2, ..., 2^{\ell-1}\}$. Therefore, $X_n^{\ell-1}$ corresponds to X_n^ℓ for $n = 0, 1, 2, ..., 2^{\ell-1}$.
For pricing lookback options with the EM scheme, as an approximation of the minimum of the process we have simply taken $\min_n X_n^\ell$. This approximation could be improved by taking

$$X_{min}^\ell = \min_n \left(X_n^\ell - \beta^* g(X_n^\ell) \Delta t_\ell^{1/2} \right).$$

Here $\beta^* \approx 0.5826$ is a constant which corrects the $\mathcal{O}(\Delta t_\ell^{1/2})$ leading order error due to the discrete sampling of the path, and thereby restores $\mathcal{O}(\Delta t_\ell)$ weak convergence [6]. However, using this approximation, the difference between the computed minimum values and the fine and coarse paths is $\mathcal{O}(\Delta t_\ell^{1/2})$, and hence the variance V_ℓ is $\mathcal{O}(\Delta t_\ell)$, corresponding to $\beta = 1$. In the previous section, this was acceptable because $\beta = 1$ was the best that could be achieved in general with the Euler path discretization which was used, but we now aim to achieve an improved convergence rate using the Milstein scheme.

In order to improve the convergence, the Brownian Bridge interpolant $\tilde{X}^\ell(t)$ defined in (11) is used. We have

$$\begin{aligned}\min_{0\leq t<T} \tilde{X}^\ell(t) &= \min_{0\leq n\leq 2^{\ell-1}-\frac{1}{2}} \left[\min_{n\Delta t_{l-1}\leq t<(n+\frac{1}{2})\Delta t_{l-1}} \tilde{X}^\ell(t) \right] \\ &= \min_{0\leq n\leq 2^{\ell-1}-\frac{1}{2}} X_{n,min}^\ell,\end{aligned}$$

where minimum of the fine approximation over the first half of the coarse time-step is given by [24]

$$X_{n,min}^\ell = \tfrac{1}{2}\left(X_n^\ell + X_{n+\frac{1}{2}}^\ell - \sqrt{\left(X_{n+\frac{1}{2}}^\ell - X_n^\ell\right)^2 - 2\, g(X_n^\ell)^2\, \Delta t_l \log U_n^\ell} \right), \qquad (13)$$

and minimum of the fine approximation over the second half of the coarse time-step is given by

$$\begin{aligned}X_{n+\frac{1}{2},min}^\ell = \tfrac{1}{2}\Big(& X_{n+\frac{1}{2}}^\ell + X_{n+1}^\ell \\ & - \sqrt{\left(X_{n+1}^\ell - X_{n+\frac{1}{2}}^\ell\right)^2 - 2\, g(X_{n+\frac{1}{2}}^\ell)^2\, \Delta t_\ell \log U_{n+\frac{1}{2}}^\ell} \Big),\end{aligned} \qquad (14)$$

where $U_n^\ell, U_{n+\frac{1}{2}}^\ell$ are uniform random variables on the unit interval. For the coarse path, in order to improve the MLMC variance a slightly different estimator is used, see (3). Using the same Brownian increments as we used on the fine path (to guarantee that we stay on the same path), equation (11) is used to define $\tilde{X}_{n+\frac{1}{2}}^{\ell-1} \equiv$

$\tilde{X}^{\ell-1}((n+\frac{1}{2})\Delta t_{\ell-1})$. Given this interpolated value, the minimum value over the interval $[n\Delta t_{\ell-1},(n+1)\Delta t_{\ell-1}]$ can then be taken to be the smaller of the minima for the two intervals $[n\Delta t_{\ell-1},(n+\frac{1}{2})\Delta t_{\ell-1})$ and $[(n+\frac{1}{2})\Delta t_{\ell-1},(n+1)\Delta t_{\ell-1})$,

$$\begin{aligned} X^{\ell-1}_{n,min} &= \tfrac{1}{2}\Big(X^{\ell-1}_{n}+\tilde{X}^{\ell-1}_{n+\frac{1}{2}} \\ &\quad -\sqrt{\Big(\tilde{X}^{\ell-1}_{n+\frac{1}{2}}-X^{\ell-1}_{n}\Big)^2-2\,(g(X^{\ell-1}_{n}))^2\,\tfrac{\Delta t_{\ell-1}}{2}\log U^{\ell}_{n}}\ \Big), \\ X^{\ell-1}_{n+\frac{1}{2},min} &= \tfrac{1}{2}\Big(\tilde{X}^{\ell-1}_{n+\frac{1}{2}}+X^{\ell-1}_{n+1} \\ &\quad -\sqrt{\Big(X^{\ell-1}_{n+1}-\tilde{X}^{\ell-1}_{n+\frac{1}{2}}\Big)^2-2\,(g(X^{\ell-1}_{n}))^2\,\tfrac{\Delta t_{\ell-1}}{2}\log U^{\ell}_{n+\frac{1}{2})}}\ \Big). \end{aligned} \tag{15}$$

Note that $g(X^{\ell-1}_{n})$ is used for both time steps. It is because we used the Brownian Bridge with diffusion term $g(X^{\ell-1}_{n})$ to derive both minima. If we changed $g(X^{\ell-1}_{n})$ to $g(\tilde{X}^{\ell-1}_{n+\frac{1}{2}})$ in $X^{\ell-1}_{n+\frac{1}{2},min}$, this would mean that different Brownian Bridges were used on the first and second half of the coarse time-step and as a consequence condition (3) would be violated. Note also the re-use of the same uniform random numbers U^{ℓ}_{n} and $U^{\ell}_{n+\frac{1}{2}}$ used to compute the fine path minimum. The $\min(X^{\ell-1}_{n,min}, X^{\ell-1}_{n+\frac{1}{2},min})$ has exactly the same distribution as $X^{\ell-1}_{n,min}$, since they are both based on the same Brownian interpolation, and therefore equality (3) is satisfied. Giles et al. [11] proved the following Theorem:

Theorem 2. *The multilevel approximation for a lookback option which is a uniform Lipschitz function of $x(T)$ and $\inf_{[0,T]} x(t)$ has $V_l = o(\Delta t_l^{2-\delta})$ for any $\delta>0$.*

3.3. *Conditional Monte Carlo*

Giles [17] and Giles et al. [11] have shown that combining conditional Monte Carlo with MLMC results in superior estimators for various financial payoffs.

To obtain an improvement in the convergence of the MLMC variance barrier and digital options, conditional Monte Carlo methods is employed. We briefly describe it here. Our goal is to calculate $\mathbb{E}[P]$. Instead, we can write

$$\mathbb{E}[P] = \mathbb{E}\big[\mathbb{E}[P \mid Z]\big],$$

where Z is a random vector. Hence $\mathbb{E}[P \mid Z]$ is an unbiased estimator of $\mathbb{E}[P]$. We also have

$$\mathrm{Var}\,[P] = \mathbb{E}\big[\mathrm{Var}\,[P \mid Z]\big] + \mathrm{Var}\,\big[\mathbb{E}[P \mid Z]\big],$$

hence $\mathrm{Var}\,\big[\mathbb{E}[P \mid Z]\big] \le \mathrm{Var}\,(P)$. In the context of MLMC we obtain a better variance convergence if we condition on different vectors on the fine and the coarse level. That is on the fine level we take $\mathbb{E}[P^f \mid Z^f]$, where $Z^f = \{X^{\ell}_{n}\}_{0\le n\le 2^{\ell}}$. On the coarse level instead of taking $\mathbb{E}[P^c \mid Z^c]$ with $Z^c = \{X^{\ell-1}_{n}\}_{0\le n\le 2^{\ell-1}}$, we take $\mathbb{E}[P^c \mid Z^c, \tilde{Z}^c]$, where $\tilde{Z}^c = \{\tilde{X}^{\ell-1}_{n+\frac{1}{2}}\}_{0\le n\le 2^{\ell-1}}$ are obtained from equation (11). Condition (3) trivially holds by tower property of conditional expectation

$$\mathbb{E}\left[\mathbb{E}[P^c \mid Z^c]\right] = \mathbb{E}[P^c] = \mathbb{E}\left[\mathbb{E}[P^c \mid Z^c, \tilde{Z}^c]\right].$$

3.4. *Barrier options*

The barrier option which is considered is a down-and-out option for which the payoff is a Lipschitz function of the value of the underlying at maturity, provided the underlying has never dropped below a value $B \in \mathbb{R}$,

$$P = f(x(T))\ \mathbf{1}_{\{\tau > T\}}.$$

The crossing time τ is defined as

$$\tau = \inf_t \{x(t) < B\}.$$

This requires the simulation of $(x(T), \mathbf{1}_{\tau>T}))$. The simplest method sets

$$\tau^{\Delta t_\ell} = \inf_n \{X^\ell_n < B\}$$

and as an approximation takes $(X^\ell_{2^{\ell-1}}, \mathbf{1}_{\{\tau^{\Delta t_\ell} > 2^{\ell-1}\}})$. But even if we could simulate the process $\{x(n\Delta t_\ell)\}_{0\le n\le 2^{\ell-1}}$ it is possible for $\{x(t)\}_{0\le t\le T}$ to cross the barrier between grid points. Using the Brownian Bridge interpolation we can approximate $\mathbf{1}_{\{\tau>T\}}$ by

$$\prod_{n=0}^{2^{\ell-1}-\frac{1}{2}} \mathbf{1}_{\{X^\ell_{n,min} \ge B\}}.$$

This suggests following the lookback approximation in computing the minimum of both the fine and coarse paths. However, the variance would be larger in this case because the payoff is a discontinuous function of the minimum. A better treatment, which is the one used in [16], is to use the conditional Monte Carlo approach to further smooth the payoff. Since the process X^ℓ_n is Markovian we have

$$\begin{aligned}
&\mathbb{E}\Big[f(X^\ell_{2^{\ell-1}}) \prod_{n=0}^{2^{\ell-1}-\frac{1}{2}} \mathbf{1}_{\{X^\ell_{n,min} \ge B\}}\Big] \\
&= \mathbb{E}\left[\mathbb{E}\Big[f(X^\ell_{2^{\ell-1}}) \prod_{n=0}^{2^{\ell-1}-\frac{1}{2}} \mathbf{1}_{\{X^\ell_{n,min} \ge B\}} \mid X^\ell_0, \ldots, X^\ell_{2^{\ell-1}}\Big]\right] \\
&= \mathbb{E}\left[f(X^\ell_{2^{\ell-1}}) \prod_{n=0}^{2^{\ell-1}-\frac{1}{2}} \mathbb{E}\Big[\mathbf{1}_{\{X^\ell_{n,min} \ge B\}} \mid X^\ell_n, X^\ell_{n+1}\Big]\right] \\
&= \mathbb{E}\left[f(X^\ell_{2^{\ell-1}}) \prod_{n=0}^{2^{\ell-1}-\frac{1}{2}} (1 - p^\ell_n)\right],
\end{aligned}$$

where from [24]

$$p_n^\ell = \mathbb{P}\left(\inf_{n\Delta t_\ell \le t < (n+\frac{1}{2})\Delta t_\ell} \tilde{X}(t) < B \mid X_n^\ell, X_{n+\frac{1}{2}}^\ell\right)$$
$$= \exp\left(\frac{-2\,(X_n^\ell - B)^+ (X_{n+\frac{1}{2}}^\ell - B)^+}{g(X_n^\ell)^2\,\Delta t_\ell}\right),$$

and

$$p_{n+\frac{1}{2}}^\ell = \mathbb{P}\left(\inf_{(n+\frac{1}{2})\Delta t_\ell \le t < (n+1)\Delta t_\ell} \tilde{X}(t) < B \mid X_{n+\frac{1}{2}}^\ell, X_{n+1}^\ell\right)$$
$$= \exp\left(\frac{-2\,(X_{n+\frac{1}{2}}^\ell - B)^+ (X_{n+1}^\ell - B)^+}{g(X_{n+\frac{1}{2}}^\ell)^2\,\Delta t_\ell}\right).$$

Hence, for the fine path this gives

$$P_\ell^f = f(X_{2^{\ell-1}}^\ell) \prod_{n=0}^{2^{\ell-2}-\frac{1}{2}} (1 - p_n^\ell), \tag{16}$$

The payoff for the coarse path is defined similarly. However, in order to reduce the variance, we subsample $\tilde{X}_{n+\frac{1}{2}}^{\ell-1}$, as we did for lookback options, from the Brownian Bridge connecting $X_n^{\ell-1}$ and $X_{n+1}^{\ell-1}$

$$\mathbb{E}\big[f(X_{2^{\ell-1}}^{\ell-1}) \prod_{n=0}^{2^{\ell-1}-1} \mathbf{1}_{\{X_{n,min}^{\ell-1} \ge B\}}\big]$$
$$= \mathbb{E}\left[\mathbb{E}\big[f(X_{2^{\ell-1}}^{\ell-1}) \prod_{n=0}^{2^{\ell-1}-1} \mathbf{1}_{\{X_{n,min}^{\ell-1} \ge B\}} \mid X_0^{\ell-1}, \tilde{X}_{\frac{1}{2}}^{\ell-1}, \ldots, \tilde{X}_{2^{\ell-1}-\frac{1}{2}}^{\ell-1}, X_{2^{\ell-1}}^{\ell-1}\big]\right]$$
$$= \mathbb{E}\left[f(X_{2^{\ell-1}}^{\ell-1}) \prod_{n=0}^{2^{\ell-1}-1} \mathbb{E}\big[\mathbf{1}_{\{X_{n,min}^{\ell-1} \ge B\}} \mid X_n^{\ell-1}, \tilde{X}_{n+\frac{1}{2}}^{\ell-1}, X_{n+1}^{\ell-1}\big]\right]$$
$$= \mathbb{E}\left[f(X_{2^{\ell-1}}^{\ell-1}) \prod_{n=0}^{2^{\ell-1}-1} (1 - p_{1,n}^{\ell-1})(1 - p_{2,n}^{\ell-1})\right],$$

where

$$p_{1,n}^{\ell-1} = \exp\left(\frac{-2\,(X_n^{\ell-1} - B)^+ (\tilde{X}_{n+\frac{1}{2}}^{\ell-1} - B)^+}{g(X_n^{\ell-1})^2\,\Delta t_\ell}\right),$$

and

$$p_{2,n}^{\ell-1} = \exp\left(\frac{-2\,(\tilde{X}_{n+\frac{1}{2}}^{\ell-1} - B)^+ (X_{n+1}^{\ell-1} - B)^+}{g(X_n^{\ell-1})^2\,\Delta t_l}\right).$$

Note that the same $g(X_n^{\ell-1})$ is used (rather than using $g(\tilde{X}_{n+\frac{1}{2}}^{\ell-1})$ in $p_{2,n}^{\ell-1}$) to calculate both probabilities for the same reason as we did for lookback options. The final estimator can be written as

$$P_{\ell-1}^c = f(X_{2^{\ell-1}}^{\ell-1}) \prod_{n=0}^{2^{\ell-1}-1} (1-p_{1,n}^{\ell-1})(1-p_{2,n}^{\ell-1}). \tag{17}$$

Giles et al. [11] proved the following theorem

Theorem 3. *Provided* $\inf_{[0,T]} |g(B)| > 0$, *and* $\inf_{[0,T]} x(t)$ *has a bounded density in the neighbourhood of* B*, then the multilevel estimator for a down-and-out barrier option has variance* $V_\ell = o(\Delta t_\ell^{3/2-\delta})$ *for any* $\delta > 0$.

The reason the variance is approximately $o(\Delta t_\ell^{3/2-\delta})$ instead of $\mathcal{O}(\Delta t_\ell^2)$ is the following: due to the strong convergence property the probability of the numerical approximation being outside $\Delta t_\ell^{1-\delta}$-neighborhood of the solution to the SDE (4) is arbitrary small, that is for any $\varepsilon > 0$

$$\begin{aligned}\mathbb{P}\left(\sup_{0\le n\Delta t_\ell \le T} \|x(n\Delta t_\ell) - X_n^\ell\| \ge \Delta t_\ell^{1-\varepsilon}\right) \\ \le \Delta t_\ell^{-p+p\varepsilon}\mathbb{E}\left[\sup_{0\le n\Delta t_\ell\le T} \|x(n\Delta t_\ell) - X_n^\ell\|^p\right] = \mathcal{O}(\Delta_\ell^{p\varepsilon}).\end{aligned} \tag{18}$$

If $\inf_{[0,T]} x(t)$ is outside the $\Delta t_\ell^{1/2}$-neighborhood of the barrier B then by (18) it is shown that so are numerical approximations. The probabilities of crossing the barrier in that case are asymptotically either 0 or 1 and essentially we are in the Lipschitz payoff case. If the $\inf_{[0,T]} x(t)$ is within the $\Delta t_\ell^{1/2}$-neighborhood of the barrier B then so are the numerical approximations. In that case it can be shown that $\mathbb{E}[(P_\ell^f - P_{\ell-1}^c)^2] = \mathcal{O}(\Delta t^{1-\delta})$ but due to the bounded density assumption, the probability that $\inf_{[0,T]} x(t)$ is within $\Delta t_\ell^{1/2}$-neighborhood of the barrier B is of order $\Delta t_\ell^{1/2-\delta}$. Therefore the overall MLMC variance is $V_\ell = o(\Delta_\ell^{3/2-\delta})$ for any $\delta > 0$.

3.5. *Digital options*

A digital option has a payoff which is a discontinuous function of the value of the underlying asset at maturity, the simplest example being

$$P = \mathbf{1}_{\{x(T)>B\}}.$$

Approximating $\mathbf{1}_{\{x(T)>B\}}$ based only on simulations of $x(T)$ by Milstein scheme will lead to an $\mathcal{O}(\Delta t_\ell)$ fraction of the paths having coarse and fine path approximations to $x(T)$ on either side of the strike, producing $P_\ell - P_{\ell-1} = \pm 1$, resulting in $V_\ell = \mathcal{O}(\Delta t_\ell)$. To improve the variance to $\mathcal{O}(\Delta t_\ell^{3/2-\delta})$ for all $\delta > 0$, the conditional Monte Carlo method is used to smooth the payoff (see section 7.2.3 in [24]). This approach was proved to be successful in Giles et al. [11] and was tested numerically in [16],

If $X^\ell_{2^\ell-1-\frac{1}{2}}$ denotes the value of the fine path approximation one time-step before maturity, then the motion thereafter is approximated as Brownian motion with constant drift $f(X^\ell_{2^\ell-1-\frac{1}{2}})$ and volatility $g(X^\ell_{2^\ell-1-\frac{1}{2}})$. The conditional expectation for the payoff is the probability that $X^\ell_{2^\ell-1} > B$ after one further time-step, which is

$$P^f_\ell = \mathbb{E}\left[\mathbf{1}_{\{X^\ell_{2^\ell-1}>B\}} \mid X^\ell_{2^\ell-1-\frac{1}{2}}\right] = \Phi\left(\frac{X^\ell_{2^\ell-1-\frac{1}{2}} + f(X^\ell_{2^\ell-1-\frac{1}{2}})\Delta t_\ell - B}{\mid g(X^\ell_{2^\ell-1-\frac{1}{2}}) \mid \sqrt{\Delta t_\ell}}\right), \tag{19}$$

where Φ is the cumulative Normal distribution.

For the coarse path, we note that given the Brownian increment $\Delta w^{\ell-1}_{2^{\ell-1}-\frac{1}{2}}$ for the first half of the last coarse time-step (which comes from the fine path simulation), the probability that $X^\ell_{2^\ell-1} > B$ is

$$\begin{aligned} P^c_{\ell-1} =& \mathbb{E}\left[\mathbf{1}_{\{X^{\ell-1}_{2^{\ell-1}}>B\}} \mid X^{\ell-1}_{2^{\ell-1}-1}, \Delta w^{\ell-1}_{2^{\ell-1}-\frac{1}{2}}\right] \\ =& \Phi\left(\frac{X^{2^{\ell-1}}_{2^{\ell-1}-1} + f(X^{\ell-1}_{2^{\ell-1}-1})\Delta t_{\ell-1} + g(X^{\ell-1}_{2^{\ell-1}-1})\Delta w^{\ell-1}_{2^{\ell-1}-\frac{1}{2}} - B}{\mid g(X^{2^{\ell-1}}_{2^{\ell-1}-1}) \mid \sqrt{\Delta t_\ell}}\right). \end{aligned} \tag{20}$$

The conditional expectation of (20) is equal to the conditional expectation of $P^f_{\ell-1}$ defined by (19) on level ℓ–ℓ, and so equality (3) is satisfied. A bound on the variance of the multilevel estimator is given by the following result:

Theorem 4. *Provided $g(B) \neq 0$, and $x(t)$ has a bounded density in the neighbourhood of B, then the multilevel estimator for a digital option has variance $V_l = o(\Delta t_l^{3/2-\delta})$ for any $\delta > 0$.*

Results of the above section were tested numerically in [17] and are summarized in the table 2.

Table 2. Orders of convergence for V_l as observed numerically and proved analytically for Milstein discretizations; δ can be any strictly positive constant.

	Milstein	
option	numerical	analysis
Lipschitz	$\mathcal{O}(Dt_l^2)$	$\mathcal{O}(Dt_l^2)$
Asian	$\mathcal{O}(Dt_l^2)$	$\mathcal{O}(Dt_l^2)$
lookback	$\mathcal{O}(Dt_l^2)$	$o(Dt_l^{2-\delta})$
barrier	$\mathcal{O}(Dt_l^{3/2})$	$o(Dt_l^{3/2-\delta})$
digital	$\mathcal{O}(Dt_l^{3/2})$	$o(Dt_l^{3/2-\delta})$

4. Greeks with MLMC

Accurate calculation of prices is only one objective of Monte Carlo simulations. Even more important in some ways is the calculation of the sensitivities of the prices to various input parameters. These sensitivities, known collectively as the "Greeks", are important for risk analysis and mitigation through hedging.

Here we follow the results by Burgos at al. [8] to present how MLMC can applied in this setting. The pathwise sensitivity approach (also known as Infinitesimal Perturbation Analysis) is one of the standard techniques for computing these sensitivities [24]. However, the pathwise approach is not applicable when the financial payoff function is discontinuous. One solution to these problems is to use the Likelihood Ratio Method (LRM) but its weaknesses are that the variance of the resulting estimator is usually $\mathcal{O}(\Delta t_l^{-1})$.

Three techniques are presented that improve MLMC variance: payoff smoothing using conditional expectations [24]; an approximation of the above technique using path splitting for the final timestep [2]; the use of a hybrid combination of pathwise sensitivity and the Likelihood Ratio Method [19]. We discuss the strengths and weaknesses of these alternatives in different multilevel Monte Carlo settings.

4.1. *Monte Carlo Greeks*

Consider the approximate solution of the general SDE (4) using Euler discretization (6). The Brownian increments can be defined to be a linear transformation of a vector of independent unit Normal random variables Z.

The goal is to efficiently estimate the expected value of some financial payoff function $P(x(T))$, and numerous first order sensitivities of this value with respect to different input parameters such as the volatility or one component of the initial data $x(0)$. In more general cases P might also depend on the values of process $\{x(t)\}_{0\leq t\leq T}$ at intermediate times.

The pathwise sensitivity approach can be viewed as starting with the expectation expressed as an integral with respect to Z:

$$V_\ell \equiv \mathbb{E}\left[P(X_n^\ell(Z,\theta))\right] = \int P(X_n^\ell(Z,\theta))\; p_Z(Z)\; dZ. \tag{21}$$

Here θ represents a generic input parameter, and the probability density function for Z is

$$p_Z(Z) = (2\pi)^{-d/2} \exp\left(-\|Z\|_2^2/2\right),$$

where d is the dimension of the vector Z.

Let $X_n^\ell = X_n^\ell(Z,\theta)$. If the drift, volatility and payoff functions are all differentiable, (21) may be differentiated to give

$$\frac{\partial V_\ell}{\partial \theta} = \int \frac{\partial P(X_n^\ell)}{\partial X_n^\ell}\, \frac{\partial X_n^\ell}{\partial \theta}\; p_Z(Z)\; \Delta Z, \tag{22}$$

with $\dfrac{\partial X_n^\ell}{\partial \theta}$ being obtained by differentiating (6) to obtain

$$\begin{aligned}\frac{\partial X_{n+1}^\ell}{\partial \theta} = \frac{\partial X_n^\ell}{\partial \theta} &+ \left(\frac{\partial f(X_n^\ell,\theta)}{\partial X_n^\ell}\,\frac{\partial X_n^\ell}{\partial \theta} + \frac{\partial f(X_n^\ell,\theta)}{\partial \theta}\right)\Delta t_l \\ &+ \left(\frac{\partial g(X_n^\ell,\theta)}{\partial X_n^\ell}\,\frac{\partial X_n^\ell}{\partial \theta} + \frac{\partial g(X_n^\ell,\theta)}{\partial \theta}\right)\Delta w_{n+1}^l.\end{aligned} \tag{23}$$

We assume that $Z \to \Delta w_{n+1}^l$ mapping does not depend on θ. It can be proved that (22) remains valid (that is we can interchange integration and differentiation) when the payoff function is continuous and piecewise differentiable, and the numerical estimate obtained by standard Monte Carlo with M independent path simulations

$$M^{-1}\sum_{m=1}^{M}\frac{\partial P(X_n^{\ell,m})}{\partial X_n^\ell}\,\frac{\partial X_n^{\ell,m}}{\partial \theta}$$

is an unbiased estimate for $\partial V/\partial\theta$ with a variance which is $\mathcal{O}(M^{-1})$, if $P(x)$ is Lipschitz and the drift and volatility functions satisfy the standard conditions [34].

Performing a change of variables, the expectation can also be expressed as

$$V_l \equiv \mathbb{E}\left[P(X_n^\ell)\right] = \int P(x)\; p_{X_n^\ell}(x,\theta)dx, \tag{24}$$

where $p_{X_n^\ell}(x,\theta)$ is the probability density function for X_n^ℓ which will depend on all of the inputs parameters. Since probability density functions are usually smooth, (24) can be differentiated to give

$$\frac{\partial V_\ell}{\partial \theta} = \int P(x)\,\frac{\partial p_{X_n^\ell}}{\partial \theta}\,dx = \int P(x)\,\frac{\partial(\log p_{X_n^\ell})}{\partial \theta}\,p_{X_n^\ell}\,dx$$

$$= \mathbb{E}\left[P(x)\,\frac{\partial(\log p_{X_n^\ell})}{\partial \theta}\right].$$

which can be estimated using the unbiased Monte Carlo estimator

$$M^{-1}\sum_{m=1}^{M} P(X_n^{\ell,m})\,\frac{\partial \log p_{X_n^\ell}(X_n^{\ell,m})}{\partial \theta}$$

This is the Likelihood Ratio Method. Its great advantage is that it does not require the differentiation of $P(X_n^\ell)$. This makes it applicable to cases in which the payoff is discontinuous, and it also simplifies the practical implementation because banks often have complicated flexible procedures through which traders specify payoffs. However, it does have a number of limitations, one being a requirement of absolute continuity which is not satisfied in a few important applications such as the LIBOR market model [24].

4.2. *Multilevel Monte Carlo Greeks*

The MLMC method for calculating Greeks can be written as

$$\frac{\partial V}{\partial \theta} = \frac{\partial \mathbb{E}(P)}{\partial \theta} \approx \frac{\partial \mathbb{E}(P_L)}{\partial \theta} = \frac{\partial \mathbb{E}(P_0)}{\partial \theta} + \sum_{\ell=1}^{L} \frac{\partial \mathbb{E}(P_\ell^f - P_{\ell-1}^c)}{\partial \theta}. \tag{25}$$

Therefore extending Monte Carlo Greeks to MLMC Greeks is straightforward. However, the challenge is to keep the MLMC variance small. This can be achieved by appropriate smoothing of the payoff function. The techniques that were presented in section 3.2 are also very useful here.

4.3. *European call*

As an example we consider an European call $P = (x(T) - B)^+$ with $x(t)$ being a geometric Brownian motion with Milstein scheme approximation given by

$$X_{n+1}^\ell = X_n^\ell + r\, X_n^\ell \Delta t_\ell + \sigma\, X_n^\ell \Delta w_{n+1}^\ell + \frac{\sigma^2}{2}((\Delta w_{n+1}^\ell)^2 - \Delta t_\ell). \tag{26}$$

We illustrate the techniques by computing delta (δ) and vega (ν), the sensitivities to the asset's initial value $x(0)$ and to its volatility σ.

Since the payoff is Lipschitz, we can use pathwise sensitivities. We observe that

$$\frac{\partial}{\partial x}(x - B)^+ = \begin{cases} 0, & \text{for } x < B \\ 1, & \text{for } x > B \end{cases}$$

This derivative fails to exists when $x = B$, but since this event has probability 0, we may write

$$\frac{\partial}{\partial x}(x - K)^+ = \mathbf{1}_{\{X > B\}}.$$

Therefore we are essentially dealing with a digital option.

4.4. *Conditional Monte Carlo for Pathwise Sensitivity*

Using conditional expectation the payoff can be smooth as we did it in Section 3.2. European calls can be treated in the exactly the same way as Digital option in Section 3.2, that is instead of simulating the whole path, we stop at the penultimate step and then on the last step we consider the full distribution of $(X_{2^l}^\ell \mid w_0^l, \ldots, w_{2^l-1}^l)$. For digital options this approach leads to (19) and (20). For the call options we can do analogous calculations. In [8] numerical results for this approach obtained, with scalar Milstein scheme used to obtain the penultimate step. They results are presented in Table 3. For lookback options conditional expectations leads to (13) and (15) and for barriers to (16) and (17). Burgos et al [8], applied pathwise sensitivity to these smoothed payoffs, with scalar Milstein scheme used to obtain the penultimate step, and obtained numerical results that we present in Table 4.

Table 3. Orders of convergence for V_ℓ as observed numerically and corresponding MLMC complexity.

	Call		Digital	
Estimator	β	MLMC Complexity	β	MLMC Complexity
Value	≈ 2.0	$\mathcal{O}(\epsilon^{-2})$	≈ 1.4	$\mathcal{O}(\epsilon^{-2})$
Delta	≈ 1.5	$\mathcal{O}(\epsilon^{-2})$	≈ 0.5	$\mathcal{O}(\epsilon^{-2.5})$
Vega	≈ 2	$\mathcal{O}(\epsilon^{-2})$	≈ 0.6	$O(\epsilon^{-2.4})$

Table 4. Orders of convergence for V_ℓ as observed numerically and corresponding MLMC complexity.

	Lookback		Barrier	
Estimator	β	MLMC Complexity	β	MLMC Complexity
Value	≈ 1.9	$\mathcal{O}(\epsilon^{-2})$	≈ 1.6	$\mathcal{O}(\epsilon^{-2})$
Delta	≈ 1.9	$\mathcal{O}(\epsilon^{-2})$	≈ 0.6	$\mathcal{O}(\epsilon^{-2.4})$
Vega	≈ 1.3	$\mathcal{O}(\epsilon^{-2})$	≈ 0.6	$O(\epsilon^{-2.4})$

4.5. *Split pathwise sensitivities*

There are two difficulties in using conditional expectation to smooth payoffs in practice in financial applications. This first is that conditional expectation will often become a multi-dimensional integral without an obvious closed-form value, and the second is that it requires a change to the often complex software framework used to specify payoffs. As a remedy for these problems the splitting technique to approximate $\mathbb{E}\left[P(X^\ell_{2^l}) \mid X^\ell_{2^\ell-1}\right]$ and $\mathbb{E}\left[P(X^{\ell-1}_{2^{\ell-1}}) \mid X^{\ell-1}_{2^{\ell-1}-1}, \Delta w^\ell_{2^\ell-2}\right]$, is used. We get numerical estimates of these values by "splitting" every simulated path on the final timestep. At the fine level: for every simulated path, a set of s final increments $\{\Delta w^{\ell,i}_{2^\ell}\}_{i\in[1,s]}$ is simulated, which can be averaged to get

$$\mathbb{E}\left[P(X^\ell_{2^\ell}) \mid X^\ell_{2^\ell-1}\right] \approx \frac{1}{s}\sum_{i=1}^{s} P(X^\ell_{2^\ell-1}, \Delta w^{\ell,i}_{2^\ell}) \tag{27}$$

At the coarse level, similar to the case of digital options, the fine increment of the Brownian motion over the first half of the coarse timestep is used,

$$\mathbb{E}\left[P(X^{\ell-1}_{2^{\ell-1}}) \mid X^{\ell-1}_{2^{\ell-1}-1}, \Delta w^\ell_{2^\ell-2}\right] \approx \frac{1}{s}\sum_{i=1}^{s} P(X^{\ell-1}_{2^{\ell-1}-1}, \Delta w^\ell_{2^\ell-2}, \Delta w^{\ell-1,i}_{2^{\ell-1}}) \tag{28}$$

This approach was tested in [8], with scalar the Milstein scheme used to obtain the penultimate step, and is presented in Table 5. As expected the values of β tend to the rates offered by conditional expectations as s increases and the approximation gets more precise.

4.6. *Optimal number of samples*

The use of multiple samples to estimate the value of the conditional expectations is an example of the splitting technique [2]. If w and z are independent random

Table 5. Orders of convergence for V_ℓ as observed numerically and the corresponding MLMC complexity.

Estimator	s	β	MLMC Complexity
Value	10	≈ 2.0	$O(\epsilon^{-2})$
	500	≈ 2.0	$O(\epsilon^{-2})$
Delta	10	≈ 1.0	$O(\epsilon^{-2}(\log\epsilon)^2)$
	500	≈ 1.5	$O(\epsilon^{-2})$
Vega	10	≈ 1.6	$O(\epsilon^{-2})$
	500	≈ 2.0	$O(\epsilon^{-2})$

variables, then for any function $P(w,z)$ the estimator

$$Y_{M,S} = M^{-1} \sum_{m=1}^{M} \left(S^{-1} \sum_{i=1}^{S} P(w^m, z^{(m,i)}) \right)$$

with independent samples w^m and $z^{m,i}$ is an unbiased estimator for

$$\mathbb{E}_{w,z}\left[P(w,z)\right] \equiv \mathbb{E}_w \left[\mathbb{E}_z[P(w,z)\,|\,w] \right],$$

and its variance is

$$\mathbb{V}[Y_{M,S}] = M^{-1}\, \mathbb{V}_w \left[\mathbb{E}_z[P(w,z)\,|\,w] \right] + (MS)^{-1}\, \mathbb{E}_w \left[\mathbb{V}_z[P(w,z)\,|\,w] \right].$$

The cost of computing $Y_{M,S}$ with variance $v_1\, M^{-1} + v_2\,(MS)^{-1}$ is proportional to

$$c_1\, M + c_2\, MS,$$

with c_1 corresponding to the path calculation and c_2 corresponding to the payoff evaluation. For a fixed computational cost, the variance can be minimized by minimizing the product

$$\left(v_1 + v_2\, s^{-1}\right)(c_1 + c_2\, s) = v_1\, c_2\, s + v_1\, c_1 + v_2\, c_2 + v_2\, c_1\, s^{-1},$$

which gives the optimum value $s_{\text{opt}} = \sqrt{v_2\, c_1 / v_1\, c_2}$.

c_1 is $\mathcal{O}(\Delta t_\ell^{-1})$ since the cost is proportional to the number of timesteps, and c_2 is $\mathcal{O}(1)$, independent of Δt_ℓ. If the payoff is Lipschitz, then v_1 and v_2 are both $\mathcal{O}(1)$ and $S_{\text{opt}} = \mathcal{O}(\Delta t_\ell^{-1/2})$.

4.7. *Vibrato Monte Carlo*

The idea of vibrato Monte Carlo is to combine pathwise sensitivity and Likelihood Ration Method. Adopting the conditional expectation approach, each path simulation for a particular set of Brownian motion increments $w^\ell \equiv (\Delta w_1^\ell, \Delta w_2^\ell, \ldots, \Delta w_{2^\ell-1}^\ell)$ (excluding the increment for the final timestep) computes a conditional Gaussian probability distribution $p_X(X_{2^\ell}^\ell | w^\ell)$. For a scalar SDE, if μ_{w^ℓ} and σ_{w^ℓ} are the mean and standard deviation for given w^ℓ, then

$$X_{2^l}^\ell(w^\ell, Z) = \mu_{w^\ell} + \sigma_{w^\ell} Z,$$

where Z is a unit Normal random variable. The expected payoff can then be expressed as

$$V_\ell = \mathbb{E}\left[\mathbb{E}\left[P(X_{2^\ell}^\ell)\,|\,w^\ell\right]\right] = \int \left\{ \int P(x)\, p_{X_{2^\ell}^\ell}(x\,|\,w^\ell)\, \mathrm{d}x \right\} p_{w^\ell}(y)\, \mathrm{d}y.$$

The outer expectation is an average over the discrete Brownian motion increments, while the inner conditional expectation is averaging over Z.

To compute the sensitivity to the input parameter θ, the first step is to apply the pathwise sensitivity approach for fixed w^l to obtain $\partial \mu_{w^l}/\partial\theta, \partial\sigma_{w^l}/\partial\theta$. We then apply LRM to the inner conditional expectation to get

$$\frac{\partial V_\ell}{\partial \theta} = \mathbb{E}\left[\frac{\partial}{\partial\theta}\mathbb{E}\left[P(X_{2^\ell}^\ell)\,|\,w^\ell\right]\right] = \mathbb{E}\left[\mathbb{E}_Z\left[P(X_{2^\ell}^\ell)\,\frac{\partial(\log p_{X_{2^\ell}^\ell})}{\partial\theta}\,|\,w^\ell\right]\right],$$

where

$$\frac{\partial(\log p_{X_{2^\ell}^\ell})}{\partial\theta} = \frac{\partial(\log p_{X_{2^\ell}^\ell})}{\partial\mu_{w^\ell}}\,\frac{\partial\mu_{w^\ell}}{\partial\theta} + \frac{\partial(\log p_{X_{2^\ell}^\ell})}{\partial\sigma_{w^\ell}}\,\frac{\partial\sigma_{w^\ell}}{\partial\theta}.$$

This leads to the estimator

$$\begin{aligned}\frac{\partial V_\ell}{\partial\theta} \approx \frac{1}{N_\ell}\sum_{m=1}^{N_\ell} &\Bigg(\frac{\partial\mu_{w^{\ell,m}}}{\partial\theta}\mathbb{E}\left[P\left(X_{2^\ell}^\ell\right)\frac{\partial(\log p_{X_{2^\ell}^\ell})}{\partial\mu_{w^\ell}}|w^{\ell,m}\right] \\ &+\frac{\partial\sigma_{\hat{w}^{\ell,m}}}{\partial\theta}\mathbb{E}\left[P\left(X_{2^\ell}^\ell\right)\frac{\partial(\log p_{X_{2^\ell}^\ell})}{\partial\sigma_{w^\ell}}|w^{\ell,m}\right]\Bigg)\end{aligned} \tag{29}$$

We compute $\dfrac{\partial\mu_{w^{\ell,m}}}{\partial\theta}$ and $\dfrac{\partial\sigma_{w^{\ell,m}}}{\partial\theta}$ with pathwise sensitivities. With $X_{2^l}^{\ell,m,i} = X_{2^\ell}^\ell(w^{\ell,m}, Z^i)$, we substitute the following estimators into (29)

$$\begin{cases}\mathbb{E}\left[P\left(X_{2^\ell}^\ell\right)\dfrac{\partial(\log p_{X_{2^\ell}^\ell})}{\partial\mu_{w^\ell}}|w^{\ell,m}\right] \approx \dfrac{1}{s}\displaystyle\sum_{i=1}^{s}\left(P\left(X_{2^\ell}^{\ell,m,i}\right)\dfrac{X_{2^\ell}^{2^\ell,m,i}-\mu_{w^{\ell,m}}}{\sigma^2_{w^{\ell,m}}}\right)\\ \mathbb{E}\left[P\left(X_{2^\ell}^\ell\right)\dfrac{\partial(\log p_{X_{2^\ell}^\ell})}{\partial\sigma_{w^\ell}}|\hat{w}^{\ell,m}\right] \approx \dfrac{1}{s}\displaystyle\sum_{i=1}^{s}P\left(X_{2^\ell}^{\ell,m,i}\right)\left(-\dfrac{1}{\sigma_{w^{\ell,m}}}+\dfrac{\left(X_{2^\ell}^{\ell,m,i}-\mu_{w^{\ell,m}}\right)^2}{\sigma^3_{w^{\ell,m}}}\right)\end{cases}$$

In a multilevel setting, at the fine level we can use (29) directly. At the coarse level, as for digital options in section 3.5, the fine Brownian increments over the first half of the coarse timestep are re-used to derive (29).

The numerical experiments for the call option with $s = 10$ was obtained [8], with scalar Milstein scheme used to obtain the penultimate step.

Estimator	β	MLMC Complexity
Value	≈ 2.0	$O(\epsilon^{-2})$
Delta	≈ 1.5	$O(\epsilon^{-2})$
Vega	≈ 2.0	$O(\epsilon^{-2})$

Although the discussion so far has considered an option based on the value of a single underlying value at the terminal time T, it can be shown that the idea extends very naturally to multidimensional cases, producing a conditional multivariate Gaussian distribution, and also to financial payoffs which are dependent on values at intermediate times.

5. MLMC for Jump-diffusion processes

Giles and Xia in [47] investigated the extension of the MLMC method to jump-diffusion SDEs. We consider models with finite rate activity using a jump-adapted discretization in which the jump times are computed and added to the standard uniform discretization times. If the Poisson jump rate is constant, the jump times are the same on both paths and the multilevel extension is relatively straightforward, but the implementation is more complex in the case of state-dependent jump rates for which the jump times naturally differ.

Merton [36] proposed a jump-diffusion process, in which the asset price follows a jump-diffusion SDE:

$$dx(t) = f(x(t-))\,\Delta t + g(x(t-))\,\Delta w(t) + c(x(t-))\,\Delta J(t), \quad 0 \le t \le T, \tag{30}$$

where the jump term $J(t)$ is a compound Poisson process $\sum_{i=1}^{N(t)}(Y_i - 1)$, the jump magnitude Y_i has a prescribed distribution, and $N(t)$ is a Poisson process with intensity λ, independent of the Brownian motion. Due to the existence of jumps, the process is a càdlàg process, i.e. having right continuity with left limits. We note that $x(t-)$ denotes the left limit of the process while $x(t) = \lim_{s\to t+} x(t)$. In [36], Merton also assumed that $\log Y_i$ has a normal distribution.

5.1. *A Jump-adapted Milstein discretization*

To simulate finite activity jump-diffusion processes, Giles and Xia [47] used the jump-adapted approximation from Platen and Bruti-Liberat [40]. For each path simulation, the set of jump times $\mathbb{J} = \{\tau_1, \tau_2, \ldots, \tau_m\}$ within the time interval $[0, T]$ is added to a uniform partition $\mathcal{P}_{\Delta t_l} := \{n\Delta t_l : n = 0, 1, 2, ..., 2^l\}$. A combined set of discretization times is then given by $\mathbb{T} = \{0 = t_0 < t_1 < t_2 < \ldots < t_M = T\}$ and we define a the length of the timestep as $\Delta t_l^n = t_{n+1} - t_n$. Clearly, $\Delta t_l^n \le \Delta t_l$.

Within each timestep the scalar Milstein discretization is used to approximate the SDE (30), and then the jump is simulated when the simulation time is equal to one of the jump times. This gives the following numerical method:

$$\begin{aligned} X_{n+1}^{\ell,-} &= X_n^\ell + f(X_n^\ell)\,\Delta t_\ell^n + g(X_n^\ell)\,\Delta w_{n+1}^\ell \\ &\quad + \tfrac{1}{2}\, g'(X_n^\ell)\, g(X_n^\ell)\,(\Delta(w_n^\ell)^2 - \Delta t_\ell^n), \\ X_{n+1}^\ell &= \begin{cases} X_{n+1}^{\ell,-} + c(X_{n+1}^{\ell,-})(Y_i - 1), & \text{when } t_{n+1} = \tau_i; \\ X_{n+1}^{\ell,-}, & \text{otherwise,} \end{cases} \end{aligned} \tag{31}$$

where $X_n^{\ell,-} = X_{t_n-}^{\ell}$ is the left limit of the approximated path, Δw_n^{ℓ} is the Brownian increment and Y_i is the jump magnitude at τ_i.

5.1.1. *Multilevel Monte Carlo for constant jump rate*

In the case of the jump-adapted discretization the telescopic sum (1) is written down with respect to Δt_ℓ rather than to Δt_ℓ^n. Therefore, we have to define the computational complexity as the expected computational cost since different paths may have different numbers of jumps. However, the expected number of jumps is finite and therefore the cost bound in assumption *iv*) will still remain valid for an appropriate choice of the constant c_3.

The MLMC approach for a constant jump rate is straightforward. The jump times τ_j, which are the same for the coarse and fine paths, are simulated by setting $\tau_j - \tau_{j-1} \sim \exp(\lambda)$.

Pricing European call and Asian options in this setting is straightforward. For lookback, barrier and digital options we need to consider Brownian bridge interpolations as we did in Section 3.2. However, due to presence of jumps some small modifications are required. To improve convergence we will be looking at Brownian bridges between time-steps coming from jump-adapted discretization. In order to obtain an interpolated value $\tilde{X}^{2^{\ell-1}}_{n+\frac{1}{2}}$ for the coarse time-step a Brownian Bridge interpolation over interval $[k_n, \hat{k}_n]$ is considered, where

$$\begin{aligned} k_n &= \max\left\{n\Delta t_{\ell-1}^n, \max\left\{\tau \in \mathbb{J} : \tau < (n+\tfrac{1}{2})\Delta t_{\ell-1}^n\right\}\right\} \\ \hat{k}_n &= \min\left\{(n+1)\Delta t_{\ell-1}^n, \min\left\{\tau \in \mathbb{J} : \tau > (n+\tfrac{1}{2})\Delta t_{\ell-1}^n\right\}\right\}. \end{aligned} \tag{32}$$

Hence

$$\begin{aligned} \tilde{X}^{\ell-1}_{n+\frac{1}{2}} &= X^{\ell-1}_{k_n} + \lambda_{\ell-1}\,(X^{\ell-1}_{\hat{k}_n} - X^{\ell-1}_{k_n}) \\ &\quad + g(X^{\ell-1}_{k_n})\left(w^\ell((n+\tfrac{1}{2})\Delta t_{\ell-1}) - w^\ell(k_n) - \lambda_{\ell-1}\,(w^\ell(\hat{k}_n) - w^\ell(k_n))\right) \end{aligned}$$

where $\lambda_{\ell-1} \equiv ((n+\frac{1}{2})\Delta t_{\ell-1}^n - k_n)/(\hat{k}_n - k_n)$.

In the same way as in Section 3.2, the minima over time-adapted discretization can be derived. For the fine time-step we have

$$X^\ell_{n,min} = \tfrac{1}{2}\left(X_n^\ell + X^{\ell,-}_{n+\frac{1}{2}} - \sqrt{\left(X^{\ell,-}_{n+\frac{1}{2}} - X_n^\ell\right)^2 - 2\,g(X_n^\ell)^2\,\Delta t_\ell^n \log U_n^\ell}\right).$$

Notice the use of the left limits $X^{\ell,-}$. Following discussion in the previous sections, the minima for the coarse time-step can be derived using interpolated value $\tilde{X}^{\ell-1}_{n+\frac{1}{2}}$. Deriving the payoffs for lookback and barrier option is now straightforward.

For digital options, due to jump-adapted time grid, in order to find conditional expectations, we need to look at relations between the last jump time and the last timestep before expiry. In fact, there are three cases:

(1) The last jump time τ happens before penultimate fixed-time timestep, i.e. $\tau < (2^{l-1} - 2)\Delta t_l$.
(2) The last jump time is within the last fixed-time timestep , i.e. $\tau > (2^{l-1} - 1)\Delta t_l$;
(3) The last jump time is within the penultimate fixed-time timestep, i.e. $(2^{l-1} - 1)\Delta t_l > \tau > (2^{l-1} - 2)\Delta t_l$.

With this in mind we can easily write down the payoffs for the coarse and fine approximations as we presented in Section 3.5.

5.1.2. *MLMC for Path-dependent rates*

In the case of a path-dependent jump rate $\lambda(x(t))$, the implementation of the multilevel method becomes more difficult because the coarse and fine path approximations may have jumps at different times. These differences could lead to a large difference between the coarse and fine path payoffs, and hence greatly increase the variance of the multilevel correction. To avoid this, Giles and Xia [47] modified the simulation approach of Glasserman and Merener [26] which uses "thinning" to treat the case in which $\lambda(x(t), t)$ is bounded. Let us recall the thinning property of Poisson processes. Let $(N_t)_{t\geq 0}$ be a Poisson process with intensity λ and define a new process Z_t by "thinning" N_t: take all the jump times $(\tau_n, n \geq 1)$ corresponding to N, keep then with probability $0 < p < 1$ or delete then with probability $1 - p$, independently from each other. Now order the jump times that have not been deleted: $(\tau_n^{'}, n \geq 1)$, and define

$$Z_t = \sum_{n\geq 1} \mathbf{1}_{t\geq \tau_n^{'}}.$$

Then the process Z is Poisson process with intensity $p\lambda$.

In our setting, first a Poisson process with a constant rate λ_{sup} (which is an upper bound of the state-dependent rate) is constructed. This gives a set of candidate jump times, and these are then selected as true jump times with probability $\lambda(x(t), t)/\lambda_{\text{sup}}$. The following jump-adapted thinning Milstein scheme is obtained

(1) Generate the jump-adapted time grid for a Poisson process with constant rate λ_{sup};
(2) Simulate each timestep using the Milstein discretization;
(3) When the endpoint t_{n+1} is a candidate jump time, generate a uniform random number $U \sim [0, 1]$, and if $U < p_{t_{n+1}} = \dfrac{\lambda(x(t_{n+1}-), t_{n+1})}{\lambda_{\text{sup}}}$, then accept t_{n+1} as a real jump time and simulate the jump.

In the multilevel implementation, the straightforward application of the above algorithm will result in different acceptance probabilities for fine and coarse level. There may be some samples in which a jump candidate is accepted for the fine path, but not for the coarse path, or vice versa. Because of the first order strong convergence, the difference in acceptance probabilities will be $\mathcal{O}(\Delta t_\ell)$, and hence there is an $\mathcal{O}(\Delta t_\ell)$ probability of coarse and fine paths differing in accepting candidate jumps. Such differences will give an $\mathcal{O}(1)$ difference in the payoff value, and hence the multilevel variance will be $\mathcal{O}(h)$. A more detailed analysis of this is given in [46].

To improve the variance convergence rate, a change of measure is used so that the acceptance probability is the same for both fine and coarse paths. This is achieved by taking the expectation with respect to a new measure Q:

$$\mathbb{E}_P[P_\ell^f - P_{\ell-1}^c] = \mathbb{E}_Q[P_\ell^f \prod_\tau R_\tau^f - P_{\ell-1}^c \prod_\tau R_\tau^c]$$

where τ are the jump times. The acceptance probability for a candidate jump under the measure Q is defined to be $\frac{1}{2}$ for both coarse and fine paths, instead of $p_\tau = \lambda(X(\tau-),\tau) \,/\, \lambda_{\text{sup}}$. The corresponding Radon-Nikodym derivatives are

$$R_\tau^f = \begin{cases} 2p_\tau^f, & \text{if } U < \dfrac{1}{2}\,; \\ 2(1-p_\tau^f), & \text{if } U \geq \dfrac{1}{2}\,, \end{cases} \qquad R_\tau^c = \begin{cases} 2p_\tau^c, & \text{if } U < \dfrac{1}{2}\,; \\ 2(1-p_\tau^c), & \text{if } U \geq \dfrac{1}{2}\,, \end{cases}$$

Since $\mathbb{V}[R_\tau^f - R_\tau^c] = \mathcal{O}(\Delta t^2)$ and $\mathbb{V}[\widehat{P}_\ell - \widehat{P}_{\ell-1}] = \mathcal{O}(\Delta t^2)$, this results in the multilevel correction variance $\mathbb{V}_Q[\widehat{P}_\ell \prod_\tau R_\tau^f - \widehat{P}_{\ell-1} \prod_\tau R_\tau^c]$ being $\mathcal{O}(\Delta t^2)$.

If the analytic formulation is expressed using the same thinning and change of measure, the weak error can be decomposed into two terms as follows:

$$\begin{aligned} \mathbb{E}_Q\left[\widehat{P}_\ell \prod_\tau R_\tau^f - P \prod_\tau R_\tau\right] &= \mathbb{E}_Q\left[(\widehat{P}_\ell - P) \prod_\tau R_\tau^f\right] \\ &+ \mathbb{E}_Q\left[P\,(\prod_\tau R_\tau^f - \prod_\tau R_\tau)\right]. \end{aligned}$$

Using Hölder's inequality, the bound $\max(R_\tau, R_\tau^f) \leq 2$ and standard results for a Poisson process, the first term can be bounded using weak convergence results for the constant rate process, and the second term can be bounded using the corresponding strong convergence results [46]. This guarantees that the multilevel procedure does converge to the correct value.

5.2. *Lévy processes*

Dereich and Heidenreich [13] analysed approximation methods for both finite and infinite activity Lévy driven SDEs with globally Lipschitz payoffs. They have derived upper bounds for MLMC variance for the class of path dependent payoffs that are Lipschitz continuous with respect to supremum norm. One of their main findings is that the rate of MLMC variance converges is closely related to Blumenthal-Getoor index of the driving Lévy process that measures the frequency of small jumps. In [13] authors considered SDEs driven by the Lévy process

$$s(t) = \Sigma\, w(t) + L(t) + b\, t,$$

where Σ is the diffusion coefficient, $L(t)$ is a compensated jump process and b is a drift coefficient. The simplest treatment is to neglect all the jumps with size smaller than h. To construct MLMC they took h^ℓ, that is at level ℓ they neglected jumps smaller than h^ℓ. Then similarly as in the previous section, a uniform time discretization Δt_ℓ augmented with jump times is used. Let us denote by $\Delta L(t) = L(t) - L(t)^-$, the jump-discontinuity at time t. The crucial observation is that for $h^{'} > h > 0$ the jumps of the process $L^{h^{'}}$ can be obtained from those of L^h by

$$\Delta L(t)^{h^{'}} = \Delta L_t^h\, \mathbf{1}_{\{|\Delta L(t)^h| > h'\}},$$

this gives the necessary coupling to obtain a good MLMC variance. We define a decreasing and invertible function $g : (0,\infty) \to (0,\infty)$ such that

$$\int \frac{|\, x\,|^2}{h^2} \wedge 1 \nu(dx) \le g(h) \quad \text{for all} \quad h > 0,$$

where ν is a Lévy measure, and for $\ell \in \mathbb{N}$ we define

$$\Delta t_\ell = 2^{-\ell} \quad \text{and} \quad h^\ell = g^{-1}(2^\ell).$$

With this choice of Δt_ℓ and h^ℓ, authors in [13] analysed the standard Euler-Maruyama scheme for Lévy driven SDEs. This approach gives good results for a Blumenthal-Getoor index smaller than one. For a Blumenthal-Getoor index bigger than one, Gaussian approximation of small jumps gives better results [12].

6. Multi-dimensional Milstein scheme

In the previous sections it was shown that by combining a numerical approximation with the strong order of convergence $\mathcal{O}(\Delta t_\ell)$ with MLMC results in reduction of the computational complexity to estimate expected values of functionals of SDE solutions with a root-mean-square error of ϵ from $\mathcal{O}(\epsilon^{-3})$ to $\mathcal{O}(\epsilon^{-2})$. However, in general, to obtain a rate of strong convergence higher than $O(\Delta t^{1/2})$ requires simulation, or approximation, of Lévy areas. Giles and Szpruch in [22] through the construction of a suitable antithetic multilevel correction estimator, showed that we can avoid the simulation of Lévy areas and still achieve an $O(\Delta t^2)$ variance for smooth payoffs, and almost an $O(\Delta t^{3/2})$ variance for piecewise smooth payoffs, even though there is only $O(\Delta t^{1/2})$ strong convergence.

In the previous sections we have shown that it can be better to use different estimators for the finer and coarser of the two levels being considered, P_ℓ^f when level ℓ is the finer level, and P_ℓ^c when level ℓ is the coarser level. In this case, we required that

$$\mathbb{E}[P_\ell^f] = \mathbb{E}[P_\ell^c] \quad \text{for } \ell = 1, \ldots, L, \tag{33}$$

so that

$$\mathbb{E}[P_L^f] = \mathbb{E}[P_0^f] + \sum_{\ell=1}^{L} \mathbb{E}[P_\ell^f - P_{\ell-1}^c],$$

still holds. For lookback, barrier and digital options we showed that we can obtain a better MLMC variance by suitable modifying the estimator on the coarse levels. By further exploiting the flexibility of MLMC, Giles and Szpruch [22] modified the estimator on the fine levels in order to avoid simulation of the Lévy areas.

6.1. *Antithetic MLMC estimator*

Based on the well-known method of antithetic variates (see for example [24]), the idea for the antithetic estimator is to exploit the flexibility of the more general MLMC estimator by defining $P_{\ell-1}^c$ to be the usual payoff $P(X^c)$ coming from a level $\ell-1$ coarse simulation X^c, and defining P_ℓ^f to be the average of the payoffs $P(X^f), P(X^a)$ coming from an antithetic pair of level ℓ simulations, X^f and X^a.

X^f will be defined in a way which corresponds naturally to the construction of X^c. Its antithetic "twin" X^a will be defined so that it has exactly the same distribution as X^f, conditional on X^c, which ensures that $\mathbb{E}[P(X^f)] = \mathbb{E}[P(X^a)]$ and hence (3) is satisfied, but at the same time

$$\left(X^f - X^c\right) \;\approx\; -\left(X^a - X^c\right)$$

and therefore

$$\left(P(X^f) - P(X^c)\right) \;\approx\; -\left(P(X^a) - P(X^c)\right),$$

so that $\frac{1}{2}\left(P(X^f) + P(X^a)\right) \approx P(X^c)$. This leads to $\frac{1}{2}\left(P(X^f) + P(X^a)\right) - P(X^c)$ having a much smaller variance than the standard estimator $P(X^f) - P(X^c)$.

We now present a lemma which gives an upper bound on the convergence of the variance of $\frac{1}{2}\left(P(X^f) + P(X^a)\right) - P(X^c)$.

Lemma 1. *If $P \in C^2(\mathbb{R}^d, \mathbb{R})$ and there exist constants L_1, L_2 such that for all $x \in \mathbb{R}^d$*

$$\left\| \frac{\partial P}{\partial x} \right\| \le L_1, \qquad \left\| \frac{\partial^2 P}{\partial x^2} \right\| \le L_2.$$

then for $p \ge 2$,

$$\begin{aligned} &\mathbb{E}\left[\left(\tfrac{1}{2}(P(X^f) + P(X^a)) - P(X^c)\right)^p\right] \\ &\quad\le\; 2^{p-1} L_1^p \, \mathbb{E}\left[\left\|\tfrac{1}{2}(X^f + X^a) - X^c\right\|^p\right] \;+\; 2^{-(p+1)} L_2^p \, \mathbb{E}\left[\left\|X^f - X^a\right\|^{2p}\right]. \end{aligned}$$

In the multidimensional SDE applications considered in finance, the Milstein approximation with the Lévy areas set to zero, combined with the antithetic construction, leads to $X^f - X^a = O(\Delta t^{1/2})$ but $\overline{X}^f - X^c = O(\Delta t)$. Hence, the variance $\mathbb{V}[\frac{1}{2}(P_l^f + P_l^a) - P_{l-1}^c]$ is $O(\Delta t^2)$, which is the order obtained for scalar SDEs using the Milstein discretization with its first order strong convergence.

6.2. *Clark-Cameron Example*

The paper of Clark and Cameron [9] addresses the question of how accurately one can approximate the solution of an SDE driven by an underlying multi-dimensional Brownian motion, using only uniformly-spaced discrete Brownian increments. Their model problem is

$$\begin{aligned} \mathrm{d}x_1(t) &= \mathrm{d}w_1(t) \\ \mathrm{d}x_2(t) &= x_1(t)\,\mathrm{d}w_2(t), \end{aligned} \tag{34}$$

with $x(0)=y(0)=0$, and zero correlation between the two Brownian motions $w_1(t)$ and $w_2(t)$. These equations can be integrated exactly over a time interval $[t_n, t_{n+1}]$, where $t_n = n\,\Delta t$, to give

$$\begin{aligned} x_1(t_{n+1}) &= x_1(t_n) + \Delta w_{1,n} \\ x_2(t_{n+1}) &= x_2(t_n) + x_1(t_n)\Delta w_{2,n} + \tfrac{1}{2}\Delta w_{1,n}\Delta w_{2,n} + \tfrac{1}{2}A_{12,n} \end{aligned} \tag{35}$$

where $\Delta w_{i,n} \equiv w_i(t_{n+1}) - w_i(t_n)$, and $A_{12,n}$ is the Lévy area defined as

$$A_{12,n} = \int_{t_n}^{t_{n+1}} \Big(w_1(t) - w_1(t_n)\Big)\,\mathrm{d}w_2(t) - \int_{t_n}^{t_{n+1}} \Big(w_2(t) - w_2(t_n)\Big)\,\mathrm{d}w_1(t).$$

This corresponds exactly to the Milstein discretization presented in (7), so for this simple model problem the Milstein discretization is exact.

The point of Clark and Cameron's paper is that for *any* numerical approximation $X(T)$ based solely on the set of discrete Brownian increments Δw,

$$\mathbb{E}[(x_2(T) - X_2(T))^2] \geq \tfrac{1}{4}\,T\,\Delta t.$$

Since in this section we use superscript f, a, c for fine X^f, antithetic X^a and coarse X^c approximations, respectively, we drop the superscript ℓ for the clarity of notation.

We define a coarse path approximation X^c with timestep Δt by neglecting the Lévy area terms to give

$$\begin{aligned} X^c_{1,n+1} &= X^c_{1,n} + \Delta w^{\ell-1}_{1,n+1} \\ X^c_{2,n+1} &= X^c_{2,n} + X^c_{1,n}\Delta w^{\ell-1}_{2,n+1} + \tfrac{1}{2}\,\Delta w^{\ell-1}_{1,n+1}\,\Delta w^{\ell-1}_{2,n+1} \end{aligned} \tag{36}$$

This is equivalent to replacing the true Brownian path by a piecewise linear approximation as illustrated in Figure 1.

Similarly, we define the corresponding two half-timesteps of the first fine path approximation X^f by

$$\begin{aligned} X^f_{1,n+\frac{1}{2}} &= X^f_{1,n} + \Delta w^{\ell}_{1,n+\frac{1}{2}} \\ X^f_{2,n+\frac{1}{2}} &= X^f_{2,n} + X^f_{1,n}\,\Delta w^{\ell}_{2,n+\frac{1}{2}} + \tfrac{1}{2}\,\Delta w^{\ell}_{1,n+\frac{1}{2}}\,\Delta w^{\ell}_{2,n+\frac{1}{2}} \\ X^f_{1,n+1} &= X^f_{1,n+1} + \Delta w^{\ell}_{1,n+1} \\ X^f_{2,n+1} &= X^f_{2,n+\frac{1}{2}} + X^f_{1,n+\frac{1}{2}}\,\Delta w^{\ell}_{2,n+1} + \tfrac{1}{2}\,\Delta w^{\ell}_{1,n+1}\,\Delta w^{\ell}_{2,n+1} \end{aligned}$$

where $\Delta w^{\ell-1}_{n+1} = \Delta w^{\ell}_{n+\frac{1}{2}} + \Delta w^{\ell}_{n+1}$. Using this relation, the equations for the two fine timesteps can be combined to give an equation for the increment over the coarse timestep,

$$\begin{aligned} X^f_{1,n+1} &= X^f_{1,n} + \Delta w^{\ell-1}_{1,n+1} \\ X^f_{2,n+1} &= X^f_{2,n} + X^f_{1,n}\,\Delta w^{\ell-1}_{2,n+1} + \tfrac{1}{2}\,\Delta w^{\ell-1}_{1,n+1}\,\Delta w^{\ell-1}_{2,n+1} \\ &\quad + \tfrac{1}{2}\left(\Delta w^{\ell}_{1,n+\frac{1}{2}}\,\Delta w^{\ell}_{2,n+1} - \Delta w^{\ell}_{2,n+\frac{1}{2}}\,\Delta w^{\ell}_{1,n+1}\right). \end{aligned} \tag{37}$$

The antithetic approximation X^a_n is defined by exactly the same discretization except that the Brownian increments δw_n and $\delta w_{n+\frac{1}{2}}$ are swapped, as illustrated in Figure 1. This gives

$$\begin{aligned} X^a_{1,n+\frac{1}{2}} &= X^a_{1,n} + \Delta w^{\ell}_{1,n+1}, \\ X^a_{2,n+\frac{1}{2}} &= X^a_{2,n} + X^a_{1,n}\,\Delta w^{\ell}_{2,n+1} + \tfrac{1}{2}\,\Delta w^{\ell}_{1,n+1}\,\Delta w^{\ell}_{2,n+1}, \\ X^a_{1,n+1} &= X^a_{1,n+\frac{1}{2}} + \Delta w^{\ell}_{1,n+\frac{1}{2}}, \\ X^a_{2,n+1} &= X^a_{2,n+\frac{1}{2}} + X^a_{1,n+\frac{1}{2}}\,\Delta w^{\ell}_{2,n+\frac{1}{2}} + \tfrac{1}{2}\,\Delta w^{\ell}_{1,n+\frac{1}{2}}\,\Delta w^{\ell}_{2,n+\frac{1}{2}}, \end{aligned}$$

and hence

$$\begin{aligned} X^a_{1,n+1} &= X^a_{1,n} + \Delta w^{\ell-1}_{1,n+1}, \\ X^a_{2,n+1} &= X^a_{2,n} + X^a_{1,n}\,\Delta w^{\ell-1}_{2,n+1} + \tfrac{1}{2}\,\Delta w^{\ell-1}_{1,n+1}\,\Delta w^{\ell-1}_{2,n+1} \\ &\quad - \tfrac{1}{2}\left(\Delta w^{\ell}_{1,n+\frac{1}{2}}\,\Delta w^{\ell}_{2,n+1} - \Delta w^{\ell}_{2,n+\frac{1}{2}}\,\Delta w^{\ell}_{1,n+1}\right). \end{aligned} \tag{38}$$

Swapping $\Delta w^{\ell}_{n+\frac{1}{2}}$ and Δw^{ℓ}_{n+1} does not change the distribution of the driving Brownian increments, and hence X^a has exactly the same distribution as X^f. Note also the change in sign in the last term in (37) compared to the corresponding term in

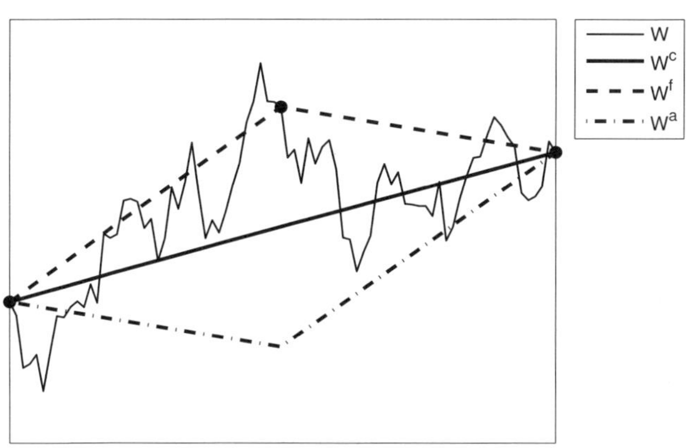

Figure 1. Brownian path and approximations over one coarse timestep

(38). This is important because these two terms cancel when the two equations are averaged.

These last terms correspond to the Lévy areas for the fine path and the antithetic path, and the sign reversal is a particular instance of a more general result for time-reversed Brownian motion, [30]. If $(w_t, 0 \le t \le 1)$ denotes a Brownian motion on the time interval $[0,1]$ then the time-reversed Brownian motion $(z_t, 0 \le t \le 1)$ defined by

$$z_t = w_1 - w_{1-t}, \tag{39}$$

has exactly the same distribution, and it can be shown that its Lévy area is equal in magnitude and opposite in sign to that of w_t.

Lemma 2. *If X^f_n, X^a_n and X^c_n are as defined above, then*

$$X^f_{1,n} = X^a_{1,n} = X^c_{1,n}, \quad \tfrac{1}{2}\left(X^f_{2,n} + X^a_{2,n}\right) = X^c_{2,n}, \quad \forall n \le N$$

and

$$\mathbb{E}\left[\left(X^f_{2,N} - X^a_{2,N}\right)^4\right] = \tfrac{3}{4}\, T\,(T+\Delta t)\,\Delta t^2.$$

In the next section we will see how this lemma generalizes to non-linear multidimensional SDEs (4).

6.3. *Milstein discretization - General theory*

Using the coarse timestep Δt, the coarse path approximation X^c_n, is given by the Milstein approximation without the Lévy area term,

$$\begin{aligned} X^c_{i,n+1} = {} & X^c_{i,n} + f_i(X^c_n)\,\Delta t_{\ell-1} + \sum_{j=1}^{m} g_{ij}(X^c_n)\,\Delta w^{\ell-1}_{j,n+1} \\ & + \sum_{j,k=1}^{m} h_{ijk}(X^c_n)\left(\Delta w_{j,n}\,\Delta w^{\ell-1}_{k,n+1} - \Omega_{jk}\,\Delta t_{\ell-1}\right). \end{aligned}$$

The first fine path approximation X^f_n (that corresponds to X^c_n) uses the corresponding discretization with timestep $\Delta t/2$,

$$\begin{aligned} X^f_{i,n+\frac{1}{2}} = {} & X^f_{i,n} + f_i(X^f_n)\,\Delta t_{\ell-1}/2 + \sum_{j=1}^{m} g_{ij}(X^f_n)\,\Delta w^{\ell}_{j,n+\frac{1}{2}} \\ & + \sum_{j,k=1}^{m} h_{ijk}(X^f_n)\left(\Delta w^{\ell}_{j,n+\frac{1}{2}}\,\Delta w^{\ell}_{k,n+\frac{1}{2}} - \Omega_{jk}\,\Delta t_{\ell-1}/2\right), \end{aligned} \tag{40}$$

$$
\begin{aligned}
X^f_{i,n+1} &= X^f_{i,n+\frac{1}{2}} + f_i(X^f_{n+\frac{1}{2}})\,\Delta t_{\ell-1}/2 + \sum_{j=1}^{m} g_{ij}(X^f_{n+\frac{1}{2}})\,\Delta w^{\ell}_{j,n+1} \\
&\quad + \sum_{j,k=1}^{m} h_{ijk}(X^f_{n+\frac{1}{2}})\left(\Delta w^{\ell}_{j,n+1}\,\Delta w^{\ell}_{k,n+1} - \Omega_{jk}\,\Delta t_{\ell-1}/2\right),
\end{aligned}
\tag{41}
$$

where $\Delta w^{\ell-1}_{n+1} = \Delta w^{\ell}_{n+\frac{1}{2}} + \Delta w^{\ell}_{n+1}$.

The antithetic approximation X^a_n is defined by exactly the same discretization except that the Brownian increments $Dw^{\ell}_{n+\frac{1}{2}}$ and Δw^{ℓ}_{n+1} are swapped, so that

$$
\begin{aligned}
X^a_{i,n+\frac{1}{2}} &= X^a_{i,n} + f_i(X^a_n)\,\Delta t_{\ell-1}/2 + \sum_{j=1}^{m} g_{ij}(X^a_n)\,\delta w_{n+\frac{1}{2}} \\
&\quad + \sum_{j,k=1}^{m} h_{ijk}(X^a_n)\left(\Delta w^{\ell}_{j,n+1}\,\Delta w^{\ell}_{k,n+1} - \Omega_{jk}\,\Delta t_{\ell-1}/2\right), \\
X^a_{i,n+1} &= X^a_{i,n+\frac{1}{2}} + f_i(X^a_{n+\frac{1}{2}})\,\Delta t_{\ell-1}/2 + \sum_{j=1}^{m} g_{ij}(X^a_{n+\frac{1}{2}})\,\Delta w^{\ell}_{j,n+\frac{1}{2}} \\
&\quad + \sum_{j,k=1}^{m} h_{ijk}(X^a_{n+\frac{1}{2}})\left(\Delta w^{\ell}_{j,n+\frac{1}{2}}\,\Delta w^{\ell}_{k,n+\frac{1}{2}} - \Omega_{jk}\,\Delta t_{\ell-1}/2\right).
\end{aligned}
\tag{42}
$$

It can be shown that [22]

Lemma 3. *For all integers $p \geq 2$, there exists a constant K_p such that*

$$
\mathbb{E}\left[\max_{0\leq n\leq N} \|X^f_n - X^a_n\|^p\right] \leq K_p\,\Delta t^{p/2}.
$$

Let's denote the average fine and antithetic path as follows

$$
\overline{X}^f_n \equiv \tfrac{1}{2}(X^f_n + X^a_n).
$$

The main results of [22] is the following theorem:

Theorem 5. *For all $p \geq 2$, there exists a constant K_p such that*

$$
\mathbb{E}\left[\max_{0\leq n\leq N} \|\overline{X}^f_n - X^c_n\|^p\right] \leq K_p\,\Delta t^{p}.
$$

This together with a classical strong convergence result for Milstein discretization allows to estimate the MLMC variance for smooth payoffs. n the case of payoff which is a smooth function of the final state $x(T)$, taking $p=2$ in Lemma 1, $p=4$ in Lemma 3 and $p=2$ in Theorem 5, immediately gives the result that the multilevel variance

$$
\mathbb{V}\left[\tfrac{1}{2}\left(P(X^f_N) + P(X^a_N)\right) - P(X^c_N)\right]
$$

has an $O(\Delta t^2)$ upper bound. This matches the convergence rate for the multilevel method for scalar SDEs using the standard first order Milstein discretization, and is much better than the $O(\Delta t)$ convergence obtained with the Euler-Maruyama discretization.

However, very few financial payoff functions are twice differentiable on the entire domain $\mathbb{R}^d$. A more typical 2D example is a call option based on the minimum of two assets,

$$P(x(T)) \equiv \max\left(0, \min(x_1(T), x_2(T)) - K\right),$$

which is piecewise linear, with a discontinuity in the gradient along the three lines (s, K), (K, s) and (s, s) for $s \geq K$.

To handle such payoffs, an assumption which bounds the probability of the solution of the SDE having a value at time T close to such lines with discontinuous gradients is needed.

Assumption 6. The payoff function $P \in C(\mathbb{R}^d, \mathbb{R})$ has a uniform Lipschitz bound, so that there exists a constant L such that

$$|P(x) - P(y)| \leq L\, |x - y|, \quad \forall\, x, y \in \mathbb{R}^d,$$

and the first and second derivatives exist, are continuous and have uniform bound L at all points $x \notin K$, where K is a set of zero measure, and there exists a constant c such that the probability of the SDE solution $x(T)$ being within a neighborhood of the set K has the bound

$$\mathbb{P}\left(\min_{y \in K} \|x(T) - y\| \leq \varepsilon\right) \leq c\, \varepsilon, \quad \forall\, \varepsilon > 0.$$

In a 1D context, Assumption 6 corresponds to an assumption of a locally bounded density for $x(T)$.

Giles and Szpruch in [22] proved the following result

Theorem 7. *If the payoff satisfies Assumption 6, then*

$$\mathbb{E}\left[\left(\tfrac{1}{2}(P(X_N^f) + P(X_N^a)) - P(X_N^c)\right)^2\right] = o(\Delta t^{3/2-\delta})$$

for any $\delta > 0$.

6.4. *Piecewise linear interpolation analysis*

The piecewise linear interpolant $X^c(t)$ for the coarse path is defined within the coarse timestep interval $[t_k, t_{k+1}]$ as

$$X^c(t) \equiv (1-\lambda)\, X_k^c + \lambda\, X_{k+1}^c, \quad \lambda \equiv \frac{t - t_k}{t_{k+1} - t_k}.$$

Likewise, the piecewise linear interpolants $X^f(t)$ and $X^a(t)$ are defined on the fine timestep $[t_k, t_{k+\frac{1}{2}}]$ as

$$X^f(t) \equiv (1-\lambda)\, X^f_k + \lambda\, X^f_{k+\frac{1}{2}}, \quad X^a(t) \equiv (1-\lambda)\, X^a_k + \lambda\, X^a_{k+\frac{1}{2}}, \quad \lambda \equiv \frac{t - t_k}{t_{k+\frac{1}{2}} - t_k},$$

and there is a corresponding definition for the fine timestep $[t_{k+\frac{1}{2}}, t_{k+1}]$. It can be shown that [22]

Theorem 8. *For all $p \geq 2$, there exists a constant K_p such that*

$$\sup_{0\leq t\leq T} \mathbb{E}\left[\, \|X^f(t) - X^a(t)\|^p \right] \leq K_p\, \Delta t^{p/2},$$

$$\sup_{0\leq t\leq T} \mathbb{E}\left[\left\|\overline{X}^f(t) - X^c(t)\right\|^p \right] \leq K_p\, \Delta t^p,$$

where $\overline{X}^f(t)$ is the average of the piecewise linear interpolants $X^f(t)$ and $X^a(t)$.

For an Asian option, the payoff depends on the average

$$x_{ave} \equiv T^{-1} \int_0^T x(t)\, \mathrm{d}t.$$

This can be approximated by integrating the appropriate piecewise linear interpolant which gives

$$\begin{aligned}
X^c_{ave} &\equiv T^{-1} \int_0^T X^c(t)\, \mathrm{d}t = N^{-1} \sum_{n=0}^{N-1} \tfrac{1}{2}(X^c_n + X^c_{n+1}),\\
X^f_{ave} &\equiv T^{-1} \int_0^T X^f(t)\, \mathrm{d}t = N^{-1} \sum_{n=0}^{N-1} \tfrac{1}{4}(X^f_n + 2X^f_{n+\frac{1}{2}} + X^f_{n+1}),\\
X^a_{ave} &\equiv T^{-1} \int_0^T X^a(t)\, \mathrm{d}t = N^{-1} \sum_{n=0}^{N-1} \tfrac{1}{4}(X^a_n + 2X^a_{n+\frac{1}{2}} + X^a_{n+1}).
\end{aligned}$$

Due to Hölder's inequality,

$$\begin{aligned}
\mathbb{E}\left[\|X^f_{ave} - X^a_{ave}\|^p \right] &\leq T^{-1} \int_0^T \mathbb{E}\left[\|X^f(t) - X^a(t)\|^p \right] \mathrm{d}t\\
&\leq \sup_{[0,T]} \mathbb{E}\left[\|X^f(t) - X^a(t)\|^p \right],
\end{aligned}$$

and similarly

$$\mathbb{E}\left[\|\tfrac{1}{2}(X^f_{ave} + X^a_{ave}) - X^c_{ave}\|^p \right] \leq \sup_{[0,T]} \mathbb{E}\left[\left\|\overline{X}^f(t) - X^c(t)\right\|^p \right].$$

Hence, if the Asian payoff is a smooth function of the average, then we obtain a second order bound for the multilevel correction variance.

This analysis can be extended to include payoffs which are a smooth function of a number of intermediate variables, each of which is a linear functional of the path $x(t)$ of the form

$$\int_0^T g^T(t)\, x(t)\, \mu(\mathrm{d}t),$$

for some vector function $g(t)$ and measure $\mu(\mathrm{d}t)$. This includes weighted averages of $x(t)$ at a number of discrete times, as well as continuously-weighted averages over the whole time interval.

As with the European options, the analysis can also be extended to payoffs which are Lipschitz functions of the average, and have first and second derivatives which exist, and are continuous and uniformly bounded, except for a set of points K of zero measure.

Assumption 9. The payoff $P \in C(\mathbb{R}^d, \mathbb{R})$ has a uniform Lipschitz bound, so that there exists a constant L such that

$$|P(x) - P(y)| \leq L\, |x-y|\,, \quad \forall\, x, y \in \mathbb{R}^d,$$

and the first and second derivatives exist, are continuous and have uniform bound L at all points $x \notin K$, where K is a set of zero measure, and there exists a constant c such that the probability of x_{ave} being within a neighborhood of the set K has the bound

$$\mathbb{P}\left(\min_{y\in K} \|x_{ave} - y\| \leq \varepsilon\right) \leq c\, \varepsilon, \quad \forall\, \varepsilon > 0.$$

Theorem 10. *If the payoff satisfies Assumption 9, then*

$$\mathbb{E}\left[\left(\tfrac{1}{2}(P(X^f_{ave}) + P(X^a_{ave})) - P(X^c_{ave})\right)^2\right] = o(\Delta t^{3/2-\delta})$$

for any $\delta > 0$.

We refer the reader to [22] for more details.

6.5. *Simulations for antithetic Monte Carlo*

Here we present numerical simulations for a European option for process simulated by X^f, X^a and X^c defined in section 6.2 with initial conditions $x_1(0) = x_2(0) = 1$. The results in Figure 2 are for a European call option with terminal time 1 and strike $K = 1$, that is $P = (x(T) - K)^+$ The top left plot shows the behavior of the variance of both P_ℓ and $P_\ell - P_{\ell-1}$. The superimposed reference slope with rate 1.5 indicates that the variance $\mathbb{V}_\ell = \mathbb{V}[P_\ell - P_{\ell-1}] = O(\Delta t_\ell^{1.5})$, corresponding to $O(\epsilon^{-2})$ computational complexity of antithetic MLMC. The top right plot shows that $\mathbb{E}[P_\ell - P_{\ell-1}] = O(\Delta t_\ell)$. The bottom left plot shows the computational complexity C (as defined in Theorem 1) with desired accuracy ϵ. The plot is of $\epsilon^2\, C$ versus ϵ, because we expect to see that $\epsilon^2 C$ is only weakly dependent on ϵ for MLMC. For standard Monte Carlo, theory predicts that $\epsilon^2 C$ should be proportional to the number of timesteps on the finest level, which in turn is roughly proportional to ϵ^{-1} due to the weak convergence order. For accuracy $\varepsilon = 10^{-4}$, the antithetic MLMC is approximately 500 times more efficient than the standard Monte Carlo. The bottom right plot shows that $\mathbb{V}[X_{1.\ell} - X_{1.\ell-1}] = O(\Delta t_\ell)$. This corresponds to standard strong convergence of order 0.5. We have also tested the algorithm presented in

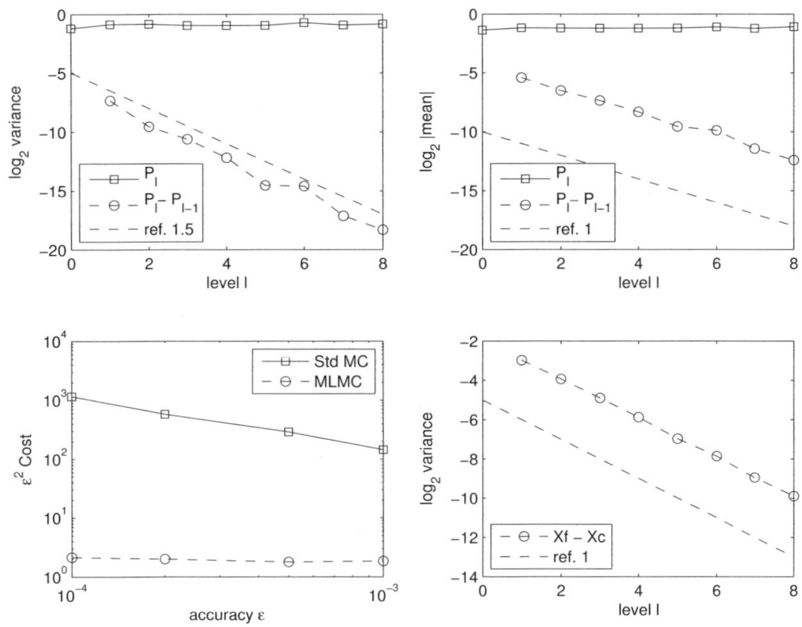

Figure 2. Call option.

[22] for approximation of Asian options. Our results were almost identical as for European options. In order to treat the lookback, digital and barrier options we found that a suitable antithetic approximation to the Lévy areas are needed. For suitable modification of the antithetic MLMC estimator we performed numerical experiments where we obtained $O(\epsilon^{-2}\log(\epsilon)^2)$ complexity for estimating barrier, digital and lookback options. Currently, we are working on theoretical justification of our results.

7. Other uses of multilevel method

7.1. *SPDEs*

Multilevel method has been used for a number of parabolic and elliptic SPDE applications [4; 10; 27] but the first use for a financial SPDE is in a new paper by Giles & Reisinger [21].

This paper considers an unusual SPDE which results from modelling credit default probabilities,

$$\Delta p = -\mu \frac{\partial p}{\partial x} \Delta t + \frac{1}{2} \frac{\partial^2 p}{\partial x^2} \Delta t - \sqrt{\rho} \frac{\partial p}{\partial x} \Delta M_t, \quad x > 0 \tag{43}$$

subject to boundary condition $p(0, t) = 0$. Here $p(x, t)$ represents the probability density function for firms being a distance x from default at time t. The diffusive term is due to idiosyncratic factors affecting individual firms, while the stochastic term due to the scalar Brownian motion M_t corresponds to the systemic movement due to random market effects affecting all firms.

Using a Milstein time discretization with uniform timestep k, and a central space discretization of the spatial derivatives with uniform spacing h gives the numerical approximation

$$\begin{aligned} p_j^{n+1} = p_j^n \; &- \frac{\mu\, k + \sqrt{\rho\, k}\, Z_n}{2h} \left(p_{j+1}^n - p_{j-1}^n\right) \\ &+ \frac{(1-\rho)\, k + \rho\, k\, Z_n^2}{2h^2} \left(p_{j+1}^n - 2p_j^n + p_{j-1}^n\right), \end{aligned} \tag{44}$$

where the Z_n are standard Normal random variables so that $\sqrt{h} Z_n$ corresponds to an increment of the driving scalar Brownian motion.

The paper shows that the requirment for mean-square stability as the grid is refined and $k, h \to 0$ is $k/h^2 \leq (1+2\rho^2)^{-1}$, and in addition the accuracy is $O(k, h^2)$. Because of this, the multilevel treatment considers a sequence of grids with $h_\ell = 2\, h_{\ell-1}$, $k_\ell = 4\, k_{\ell-1}$.

The multilevel implementation is very straightforward, with the Brownian increments for the fine path being summed pairwise to give the corresponding Brownian increments for the coarse path. The payoff corresponds to different tranches of a credit derivative that depends on a numerical approximation of the integral

$$\int_0^\infty p(x, t)\; \Delta x.$$

The computational cost increases by factor 8 on each level, and numerical experiments indicate that the variance decreases by factor 16. The MLMC Theorem still applies in this case, with $\beta = 4$ and $\gamma = 3$, and so the overall computational complexity to achieve an $O(\epsilon)$ RMS error is again $\mathcal{O}(\epsilon^{-2})$.

7.2. *Nested simulation*

The pricing of American options is one of the big challenges for Monte Carlo methods in computational finance, and Belomestny & Schoenmakers have recently written a very interesting paper on the use of multilevel Monte Carlo for this purpose [5]. Their method is based on Anderson & Broadie's dual simulation method [1] in which a key component at each timestep in the simulation is to estimate a conditional expectation using a number of sub-paths.

In their multilevel treatment, Belomestny & Schoenmakers use the same uniform timestep on all levels of the simulation. The quantity which changes between different levels of simulation is the number of sub-samples used to estimate the conditional expectation.

To couple the coarse and fine levels, the fine level uses N_ℓ sub-samples, and the coarse level uses $N_{\ell-1} = N_\ell/2$ of them. Similar research by N. Chen * found the multilevel correction variance is reduced if the payoff on the coarse level is replaced by an average of the payoffs obtained using the first $N_\ell/2$ and the second $N_\ell/2$ samples. This is similar in some ways to the antithetic approach described in section 6.

In future research, Belomestny & Schoenmakers intend to also change the number of timesteps on each level, to increase the overall computational benefits of the multilevel approach.

7.3. *Truncated series expansions*

Building on earlier work by Broadie and Kaya [7], Glasserman and Kim have recently developed an efficient method [25] of exactly simulating the Heston stochastic volatility model [19].

The key to their algorithm is a method of representing the integrated volatility over a time interval $[0,T]$, conditional on the initial and final values, v_0 and v_T as

$$\left(\int_0^T V_s\, ds \;\middle|\; V_0 = v_0, V_T = v_T \right) \overset{d}{=} \sum_{n=1}^{\infty} x_n + \sum_{n=1}^{\infty} y_n + \sum_{n=1}^{\infty} z_n$$

where x_n, y_n, z_n are independent random variables.

In practice, they truncate the series expansions at a level which ensures the desired accuracy, but a more severe truncation would lead to a tradeoff between accuracy and computational cost. This makes the algorithm a candidate for a multilevel treatment in which level ℓ computation performs the truncation at N_ℓ (taken to be the same for all three series, for simplicity).

To give more details, the level ℓ computation would use

$$\sum_{n=1}^{N_\ell} x_n + \sum_{n=1}^{N_\ell} y_n + \sum_{n=1}^{N_\ell} z_n$$

while the level $\ell-1$ computation would use

$$\sum_{n=1}^{N_{\ell-1}} x_n + \sum_{n=1}^{N_{\ell-1}} y_n + \sum_{n=1}^{N_{\ell-1}} z_n$$

with the same random variables x_n, y_n, z_n.

This kind of multilevel treatment has not been tested experimentally, but it seems that it might yield some computational savings even though Glasserman

*unpublished, but presented at the MCQMC12 conference.

and Kim typically retain only 10 terms in their summations through the use of a carefully constructed estimator for the truncated remainder. In other circumstances requiring more terms to be retained, the savings may be larger.

7.4. *Mixed precision arithmetic*

The final example of the use of multilevel is unusual, because it concerns the computer implementation of Monte Carlo algorithms.

In the latest CPUs from Intel and AMD, each core has a vector unit which can perform 8 single precision or 4 double precision operations with one instruction. Together with the obvious fact that double precision variables are twice as big as single precision variables and so require twice as much time to transfer, in bulk, it leads to single precision computations being twice as fast as double precision computations. On GPUs (graphics processors) the difference in performance can be even larger, up to a factor of eight in the most extreme cases.

This raises the question of whether single precision arithmetic is sufficient for Monte Carlo simulation. In general, our view is that the errors due to single precision arithmetic are much smaller than the errors due to

- statistical error due to Monte Carlo sampling;
- bias due to SDE discretization;
- model uncertainty.

We have just two concerns with single precision accuracy:

- there can be significant errors when averaging the payoffs unless one uses binary tree summation to perform the summation;
- when computing Greeks using "bumping", the single precision inaccuracy can be greatly amplified if a small bump is used.

Our advice would be to always use double precision for the final accumulation of payoff values, and pathwise sensitivity analysis as much as possible for computing Greeks, but if there remains a need for the path simulation to be performed in double precision then one could use two-level approach in which level 0 corresponds to single precision and level 1 corresponds to double precison.

On both levels one would use the same random numbers. The multilevel analysis would then give the optimal allocation of effort between the single precision and double precision computations. Since it is likely that most of the calculations would be single precision, the computational savings would be a factor two or more compared to standard double precision calculations.

8. Multilevel Quasi-Monte Carlo

In Theorem 1, if $\beta > \gamma$, so that rate at which the multilevel variance decays with increasing grid level is greater than the rate at which the computational cost increases, then the dominant computational cost is on the coarsest levels of approximation.

Since coarse levels of approximation correspond to low-dimensional numerical quadrature, it is quite natural to consider the use of quasi-Monte Carlo techniques. This has been investigated by Giles & Waterhouse [23] in the context of scalar SDEs with a Lipschitz payoff. Using the Milstein approximation with a doubling of the number of timesteps on each level gives $\beta = 2$ and $\gamma = 1$. They used a rank-1 lattice rule to generate the quasi-random numbers, randomisation with 32 independent offsets to obtain confidence intervals, and a standard Brownian Bridge construction of the increments of the driving Brownian process.

Their empirical observation was that MLMC on its own was better than QMC on its own, but the combination of the two was even better. The QMC treatment greatly reduced the variance per sample for the coarsest levels, resulting in significantly reduced costs overall. In the simplest case of a European call option, shown in Figure 3, the top left plot shows the reduction in the variance per sample as the number of QMC points is increased.

The benefit is much greater on the coarsest levels than on the finest levels. In the bottom two plots, the number of QMC points on each level is determined automatically to obtain the required accuracy; see [23] for the precise details. Overall, the computational complexity appears to be reduced from $O(\epsilon^{-2})$ to approximately $O(\epsilon^{-1.5})$.

Giles & Waterhouse interpreted the fact that the variance is not reduced on the finest levels as being due to a lack of significant low-dimensional content. i.e. the difference in the two payoffs due to neighboring grid levels is due to the difference in resolution of the driving Brownian path, and this is inherently of high dimensionality. This suggests that in other applications with $\beta < \gamma$, which would lead to the dominant cost being on the finest levels, then the use of quasi-Monte Carlo methods is unlikely to yield any benefits.

Further research is needed in this area to investigate the use of other low-discrepancy sequences (e.g. Sobol) and other ways of generating the Brownian increments (e.g. PCA). We also refer the reader to [15] for some results for randomized multilevel quasi-Monte Carlo.

9. Conclusions

In the past 6 years, considerable progress has been achieved with the multilevel Monte Carlo method for financial options based on underlying assets described by Brownian diffusions, jump diffusions, and more general Lévy processes.

The multilevel approach is conceptually very simple. In essence it is a recursive control variate strategy, using a coarse path simulation as a control variate for a fine path simulation, relying on strong convergence properties to ensure a very strong correlation between the two.

In practice, the challenge is to couple the coarse and fine path simulations as

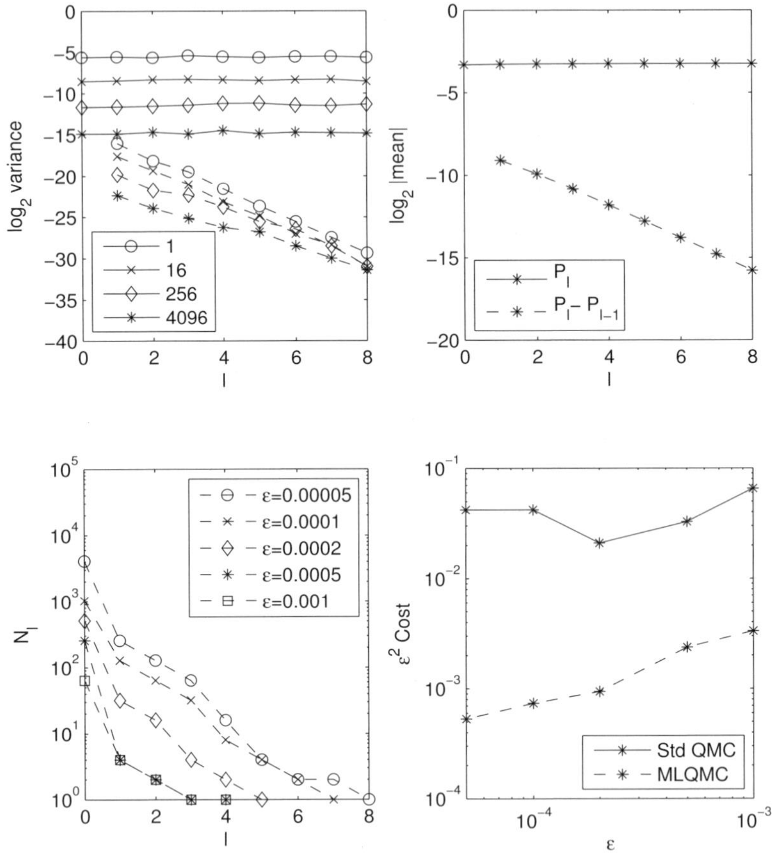

Figure 3. European call option (Fig 6.1 from [23])

tightly as possible, minimizing the difference in the payoffs obtained for each. In doing this, there is considerable freedom to be creative, as shown in the use of Brownian Bridge constructions to improve the variance for lookback and barrier options, and in the antithetic estimators for multi-dimensinal SDEs which would require the simulation of Lévy areas to achieve first order strong convergence. Another challenge is avoiding large payoff differences due to discontinuous payoffs; here one can often use either conditional expectations to smooth the payoff, or a change of measure to ensure that the coarse and fine paths are on the same side of the discontinuity.

Overall, multilevel methods are being used for an increasingly wide range of applications. This biggest savings are in situations in which the coarsest approximation is very much cheaper than the finest. If the finest level of approaximation has only 32 timesteps, then there are very limited savings to be achieved, but if the finest level has 256 timesteps, then the potential savings are much larger.

Looking to the future, exciting areas for further research include:

- more research on multilevel techniques for American and Bermudan options;
- more investigation of multilevel Quasi Monte Carlo methods;
- use of multilevel ideas for completely new financial applications, such as Gaussian copula and new SPDE models.

Bibliography

[1] Andersen, L. and Broadie, M.: A primal-dual simulation algorithm for pricing multidimensional American options. *Management Science*, 50(9):1222–1234, 2004.

[2] Asmussen, A. and Glynn, P.: *Stochastic Simulation*. Springer, New York, 2007.

[3] Avikainen, R.: On irregular functionals of SDEs and the Euler scheme. *Finance and Stochastics*, 13(3):381–401, 2009.

[4] Barth, A., Schwab, C. and Zollinger, N.: Multi-level Monte Carlo finite element method for elliptic PDEs with stochastic coefficients. *Numerische Mathematik*, 119(1):123–161, 2011.

[5] Belomestny, D. and Schoenmakers, J.: Multilevel dual approach for pricing American style derivatives. Preprint 1647, WIAS, 2011.

[6] Broadie, M., Glasserman, P. and Kou, S.. A continuity correction for discrete barrier options. *Mathematical Finance*, 7(4):325–348, 1997.

[7] Broadie, M. and Kaya, O.: Exact simulation of stochastic volatility and other affine jump diffusion processes. 54(2):217–231, 2006.

[8] Burgos, S. and Giles, M.B.: Computing Greeks using multilevel path simulation. In L. Plaskota and H. Woźniakowski, editors, *Monte Carlo and Quasi-Monte Carlo Methods 2010*. Springer-Verlag, 2012.

[9] Clark, J.M.C. and Cameron, R.J.: The maximum rate of convergence of discrete approximations for stochastic differential equations. In B. Grigelionis, editor, *Stochastic Differential Equations, no. 25 in Lecture Notes in Control and Information Sciences*. Springer-Verlag, 1980.

[10] Cliffe, K.A., Giles, M.B., Scheichl, R. and Teckentrup, A.: Multilevel Monte Carlo methods and applications to elliptic PDEs with random coefficients. *Computing and Visualization in Science*, 14(1):3–15, 2011.

[11] Debrabant, K., Giles, M.B. and Rossler, A.: Numerical analysis of multilevel monte carlo path simulation using milstein discretization: scalar case. *Technical report*, 2011.

[12] Dereich, S.: Multilevel Monte Carlo algorithms for Lévy-driven SDEs with Gaussian correction. *Annals of Applied Probability*, 21(1):283–311, 2011.

[13] Dereich, S. and Heidenreich, F. A multilevel monte carlo algorithm for lévy-driven stochastic differential equations. *Stochastic Processes and their Applications*, 2011.

[14] Gaines, J.G. and Lyons, T.J.: Random generation of stochastic integrals. *SIAM Journal of Applied Mathematics*, 54(4):1132–1146, 1994.

[15] Gerstner, T. and Noll, M.: Randomized multilevel quasi-Monte Carlo path simulation. *Chapter in this book*, 2012.

[16] Giles, M.B.: Monte Carlo evaluation of sensitivities in computational finance. Technical Report NA07/12, 2007.

[17] Giles, M.B.: Improved multilevel Monte Carlo convergence using the Milstein scheme. In A. Keller, S. Heinrich, and H. Niederreiter, editors, *Monte Carlo and Quasi-Monte Carlo Methods 2006*, pages 343–358. Springer-Verlag, 2008.

[18] Giles, M.B.: Multilevel Monte Carlo path simulation. *Operations Research*, 56(3):607–617, 2008.

[19] Giles, M.B.: Multilevel Monte Carlo for basket options. In *Proceedings of the Winter Simulation Conference 2009*, 2009.

[20] Giles, M.B., Higham, D.J. and Mao, X.: Analysing multilevel Monte Carlo for options with non-globally Lipschitz payoff. *Finance and Stochastics*, 13(3):403–413, 2009.

[21] Giles, M.B. and Reisinger, C.: Stochastic finite differences and multilevel Monte Carlo for a class of SPDEs in finance. *SIAM Journal of Financial Mathematics*, to appear, 2012.

[22] Giles, M.B. and Szpruch, L.: Antithetic Multilevel Monte Carlo estimation for multi-dimensional SDEs without Lévy area simulation. *Arxiv preprint arXiv:1202.6283*, 2012.

[23] Giles, M.B. and Waterhouse, B.J.: Multilevel quasi-Monte Carlo path simulation. In *Advanced Financial Modelling*, Radon Series on Computational and Applied Mathematics, pages 165–181. de Gruyter, 2009.

[24] Glasserman, P.: *Monte Carlo Methods in Financial Engineering*. Springer, New York, 2004.

[25] Glasserman, P. and Kim, K.-K.: Gamma expansion of the Heston stochastic volatility model. *Finance and Stochastics*, 15(2):267–296, 2011.

[26] Glasserman, P. and Merener, N.: Convergence of a discretization scheme for jump-diffusion processes with state-dependent intensities. *Proc. Royal Soc. London A*, 460:111–127, 2004.

[27] Graubner, S.: Multi-level Monte Carlo Methoden für stochastiche partial Differentialgleichungen. Diplomarbeit, TU Darmstadt, 2008.

[28] Heinrich, S.: *Multilevel Monte Carlo Methods*, volume 2179 of *Lecture Notes in Computer Science*, pages 58–67. Springer-Verlag, 2001.

[29] Heston, S.I.: A closed-form solution for options with stochastic volatility with applications to bond and currency options. *Review of Financial Studies*, 6:327–343, 1993.

[30] Karatzas, I. and Shreve, S.E.: *Brownian Motion and Stochastic Calculus*. Graduate Texts in Mathematics, Vol. 113. Springer, New York, 1991.

[31] Kebaier, A.: Statistical Romberg extrapolation: a new variance reduction method and applications to options pricing. *Annals of Applied Probability*, 14(4):2681–2705, 2005.

[32] Kloeden, P.E. and Neuenkirch, A.: Convergence of numerical methods for stochastic differential equations in mathematical finance. *Chapter in this book*, 2012.

[33] Kloeden, P.E., Neuenkirch, A. and Pavani, R.: Multilevel monte carlo for stochastic differential equations with additive fractional noise. *Annals of Operations Research*, pages 1–22, 2011.

[34] Kloeden, P.E. and Platen, E.: *Numerical Solution of Stochastic Differential Equations*. Springer, Berlin, 1992.

[35] Mao, X. and Szpruch, L.: Strong convergence rates for backward euler–maruyama method for non-linear dissipative-type stochastic differential equations with super-linear diffusion coefficients. *Stochastics*, 2012.

[36] Merton, R.C.: Option pricing when underlying stock returns are discontinuous. *Journal of Finance*, 3:125–144, 1976.

[37] Müller-Gronbach, T.: The optimal uniform approximation of systems of stochastic differential equations. *The Annals of Applied Probability*, 12(2):664–690, 2002.

[38] Müller-Gronbach, T.: *Strong approximation of systems of stochastic differential equations*. Habilitation thesis, TU Darmstadt, 2002.

[39] Pagès, G.: Multi-step Richardson-Romberg extrapolation: remarks on variance con-

trol and complexity. *Monte Carlo Methods and Applications*, 13(1):37–70, 2007.

[40] Platen, E. and Bruti-Liberati, N.: *Numerical Solution of Stochastic Differential Equations With Jumps in Finance*. Springer, 2010.

[41] Rydén, T. and Wiktorsson, M.: On the simulation of iterated Itô integrals. *Stochastic Processes and their Applications*, 91(1):151–168, 2001.

[42] Speight, A.L.: A multilevel approach to control variates. *Journal of Computational Finance*, 12:1–25, 2009.

[43] Speight, A.L.: Multigrid techniques in economics. *Operations Research*, 58(4):1057–1078, 2010.

[44] Szpruch, L., Mao, X., Higham, D.J. and Pan, J.: Numerical simulation of a strongly nonlinear ait-sahalia-type interest rate model. *BIT Numerical Mathematics*, 51(2):405–425, 2011.

[45] Wiktorsson, M.: Joint characteristic function and simultaneous simulation of iterated Itô integrals for multiple independent Brownian motions. *Annals of Applied Probability*, 11(2):470–487, 2001.

[46] Xia, Y.: Multilevel Monte Carlo method for jump-diffusion SDEs. *Arxiv preprint arXiv:1106.4730*, 2011.

[47] Xia, Y. and Giles, M.B.: Multilevel path simulation for jump-diffusion SDEs. In L. Plaskota and H. Woźniakowski, editors, *Monte Carlo and Quasi-Monte Carlo Methods 2010*. Springer-Verlag, 2012.

Chapter 2

Convergence of numerical methods for stochastic differential equations in mathematical finance

Peter Kloeden and Andreas Neuenkirch

Institut für Mathematik, Goethe-Universität, Robert-Mayer-Strasse 10, D-60325 Frankfurt am Main, Germany

Institut für Mathematik, Universität Mannheim, A5,6, D-68131 Mannheim, Germany

Abstract Many stochastic differential equations that occur in financial modelling do not satisfy the standard assumptions made in convergence proofs of numerical schemes that are given in textbooks, i.e., their coefficients and the corresponding derivatives appearing in the proofs are not uniformly bounded and hence, in particular, not globally Lipschitz. Specific examples are the Heston and Cox-Ingersoll-Ross models with square root coefficients and the Ait-Sahalia model with rational coefficient functions. Simple examples show that, for example, the Euler-Maruyama scheme may not converge either in the strong or weak sense when the standard assumptions do not hold. Nevertheless, new convergence results have been obtained recently for many such models in financial mathematics. These are reviewed here. Although weak convergence is of traditional importance in financial mathematics with its emphasis on expectations of functionals of the solutions, strong convergence plays a crucial role in Multi Level Monte Carlo methods, so it and also pathwise convergence will be considered along with methods which preserve the positivity of the solutions.

1. Introduction

Consider the Itô stochastic differential equation (SDE) in $\mathbb{R}^d$

$$dX_t = a(X_t)dt + \sum_{j=1}^{m} b_j(X_t)dW_t^{(j)}, \quad t \in [0,T], \qquad X_0 = x_0 \in \mathbb{R}^d \tag{1}$$

with drift and diffusion coefficients $a,\ b_j : \mathbb{R}^d \to \mathbb{R}^d$ for $j = 1,\dots,m$. Here $W_t = (W_t^{(1)},\dots,W_t^{(m)})$, $t \geq 0$, is an m-dimensional Brownian motion on a probability space $(\Omega, \mathcal{F}, \mathbf{P})$ and superscripts in brackets label components of vectors. Throughout this article it will always be assumed that equation (1) has a unique strong solution.

Explicit solutions of such equations are rarely known, thus one has to rely on numerical methods to simulate their sample paths $X_t(\omega)$ or to estimate functionals $\mathbf{E}\Phi(X)$ for some $\Phi : C([0,T];\mathbb{R}^d) \to \mathbb{R}$. Typically, such a numerical method relies on a discretization

$$0 \le t_1 \le t_2 \le \ldots \le t_n = T$$

and a global approximation on $[0,T]$ is obtained by interpolation.

In the case of the classical weak approximation the error of an approximation $\overline{X}$ to X is measured by the quantity

$$|\mathbf{E}\phi(X_T) - \mathbf{E}\phi(\overline{X}_T)|$$

for smooth functions $\phi : \mathbb{R}^d \to \mathbb{R}$. The test functions ϕ are a particular case of the general (path-dependent) functionals Φ. In the strong approximation problem the p-th mean of the difference between X and $\overline{X}$ is analyzed, i.e.

$$\left(\mathbf{E} \sup_{k=0,\ldots,n} |X_{t_k} - \overline{X}_{t_k}|^p\right)^{1/p}$$

for the maximal error in the discretization points or

$$\left(\mathbf{E} \sup_{t\in[0,T]} |X_t - \overline{X}_t|^p\right)^{1/p}$$

for the global error, where $p \geq 1$ and $|\cdot|$ denotes the Euclidean norm. Here the mean-square error, i.e. $p = 2$, is usually studied. The recent development of the Multi-level Monte Carlo method for SDEs [19; 20] has revealed that strong error bounds are crucial for the efficient computation of functionals $\mathbf{E}\Phi(X)$.

While the strong error measures the error of the approximate sample paths $\overline{X}$ on average, the pathwise error is the random quantity

$$\sup_{k=0,\ldots,n} |X_{t_k}(\omega) - \overline{X}_{t_k}(\omega)|, \qquad \omega \in \Omega$$

and

$$\sup_{t\in[0,T]} |X_t(\omega) - \overline{X}_t(\omega)|, \qquad \omega \in \Omega$$

respectively. Here the error is analyzed for a fixed $\omega \in \Omega$ without averaging. This quantity thus gives the error of the actually calculated approximation $\overline{X}_{t_k}(\omega)$, $k = 0, \ldots, n$, respectively $\overline{X}(\omega)$.

The traditional weak and strong convergence analysis for numerical methods for stochastic differential equations (SDEs) relies on the global Lipschitz assumption, i.e. the SDE coefficients satisfy

$$|a(x) - a(y)| + \sum_{j=1}^{m} |b_j(x) - b_j(y)| \le L \cdot |x - y|, \qquad x, y \in \mathbb{R}^d$$

for some $L > 0$. However, in many SDEs used for modelling in mathematical finance this assumption is violated, so the standard results (see [39; 27]) do not apply.

The Constant Elasticity of Variance Model for asset prices [12] , which was introduced by Cox in 1975, is given by the SDE

$$dS_t = \mu S_t \, dt + \sigma S_t^\gamma \, dW_t, \qquad S_0 = s_0 > 0$$

where $\mu \in \mathbb{R}$, $\sigma > 0$ and $\gamma \in (0, 1]$ and $W_t, t \geq 0$, is a one-dimensional Brownian motion. For $\gamma = 1$ this is the standard Black-Scholes model (i.e. a geometric Brownian motion), while for $\gamma \in (0, 1)$ the diffusion coefficient of this SDE is clearly not globally Lipschitz continuous. This SDE has a unique strong solution if and only if $\gamma \in [1/2, 1]$ and takes values in $[0, \infty)$.

The Ait-Sahalia model and its generalization [1; 45], which are stochastic interest rate models, follow the dynamics

$$dX_t = \big(\alpha_{-1} X_t^{-1} - \alpha_0 + \alpha_1 X_t - \alpha_2 X_t^r\big) dt + \sigma X_t^\rho dW_t, \qquad X_0 = x_0 > 0$$

where $\alpha_i, \sigma, r, \rho > 0$, $i = -1, \ldots, 2$. Under certain conditions on the parameters (see [45]), this SDE has a unique strong solution with values in $(0, \infty)$. Note that here the diffusion coefficient grows superlinearly for large values of x while the drift coefficient has a singularity at $x = 0$.

The Heston model [26], which is an asset price model with stochastic volatility, is another example for an SDE with non-Lipschitz coefficients. This SDE takes non-negative values only and contains square root coefficients:

$$\begin{aligned} dS_t &= \mu S_t \, dt + \sqrt{V_t} S_t \left(\sqrt{1 - \rho^2} \, dW_t^{(1)} + \rho \, dW_t^{(2)}\right), && S_0 = s_0 > 0 \\ dV_t &= \kappa(\lambda - V_t) \, dt + \theta \sqrt{V_t} \, dW_t^{(2)}, && V_0 = v_0 > 0. \end{aligned}$$

The parameters satisfy $\mu \in \mathbb{R}$, $\kappa, \lambda, \theta > 0$ and $\rho \in (-1, 1)$. The second component of this SDE is the Cox-Ingersoll-Ross process, which is also used as a short rate model [13].

Finally, the use of the inverse of the CIR process as volatility process leads to the so-called 3/2-model

$$\begin{aligned} dS_t &= \mu S_t \, dt + \sqrt{V_t} S_t \left(\sqrt{1 - \rho^2} \, dW_t^{(1)} + \rho \, dW_t^{(2)}\right), && S_0 = s_0 > 0 \\ dV_t &= c_1 V_t (c_2 - V_t) \, dt + c_3 V_t^{3/2} \, dW_t^{(2)}, && V_0 = v_0 > 0 \end{aligned}$$

where $c_1, c_2, c_3 > 0$, see e.g. [27].

Motivated by these and other examples, the investigation of numerical methods for SDEs with non-Lipschitz coefficients has been an active field of research in recent years. This article, provides an overview of the new developments using the above equations as illustrative examples and discussing, in particular, Euler-type schemes. For some of the above equations exact simulation methods exist, see e.g. [11; 22] and also [9] for a class of one-dimensional equations, which are superior for the simulation of the SDEs at a single or a few time points. However, if a full sample path of the SDE has to be simulated or if the SDEs under consideration are part of a larger SDE system, then discretization schemes are typically more efficient.

2. Pathwise Convergence Rates of the Euler Scheme and general Itô-Taylor Methods

The pathwise error criteria are very robust with respect to the global Lipschitz assumption. One of the simplest approximation schemes for equation (1) is the Euler scheme

$$\overline{X}_{t_{k+1}} = \overline{X}_{t_k} + a(\overline{X}_{t_k})\Delta + \sum_{j=1}^{m} b_j(\overline{X}_{t_k})\Delta_k W^{(j)}, \qquad k = 0, 1, \ldots,$$

with $\overline{X}_0 = x_0$, where $\Delta = T/n$, $t_k = k\Delta$ and $\Delta_k W = W_{t_{k+1}} - W_{t_k}$. The Euler scheme (and all other approximation methods that will be introduced below) depend on the stepsize $\Delta > 0$, hence on $n \in \mathbb{N}$, but this dependence will be omitted whenever it is clear from the context.

From the results of Gyöngy [23] it follows that the Euler scheme has pathwise convergence order $1/2 - \varepsilon$ also if the SDE coefficients are only locally Lipschitz continuous: for all $\varepsilon > 0$

$$\sup_{k=0,\ldots,n} |X_{t_k} - \overline{X}_{t_k}| \leq \eta_\varepsilon^E \cdot n^{-1/2+\varepsilon}$$

almost surely for a finite and non-negative random variable η_ε^E under the assumption that for all $N \in \mathbb{N}$ there exist constants $L_N > 0$ such that

$$|a(x) - a(y)| + \sum_{j=1}^{m} |b_j(x) - b_j(y)| \leq L_N \cdot |x - y|, \qquad |x|, |y| \leq N.$$

Thus, the pathwise convergence rate of the Euler scheme coincides up to an arbitrarily small $\varepsilon > 0$ with its strong convergence rate 1/2, but for the pathwise convergence rate no global Lipschitz assumption is required.

Jentzen, Kloeden & Neuenkirch [37] observed that this is not a specific feature of the Euler scheme but, in fact, holds for general Itô-Taylor schemes of order $\gamma = 0.5, 1.0, 1.5, \ldots$. For the definition of these schemes, see e.g. [39]. The Euler scheme corresponds to $\gamma = 0.5$, while $\gamma = 1.0$ yields the Milstein scheme

$$\begin{aligned} \overline{X}_{t_{k+1}} &= \overline{X}_{t_k} + a(\overline{X}_{t_k})\Delta + \sum_{j=1}^{m} b_j(\overline{X}_{t_k})\Delta_k W^{(j)} \\ &\quad + \sum_{j_1,j_2=1}^{m} L^{j_1} b_{j_2}(\overline{X}_{t_k}) I_{j_1,j_2}(t_k, t_{k+1}) \end{aligned}$$

with the differential operators

$$L^j = \sum_{k=1}^{d} b_j^{(k)} \frac{\partial}{\partial x^k}, \qquad j = 1, \ldots, m$$

and the iterated Itô-integrals

$$I_{j_1,j_2}(s,t) = \int_s^t \int_s^{\tau_2} dW_{\tau_1}^{(j_1)}\, dW_{\tau_2}^{(j_2)}, \qquad j_1, j_2 = 1, \ldots, m.$$

The Itô-Taylor scheme of order 1.5 is usually called the Wagner-Platen scheme.

Theorem 1. *Let $\gamma = 0.5, 1.0, 1.5, \ldots$. Assume that $a, b_1, \ldots, b_m \in C^{2\gamma+1}(\mathbb{R}^d; \mathbb{R}^d)$ and moreover let $\overline{X}^{\gamma,n}$ be the Itô-Taylor scheme of order γ with stepsize $\Delta = T/n$. Then for every $\varepsilon > 0$ there exists a non-negative random variable η_ε^γ such that*

$$\sup_{k=0,\ldots,n} \left| X_{t_k}(\omega) - \overline{X}_{t_k}^{\gamma,n}(\omega) \right| \leq \eta_\varepsilon^\gamma(\omega) \cdot n^{-\gamma+\varepsilon}$$

for almost all $\omega \in \Omega$.

The main ingredients to obtain this result are the Burkholder-Davis-Gundy inequality, which implies that all moments of an Itô-integral are equivalent, the following Borel-Cantelli-type Lemma, and a localization procedure.

Lemma 1. *(see [38]) Let $\alpha > 0$, $c_p \geq 0$ for $p \geq 1$ and let $(Z_n)_{n \in \mathbb{N}}$ be a sequence of random variables with*

$$(\mathbf{E}|Z_n|^p)^{1/p} \leq c_p \cdot n^{-\alpha}$$

for all $p \geq 1$ and $n \in \mathbb{N}$. Then for every $\varepsilon > 0$ there exists a finite and non-negative random variable η_ε such that

$$|Z_n| \leq \eta_\varepsilon \cdot n^{-\alpha+\varepsilon}$$

almost surely for all $n \in \mathbb{N}$.

The Burkholder-Davis-Gundy inequality and the Borel-Cantelli-type Lemma allow one to show that the Itô-Taylor scheme of order γ has pathwise convergence rate $\gamma - \varepsilon$ for smooth and bounded coefficients with bounded derivatives, thereby extending the classical mean-square convergence analysis in [39]. Then a localization argument is applied to avoid the boundedness assumptions. Roughly speaking, this localization argument works as follows: A fixed sample path $X_t(\omega)$, $t \in [0,T]$, of the SDE solution is bounded, i.e. stays in some open set $B(\omega)$. However for the SDE

$$dY_t = \widetilde{a}(Y_t)\, dt + \sum_{j=1}^{m} \widetilde{b}_j(Y_t)\, dW_t^{(j)}, \qquad Y_0 = x_0$$

with smooth and bounded coefficients $\widetilde{a}$, $\widetilde{b}_j$ with bounded derivatives, which coincide with the ones of the original SDE on $B(\omega)$, the solution sample path $Y_t(\omega), t \in [0,T]$, coincides with $X_t(\omega), t \in [0,T]$. Asymptotically this also holds for the corresponding sample paths of the γ-Itô-Taylor schemes, so the pathwise convergence rates carry over.

Note that all the examples of SDEs given in the introduction take non-negative values only, so good approximation schemes should preserve this structural property. The (explicit) Euler scheme is, in general, not such a scheme, since its increments are conditionally Gaussian. For example, in case of the CIR process

$$dX_t = \kappa(\lambda - X_t)\,dt + \theta\sqrt{X_t}\,dW_t, \qquad X_0 = x_0 > 0$$

the transition density of the Euler scheme reads as

$$p(y;x) = \frac{1}{\sqrt{2\pi\theta^2 x\Delta}} \exp\left(-\frac{\left(y-(x+\kappa(\lambda-x)\Delta)\right)^2}{2\theta^2 x\Delta}\right), \qquad y \in \mathbb{R},\, x>0,$$

so negative values can be obtained with positive probability even in the first step. This has lead to many ad-hoc corrections to prevent termination of the Euler scheme. The truncated Euler scheme

$$\overline{X}_{t_{k+1}} = \overline{X}_{t_k} + \kappa(\lambda - \overline{X}_{t_k})\,\Delta + \theta\sqrt{\overline{X}_{t_k}^{+}}\,\Delta_k W, \qquad k = 0,1,\dots \tag{2}$$

was proposed in [14], while the scheme

$$\overline{X}_{t_{k+1}} = \overline{X}_{t_k} + \kappa(\lambda - \overline{X}_{t_k})\Delta + \theta\sqrt{|\overline{X}_{t_k}|}\,\Delta_k W, \qquad k = 0,1,\dots \tag{3}$$

was studied in [28]. Both approaches extend the mapping $[0,\infty) \ni x \mapsto \sqrt{x} \in [0,\infty)$ suitably to negative values of x. For the CIR process this idea was taken further by Lord, Koekkoek & van Dijk [40], who also proposed modifications of the drift coefficient for negative values of the state space.

Example 1. The following table shows the average number of negative steps per path for the above Euler approximations of the CIR process. Scenario I (taken from [2]), corresponds to the parameters

$$x_0 = 0.05, \qquad \kappa = 5.07, \qquad \lambda = 0.0457, \qquad \theta = 0.48, \qquad T = 5$$

while Scenario II (taken from [11]) uses

$$x_0 = 0.09, \qquad \kappa = 2, \qquad \lambda = 0.09, \qquad \theta = 1, \qquad T = 5.$$

The stepsize for the Euler schemes is given by $\Delta = T/n$ with $n = 512$.

average negative steps of / for	Scenario I	Scenario II
Euler scheme (2)	0.9141	64.8611
Euler scheme (3)	1.0590	74.5017

The empirical frequency of negative paths is 0.4913 in Scenario I and 0.9990 in Scenario II. These results were obtained by a Monte Carlo simulation with $N = 10^6$ repetition. They clearly indicate that the Euler scheme (3) has a tendency for negative "excursions". This can also be seen in Figure 1, which shows a sample path of the (linearly interpolated) Euler schemes (2) and (3) using the same path of the driving Brownian motion. The parameters used in this figure correspond to Scenario II.

◇

For general SDEs the procedure of modifying the coefficients outside the support of the solution has been introduced systematically in [37]. For an SDE

$$dX_t = a(X_t)\,dt + \sum_{j=1}^{m} b_j(X_t)\,dW_t^{(j)}, \qquad X_0 = x_0 \tag{4}$$

which takes values in a domain $D \subset \mathbb{R}^d$, i.e.

$$\mathbf{P}(X_t \in D,\ t \geq 0) = 1, \tag{5}$$

the auxiliary coefficients

$$\widetilde{a}(x) = a(x) \cdot \mathbf{1}_D(x) + f(x) \cdot \mathbf{1}_E(x), \qquad x \in \mathbb{R}^d$$

$$\widetilde{b}_j(x) = b_j(x) \cdot \mathbf{1}_D(x) + g_j(x) \cdot \mathbf{1}_E(x), \qquad x \in \mathbb{R}^d, \quad j = 1, \ldots, m$$

with $E = \mathbb{R}^d \setminus D$ are introduced there. A modified Itô-Taylor scheme of order γ based on the auxiliary functions f and g is then the corresponding standard Itô-Taylor scheme for the SDE

$$dX_t = \widetilde{a}(X_t)\,dt + \sum_{j=1}^{m} \widetilde{b}_j(X_t)\,dW_t^{(j)},$$

with a suitable definition of the derivatives of the coefficients on ∂D, see [37] for details. This method is well-defined as long as the coefficients of the equation are $(2\gamma + 1)$-times differentiable on D and the auxiliary functions are $(2\gamma - 1)$-times differentiable on E. The purpose of the auxiliary functions is twofold: to obtain a well-defined approximation scheme and to bring the numerical scheme back to D if it leaves D. In particular, the auxiliary functions can always be chosen to be affine or even constant. It was shown by Jentzen, Kloeden & Neuenkirch [37] that Theorem 1 adapts to modified Itô-Taylor schemes for SDEs on domains $D \subset \mathbb{R}^d$.

Theorem 2. *Let X be the solution of SDE* (4) *satisfying condition* (5). *Moreover let $\gamma = 0.5, 1.0, 1.5, \ldots$ and assume that*

$$a \in C^{2\gamma+1}(D; \mathbb{R}^d), \qquad b \in C^{2\gamma+1}(D; \mathbb{R}^{d,m})$$

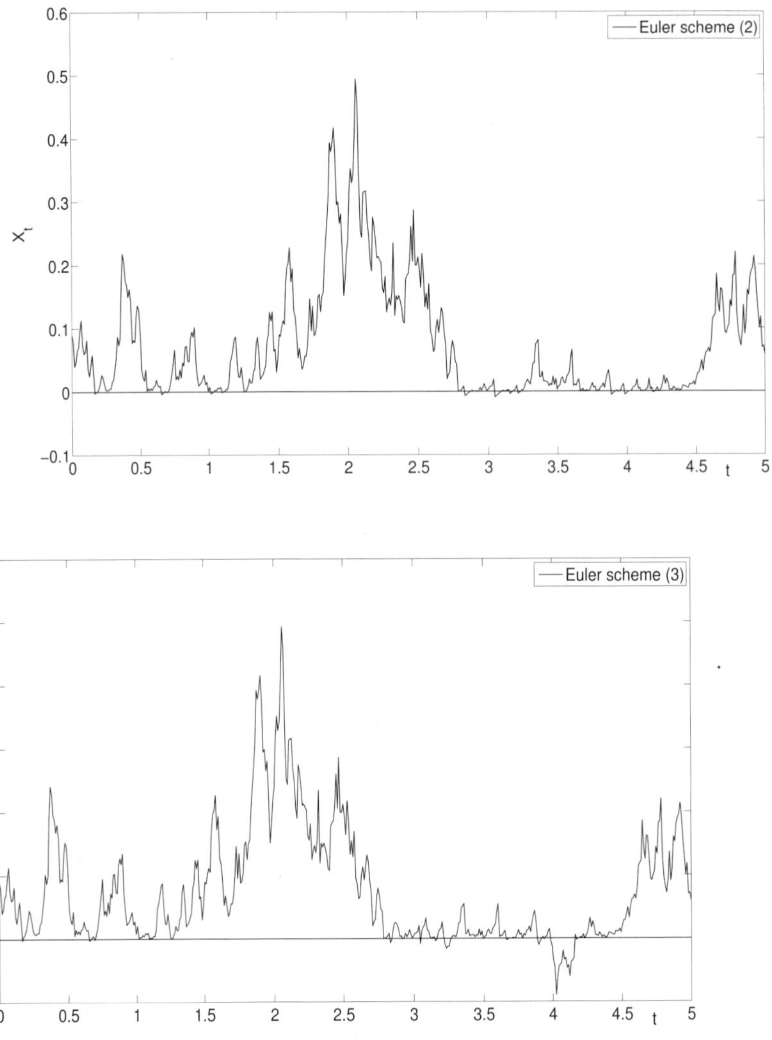

Figure 1. A path of Euler scheme (2) vs. Euler scheme (3) for the CIR process and Scenario II

and

$$f \in C^{2\gamma-1}(E;\mathbb{R}^d), \qquad g \in C^{2\gamma-1}(E;\mathbb{R}^{d,m}).$$

Finally let $\widetilde{X}^{\gamma,n}$ *be the modified Itô-Taylor method for* X *based on the auxiliary functions* f *and* g *with stepsize* $\Delta = T/n$. *Then for every* $\varepsilon > 0$ *there exists a finite and non-negative random variable* $\eta^{f,g}_{\gamma,\varepsilon}$ *such that*

$$\sup_{k=0,\ldots,n} \left| X_{t_k}(\omega) - \widetilde{X}^{\gamma,n}_{t_k}(\omega) \right| \leq \eta^{f,g}_{\gamma,\varepsilon}(\omega) \cdot n^{-\gamma+\varepsilon}$$

for almost all $\omega \in \Omega$ *and all* $n \in \mathbb{N}$.

In the case of the Euler scheme, i.e. $\gamma = 0.5$, the assumptions on a and b can be weakened to the assumption that a and b are locally Lipschitz continuous on D. For SDEs on domains in mathematical finance this condition is typically satisfied. In fact, in most cases the coefficients are infinitely differentiable.

The CIR process satisfies

$$\mathbf{P}(X_t > 0 \text{ for all } t \geq 0) = 1$$

if and only if $2\kappa\lambda \geq \theta^2$. The latter assumption is typically satisfied in interest rate applications of the CIR process. Hence, modified Taylor schemes can be used here with $D = (0, \infty)$. The truncated Euler scheme (2) corresponds to the auxiliary functions $f(x) = a(x)$, $g(x) = 0$, $x \leq 0$, while the scheme (3) uses the auxiliary functions $f(x) = a(x)$, $g(x) = \sqrt{-x}$, $x \leq 0$. Note that $2\kappa\lambda \geq \theta^2$ is satisfied in Scenario I, but not in Scenario II.

For structure preserving integration of the CIR process also the symmetrized Euler method

$$\widetilde{X}_{t_{k+1}} = \left| \widetilde{X}_{t_k} + \kappa(\lambda - \widetilde{X}t_k)\Delta + \theta\sqrt{\widetilde{X}_{t_k}}\, \Delta_k W \right|, \qquad k = 0, 1, \ldots \tag{6}$$

was proposed in [10; 8]. While the modified Euler schemes (2) and (3) may leave $(0, \infty)$ and are then forced back in the next steps, this scheme is always non-negative. Adapting this to general SDEs, which take values in a domain D, leads to the reflected Euler schemes, see e.g. [44], which are given by

$$\widetilde{X}^{\psi}_{t_{k+1}} = H^{\psi}_{t_{k+1}} \cdot \mathbf{1}_D(H^{\psi}_{t_{k+1}}) + \psi(H^{\psi}_{t_{k+1}}) \cdot \mathbf{1}_{\mathbb{R}^d \setminus D}(H^{\psi}_{t_{k+1}})$$

with $\widetilde{X}^{\psi}_0 = x_0$, where

$$H^{\psi}_{t_{k+1}} = \widetilde{X}^{\psi}_{t_k} + a(\widetilde{X}^{\psi}_{t_k})\Delta + \sum_{j=1}^{m} b_j(\widetilde{X}^{\psi}_{t_k})\Delta_k W^{(j)}$$

and a measurable projection function $\psi : \mathbb{R}^d \setminus D \to D \cup \partial D$. A straightforward modification of the above theorem yields a pathwise convergence order $1/2 - \varepsilon$ for these reflected Euler schemes if the SDE coefficients are twice continuously differentiable on D. In the same way reflected Itô-Taylor schemes of arbitrary order can be constructed and analyzed.

The symmetrized Euler scheme (6) corresponds to the reflection function $\psi(x) = |x|$. The results on modified Itô-Taylor schemes and reflected Euler methods apply also to the generalized Ait-Sahalia model with $D = (0, \infty)$ if $r > 1, \rho < (1 + r)/2$, to the Heston model with $D = (0, \infty)^2$ if $2\kappa\lambda \geq \theta^2$ and to the 3/2-model with $D = (0, \infty)^2$ and no further restrictions on the parameter.

Example 2. To illustrate the above results consider Scenario I for the Cox-Ingersoll-Ross process. Figure 2 shows for two different sample paths $\omega \in \Omega$ the maximum error in the discretization points, i.e.

$$\sup_{k=0,\dots,n} |X_{t_k}(\omega) - \overline{X}_{t_k}(\omega)|,$$

of

(i) the truncated Euler scheme (2)
(ii) the symmetrized Euler scheme (6)
(iii) the modified Milstein scheme with auxiliary functions $f(x) = \kappa(\lambda - x)$, $g(x) = 0$, i.e. a truncated Milstein scheme.

To estimate the pathwise maximum error for the above approximation schemes the Cox-Ingersoll-Ross process have been discretized with a very small step size using scheme (2). In Figure 2 log-log-coordinates are used, so the dots indicate the convergence orders 0.5 and 1. The pathwise convergence rates of all three approximation schemes are in good accordance with the theoretically predicted rates for moderate and small step sizes. For small step sizes both Euler schemes do not take negative values and hence coincide. Moreover, for small step sizes the Milstein scheme is superior due to its first order convergence.

◇

Numerical methods with pathwise convergence rates of high order are thus available also for SDEs with non-globally Lipschitz coefficients. However, while pathwise convergence rates are very important for the analysis of random dynamical systems [7; 17], one of the main objectives in mathematical finance is the pricing of (path-dependent) European-type derivatives, which means to compute real numbers $\mathbf{E}\Phi(X)$ where $\Phi : C([0,T];\mathbb{R}^d) \to \mathbb{R}$ is the discounted payoff of the derivative. Since the integrability of the random constants in the error bounds is an open problem, the above pathwise convergence rates do not imply weak or strong convergence rates. Nevertheless, if Φ is bounded and continuous and if $\overline{X}^{\gamma} = (\overline{X}_t^{\gamma})_{t\in[0,T]}$ is the piecewise linear interpolation of the γ-Itô-Taylor scheme (standard, modified or reflected) then

$$\mathbf{E}\Phi(\overline{X}^{\gamma}) \longrightarrow \mathbf{E}\Phi(X)$$

for $n \to \infty$, so for bounded and continuous pay-offs (e.g. put options) one obtains at least the convergence of the corresponding standard Monte Carlo estimators for the option price. The same is true for barrier options with payoff of the form

$$\Phi(X) = \phi(X_T)\mathbf{1}_{\{K_1 \le |X_t| \le K_2,\ t\in[0,T]\}}$$

with $0 \le K_1 \le K_2 < \infty$, if ϕ is bounded and continuous and the law of $\sup_{t\in[0,T]} |X_t|$ and $\inf_{t\in[0,T]} |X_t|$ has a density with respect to the Lebesgue measure.

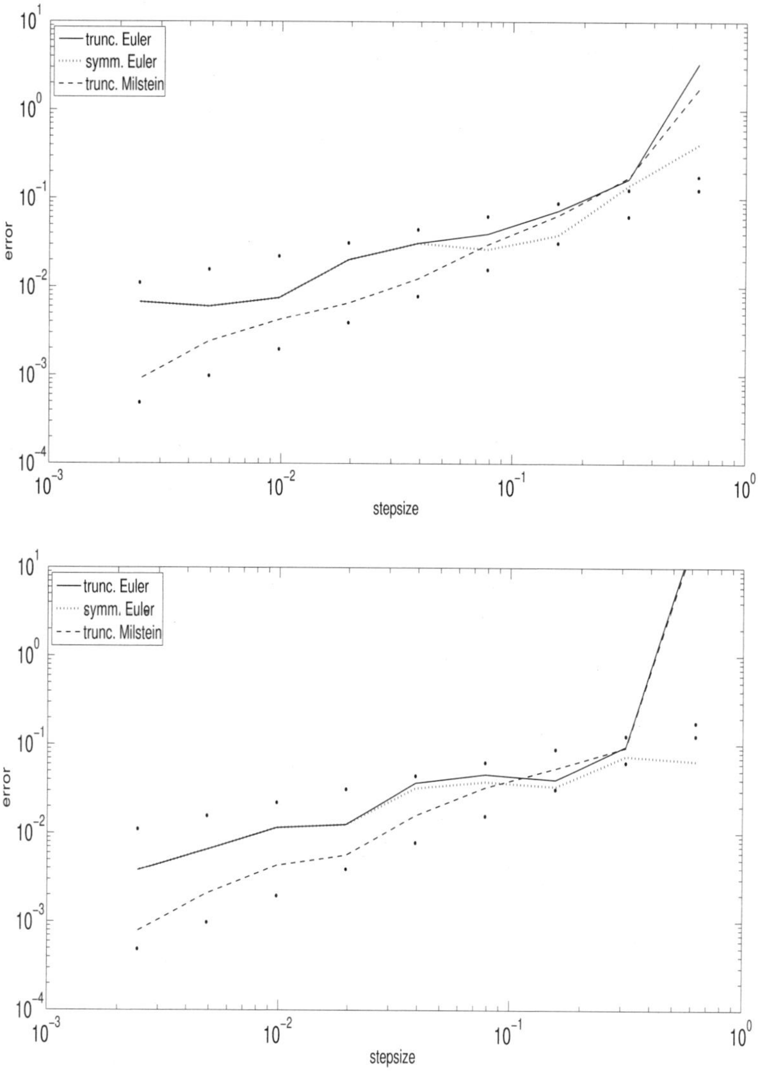

Figure 2. Pathwise maximum error vs. step size for two sample paths for the Cox-Ingersoll-Ross model for Scenario I

3. The Explicit Euler Scheme: Criteria for Weak and Strong Convergence

It was shown by Higham, Mao & Stuart in [29] that the explicit Euler scheme

$$\overline{X}_{t_{k+1}} = \overline{X}_{t_k} + a(\overline{X}_{t_k})\Delta + \sum_{j=1}^{m} b_j(\overline{X}_{t_k})\Delta_k W^{(j)}, \qquad k = 0, 1, \ldots,$$

$$\overline{X}_0 = x_0$$

is strongly convergent if the coefficients are locally Lipschitz continuous on $\mathbb{R}^d$ and a moment condition for the SDE and its Euler approximation is satisfied. This result can be extended to SDE on domains and the modified or reflected Euler scheme.

Theorem 3. *Let X be the solution of SDE (4) satisfying condition (5). Moreover, let $\widetilde{X}^n$ be the modified Euler scheme based on the auxiliary functions $f \in C(E;\mathbb{R}^d)$, $g \in C(E;\mathbb{R}^{d,m})$ with stepsize $\Delta = T/n$ or let $\widetilde{X}^n$ be the reflected Euler scheme based on the projection function $\psi : E \to D \cup \partial D$ with stepsize $\Delta = T/n$. Assume that*

$$a \in C^2(D;\mathbb{R}^d), \qquad b \in C^2(D;\mathbb{R}^{d,m})$$

and furthermore, assume that for some $p > 2$

$$\sup_{n\in\mathbb{N}} \mathbf{E} \max_{k=0,\dots n} |\widetilde{X}^n_{t_k}|^p + \mathbf{E} \sup_{t\in[0,T]} |X_t|^p < \infty. \tag{7}$$

Then

$$\lim_{n\to\infty} \mathbf{E} \max_{k=0,\dots,n} |X_{t_k} - \widetilde{X}^n_{t_k}|^2 = 0.$$

Proof. From the results of the previous section

$$\lim_{n\to\infty} \max_{k=0,\dots,n} |X_{t_k} - \widetilde{X}^n_{t_k}| = 0$$

hold, almost surely. However, assumption (7) implies the uniform integrability of

$$\max_{k=0,\dots n} |X_{t_k} - \widetilde{X}^n_{t_k}|^2, \quad n \in \mathbb{N}.$$

The assertion now follows, since uniform integrability allows integration to the limit. □

Note that assumption (7) is easily verified if the SDE coefficients have linear growth on D, i.e.

$$|a(x)| + \sum_{j=1}^{m} |b_j(x)| \le C \cdot (1 + |x|), \qquad x \in D,$$

for some $C > 0$. Turning back to the Cox-Ingersoll-Ross process this gives us strong convergence of the Euler schemes (2), (3) and (6) under the assumption $2\kappa\lambda \ge \theta^2$. Note that for the Euler schemes (2) and (3) strong convergence without a restriction on the parameter has been shown in [14] and [28] using a Yamada function technique. This technique has also been applied by Gyöngy & Rásonyi in [24] to obtain the following result:

Theorem 4. *Let $a_1, a_2, b : \mathbb{R} \to \mathbb{R}$. Consider the one-dimensional SDE*

$$dX_t = (a_1(X_t) + a_2(X_t))\,dt + b(X_t)\,dW_t, \quad t \in [0,T], \qquad X_0 = x_0 \in \mathbb{R}$$

and let $\overline{X}^n$ *be the corresponding Euler scheme with stepsize* $\Delta = T/n$. *Moreover, let* a_2 *be monotonically decreasing and assume that there exists constants* $\alpha \in [0, 1/2]$, $\beta \in (0, 1]$ *and* $C > 0$ *such that*

$$|a_1(x) - a_1(y)| \leq C \cdot |x - y|, \qquad |a_2(x) - a_2(y)| \leq C \cdot |x - y|^{\beta},$$

$$|b(x) - b(y)| \leq C \cdot |x - y|^{\frac{1}{2}+\alpha}$$

for all $x, y \in \mathbb{R}$. *Then, for all* $p \in \mathbb{N}$, *there exist constants* $K_p^{\alpha,\beta} > 0$ *such that*

$$\mathbf{E} \max_{k=0,\ldots,n} |X_{t_k} - \overline{X}^n_{t_k}|^p \leq \begin{cases} K_p^{0,\beta} \cdot \frac{1}{\log(n)} & \text{for } \alpha = 0 \\ K_p^{\alpha,\beta} \cdot \left(\frac{1}{n^{\alpha}} + \frac{1}{n^{\beta/2}}\right) & \text{for } \alpha \in (0, 1/2) \\ K_p^{\alpha,\beta} \cdot \left(\frac{1}{n^{p/2}} + \frac{1}{n^{\beta p/2}}\right) & \text{for } \alpha = 1/2 \end{cases}$$

This result can be applied to the CEV model

$$dX_t = \mu X_t \, dt + \sigma X_t^{\gamma} \, dW_t,$$

if the mapping $[0, \infty) \ni x \mapsto x^{\gamma} \in [0, \infty)$ is extended to $(-\infty, 0)$, e.g. as $(x^+)^{\gamma}$ or $|x|^{\gamma}$. Theorem 4 then yields strong convergence of the corresponding Euler schemes.

Example 3. Whether the convergence rates predicted from Theorem 4 are sharp for the CEV model remains an open problem. The following simulation study suggests that the Euler scheme has strong convergence order 1/2, at least for some parameter constellations. To better preserve the positivity of the CEV process, the Euler scheme is applied to the SDE

$$dX_t = \mu |X_t| \, dt + \sigma (X_t^+)^{\gamma} \, dW_t$$

which still fulfills the assumptions of Theorem 4 with $a_2 = 0$, i.e. $\beta = 1$, and $\alpha = \gamma - 1/2$. Its solution coincides with the CEV process.

Figure 3 shows the empirical root mean square maximum error in the discretization points versus the step size for the parameters

Set I:	$\mu = 0.1$,	$\sigma = 0.3$,	$\gamma = 0.75$,	$T = 1$,	$x_0 = 0.2$
Set II:	$\mu = 0.2$,	$\sigma = 0.5$,	$\gamma = 0.55$,	$T = 1$,	$x_0 = 0.5$

The empirical mean square maximum error in the discretization points is estimated by

$$\left(\frac{1}{N} \sum_{i=1}^{N} \max_{k=0,\ldots,n} |X_{t_k}^{*,(i)} - \overline{X}_{t_k}^{n,(i)}|^2 \right)^{1/2}$$

with $N = 5 \cdot 10^4$. Here X^* is the numerical reference solution obtained by using the same Euler scheme with very small step size and $X^{*,(i)}, \overline{X}^{n,(i)}$ are independent copies of $X^*, \overline{X}^n$. For both sets of parameter a good accordance with the convergence order 1/2 is obtained. (The dots in the figure indicate convergence order 1/2). A regression of the numerical data yields moreover the empirical convergence order 0.493923 for set I, respectively 0.509903 for set II.

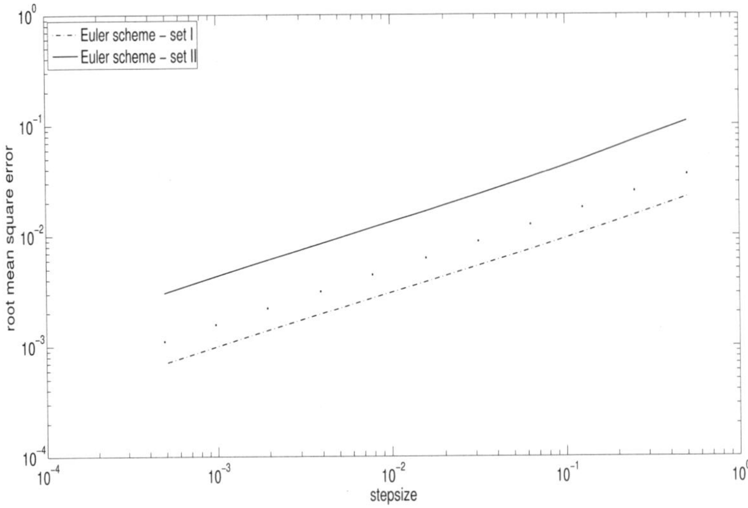

Figure 3. Root mean square error of the Euler scheme vs. step size for the CEV process for the parameter sets I and II

◇

But do Theorems 3 and 4 have any consequences for the other examples? Unfortunately not: for the Heston, Ait-Sahalia and 3/2-models, no linear growth condition is satisfied. Even worse, for the Ait-Sahalia model and the 3/2-model the moments of the Euler scheme explode! In the case of the 3/2-model the latter can be deduced from the following Theorem, which was obtained by Hutzenthaler, Jentzen & Kloeden in [34].

Theorem 5. *Let $a, b : \mathbb{R} \to \mathbb{R}$ and assume that the one-dimensional SDE*

$$dX_t = a(X_t)\,dt + b(X_t)\,dW_t, \quad t \in [0,T], \qquad X_0 = x_0 \in \mathbb{R}$$

has a unique strong solution with

$$\sup_{t\in[0,T]} \mathbb{E}|X_t|^p < \infty$$

for one $p \in [1,\infty)$. Moreover, let $b(x_0) \neq 0$ and let $C \geq 1$, $\beta > \alpha > 1$ be constants such that

$$\max\big(|a(x)|, |b(x)|\big) \geq \frac{1}{C} \cdot |x|^{\beta} \quad \text{and} \quad \min\big(|a(x)|, |b(x)|\big) \leq C \cdot |x|^{\alpha}$$

for all $|x| \geq C$. Then, the corresponding Euler scheme $\overline{X}^n$ with stepsize $\Delta = T/n$ satisfies

$$\lim_{n\to\infty} \mathbb{E}|X_T - \overline{X}_T^n|^p = \infty \quad \text{and} \quad \lim_{n\to\infty} \big|\mathbb{E}|X_T|^p - \mathbb{E}|\overline{X}_T^n|^p\big| = \infty. \qquad (8)$$

In the case of the 3/2-model, which has finite moments up to order $p < 2 + \frac{2c_1}{c_3^2}$, the coefficients are

$$a(x) = -c_2 x^2 + c_1 c_2 x, \qquad b(x) = c_3 (x^+)^{3/2}, \qquad x \in \mathbb{R},$$

so the assumptions of the above Theorem are satisfied for $\alpha = 3/2$, $\beta = 2$ and C sufficiently large.

Concerning the Ait-Sahalia model, the moments of the Euler scheme already explode in the second step. Here the first step of the Euler scheme has a Gaussian distribution with mean $x_0 + (\alpha_{-1} x_0^{-1} - \alpha_0 + \alpha_1 x_0 - \alpha_2 x_0^r)\Delta$ and variance $\alpha_3^2 x_0^{2\rho} \Delta$. The inverse of the first step must be computed for the second step of the Euler scheme, so the moments of the second step are infinite, since inverse moments of a Gaussian random variable do not exist.

Why the moments of the Euler scheme diverge for superlinearly growing coefficients – even without a singularity – can be nicely illustrated by considering the SDE

$$dX_t = -X_t^3\, dt + \sigma dW_t, \qquad X_0 = x_0 \tag{9}$$

with $\sigma \geq 0$ for which the Euler scheme reads as

$$\overline{X}^n_{t_{k+1}} = \overline{X}^n_{t_k} \left(1 - |\overline{X}^n_{t_k}|^2 \Delta\right) + \sigma \Delta_k W. \tag{10}$$

In the deterministic case, i.e., (9) and (10) with $\sigma = 0$, the Euler approximation of the deterministic equation is known to be unstable if the initial value is large (see e.g. Chapter 6 in [16]). For example, if $x_0 = n$, $T = 1$ and $\Delta = n^{-1}$ then

$$\overline{X}^n_{t_1} = n\left(1 - \frac{n^2}{n}\right) \approx -n^2$$

and therefore

$$\overline{X}^n_{t_2} = \overline{X}^n_{t_1}\left(1 - |\overline{X}^n_{t_1}|^2 \Delta\right) \approx n^5.$$

Iterating this further, one obtains

$$\left|\overline{X}^n_{t_k}\right| \gtrapprox n^{(2^k)}$$

for $k = 0, 1, \ldots, n$. Thus, $\overline{X}^n_{t_n}$ grows double-exponentially fast in n. In the presence of noise ($\sigma > 0$) there is an exponentially small event that the Brownian motion leaves the interval $[-2n, 2n]$ and on this event the approximations grow double-exponentially fast due to the deterministic dynamics. Consequently this double-exponentially growth can not be compensated by the exponentially small probability of this event, which leads to the moment explosion of the Euler approximation.

Example 4. That rare events lead to the explosion of the moments of the Euler scheme can be also seen from the following numerical example. Consider the volatility process in the 3/2-model

$$dV_t = c_1 V_t (c_2 - V_t)\, dt + c_3 V_t^{3/2}\, dW_t, \quad V_0 = v_0 > 0$$

with

$$c_1 = 1.2, \qquad c_2 = 0.8, \qquad c_3 = 1, \qquad T = 4, \qquad v_0 = 0.5$$

and try to compute

$$\mathbf{E}|X_T| = 0.566217$$

using the standard Monte Carlo estimator

$$\frac{1}{N}\sum_{i=1}^{N} |\overline{X}_T^{n,(i)}|$$

where $\overline{X}_T^{n,(1)}, \ldots, \overline{X}_T^{n,(N)}$ are iid copies of $\overline{X}_T^n$. The exact value for $\mathbf{E}|X_T|$ is computed using the inverse moments of the CIR process, see e.g. [31]. While for a moderate number of repetitions the estimator seems to converge for small step sizes (and the 'Inf'-outputs seem to be some numerical instabilities due to the large step sizes), the estimator explodes even for small step sizes when increasing the number of repetitions – as predicted by Theorem 5. Despite of this the Euler scheme for this SDE converges pathwise with rate $1/2 - \varepsilon$ due to Theorem 2.

Repetitions N / stepsize Δ	2^0	2^{-2}	2^{-4}	2^{-6}	2^{-8}	2^{-10}
10^3	6.327232	Inf	Inf	0.550185	0.553499	0.555069
10^4	6.894698	Inf	Inf	Inf	0.562716	0.563352
10^5	7.430606	Inf	Inf	Inf	0.566218	0.567106
10^6	7.227379	Inf	Inf	Inf	Inf	0.565750
10^7	7.279187	Inf	Inf	Inf	Inf	Inf

A similar moment explosion arises if a Multi-level Monte Carlo method is used to estimate $\mathbf{E}|X_T|$. This is shown, also for more general SDEs, in [36].

◇

However, in some cases using the Euler scheme one still obtains a convergent Monte Carlo estimator for functionals of the type $\mathbf{E}\phi(X_T)$. The standard Euler-based estimator for the latter quantity is

$$\frac{1}{N}\sum_{i=1}^{N} \phi(\overline{X}_T^{n,(i)}). \tag{11}$$

In the classical case, i.e. if $a, b, \phi \in C^4(\mathbb{R};\mathbb{R})$ with at most polynomially growing derivatives and a, b globally Lipschitz, one has

$$\mathbf{E}\left|\frac{1}{N}\sum_{i=1}^{N} \phi(\overline{X}_T^{n,(i)}) - \mathbf{E}\phi(X_T)\right|^2 \leq K_{Bias} \cdot \frac{1}{n^2} + K_{MC} \cdot \frac{1}{N},$$

see e.g. [39]. The first term on the right hand side corresponds to the squared bias of the Euler scheme, while the second term corresponds to the variance of the Monte Carlo simulation. It is thus optimal to choose $N = n^2$ for balancing both terms with respect to the computational cost (number of arithmetic operations, function evaluations and random numbers used), see [18]. The corresponding Monte Carlo estimator has then convergence order 1/3 in terms of the computational cost.

Hutzenthaler & Jentzen could show in [32] that if the global Lipschitz assumption on the drift-coefficient is weakened to

$$(x-y)(a(x)-a(y)) \leq L\,(x-y)^2\,, \qquad x, y \in \mathbb{R} \tag{12}$$

for some $L > 0$, then one still has

$$\left| \frac{1}{N^2} \sum_{i=1}^{N^2} \phi(\overline{X}_T^{N,(i)}) - \mathbf{E}\phi(X_T) \right| \leq \eta_\varepsilon \cdot N^{-(1-\varepsilon)}$$

almost surely for all $\varepsilon > 0$ and almost-surely finite and non-negative random variables η_ε.

Weak approximation under non-standard assumptions is also studied by Milstein & Tretyakov in [42]. In their approach, simulations which leave a ball with sufficiently large radius are discarded. In the context of the Euler scheme with equidistant stepsize this estimator reads as

$$\frac{1}{N} \sum_{i=1}^{N} \phi(\overline{X}_T^{n,(i)}) \cdot \mathbf{1}_{\{\sup_{k=0,\ldots,n} |X_{t_k}^{n,(i)}| \leq R\}}.$$

For coefficients a, b and functions ϕ satisfying a Lyapunov-type condition still a convergent Monte Carlo estimator is obtained, when matching the discarding radius R appropriately to the number of repetitions N and the stepsize of the discretization n.

Condition (12) on the drift coefficient is the so-called one-sided Lipschitz condition. This condition is also very useful to obtain strong convergence results for implicit Euler methods and tamed Euler schemes, which will be explained in the next section. Very recently a unifying framework for the analysis of Euler-type methods has been provided in [33].

4. Strong convergence of implicit and tamed Euler schemes

The condition in Theorem 3 for the strong convergence of the Euler scheme which is usually difficult to verify is the finiteness of its moments, i.e.

$$\sup_{n \in \mathbb{N}} \mathbf{E} \max_{k=0,\ldots,n} |\overline{X}_{t_k}^n|^p < \infty$$

for some $p > 2$. Moreover, this condition may even fail to hold for specific equations, see Theorem 5. However, both problems can be overcome in some situations if appropriate drift-implicit Euler schemes are used. The split-step backward Euler scheme is defined as

$$X_{t_k}^* = \overline{X}_{t_k} + a(X_{t_k}^*)\Delta, \qquad \overline{X}_{t_{k+1}} = X_{t_k}^* + \sum_{j=1}^m b_j(X_{t_k}^*)\Delta_k W^{(j)} \tag{13}$$

for $k = 0, 1, \ldots$ with $\overline{X}_0 = x_0$, while the backward or drift-implicit Euler scheme reads as

$$\overline{X}_{t_{k+1}} = \overline{X}_{t_k} + a(\overline{X}_{t_{k+1}})\Delta + \sum_{j=1}^m b_j(\overline{X}_{t_k})\Delta_k W^{(j)}. \tag{14}$$

Both schemes are defined via an implicit equation, whose solvability relies on the properties of the drift-coefficient a. The following result has been obtained by Higham, Mao & Stuart in [29].

Theorem 6. *Let $a, b_j \in C^1(\mathbb{R}^d; \mathbb{R}^d)$, $j = 1, \ldots, m$, and assume that there exist constants $L_1, L_2 > 0$ such that*

$$\langle x - y, a(x) - a(y)\rangle \leq L_1 \cdot |x - y|^2, \qquad x, y \in \mathbb{R}^d,$$

$$\sum_{j=1}^m |b_j(x) - b_j(y)|^2 \leq L_2 \cdot |x - y|^2, \qquad x, y \in \mathbb{R}^d.$$

Then, the split-step backward Euler scheme given by (13) *with stepsize $\Delta = T/n$ is well defined for $\Delta < \Delta_* := 1/\max\{1 + 2L_1, 4L_2\}$ and satisfies*

$$\lim_{n\to\infty} \mathbf{E} \max_{k=0,\ldots,n} |X_{t_k} - \overline{X}_{t_k}^n|^2 = 0.$$

The conditions on the coefficients imply that the SDE has bounded moments of any order, and also allow one to show that the split-step Euler method has moments of any order. The implicitness of the method is crucial for the latter. Furthermore, the split-step Euler method coincides with the explicit Euler method for the perturbed SDE

$$dX_t^\Delta = a(h_\Delta(X_t^\Delta))\, dt + \sum_{j=1}^m b_j(h_\Delta(X_t^\Delta))\, dW^{(j)}(t), \qquad X_0^\Delta = x_0. \tag{15}$$

Here the function $h_\Delta : \mathbb{R}^d \to \mathbb{R}^d$ is defined as the unique solution of the equation

$$h_\Delta(x) = x + a(h_\Delta(x))\Delta, \qquad x \in \mathbb{R}^d$$

with $\Delta < \Delta_*$. Since h_Δ converges to the identity for $\Delta \to 0$ this perturbed SDE is close to original SDE. To establish Theorem 6, it thus remains to show that the split-step backward Euler scheme is close to (15), which can be done along the lines of the proof of Theorem 3.

If the drift-coefficient is additionally also polynomially Lipschitz, then the standard strong convergence rate 1/2 can even be recovered.

Theorem 7. *Let the assumptions of Theorem 6 hold and assume additionally that there exist $C, q > 0$ such that*

$$|a(x) - a(y)| \leq C \cdot (1 + |x|^q + |y|^q) \cdot |x - y|, \qquad x, y \in \mathbb{R}^d.$$

Then, the split-step backward Euler scheme given by (13) and the backward Euler scheme given by (14) are well defined for $\Delta < \Delta^*$ and have strong convergence order 1/2, i.e. for both schemes there exists a constant $K > 0$ such that

$$\mathbf{E} \max_{k=0,\dots,n} |X_{t_k} - \overline{X}^n_{t_k}|^2 \leq K \cdot n^{-1}.$$

As pointed out above, in each step of both schemes an implicit equation has to be solved. If the function h_Δ is not known explicitly, this has to be done numerically and may be time-consuming. Solving implicit equations can be avoided by using the so-called tamed Euler method, which has been proposed by Hutzenthaler, Jentzen & Kloeden in [35]:

$$\overline{X}_{t_{k+1}} = \overline{X}_{t_k} + \frac{1}{1 + |a(\overline{X}_{t_k})|\Delta} a(\overline{X}_{t_k})\Delta + \sum_{j=1}^{m} b_j(\overline{X}_{t_k})\Delta_k W^{(j)}. \tag{16}$$

Here the drift-term is "tamed" by the factor $\frac{1}{1+|a(\overline{X}_{t_k})|\Delta}$ in the k-th step, which prevents a possible explosion of the scheme.

Theorem 8. *Let the assumptions of Theorem 7 hold. Then, there exists a constant $K > 0$ such that the tamed Euler scheme given by (16) satisfies*

$$\mathbf{E} \max_{k=0,\dots,n} |X_{t_k} - \overline{X}^n_{t_k}|^2 \leq K \cdot n^{-1}.$$

Here, the difficulty is again to control the moments of the approximation scheme. For this appropriate processes are used that dominate the tamed Euler scheme on subevents whose probabilities converge sufficiently fast to one.

The Theorems given so far in this section require the diffusion coefficient to be globally Lipschitz, which is often not fulfilled in SDEs arising from mathematical finance. However, the backward Euler method can be also successfully applied to the Ait-Sahalia interest rate model

$$dX_t = \left(\alpha_{-1}X_t^{-1} - \alpha_0 + \alpha_1 X_t - \alpha_2 X_t^r\right)dt + \sigma X_t^\rho dW_t \tag{17}$$

where $\alpha_i, \sigma > 0$, $i = -1, \dots, 2$ and $r, \rho > 1$. The following result has been obtained by Szpruch et al. in [45]:

Theorem 9. *Consider the SDE (17) and assume that*

$$r + 1 > 2\rho.$$

Then the corresponding backward Euler method (14) *with stepsize* $\Delta = T/n$ *is well defined if* $\Delta \leq 1/\alpha_1$, *and*

$$\lim_{n\to\infty} \mathbf{E} \max_{k=0,\ldots,n} |X_{t_k} - \overline{X}_{t_k}^n|^2 = 0.$$

Here the drift coefficient is still one-sided Lipschitz on the domain of the SDE, i.e.

$$(x-y)(a(x)-a(y)) \leq \alpha_1 |x-y|^2, \qquad x, y > 0,$$

and, moreover, $-a$ is coercive on $(0,\infty)$, i.e.

$$\lim_{x\to 0} a(x) = \infty \qquad \lim_{x\to\infty} a(x) = -\infty.$$

These two properties ensure that the drift-implicit Euler scheme for (17) is well-defined and, in particular, takes only strictly positive values.

The drift coefficient in the volatility process

$$dV_t = c_1 V_t (c_2 - V_t)\, dt + c_3 V_t^{3/2}\, dW_t, \quad V_0 = v_0 > 0$$

in the 3/2-model is also one-sided Lipschitz on $(0,\infty)$. It does not, however, satisfy the coercivity assumption. Consequently, the drift-implicit Euler scheme cannot be applied here, since the implicit equation may not be solvable. Note that very recently, Higham et al. introduced in [30] a double-implicit Milstein scheme, which is strongly convergent for the 3/2-model and similar SDEs.

5. Strong Convergence Rates for the approximation of the Cox-Ingersoll-Ross process and the Heston model

Strong convergence rates for the approximation of the CIR process

$$dX_t = \kappa(\lambda - X_t)\, dt + \theta\sqrt{X_t}\, dW_t, \quad t \in [0,T], \qquad X_0 = x_0 > 0$$

with $\kappa, \lambda, \theta > 0$ have been a long standing open problem, even in the regime where the CIR process does not hit zero, i.e. when $2\kappa\lambda \geq \theta^2$.

The first non-logarithmic rates were derived by Berkaoui, Bossy & Diop for the symmetrized Euler scheme (6), i.e.

$$\overline{X}_{t_{k+1}} = \left| \overline{X}_{t_k} + \kappa(\lambda - \overline{X}_{t_k})\Delta + \theta\sqrt{\overline{X}_{t_k}}\, \Delta_k W \right|.$$

They showed in [8] that

$$\mathbf{E} \max_{k=0,\ldots,n} |X_{t_k} - \overline{X}_{t_k}|^{2p} \leq C_p \cdot \Delta^p$$

under the assumption

$$\frac{2\kappa\lambda}{\theta^2} > 1 + \sqrt{8} \max\left\{ \frac{\sqrt{\kappa}}{\theta}\sqrt{16p-1}, 16p-2 \right\},$$

where the constant $C_p > 0$ depends only on $p, \kappa, \lambda, \theta, x_0$ and T. Strong convergence rates for a drift-implicit Euler-type scheme were recently obtained under mild assumptions by Dereich, Neuenkirch & Szpruch in [15]. Their key tool is the use of the Lamperti-transformation: by the Itô formula, the transformed process $Y_t = \sqrt{X_t}$ satisfies the SDE

$$dY_t = \frac{\alpha}{Y_t}\,dt + \beta Y_t\,dt + \gamma\,dW_t, \quad t \geq 0, \qquad Y_0 = \sqrt{x_0} \tag{18}$$

with

$$\alpha = \frac{4\kappa\lambda - \theta^2}{8}, \qquad \beta = -\frac{\kappa}{2}, \qquad \gamma = \frac{\theta}{2}.$$

At first glance this transformation does not help at all, since the drift coefficient of the arising SDE is singular. However,

$$a(x) = \frac{\alpha}{x} + \beta x, \qquad x > 0,$$

satisfies for $\alpha > 0$, $\beta \in \mathbb{R}$ the restricted one-sided Lipschitz condition

$$(x-y)(a(x)-a(y)) \leq \beta(x-y)^2, \qquad x, y > 0$$

The drift-implicit Euler method with stepsize $\Delta > 0$ in this case is

$$\overline{Y}_{t_{k+1}} = \overline{Y}_{t_k} + \left(\frac{\alpha}{\overline{Y}_{t_{k+1}}} + \beta\overline{Y}_{t_{k+1}}\right)\Delta + \gamma\Delta_k W, \qquad k = 0, 1, \ldots$$

with $\overline{Y}_0 = \sqrt{x_0}$, which has the explicit solution

$$\overline{Y}_{t_{k+1}} = \frac{\overline{Y}_{t_k} + \gamma\Delta_k W}{2(1-\beta\Delta)} + \sqrt{\frac{(\overline{Y}_{t_k} + \gamma\Delta_k W)^2}{4(1-\beta\Delta)^2} + \frac{\alpha\Delta}{1-\beta\Delta}}.$$

Setting

$$\overline{X}_{t_k} = \overline{Y}^2_{t_k}, \qquad k = 0, 1, \ldots, \tag{19}$$

gives a positivity preserving approximation of the CIR process, which is called drift-implicit square-root Euler method. This scheme had already been proposed in [4], but without a convergence analysis. Piecewise linear interpolation, i.e.

$$\overline{X}_t = \frac{t_{k+1}-t}{\Delta}\overline{X}_{t_k} + \frac{t-t_k}{\Delta}\overline{X}_{t_{k+1}}, \qquad t \in [t_k, t_{k+1}],$$

gives a global approximation $(\overline{X}_t)_{t\in[0,T]}$ of the CIR process on $[0, T]$. The main result of [15] is:

Theorem 10. *Let $2\kappa\lambda > \theta^2$, $x_0 > 0$ and $T > 0$. Then, for all*

$$1 \leq p < \frac{2\kappa\lambda}{\theta^2}$$

there exists a constant $K_p > 0$ *such that*

$$\left(\mathbf{E} \max_{t\in[0,T]} |X_t - \overline{X}_t|^p\right)^{1/p} \leq K_p \cdot \sqrt{|\log(\Delta)|} \cdot \sqrt{\Delta},$$

for all $\Delta \in (0, 1/2]$.

The restriction on p arises in the proof of the convergence rate when controlling the inverse p-th moments of the CIR process, which are infinite for $p \geq 2\kappa\lambda/\theta^2$. For further details, see [15]. Note that for SDEs with Lipschitz coefficients the convergence rate $\sqrt{|\log(\Delta)|} \cdot \sqrt{\Delta}$ is best possible with respect to the above global error criterion, see [43]. So the convergence rate given in Theorem 10 matches the rate that is optimal under standard assumptions.

Other approximation schemes for the strong approximation of the CIR process can be found in [4; 21; 25]. Among them is the drift-implicit Milstein scheme

$$\overline{Z}_{t_{k+1}} = \overline{Z}_{t_k} + \kappa(\lambda - \overline{Z}_{t_{k+1}})\Delta + \theta\sqrt{\overline{Z}_{t_k}}\Delta_k W + \frac{\theta^2}{4}((\Delta_k W)^2 - \Delta)$$

with $Z_0 = x_0$, see [21]. It can be rewritten as

$$\overline{Z}_{t_{k+1}} = \frac{1}{1+\kappa\Delta}\left(\sqrt{\overline{Z}_{t_k}} + \frac{\theta}{2}\Delta_k W\right)^2 + \frac{1}{1+\kappa\Delta}\left(\kappa\lambda - \frac{\theta^2}{4}\right)\Delta, \tag{20}$$

so this scheme preserves the positivity of the CIR process if $4\kappa\lambda \geq \theta^2$. It coincides up to a term of second order with the drift-implicit square-root Euler method, since the latter can be written as

$$\begin{aligned}\overline{X}_{t_{k+1}} =& \frac{1}{1+\kappa\Delta}\left(\sqrt{\overline{X}_{t_k}} + \frac{\theta}{2}\Delta_k W\right)^2 + \frac{1}{1+\kappa\Delta}\left(\kappa\lambda - \frac{\theta^2}{4}\right)\Delta \\ &- \frac{1}{1+\kappa\Delta}\left(\frac{4\kappa\lambda - \theta^2}{8\sqrt{\overline{X}_{t_{k+1}}}} - \frac{\kappa}{2}\sqrt{\overline{X}_{t_{k+1}}}\right)^2 \Delta^2.\end{aligned}$$

Moreover the drift-implicit Milstein scheme dominates the drift-implicit square-root Euler method:

Lemma 2. *Let* $2\kappa\lambda > \theta^2$, $x_0 > 0$ *and* $T > 0$. *Then*

$$\mathbf{P}(\overline{Z}_{t_k} \geq \overline{X}_{t_k},\ k = 0, 1, \ldots) = 1.$$

Proof. The numerical flow for the drift-implicit Milstein scheme is given by

$$\varphi_{\overline{Z}}(x; k, \Delta) = \frac{1}{1+\kappa\Delta}\left(\sqrt{x} + \frac{\theta}{2}\Delta_k W\right)^2 + \frac{1}{1+\kappa\Delta}\left(\kappa\lambda - \frac{\theta^2}{4}\right)\Delta$$

and for the drift-implicit square-root Euler method it satisfies

$$\varphi_{\overline{X}}(x;k,\Delta) + \frac{1}{1+\kappa\Delta}\left(\frac{4\kappa\lambda-\theta^2}{8\sqrt{\varphi_{\overline{X}}(x;k,\Delta)}} - \frac{\kappa}{2}\sqrt{\varphi_{\overline{X}}(x;k,\Delta)}\right)^2 \Delta^2$$
$$= \frac{1}{1+\kappa\Delta}\left(\sqrt{x}+\frac{\theta}{2}\Delta_k W\right)^2 + \frac{1}{1+\kappa\Delta}\left(\kappa\lambda - \frac{\theta^2}{4}\right)\Delta.$$

From [4] it is known that $\varphi_{\overline{X}}$ is monotone, i.e.

$$\varphi_{\overline{X}}(x_1;k,\Delta) \geq \varphi_{\overline{X}}(x_2;k,\Delta)$$

for $x_1 \geq x_2$. Thus it remains to show that

$$\varphi_{\overline{Z}}(x;k,\Delta) \geq \varphi_{\overline{X}}(x;k,\Delta)$$

for arbitrary $\Delta > 0$, $k = 0, 1, \ldots$, $x > 0$. However, this follows directly by comparing both flows. □

The above property allows one to show the strong convergence of the drift-implicit Milstein scheme, which seems not to have been established yet in the literature.

Proposition 2.1. *Let* $2\kappa\lambda > \theta^2$, $x_0 > 0$ *and* $T > 0$. *Then*

$$\lim_{n\to\infty} \mathbf{E} \max_{k=0,\ldots,n} |X_{t_k} - \overline{Z}^n_{t_k}|^2 = 0.$$

Proof. First note that the drift-implicit square-root Euler method can be rearranged as

$$X_{t_{k+1}} = \varphi_{\overline{Z}}(X_{t_k};k,\Delta) - \kappa\int_{t_k}^{t_{k+1}} (X_t - X_{t_{k+1}})\,dt$$
$$+\theta\int_{t_k}^{t_{k+1}} (\sqrt{X_t} - \sqrt{X_{t_k}})\,dW_t - \frac{\theta^2}{4}(\Delta_k W^2 - \Delta)$$

where $\varphi_{\overline{Z}}$ is the numerical flow of the drift-implicit Milstein scheme defined in the proof of the above Lemma. Thus the error $e_k = X_{t_k} - \overline{Z}_{t_k}$ satisfies the recursion

$$e_{k+1} = e_k - \kappa e_{k+1}\Delta + \theta\Big(\sqrt{X_{t_k}} - \sqrt{\overline{Z}_{t_k}}\Big)\Delta_k W + \rho_{k+1} \tag{21}$$

with $e_0 = 0$, where

$$\rho_{k+1} = -\kappa\int_{t_k}^{t_{k+1}} (X_s - X_{t_{k+1}})\,ds + \theta\int_{t_k}^{t_{k+1}} (\sqrt{X_s} - \sqrt{X_{t_k}})\,dW_s.$$

Now (21) gives

$$e_{k+1} = \frac{1}{1+\kappa\Delta}\left(e_k + \theta\Big(\sqrt{X_{t_k}} - \sqrt{\overline{Z}_{t_k}}\Big)\Delta_k W + \rho_{k+1}\right),$$

so

$$e_k = \sum_{\ell=0}^{k-1} \frac{\theta}{(1+\kappa\Delta)^{k-\ell}} \left(\sqrt{X_{t_\ell}} - \sqrt{\overline{Z}_{t_\ell}} \right) \Delta_\ell W + \sum_{\ell=0}^{k-1} \frac{1}{(1+\kappa\Delta)^{k-\ell}} \rho_{\ell+1}.$$

Straightforward calculations using (20) yield

$$\sup_{n\in\mathbb{N}} \sup_{k=0,\dots,n} \mathbf{E}\overline{Z}_{t_k} < \infty.$$

Then applying the Burkholder-Davis-Gundy inequality to the martingale

$$M_k = \sum_{\ell=0}^{k-1} (1+\kappa\Delta)^\ell \left(\sqrt{X_{t_\ell}} - \sqrt{\overline{Z}_{t_\ell}} \right) \Delta_\ell W, \qquad k = 0, 1, \dots$$

gives

$$\begin{aligned} \mathbf{E} \sup_{k=0,\dots,n} e_k^2 \leq c \sum_{\ell=0}^{n-1} (1+\kappa\Delta)^{2\ell} \mathbf{E} \left| \sqrt{X_{t_\ell}} - \sqrt{\overline{Z}_{t_\ell}} \right|^2 \Delta \\ + c\,\mathbf{E} \sup_{k=1,\dots,n} \left| \sum_{\ell=0}^{k-1} \frac{1}{(1+\kappa\Delta)^{k-\ell}} \rho_{\ell+1} \right|^2 . \end{aligned} \tag{22}$$

Here and below constants whose particular value is not important will be denoted by c regardless of their value.

It remains to estimate the terms on the right side of the equation (22). The previous Lemma implies that

$$\mathbf{E}|\overline{Z}_{t_k} - \overline{X}_{t_k}| = \mathbf{E}(\overline{Z}_{t_k} - \overline{X}_{t_k}),$$

so

$$\mathbf{E}|\overline{Z}_{t_k} - X_{t_k}| \leq 2\,\mathbf{E}|\overline{X}_{t_k} - X_{t_k}| + |\mathbf{E}(\overline{Z}_{t_k} - X_{t_k})|.$$

Clearly, Theorem 10 yields

$$\max_{k=0,\dots,n} \mathbf{E}|\overline{X}_{t_k} - X_{t_k}| \leq c \cdot \sqrt{|\log(\Delta)|} \cdot \sqrt{\Delta}.$$

Moreover,

$$\mathbf{E}\overline{Z}_{t_{k+1}} = \mathbf{E}\overline{Z}_{t_k} + \kappa(\lambda - \mathbf{E}\overline{Z}_{t_{k+1}})\Delta,$$

which is the drift-implicit Euler approximation of

$$\mathbf{E}X_t = x_0 + \int_0^t \kappa(\lambda - \mathbf{E}X_s)\, ds, \qquad t \in [0,T],$$

at the discretization points $t_k = k\Delta$, so

$$\max_{k=0,\dots,n} |\mathbf{E}(X_{t_k} - \overline{Z}_{t_k})| \leq c \cdot \Delta.$$

Hence

$$\max_{k=0,\dots,n} \mathbf{E}|\overline{X}_{t_k} - X_{t_k}| \leq c \cdot \sqrt{|\log(\Delta)|} \cdot \sqrt{\Delta}$$

which gives

$$\sum_{k=0}^{n}(1+\kappa\Delta)^{2k}\mathbf{E}\left|\sqrt{X_{t_k}}-\sqrt{Z_{t_k}}\right|^2\Delta \le c\cdot\sqrt{|\log(\Delta)|}\cdot\sqrt{\Delta} \tag{23}$$

since $|\sqrt{x}-\sqrt{y}|\le\sqrt{|x-y|}$ for $x,y>0$ and $\sup_{n\in\mathbb{N}}\sup_{k=0,\ldots,n}(1+\kappa\Delta)^{2k}<\infty$.

For the second term, applying the Burkholder-Davies-Gundy inequality and Jensen's inequality yield

$$\begin{aligned}&\mathbf{E}\sup_{k=1,\ldots,n}\left|\sum_{\ell=0}^{k-1}\frac{1}{(1+\kappa\Delta)^{k-\ell}}\rho_{\ell+1}\right|^2\\ &\le c\cdot\frac{1}{\Delta}\sum_{k=0}^{n-1}\mathbf{E}\left|\int_{t_k}^{t_{k+1}}(X_t-X_{t_{k+1}})\,dt\right|^2+c\sum_{k=0}^{n-1}\mathbf{E}\int_{t_k}^{t_{k+1}}\left|\sqrt{X_t}-\sqrt{X_{t_k}}\right|^2dt.\end{aligned}$$

Now

$$\mathbf{E}|X_t-X_s|^2\le c\cdot|t-s|,\qquad s,t\in[0,T],$$

so it follows that

$$\mathbf{E}\sup_{k=1,\ldots,n}\left|\sum_{\ell=0}^{k-1}\frac{1}{(1+\kappa\Delta)^{k-\ell}}\rho_{\ell+1}\right|^2\le c\cdot\sqrt{\Delta}, \tag{24}$$

which completes the proof of the proposition. □

Alternatively, Proposition 2.1 could have been obtained by deriving the pathwise convergence of the drift-implicit Milstein scheme and establishing the uniform integrability of the squared maximum error. Note that the above proof gives also the convergence order 1/4 up to a logarithmic term. However this rate seems to be suboptimal, see the following numerical example.

Example 5. The Figures 4 and 5 show the empirical root mean square maximum error in the discretization points, i.e.

$$\left(\frac{1}{N}\sum_{i=1}^{N}\max_{k=0,\ldots,n}|X_{t_k}^{*,(i)}-\overline{X}_{t_k}^{n,(i)}|^2\right)^{1/2},$$

versus the step size for the approximation of the CIR process. Consider the

(i) truncated Euler scheme (2)
(ii) drift-implicit square-root Euler (19)
(iii) drift-implicit Milstein scheme (20)

for the Scenarios I and II (see Example 1). Scenario I satisfies the condition of Theorem 10 and Proposition 2.1 since $2\kappa\lambda/\theta^2=2.011276\ldots$.This condition is violated in Scenario II where $2\kappa\lambda/\theta^2=0.36$. In the latter scenario, the truncation $\sqrt{x^+}$ in the definition of the schemes (ii) and (iii) is used, since both discretization schemes may take negative values here. The numerical reference solution X^* is computed in Scenario I using scheme (ii) with very small stepsize and in Scenario II with scheme (i) with a very small stepsize. The number of repetitions of the Monte Carlo simulation is $N = 5 \cdot 10^4$. In the log-log coordinates here, the dots

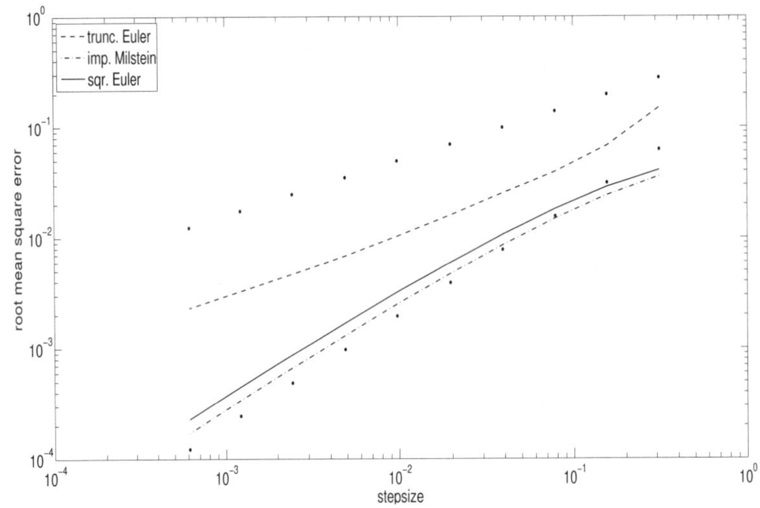

Figure 4. Root mean square errors vs. step size for Scenario I of the CIR process

indicate the convergence orders 0.5 and 1.0 in Figure 4 and 0.25 and 0.5 in Figure 5, respectively. For Scenario I the empirical mean square error for the truncated Euler scheme seems to decay with the order 0.5, while the other schemes seem to have an empirical convergence order close to 1.0. (For smooth and Lipschitz coefficients the Milstein scheme is of order one for the maximum error in the discretization points.)

For Scenario II, these convergence orders deteriorate and for all schemes are significantly lower than one half, see also the following table, where the convergence orders have been estimated by a linear regression.

empirical conv. order / for	Sc. I "part"	Sc. II "part"	Sc. I "full"	Sc. II "full"
truncated Euler	0.5739	0.3193	0.6446	0.2960
drift-imp. square-root Euler	0.9281	0.2734	0.8491	0.2837
drift-imp. Milstein	0.9447	0.3096	0.8719	0.2871

Here "part" denotes the results for the linear regression using only the step sizes $\Delta = 5 \cdot 2^{-j}$, $j = 7, \ldots, 13$, while "full" uses the full data set, i.e. the step sizes $\Delta = 5 \cdot 2^{-j}$, $j = 4, \ldots, 13$.

◇

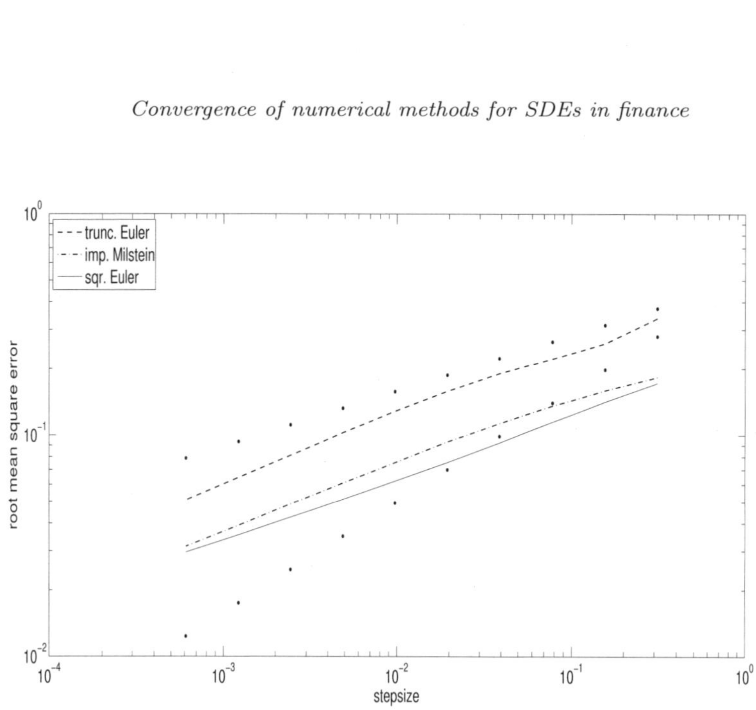

Figure 5. Root mean square errors vs. step size for Scenario II of the CIR process

Applying the Lamperti-transformation also to the asset price in the Heston model gives the log-Heston model

$$d\log(S_t) = \left(\mu - \frac{1}{2}Y_t^2\right)dt + Y_t\left(\sqrt{1-\rho^2}dW_t^{(1)} + \rho dW_t^{(2)}\right), \qquad S_0 = s_0 > 0$$

$$dY_t = \left(\frac{4\kappa\lambda - \theta^2}{8}\frac{1}{Y_t} - \frac{\kappa}{2}Y_t\right)dt + \frac{\theta}{2}dW_t^{(2)}, \qquad Y_0 = y_0 > 0.$$

The approximation of the log-Heston price is then a simple integration problem. Using the Euler scheme for the log-price equation and the drift-implicit square-root Euler scheme for the volatility process yields an approximation $\overline{H}_{t_k}$ of $\log(S_{t_k})$ given by

$$\overline{H}_{t_k} = \log(s_0) + \sum_{\ell=0}^{k-1}\left(\mu - \frac{1}{2}\overline{Y}_{t_\ell}^2\right)\Delta + \sum_{\ell=0}^{k-1}\overline{Y}_{t_\ell}\left(\sqrt{1-\rho^2}\Delta_\ell W^{(1)} + \rho\Delta_\ell W^{(2)}\right).$$

This is extended by piecewise linear interpolation to $[0, T]$.

Corollary 2.1. *Let $2\kappa\lambda > \theta^2$, $x_0 > 0$ and $T > 0$. Then, for all*

$$1 \leq p < \frac{2\kappa\lambda}{\theta^2}$$

there exists a constant $K_p > 0$ such that

$$\left(\mathbf{E}\max_{t\in[0,T]}|\log(S_t) - \overline{H}_t|^p\right)^{1/p} \leq K_p \cdot \sqrt{|\log(\Delta)|}\cdot\sqrt{\Delta},$$

for all $\Delta \in (0, 1/2]$.

Note that in the Heston model moment explosions may appear according to the parameters of the SDE. In particular, for $p > 1$ one has $\mathbf{E}S_t^p < \infty$ for all $t > 0$ if and only if

$$\rho \leq -\frac{\sqrt{p-1}}{\sqrt{p}} + \frac{\kappa}{\theta p}.$$

For more details see e.g. [6]. Whether this phenomenon also arises for discretization schemes for the Heston model is unknown at the time of writing.

Example 6. In this example we test the efficiency of the Multi-level Monte Carlo estimator $\widehat{P}_{ml}$ see [19; 20], based on the above approximation scheme for the valuation of a European Call option, i.e. for

$$p = e^{-rT}\mathbf{E}(S_T - K)^+.$$

The parameters for the Heston model are

$$\begin{aligned} &v_0 = 0.05, \quad \kappa = 5.07, \quad \lambda = 0.0457, \quad \theta = 0.48, \quad T = 1 \\ &s_0 = 100, \quad \mu = r = 0.0319, \quad \rho = -0.7, \quad K = 105. \end{aligned}$$

(Since the riskfree measure is used for the valuation we have $\mu = r$.) In view of the above convergence result for the log-Heston model, we use the number of levels $L = \lceil \log_2(T\varepsilon^{-1}) \rceil$ and the number of repetitions $N_l = \lceil L\varepsilon^{-2}T2^{-\ell} \rceil$, $\ell = 0, \ldots, L$, for a given input accuracy $\varepsilon > 0$, see [19].

The table below shows the empirical root mean square error

$$\text{rmsq} = \sqrt{\frac{1}{M}\sum_{i=1}^{M} |p - \widehat{P}_{ml}^{(i)}|^2}$$

for the Multi-level estimator versus the required number of total Euler steps. The latter is proportional to the overall computational cost of the estimator, i.e. the number of used random numbers, number of function evaluations and number of arithmetic operations. The $\widehat{P}_{ml}^{(i)}$ are iid copies of the Multi-level estimator $\widehat{P}_{ml}$ and we use $M = 5 \cdot 10^4$. The reference value $p = 7.46253$ was obtained by a numerical evaluation of its Fourier transform representation, see e.g. [7].

For comparison, we also provide the corresponding numerical data for the standard Monte Carlo estimator $\widehat{P}_{st}$, see (11), for which we use the relation $\Delta^2 = T/N$ to match stepsize Δ and numbers of repetitions N. For the same parameters as above the empirical root mean square error of $\widehat{P}_{st}$ is again estimated using $M = 5 \cdot 10^4$ repetitions.

ε	Euler steps of $\widehat{P}_{ml}$	rmsq_{emp} of $\widehat{P}_{ml}$	Euler steps of $\widehat{P}_{st}$	rmsq_{emp} of $\widehat{P}_{st}$
2^{-3}	1056	1.369616	512	1.444497
2^{-4}	7168	0.685299	4096	0.714207
2^{-5}	43520	0.352762	32768	0.357962
2^{-6}	245760	0.181384	262144	0.179231
2^{-7}	1318912	0.093485	2097152	0.089618
2^{-8}	6815744	0.047139	16777216	0.044821

The numerical data are in good accordance with the predicted convergence behavior, that is

- for the Multi-level estimator a root mean square error of order ε for a computational cost of order $\varepsilon^{-2}|\log(\varepsilon)|^2$
- and for the standard estimator a root mean square error of order ε for a computational cost of order ε^{-3}.

In particular halving the input accuracy leads for both estimators (approximately) to a halving of the empirical root mean square error. Moreover, these results illustrate nicely the superiority of the Multi-level estimator for small input accuracies.

6. Summary and Outlook

In this article we gave a survey on recent results on the convergence of numerical methods for stochastic differential equations in mathematical finance. The presented results include:

- the pathwise convergence of general Itô-Taylor schemes for strictly positive SDEs with smooth but not globally Lipschitz coefficients (Section 2);
- the construction of structure, i.e. positivity, preserving approximation schemes (Sections 2 and 5);
- the strong convergence of Euler-type methods for the CEV model and the CIR process (Section 3);
- the explosion of the moments of the Euler scheme for SDEs for the 3/2-model (Section 3);
- the strong convergence of the drift-implicit Euler scheme for the Ait-Sahalia model (Section 4);
- strong convergence rates for the approximation of the CIR and the log-Heston model using a drift-implicit Euler-type method (Section 5).

However many unsettled questions are remaining: the exact strong convergence rate of the Euler scheme for the CEV and CIR processes, the existence or non-existence of moment explosions for approximation schemes of the Heston model, how to prevent moment explosions (if they happen) by simple modifications of the scheme etc. And even if these questions are answered, the question remains whether there is a "general theory" for numerical methods for SDEs from mathematical finance or do these SDEs have to analysed one by one. So, the numerical analysis of SDEs arising in finance will be still an active and challenging field of research in the future.

Acknowledgements. The authors would like to thank Martin Altmayer, Martin Hutzenthaler and Arnulf Jentzen for valuable comments and remarks on an earlier version of the manuscript. Moreover, the authors would like to thank Mike Giles for a helpful discussion concerning the numerical evaluation of Fourier transforms.

Bibliography

[1] Ait-Sahalia, A.: Testing continuous-time models of the spot interest rate. *Rev. Financ. Stud.* **9**, no. 2, 385–426 (1996)

[2] Ait-Sahalia, A., Kimmel, R.: Maximum Likelihood Estimation of Stochastic Volatility Models. *J. Financ. Econ.* **83**, 413–452 (2007)

[3] Albrecher, H., Mayer, Ph., Schoutens, W., Tistaert, J.: The Little Heston Trap. *Wilmott Magazine*, January Issue, 83–92. (2007)

[4] Alfonsi, A.: On the discretization schemes for the CIR (and Bessel squared) processes. *Monte Carlo Methods Appl.* **11**, 355–384 (2005)

[5] Alfonsi, A.: High order discretization schemes for the CIR process: Application to affine term structure and Heston models. *Math. Comput.* **79**, no. 269, 209–237 (2010)

[6] Andersen, L., Piterbarg, V.: Moment explosions in stochastic volatility models. *Finance Stoch.* **11**, no. 1, 29–50 (2007)

[7] Arnold, L.: *Random Dynamical Systems*, Springer, Berlin (1998)

[8] Berkaoui, A., Bossy, M., Diop, A.: Euler scheme for SDEs with non-Lipschitz diffusion coefficient: strong convergence. *ESAIM, Probab. Stat.* **12**, 1–11 (2008)

[9] Beskos, A., Papaspiliopoulos, O., Roberts G.: Retrospective exact simulation of diffusion sample paths with applications. *Bernoulli* **12**, no. 6, 1077–1098 (2006)

[10] Bossy, M., Diop, A.: An efficient discretization scheme for one dimensional SDEs with a diffusion coefficient function of the form $|x|^a, a \in [1/2, 1)$. Working paper, INRIA (2007)

[11] Broadie, M., Kaya, Ö.: Exact simulation of stochastic volatility and other affine jump diffusion processes. *Oper. Res.* **54**, 217–231 (2006)

[12] Cox, J.: Notes on option pricing I: Constant elasticity of variance diffusions. Working paper, Stanford University (1975)

[13] Cox, J., Ingersoll, J., Ross, S.: A theory of the term structure of interest rates. *Econometrica* **53**, 385–408 (1985)

[14] Deelstra, G., Delbaen, F.: Convergence of discretized stochastic (interest rate) processes with stochastic drift term. *Appl. Stochastic Models Data Anal.* **14**, 77–84 (1998)

[15] Dereich, S., Neuenkirch, A., Szpruch, L.: An Euler-type method for the strong approximation of the Cox-Ingersoll-Ross process. *Proc. Roy. Soc. A*, to appear

[16] Deuflhard, P., Bornemann, F.: *Scientific computing with ordinary differential equations.* Springer, New York (2002)

[17] Garrido-Atienza, M.J., Kloeden, P.E., Neuenkirch, A.: Discretization of stationary solutions of stochastic systems driven by fractional Brownian motion, *J. Appl Math. Optim.* **60**, 151–172 (2009)

[18] Duffie, D., Glynn, P.: Efficient Monte Carlo simulation of security prices. *Ann. Appl. Probab.* **5**, no. 4, 897–905 (1995)

[19] Giles, M.: Multi-level Monte Carlo path simulation. *Oper. Res.* **56**, no. 3, 607–617 (2008)

[20] Giles, M.: Improved multilevel Monte Carlo convergence using the Milstein scheme. In: Keller, A. (ed.) et al., *Monte Carlo and Quasi-Monte Carlo Methods 2006, Proceedings.* Berlin, Springer, 343–354 (2007)

[21] Günther, M., Kahl, C., Roßberg, T.: Structure preserving stochastic integration schemes in interest rate derivative modeling. *Appl. Numer. Math.* **58**, no. 3, 284–295 (2008)

[22] Makarov, R., Glew, D.: Exact simulation of Bessel diffusions. *Monte Carlo Methods Appl.* **16**, no. 3-4, 283–306 (2010)

[23] Gyöngy, I.: A note on Euler's approximations, *Potential Anal.* **8**, 205–216 (1998)

[24] Gyöngy, I., Rásonyi, M.: A note on Euler approximations for SDEs with Hölder continuous diffusion coefficients, *Stochastic Processes Appl.* **121**, no. 10, 2189–2200 (2011)

[25] Halidias, N.: Semi discrete approximations for stochastic differential equations and applications. *Int. J. Comput. Math.* **89**, no. 6, 780—794 (2012)

[26] Heston, S.: A closed form solution for options with stochastic volatility, with applications to bonds and currency options. *Rev. Financial Stud.* **6**, 327–343 (1993)

[27] Heston, S.: A simple new formula for options with stochastic volatility. Working paper, Washington University of St. Louis (1997)

[28] Higham, D., Mao, X.: Convergence of Monte Carlo simulations involving the mean-reverting square root process. *J. Comp. Fin.* **8**, 35–62 (2005)

[29] Higham, D., Mao, X., Stuart, A.: Strong convergence of Euler-type methods for nonlinear stochastic differential equations. *SIAM J. Numer. Anal.* **40**, 1041–1063 (2002)

[30] Higham, D., Mao, X., Szpruch, L.: Convergence, Non-negativity and Stability of a New Milstein Scheme with Applications to Finance. Working paper (2012) `arXiv:1204.1647`

[31] Hurd, T.R., Kuznetsov, A.: Explicit formulas for Laplace transforms of stochastic integrals. *Markov Process. Relat. Fields* **14**, no. 2, 277–290 (2008)

[32] Hutzenthaler, M., Jentzen, A.: Convergence of the stochastic Euler scheme for locally Lipschitz coefficients. *Found. Comput. Math.* **11**, no. 6, 657–706 (2011)

[33] Hutzenthaler, M., Jentzen, A.: Numerical approximations of stochastic differential equations with non-globally Lipschitz continuous coefficients. Ann Appl.

[34] Hutzenthaler, M., Jentzen A., Kloeden, P.: Strong and weak divergence in finite time of Euler's method for stochastic differential equations with non–globally Lipschitz coefficients. *Proc. Roy. Soc. London A* **467**, no. 2130, 1563–1576 (2011)

[35] Hutzenthaler, M., Jentzen A., Kloeden, P.: Strong convergence of an explicit numerical method for SDEs with non–globally Lipschitz continuous coefficients. *Ann. Appl. Probab.* 22 (2012), 1611-1641.

[36] Hutzenthaler, M., Jentzen A., Kloeden, P.: Divergence of the multilevel Monte Carlo

Euler method for nonlinear stochastic differential equations. Working paper (2011) `arXiv:1105.0226`

[37] Jentzen, A., Kloeden, P., Neuenkirch, A.: Pathwise approximation of stochastic differential equations on domains: Higher order convergence rates without global Lipschitz coefficients. *Numer. Math.* **112**, no. 1, 41–64 (2009)

[38] Kloeden, P., Neuenkirch, A.: The pathwise convergence of approximation schemes for stochastic differential equations. *LMS J. Comput. Math.* **10**, 235–253 (2007)

[39] Kloeden, P., Platen, E.: *Numerical Solution of Stochastic Differential Equations.* 3rd edn, Springer, Berlin (1999)

[40] Lord, R., Koekkoek, R., van Dijk, D.: A comparison of biased simulation schemes for stochastic volatility models. *Quant. Finan.* **10**, no. 2, 177–194 (2010)

[41] Milstein, G.: *Numerical Integration of Stochastic Differential Equations.* Kluwer, Doordrecht (1995)

[42] Milstein, G., Tretyakov, M.: Numerical integration of stochastic differential equations with nonglobally Lipschitz coefficients. *SIAM J. Numer. Anal.* **43**, 1139–1154 (2005)

[43] Müller-Gronbach, T.: The optimal uniform approximation of systems of stochastic differential equations. *Ann. Appl. Probab.* **12**, no. 2, 664–690 (2002)

[44] Neuenkirch, A., Zähle, H.: Asymptotic error distribution of the Euler method for SDEs with non-Lipschitz coefficients. *Monte Carlo Methods Appl.* **15**, no.4, 335–353 (2009)

[45] Szpruch, L., Mao, X., Higham, D., Pan, J.: Numerical simulation of a strongly nonlinear Ait-Sahalia-type interest rate model. *BIT* **51**, 405–425 (2011)

Chapter 3

Inverse problems in finance

J. Baumeister

Department of Mathematics, Goethe University, Robert-Mayer Str. 6-10, D-60054 Frankfurt Main, Germany

Abstract We give a survey of the mathematical methods for the computational analysis of inverse problems in finance. In the first subsection we present the forward/direct problem and the inverse problems in finance. As we know, inverse problems are ill-posed in general. Therefore we give an introduction to ill-posedness to the regularization theory in the second subsection. We sketch the main recent developments in the linear and nonlinear theory. The simplest form of the inverse problem in finance is to reconstruct the constant volatility in the Black-Scholes model, in order to understand the smile in option prices. This problem and related ones are considered in the third subsection. The mathematics is elementary and the algorithms are more or less recipes. In subsection four we consider the uniqueness in the reconstruction of local volatilities (Identifiability). This uniqueness is based more or less on the equation of Dupire. The last subsection is devoted to the calibration problem: determination of the local volatility from market data. We do this for vanilla options only since the analysis of other complex models (like American options, ...) is on the one side similar and on the other side not in the focus of the financial market. We sketch various gradient methods and at the end we consider a Kaczmarz-like method.

Almost all results presented in this survey suffer fromm the fact that the nonlinearity leads to a lot of difficulties to make the presentation as smooth as desired. We see this mainly in Section 5 where we consider gradient type methods: the adaptation to the Hilbert space setting is not as perfect as necessary.

1. Forward and inverse problems in finance

In this section, we present the main concepts of financial instruments and we give an introduction to theoretical, numerical and empirical aspects of models in finance. On this basis we formulate inverse problems of finance. Especially, for the computation of solutions in inverse problems (of finance) we sketch a few methods to solve the forward problem (in finance). We follow mainly [12; 72; 129]. Some aspects presented here are also considered in [124].

1.1. *Classification of problems*

A *model* is an image of reality formulated in terms which are well defined and can be handled by theoretical investigations. Usually, a model is developed by considering the relation between causes and effects in the process under consideration.

Suppose that we have a mathematical model of a (physical, biological, social, ...) process. We assume that this model gives a description of the *system behind the process* and its *operating conditions* and explains the *principal quantities* of the model:

input, system parameters, output

The analysis of the given process via the mathematical model may then be separated into three distinct types of problems.

(A) **The direct problem.** Given the input and the system parameter, find out the output of the model.

(B) **The reconstruction problem.** Given the system parameters and the output, find out which input has led to this output.

(C) **The identification problem.** Given the input and the output, determine the system parameters which are in agreement with the relation between input and output.

We call a problem of type (A) a direct (or forward) problem since it is oriented towards a cause-effect sequence. In this sense problems of type (B) and (C) are called inverse problems because they are problems of finding out unknown causes of known consequences. It is immediately clear that the solution of one of the problems above involves a treatment of the other problems as well. A complete discussion of the model by solving the inverse problems is the main objective of *inverse modelling*.

Inverse modelling involves the estimation of the solution of an equation from a set of observed data represented by y, p in (B) and y, x in (C) respectively. In applications, the knowledge of just one observation y is not sufficient to determine p in problem (B). It is the subject of the identification theory to find out how many data y are sufficient to determine p in a uniquely way: *identifiability*.

Usually, the system is governed by a process operator transforming the input into the output. Moreover, the laws of nature are often expressed as systems of differential equations containing parameters describing the concrete situation; in a dynamical consideration the parameters are functions of time. These equations are *local* in the sense that the quantities (velocity, pressure, current in a physical situation) depend on a neighborhood of the current time only. Another typical feature of the laws is *causality*: later conditions depend on previous ones. Locality and causality are features typically associated to direct/forward problems. Inverse problems on the other hand are most often *nonlocal* and *noncausal*. These properties contribute to the fact that inverse problems are sensitive to the input data and parameters.

Let us give a description of the input, the output, the parameters and the systems by mathematical terms:

X	space of input quantities;
Y	space of output quantities;
P	space of system parameters;
$A(p)$	system operator from X into Y associated to the admissible parameter $p \in P_{\text{ad}} \subset P$.

In these terms we may formulate the problems (A), (B), (C) above in the following way:

(A) Given $x \in X$ and $p \in P_{\text{ad}}$, find $y := A(p)x$.
(B) Given $y \in Y$ and $p \in P_{\text{ad}}$, find $x \in X$ such that $A(p)x = y$.
(C) Given $x \in X$ and $y \in Y$, find $p \in P_{\text{ad}}$ such that $A(p)x = y$.

For the description of $X, Y, P, A(p)$ and P_{ad} we use methods and results from analysis, linear algebra and functional analysis.

At first glance, the direct problem (A) seems to be solved much easier than the inverse problems (B), (C). However, for the computation of $y := A(p)x$ it may be necessary to solve a differential or integral equation, a task which may be of the same complexity as the solution of the equations in the inverse problems.

The theory to solve (B) or (C) falls into two distinct parts. One deals with the ideal case in which the data are supposed to be known exactly and completely (*perfect data*). The other treats the practical problems that are created by incomplete and imprecise data (*imperfect data*). It might be thought that an exact solution to an inverse problem with perfect data would prove also useful for the practical case. But in general, it turns out in inverse problems that the solution obtained by the analytic formula obtained from the analysis of the forward problem is very sensitive to the way in which the data set is completed and to errors in it. Moreover, when the inverse problem is described by an operator equation, the naive way to solve the equation by a fit of the data to the range of the related operator does not give necessarily good results: As a rule of thumb: *a good fit to the data does not mean a good fit to the unknown solution.*

The list of inverse problems other than in finance is huge. We refer for instance to [12; 47; 61; 77; 97; 120]. Over the past 40 years, the number of publications on inverse problems has grown rapidly. Nowadays there are several mathematical journals devoted to this topic. Especially, the subject of image processing and non-invasively, non-destructively mapping have become very important: *Making the unseen visible with mathematics.* In the context of finance we are interested mainly in determination of the volatility from market data which is the most important parameter in models finance instruments. Thus, we are concerned with an inverse problem of the type (C) and, in general, all theoretical and numerical results concerning parameter identification may be applied; see for instance [11; 15; 78; 79; 105].

There are concepts that are often confused by modellers, in particular in financial and economic modeling: identification, simulation and extrapolation:

- **Identification** is building a model from the available data.
- **Simulation** is experimenting with the model to produce new and additional data.
- **Extrapolation** is producing new data with a model beyond the reference period of the available data.

With all these concepts we are concerned in this article,

1.2. *Financial market and financial products*

Here is a very short introduction to derivatives in order to fix assumptions which are needed for our considerations; see for instance [51]. A critical discussion of the problems with the financial products in the last years can be found in [50].

Fiancial market

We consider a financial market where assets (stocks, equities, interest rates, ...) are traded in continuous time and we assume that there are d risky assets. Their prices $S^1, \ldots, S^d$ are assumed to be stochastic processes. Additionally, we assume that there is a riskless asset with price S^0. To describe the dynamics of the assets $S^0, S^1, \ldots, S^d$, we have to model the dynamics of their prices. Usually, the price-dynamics of the risky assets is governed by a system of d stochastic differential equations associated with Brownian motions $dW^1, \ldots, dW^d$: The dynamics of the riskless asset S^0 may be modelled by a stochastic differential equation too. If the process is deterministic it may be modelled by a simple first order differential equation; see the third section.

A *portfolio* associated to this financial market is a stochastic process $\pi_t = (\pi_t^0, \pi_t^1, \ldots, \pi_t^d), t \geq 0$, where π_t^i denotes the number of shares of asset i held at time t. Its value is

$$V_t(\pi) := \sum_{i=0}^{d} \pi_t^i S_t^i \tag{1}$$

The family $(V_t(\pi))_{t\geq 0}$ should be considered as a family of random variables. Therefore the expectation $\mathbb{E}^{\mathbb{P}}(V_t(\pi)), t \geq 0$, is well defined (under reasonable assumptions). To analyze portfolios we have to state some conditions which have to fulfill the finance market from the theoretical point of view. Here is the list of these conditions:

- There are market participants.
- There are no *transactions costs.*
- The market is *perfectly liquid* and it is possible to purchase or sell any amount of shares or their fractions at any given time.
- *Short-selling* is allowed, i.e. values $\pi_t^i < 0$ in a portfolio are allowed.
- Borrowing as well as lending is both possible at the risk-free interest rate.
- The interest rate is known.
- There are no arbitrage opportunities. *Arbitrage* is defined as the possibility to do a set of trades that leaves you with no position, but money in the pocket; see below.

In the following, we assume without further mention that these assumptions are satisfied. Later on we shall add the assumption of *completeness.*

The assumption about the arbitrage-freeness is the most important one. In terms of a portfolio this means, roughly speaking, no portfolio π with $V_0(\pi) = 0$ and $V_T(\pi) \geq 0, \mathbb{E}(V_T(\pi)) > 0$, is possible. There is a famous expression often used in financial literature: *There is no such thing as a free lunch.* From the theoretical point of view, arbitrage-freeness ensures that a probability measure exists equivalent to the given one associated to the processes $(S_t^i)_{t\geq 0}$ with very helpful additional properties.

Derivatives

Derivatives are financial products, such as futures, options and mortage-backed securities. In general, these products are based on one or more underlying assets traded in the financial market. Thus, the underlying is usually a stock, an interest rate or an equity or a basket of these different financial objects. Here we consider the case of one underlying and an deterministic riskless asset.

A *derivative* is a contract between two parties. Typically, the seller receives money in exchange for an aggreement to purchase or sell some good or service at some future time $T > 0$ called the *maturity* of the contract. For each time $t \in [0, T]$ this contract has its price $C(t)$. The main problem is to find the price of the contract at the initial time $t = 0$. Since the payoff of the contract depends usually on the evolution of the underlying which is unknown it is difficult to determine the price of the contract/derivative.

An easy to understand example of an option is the *European call option* which gives the holder of the option the right but not the obligation to buy a stock for the *strike price* K at maturity $T > 0$. Mathematically, this means that the payoff function of the option is

$$(S_T - K)^+ := \max(0, S_T - K) \tag{2}$$

where S_t is the value of the underlying at time $t \in [0, T]$. Usually, the price of such an option will be determined under the assumption of a deterministic constant interest rate r.

An European *put option* is the same, but this time the buyer of the option has the right but not the obligation to sell the underlying. The payoff function of such a contract is

$$(K - S_T)^+ := \max(0, K - S_T) \tag{3}$$

A call or put option that can be exercised anytime during its life $[0, T]$ is called an *American option.* A huge amount of European and American options are observable on the financial markets by the notion of their price round about the world.

Once we have at hand the family of stochastic variables $S_t, t \in [0, T]$, the price $C(t)$ of an European option is given by the discounted expectation with respect to a measure Q:

$$C(t) = e^{-r(T-t)}\mathbb{E}^Q((S_T - K)^+) \tag{4}$$

Here Q is the so called *risk-neutral measure* which is a probability measure depending on the model for the evolution of the risky asset of the financial market with helpful conditions. Moreover, r is the constant interest short rate for the riskless asset.

1.3. *The forward problem of option pricing*

In this section we sketch the forward problem in order to prepare the formulation of inverse problems in the following sections. Most of the aspects considered here can be found in [36; 58].

The Black-Scholes-Merton model

Let us consider the modelling for an European option when the underlying is a stock. Thus, we have one risky asset on our financial market. The interest short rate is assumed to be deterministic and constant. The associated theory is called *Arbitrage-Free Option pricing* and was among others developed by Black, Scholes and Merton in the early 70s; see [21; 109]. A first pricing formula for options appears in thesis of Bachelier (see [8]). In this model the dynamics of the underlying is governed by a drifted Brownian motion.

As a paradigmatic example, we consider the problem of an European call option with maturity T and strike K on the underlying asset $S = (S_t)_{t \geq 0}$ whose dynamics is described by the stochastic differential equation

$$\frac{dS}{S} = \underbrace{(\mu - q)\, dt}_{\text{deterministic component}} + \underbrace{\sigma\, dW_t}_{\text{random component}} \tag{5}$$

Here $(W_t)_{t \geq 0}$ denotes the standard Brownian motion/Wiener process. The stochastic differential equation is called the (drifted) *geometric Brownian motion.* μ is the so called *drift constant*, $q \geq 0$ is the *rate of a continuous dividend.* The increments $W_t - W_s, s < t$, of the Wiener process are independent Gaussian random variables with expectation zero and unit variance $\sqrt{t-s}$.

Throughout, we do not consider the modelling of dividents. This means that we choose $q = 0$ in equation (5). This restriction is a technical simplification only.

Under the assumptions of our finance market (liquidity, absence of arbitrage and transaction costs, completeness,...) the call price $C = C(S,t) = C(S,t;K,T,r,q,\sigma)$ has the probabilistic representation

$$C(S,t;K,T,r,\sigma) = \exp(-r(T-t))\mathbb{E}^{Q}_{t,S}((S_T - K)^+)\,, \tag{6}$$

where $\mathbb{E}^{Q}_{t,S}$ is the expected value of the *risk-neutral* probability measure Q. Under this probability measure the drift constant μ is eliminated, or in other terms: the drift constant μ is replaced by the interest rate r. The risk-neutral probability measure differs from so called *subjective measure* $\mathbb{P}$ in the sense that it is the one for which the discounted process $(e^{-r(T-t)}S_t)_{t\geq 0}$ is a martingal; see [93; 101] for more details.

An interpretation of the representation (6) would be: for each realization ω of the market, the payoff $(S_T(\omega) - K)^+$ should be brought to its present value $e^{-r(T-t)}(S_T(\omega) - K)^+$ by means of discounting by the interest rate r. Then the average over all the possible realizations with respect to the risk-neutral measure gives the option price.

There are five factors that affect an option's price: strike price, underlying stock price, maturity time, interest rate, rate of dividend, volatility. The variance σ of the random component in the stochastic differential equation (5) is the major determinant. In the context of finance it is called *volatility.* For short, there are three different types of volatility: implied volatility, stochastic volatility and local volatility; see below.

Pricing formula

Let us surpress the parameter r, σ, K, T in the representation (6), i.e. $C = C(S,t)$. Using Itô's formula (see [82; 92]) the fair price $C = C(S,t)$ for the European call option satisfies the so called *Black-Scholes equation* with boundary and final values:

$$\frac{\partial C}{\partial t} + \frac{1}{2}\sigma^2 S^2 \frac{\partial^2 C}{\partial S^2} + rS\frac{\partial C}{\partial S} - rC = 0 \;\; (S \in (0,\infty), t \in [0,T))\,, \tag{7}$$

$$C(0,t) = 0\,,\; \lim_{S\to\infty}(C(S,t) - S) = 0\,,\; t \in (0,T)\,, \tag{8}$$

$$C(S,T) = (S-K)^+\,,\; S > 0\,. \tag{9}$$

Since the problem (7),(8),(9) can be transformed to a boundary value problem for the heat equation – we will see this transformation later on – the solution can be given by the following formula:

$$C(S,t) = S\mathcal{N}(d_+(\sigma)) - Ke^{-r(T-t)}\mathcal{N}(d_-(\sigma))\,, \tag{10}$$

where we use

$$\mathcal{N}(a) := \int_{-\infty}^{a} \frac{1}{\sqrt{2\pi}} \exp(-\frac{s^2}{2})ds\,,\ a \in \mathbb{R}$$

$$d_\pm(\sigma) = \frac{\ln(\frac{S}{K}) + (r \pm \frac{\sigma^2}{2})(T-t)}{\sigma\sqrt{T-t}}\,,\ \sigma \geq 0\,;$$

here $\mathcal{N}$ denotes the cumulative normal distribution function. Especially, $C(S_0, t_0)$ is the price which the emittent of an European call option should demand if the price of the underlying asset is S_0 at time t_0.

The pricing of an European put option can be done in the same way. Under the assumptions for the finance market (see below) the so called *put-call-parity* holds:

$$P(S,t) + S = C(S,t) + Ke^{-r(T-t)}\,,\ S > 0, t \in [0,T]\,, \tag{11}$$

where P, C is the price of the put and call, respectively. Therefore the price of a put option can be computed from the price of a call option. This is important in section 5 when we shall consider weak solutions.

Here are some properties which satisfy the price $C = C(S,t;K,T), P = P(S,t;K,T)$ of an European call option and a put option, respectively. For a proof see [36].

Properties 1.1.

- $0 \leq (S - Ke^{-r(T-t)})^+ \leq C(S,t;K,T) \leq S\,,\ t \in [0,T]\,.$
- $0 \leq (Ke^{-r(T-t)} - S)^+ \leq P(S,t;K,T) \leq Ke^{-r(T-t)}\,,\ t \in [0,T]\,.$
- $C(S,t;K,\cdot)$ monotone nondecreasing.
- $C(S,t,\cdot,T)$ is monotone nonincreasing and convex.
- $C(\cdot,t;K,T)$ is monotone nondecreasing and convex.
- $\lim_{K\to\infty} C(S,t;K,T) = 0\,.$

These properties are important for the completion of the market data in a concise form. We come back to this properties in the next sections.

Sensitivity

By showing how the model behavior responds to changes in parameter values, sensitivity analysis is a useful tool in model building as well as in model evaluation and it helps to build confidence in the model. *Sensitivity analysis* is used to determine how "sensitive" a model is to changes in the value of the parameters of the model and to changes in the structure of the model. Sensitivity analysis in finance is a wide field; see [53; 54] for instance.

Sensitivity in finance is mainly related to the so called *greeks*.* These are quantities which measure the dependency of option prices C on parameters of the models.

Delta $\Delta := \frac{\partial C}{\partial S}$ is a quantity which is used as a hedging-parameter.

Gamma $\Gamma := \frac{\partial^2 C}{\partial S^2}$ can be used to measure the sensitivity of the option price with respect to large variations of the price of the underlying.

Rho $\mathcal{P} := \frac{\partial C}{\partial r}$.

Vega $\mathcal{V} := \frac{\partial C}{\partial \sigma}$ is the most important sensitivity quantity.

Theta $\Theta := \frac{\partial C}{\partial t}$ can be used as a measure concerning the dependence of the option price on the time to the maturity.

These greeks are defined independently of the model for the option prices. The computation of the greeks is a very important task in using models for option pricing. In particular, the determination of the Vega is of high interest.

1.4. *Volatility*

The Black-Scholes formula (10) worked well before the huge crashes 1987 and 1989, but since then, it has been observed that market prices contradict the assumption of constant volatility: option prices for different maturities change drastically, and option prices for different strikes also experience significant variations. We conclude that the Black-Scholes-Merton model does not describe the market correctly; one calls this effect the *smile*. There are mainly three different strategies to remedy this observation:

- Implied volatility
- Stochastic volatility
- Local volatility

Implied volatility

Here we state the question: what volatility is implied in observed option prices, if the Black-Scholes model is a valid description of market conditions? Thus, the *implied volatility* is the value of σ such that the price of a European call is equal to the value obtained by applying the Black-Scholes formula. On the market, the implied volatility is the volatility which use the traders, since then by the Black-Scholes formula the price of the option can be computed. We will consider the

*These numbers are called *greeks* since they are denoted by greek letters. An exceptional case is the sensitivity with respect to the underlying: Vega is no greek letter.

computation of the implied volatility in the next chapter.

According to the Black-Scholes model, we should expect implied volatilities for different calls on the same underlying to be identical. But it can be observed that this is not the case: there is a dependence on the strike price K, on the *moneyness* (nearness of the stock price and the strike price) and on the remaining lifetime of the option. This is confirmed by a series of empirical studies; see [39] for instance.

The collection of implied volatilities for different strikes and maturities is called the *implied volatility surface* because if we plot that table, we can see a *surface*. But this surface cannot be used for pricing options for strikes and maturities not quoted on the market. Standard interpolation techniques may give rise to arbitrage in the interpolated volatility surface even if there is no arbitrage in the original set.

Stochastic volatility

To overcome the smile-effect, one may extend the model. One widely used approach is to consider that the volatility also follows a stochastic diffusion process: A concrete example of this kind is the so called *Heston-modell* (see [76]). Stochastic volatility models are useful because they explain in a selfconsistent way why it is that options with different strikes and expirations have different implied volatilities.

Local volatility

Another case is when the volatility σ is not a constant any more but rather a deterministic function of the underlying asset and the time. These types of models are called *local volatility models*. For such models, the dynamics of the underlying asset is governed by the stochastic differential equation

$$\frac{dS}{S} = \mu\, dt + \sigma(S,t)\, dW_t. \tag{12}$$

Here σ is a function of the underlying asset and of time. Now, we are forced to use a method for the computation of option prices where the volatility σ is a function of the underlying and time:

$$\sigma : [0,\infty) \times [0,T] \longrightarrow \mathbb{R}.$$

Again, one can show that the price of an European call option whose dynamic of the underlying is governed by the equation (12) is implicitely given as a solution of the following system of equations:

$$\frac{\partial C}{\partial t} + \frac{1}{2}\sigma(S,t)^2 S^2 \frac{\partial^2 C}{\partial S^2} + rS\frac{\partial C}{\partial S} - rC = 0 \quad (S \in (0,\infty), t \in [0,T)), \tag{13}$$

$$C(0,t) = 0\,,\ \lim_{S\to\infty} (C(S,t) - S) = 0\,,\ t \in (0,T)\,, \tag{14}$$

$$C(S,T) = (S-K)^+\,,\ S > 0\,. \tag{15}$$

In general, there is no closed form formula for the solution.

Existence and logarithmic variables

Let us consider the Black-Scholes equation with boundary and final conditions and local volatility; see (13),(14),(15). Let us fix the parameters r and the volatility function σ in the Black-Scholes equation. Concerning the local volatility we assume*

$$0 < \sigma_{\min} \le \sigma(x,t) \le \sigma_{\max}\,,\ x,t > 0\,,\ \text{whith constants } \sigma_{\min}, \sigma_{\max} \tag{16}$$

$$\begin{aligned}&\sigma \text{ is Hölder-continuous },\ i.e.\ \sigma \in C^{\gamma,\gamma/2}([0,\infty) \times (t_*,T])(0 < t_* \le T)\\ &\quad \text{with } 0 < \gamma < 1\,.\end{aligned} \tag{17}$$

Theorem 1.1 (Existence for the forward problem).
The initial-boundary value problem (13), (14), (15) *posesses (under the assumptions* (16), (17)*) for each pair* (K,T) *a continuous solution* $C(\cdot,\cdot,K,T,r,\sigma)$ *and we have the growth-bound*

$$|C(S,t;K,T,r,\sigma)| \le c_1 S^{c_2 \ln(S)}\,,\ S,t > 0 \quad (c_1,c_2 \ge 0\,. \tag{18}$$

Proof:
We set for a fixed pair (K,T)

$$y := \ln(S/K), \tau := T-t, a(y,\tau) := \frac{1}{2}\sigma(S,t)^2\,,\ c(y,\tau) := C(S,t;K,T,r,\sigma)\,. \tag{19}$$

Then we obtain from the equations (13), (14), (15)

$$\mathcal{L}_{BS,log}c := \frac{\partial c}{\partial \tau} - a(y,\tau)\frac{\partial^2 c}{\partial y^2} - (r - a(y,\tau))\frac{\partial c}{\partial y} - rc = 0\,, \quad y \in \mathbb{R}, \tau \in (0,T)\,. \tag{20}$$

$$c(-\infty,\tau) = 0\,,\ \lim_{y\to\infty}(c(y,\tau) - Ke^y) = 0\,,\ \tau \in (0,T)\,, \tag{21}$$

$$c(y,0) = K(e^y - 1)^+\,,\ y \in \mathbb{R}\,. \tag{22}$$

From [57], Chapter 1, and [24], Chapter 2, we obtain a classical solution of (20),..., (22). A transformation back to the S-t–variables gives the result. For additional details see [72].

Of course, one can say more on the quality of the solution in Theorem 1.1 (boundedness, Hölder-continuity). In general, finite difference methods are used to solve the initial boundary value problem† problem numerically.

Should one prefer the S-t–variables or the (artificial) y-τ–variables in the analysis of the Black-Scholes-model? The answer is, it does not matter in theoretical questions concerning existence, uniqueness and stability but it is important in numerical considerations: the discretization in different variables in a uniform manner

*For Hölder continuity consult [57; 128] for instance.

†Clearly, (15) is a condition on the end of the interval $[0,T]$ but by time reversing a parabolic equation with an initial condistion results.

has serious consequences. A uniform grid in the y-τ–variables means a very dense grid in the S-t–variables at $S=0$, a uniform grid in the S-t-variables at "$S=\infty$" is essential a dense grid in the y-τ-variables at "$y=\infty$". Therefore for both choices of variables numerical methods have to refine the grid appropriatly.

1.5. *Inverse problems of finance*

Suppose that we have a model to list the option prices as function of the parameters already introduced:

$$C\,:\,(0,\infty)\times(0,T)\ni(S,t)\;\longmapsto\;C(S,t;K,T,r,\sigma)\in\mathbb{R} \tag{23}$$

Direct problem of finance

Given $S_*, t_*, r, T_{\max}$, a (continuous) function $\sigma:(0,\infty)\times[t_*,T_{\max})\longrightarrow(0,\infty)$ and a set $R\subset(0,\infty)\times[t_*,T_{\max})$.
Determine the operator $G(\sigma):R\ni(K,T)\;\longmapsto\;C(S_*,t_*;K,T,r,\sigma)\in[0,\infty)$.

For the evaluation of the operator $G(\sigma)$ along the Black Scholes model in $(K,T)\in R$ we have to solve the partial differential equation (13) with boundary conditions (14) and final condition (15) or to solve the stochastic differential equation (5). This can be done as follows:

- Finite difference and finite element methods in computing solutions of (7), (8), (9); see [1; 72]
- Simulation methods using discrete models of the stochastic differential equation (5) in combination with Monte Carlo methods, especially multilevel Monte Carlo methods; see for instance [75; 83; 100].

In any case, a huge amount of computational work is necessary. We will see in section 4 and 5 that this can be avoided by a clever idea (Dupire's equation or dual Black-Scholes equation).

Market prices of financial derivatives such as options are directly observable. In the context above, this means that values in the range of C (see (23)) are observable in the market. Since the values of S and the constants K,T,r are also known from the market the only parameter which is unknown is the volatility. According to the formulation of the direct problem in finance using the operator $G(\sigma)$ the inverse problem should be:

Inverse problem of finance/Identifiability

Given $S_*, t_*, r, T_{\max}$ and sets $R,I\subset(0,\infty)\times[t_*,T_{\max})$.
Prove that
$G(\sigma)(K,T)=G(\tilde\sigma)(K,T)$ for all $(K,T)\in R$ implies $\sigma(S,t)=\tilde\sigma(S,t)$, $(S,t)\in I$.

Clearly, the chance that identifiability can be proved, is related to the "size" and the relation of the sets R, I . The main question is whether a given subset R is sufficient for proving the identifiability in the set I . In particular, R and I can be the whole set $[0, \infty) \times [0, T_{\max})$. The main rule of thumb in this identifiability problem says that for a stable reconstruction of the unknown properties of the system in a certain subset I it is necessary to have data in a subset R which is in size and quality related to I .

The *calibration of a financial models* is the process of tuning the model parameters to fit market data to the results of the model. Each calibration of a model needs to choose a calibration set; see below. Obviously, the choice of the calibration set may influence the outcome of the calibration. The calibration problem is the inverse of the pricing problem. Instead of computing prices in a model with given values for its parameters, one wishes to compute the values of the model parameters that are consistent with observed prices (up to the bid-ask spread).

Inverse problem of finance/Calibration

Given $S_*, t_*, r, T_{\max}$, a set $M \subset (0, \infty) \times [t_*, T_{\max})$ and a function $v : M \longrightarrow \mathbb{R}$.
Find $\sigma : 0, \infty) \times [t_*, T_{\max}) \longrightarrow (0, \infty)$ such that

$$C(S_*, t_*; K, T, r, \sigma) = v(K, T) \text{ for } (K, T) \in M .$$

In the real situation, M is a discrete set and numerical methods have to take into account this fact. The situation that M is a singleton is a special case of calibration when the Black-Scholes-formula is used as the pricing model: that is the case of the implied volatility. Then we have to solve a nonlinear equation. This (numerical) problem is discussed in the third section.

As a rule, the calibration problem is ill-posed: existence and uniqueness are in question, a continuous dependence of the solution on the function v is in doubt. This is a challenge for a numerical solution of the calibration problem. The remedy is regularization.

There are three factors which have to be taken into account when we consider a method for the solution of the calibration problem above:

(1) One has to consider the **bid-ask-spread**. The bid/ask spread (also known as bid/offer or buy/sell spread) for securities (such as stocks, futures, options) is the difference between the prices quoted for an immediate sale (ask) and an immediate purchase (bid).
(2) **Noise** There is always a degree of randomness in the level of stock prices in the market. So, we have to solve the calibration problem for erroneous data.
(3) **Unmodelled dynamics** We suspect that investors have shifted their forward-looking focus to a specific point in the future beyond what we

can see in the available data.

2. An introduction into regularization of ill-posed problems

In this chapter we sketch the inverse terminology and the problems which result from the fact that inverse problems are ill-posed in general. In the focus are aspects of the regularization theory for ill-posed inverse problems, a very active branch of applied mathematics. This theory is of enormous help to understand the problems and questions atising in studying inverse problems of finance. We follow in this section the monographs [12; 47; 97; 90].

2.1. *Ill-posedness/Well-posedness*

Here we explain basics of for classificying (systems of) equations whose solution do not depend continously on the involved data.

Definition

In a complete analysis of an inverse problems – we denote the collection of methods and results in this field by "inversion theory"- the questions of *existence, uniqueness, stability* and *construction* of objects have to be considered.

The question of existence and uniqueness is of great importance in testing the assumption behind any mathematical model. If the answer in the uniqueness question is no, then we know that even perfect data do not contain enough information to recover the physical quantity to be estimated. By questioning for stability we have to decide whether the solution depends continuously on the data. Stability is necessary if we want to be sure that a variation of the given data in a sufficiently small range leads to an arbitrarily small change in the solution. This concept was introduced by *Hadamard* in 1902 in connection with the study of boundary value problems for partial differential equations and he designated unstable problems *ill-posed** otherwise *well-posed*; see [64]. The nature of inverse problems (irreversibility, causality, unmodelled structures, ...) leads to ill-posedness as a characteristic property of these problems.

The prototype of an inverse problem will be an equation of the form

$$F(x) = y \quad (x \in X, y \in Y) \tag{24}$$

with a mapping F from the Banach space X into the Banach space Y; in the formulation above we have $F(x) = A(p)x$ in the case (B). For such an equation, the unknown is x and the data are usually the right-hand side y. If the stability

*Hadamard believed – many mathematicians still do – that ill-posed problems are actually incorrectely posed and artificial in that they would not describe physical systems. He was wrong! Nowadays we know that such problems arise in a fundamental way in the modelling of complex (physical) systems.

condition is violated, the numerical solution of the inverse problem by standard methods is difficult and often yields instability, even if the data are exact (since any numerical method has internal errors acting like noise). Therefore, special techniques, so called *regularization methods* have to be used in order to obtain a stable approximation of the solution. The appropriate construction and analysis of regularization methods and subsequently (or simultaneously) of numerical schemes is the major issue in the solution of inverse problems.

When solving ill-posed problems (numerically), we must certainly expect some difficulties, since any error acts as a perturbation on the original equation and so may cause arbitrarily large variations in the solution. Since errors cannot be completely avoided, there may be a range of plausible solutions and we have to find out a reasonable solution. These ambiguities in the solution of inverse problems which are unstable can be reduced by incorporating some sort of *a-priori information* that limits the class of allowable solutions. By a-priori information we mean an information which has been obtained independently of the observed values of the data. This a-priori information may be given as a deterministic or a statistical information. We shall restrict ourselves mainly to deterministic considerations, only in the last section we strive the case of the stochastic regularization.

As we already have seen, an inverse problem may be formulated as the problem to solve an equation governed by an operator, let's say A. Ill-posedness tells us that the inverse A^{-1} does not exist and/or is not continuous. The remedy is *regularization*. The idea of the regularization theory is to replace the inverse of such an operator by a one-parameter family of continuous operators and to choose the "best approximation" in this family by a clever strategy. Such a *regularization strategy* uses two main ingredients for solving an inverse problem in a stable way: a-priori information and signal to noise ratio (SNR). The *signal to noise ratio* is a quantity which describes the relation of the size of the true solution (signal, image, ...) to the size of the noise contained in the measured quantity (right-hand side of the equation, ...).

Tutorial example: numerical differentiation

Let us present a first example of an ill-posed problem: (numerical) differentiation of a function. The differentiation of a (measured) function is involved in many inverse problems. In a mechanical system one may ask for hidden forces and since Newton's law relates forces to velocities and accelerations one has to differentiate observed data when we want to compute the forces. Moreover, one can see that numerical differentiation is implicitly present in the problem of X-ray tomography, in parameter identification, in the determination of heat flow, and edge detection. We will see that this problem is important too in finding the volatility from market data and of sensitivities (greeks) for option prices.

Let us consider the problem of finding the integral of a given function. This can be done analytically and numerically in a very stable way. When this problem is considered as a direct (forward) problem then to differentiate a given function is the related inverse problem. A mathematical description is given as follows:

Direct Problem: With a continuous function $x:[0,1]\longrightarrow\mathbb{R}$ compute

$$y(t) := \int_0^t x(s)ds\,,\; t\in[0,1]\,.$$

Inverse Problem: Given a differentiable function $y:[0,1]\longrightarrow\mathbb{R}$ determine $x := y'$.

We are interested in the inverse problem. Since y should be considered as the result of measurements the data y are noisy and we may not expect that the noisy data $\tilde{y}$ of y is presented by a continuously differentiable function. Therefore, the inverse problem has no obvious solution, especially when $\tilde{y}(0)\neq 0$. (In practice we simply subtract $\tilde{y}(0)$ from $\tilde{y}$.) Moreover, the problem should not be formulated in the space of continuous functions since perturbations due to noise lead to functions which are not continuous. But the analysis and the message of the results is not very different from the following in the more elementary case of continuous perturbations.

Suppose that instead of the continuous function $y:[0,1]\longrightarrow\mathbb{R}$ a "measured" function $\tilde{y}$ denoted by $y^\varepsilon:[0,1]\longrightarrow\mathbb{R}$ is available only. We assume:

$$|y^\varepsilon(t)-y(t)|\le\varepsilon \text{ for all } t\in[0,1]\,.$$

Clearly, ε is the level of noise in the measurement of y^ε. It is reasonable to try to reconstruct the derivative $x:=y'$ of y at $\tau\in(0,1)$ by

$$x^{\varepsilon,h}(\tau) := D_h y^\varepsilon(\tau) := \frac{y^\varepsilon(\tau+h)-y^\varepsilon(\tau)}{h}$$

where the discretization parameter $h\neq 0$ has to be chosen such that $\tau+h\in[0,1]$. We obtain

$$\begin{aligned}|x^{\varepsilon,h}(\tau)-x(\tau)| \le & \left|\frac{y(\tau+h)-y(\tau)}{h}-x(\tau)\right| \\ & + \left|\frac{(y^\varepsilon-y)(\tau+h)-(y^\varepsilon-y)(\tau)}{h}\right|.\end{aligned}$$

Under the assumption that the unknown solution x is continuously differentiable we have

$$\frac{y(\tau+h)-y(\tau)}{h}-x(\tau)=\frac{1}{2}y''(\eta)h \text{ for some } \eta\in[0,1]\,.$$

When we know a bound (*a priori* bound)

$$|x'(t)|\le E \text{ for all } t\in[0,1]\,, \tag{25}$$

then the estimate

$$|x^{\varepsilon,h}(\tau)-x(\tau)|\le\frac{1}{2}hE+2\frac{\varepsilon}{h} \tag{26}$$

follows. Now it is clear that the best what we can do is to balance the terms on the right hand side in (26), i.e. to choose the quantity h such that both terms of the right hand side in (26) are equal. This is done by the choice

$$h := h_{\rm opt} := 2\sqrt{\frac{\varepsilon}{E}}\,. \tag{27}$$

This gives

$$|x^{\varepsilon,h(\varepsilon)}(\tau) - x(\tau)| \le 2\sqrt{E\,\varepsilon}\,. \tag{28}$$

The Diagram 1 which is a graphical presentation of the bound (26) is typical for approximations in ill-posed problems: there are two terms in the error estimate, a term due to approximation of the inverse mapping and a term due to measurement error. The balance of these two terms gives an "optimal" reconstruction result. Thus, in contrast to well-posed problems, it is not the best to discretize finer and finer. Moreover, the bound in (28) shows that halving the (measurement) error ε does not lead to a halving of the absolute error in the result (as it is usually the case in well posed problems). One may consider ill-posed problems under the motto *"When the imprecise is preciser"*.*

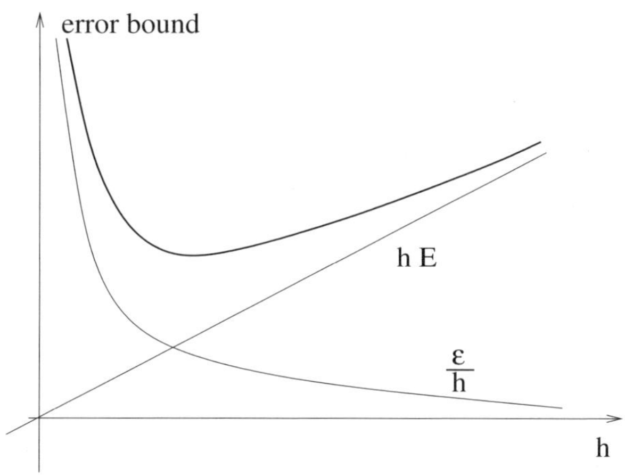

Figure 1. Error balance

The requirement (25) describes an information concerning the solution we want to find. Therefore this information is called *a priori information/a priori knowledge.* The quantity

$$\mathrm{SNR} := \frac{E}{\varepsilon}$$

may be considered as the signal to noise ratio. As a so called *merit function* we have used the norm of the deviation of the approximation from the (unknown) solution.

*This is the title of [96].

The regularization strategy to balance the merit function with respect to the *a priori* information and the signal to noise ratio SNR is called an *a priori strategy* since it can be done without calculating some approximations.

Numerical differentiation is studied in a huge amount of papers. See for instance [4; 5; 70; 107; 114; 118].

Numerical methods: inverse crime

Often, tests of the exact or approximate theoretical models employed in inversion schemes are made with synthetic data. Generating the latter also requires a theoretical model which, mathematically speaking, can be identical to, or different from, the one employed in the inversion scheme. In [34] the authors coin the expression *inverse crime* to denote the act of employing the same model to generate, as well as to invert, synthetic data. Moreover, they warn against committing the inverse crime, *in order to avoid trivial inversion* and go on to state: *it is crucial that the synthetic data be obtained by a forward solver which has no connection to the inverse solver.* Shortly: The term *"inverse crime"* is saying that sampling the data in the forward problem and solving the inverse problem are done in the same manner. Inverse crime arises when

- the numerically produced simulated data is produced by the same model that is used to invert the data;
- the discretization in the numerical simulation is the same as the one used in the inversion.

When inverse crimes are not avoided a numerical algorithm may lead to unreallistically optimistic results. Note that inverse crimes are not possible in situations where actual real-world measured data are used, they are only a problem of computational simulation studies. In the context of mathfinance it is important that there is huge amount of numerical methods to solve the forward problem which makes it possible to choose different methods in the solution of the forward and inverse problems; see [99; 100; 124] for numerical methods.

Let us revisite the problem of numerical differentiation. In order to check a method of numerical differention which uses synthetic data of y on a grid $0 = t_0 < t_1 < \cdots < t_N = 1$ we can procede as follows: use a finer grid $0 = s_0 < s_1 < \cdots < s_M = 1$ – M should be not a multiple of N – and interpolate the data on the finer grid such that data are available on the coarser grid and add a noise to the data obtained. Then apply the method to the noisy data.

2.2. *Regularization*

Here we discuss the methods which are used to solve an ill-posed problem in a stable way. We restrict ourselves to the linear case. Later on we shall show how the following considerations may be extended to the nonlinear case.

The linear equation

Let us introduce some notations from functional analysis; for a detailed introduction the reader may consult [6]. Let X, Y be Banach spaces, i.e. complete normed vector spaces over the scalar field $\mathbb{K} \in \{\mathbb{R}, \mathbb{C}\}$; the norm in these spaces is denoted by $\|\cdot\|_X$ and $\|\cdot\|_Y$, respectively. For every linear and bounded mapping A from the Banach space X into Y we use:

$\operatorname{ran}(A) := \{y \in Y | y = Ax \text{ for some } x \in X\}$ is the range of A,
$\ker(A) := \{x \in X | Ax = \theta\}$ is the null space of A,
$\sigma(A)$ is the spectrum of A,
$\rho(A) := \mathbb{K} \backslash \sigma(A)$ is the resolvent set of A,
$\operatorname{rad}(A) := \sup_{\lambda \in \sigma(A)} |\lambda|$ is the spectral radius of A.

The adjoint operator of A is denoted by A^*. Moreover, if $S \subset X$, then $A_{|S}$ means A restricted to S.

Let us consider the equation

$$Ax = y \quad (x \in X, y \in Y) \tag{29}$$

where A is a linear and continuous mapping from the Banach space X into Y. We consider the equation (29) under the following assumptions:

(A0) X and Y are Hilbert spaces with inner products $\langle \cdot, \cdot \rangle_X$ and $\langle \cdot, \cdot \rangle_Y$.
(A1) $\operatorname{ran}(A)$ is dense in Y but not closed

Notice that condition **(A1)** indicates that $\operatorname{ran}(A)$ has infinite dimension. Moreover, it has the consequence that the equation (29) is ill-posed due to the lack of continuity of the inverse of A even when it exists.

Clearly, when $y \notin \operatorname{ran}(A)$ the equation (29) has no solution. Then, instead of (29) we may consider the so called *normal equation* (see below):

$$A^*Ax = A^*y \quad (x \in X, y \in Y) \tag{30}$$

A solution x of equation (30) is called a *least squares solution* or a *normal solution* of (29). A normal solution of (29) exists for each $y \in \operatorname{ran}(A) + \operatorname{ran}(A)^\perp$ where $\operatorname{ran}(A)^\perp$ is the orthogonal space of $\operatorname{ran}(A)$ with respect to the inner product $\langle \cdot, \cdot \rangle_Y$. There exists no normal solution of (29) for $y \notin \operatorname{ran}(A) + \operatorname{ran}(A)^\perp$.

Since for every $y^0 \in Y$ the set $\{x \in X | A^*Ax^* = A^*y^0\}$ is convex and closed the following definition makes sense:

Definition 2.1. Consider $y^0 \in \operatorname{ran}(A) + \operatorname{ran}(A)^\perp$. Then $x^{0,+}$ with

$$\|x^{0,+}\|_X = \min\{\|x^*\|_X | A^*Ax^* = A^*y^0\} \tag{31}$$

is called the ***minimal least squares solution*** of the equation $Ax = y^0$. □

It can be shown that $x^{0,+}$ is the unique solution to the equation (30) in the closure of $\mathrm{ran}(A)$. As a consequence, we may define the ***(Moore-Penrose-) pseudo-inverse*** $A^\dagger$:

$$A^\dagger : \mathrm{ran}(A) + \mathrm{ran}(A)^\perp \ni y^0 \longmapsto x^{0,+} \in X \,.$$

One can show that $A^\dagger$ is a linear operator. Notice that $A^\dagger$ is not bounded under the condition **(A1)**.

A comprehensive presentation of pseudoinverses of linear operators may be found in [59].

Regularization scheme

The key idea to solve an ill-posed linear equation (29) in a stable way is to find a good approximation R for the pseudo-inverse operator $A^\dagger$; R should be a continuous operator from Y nach X . Since $\mathrm{ran}(A)$ is not closed we cannot require $R_{|\mathrm{ran}(A)} = A^\dagger$ since this would imply that $A^\dagger$ is continuous.

Definition 2.2. A family $(R_h)_{h>0}$ of linear bounded operators from Y into X is called a ***regularizing family*** for A if

$$\lim_{h\downarrow 0} R_h A x = x \text{ for all } x \in X \,. \tag{32}$$

□

Obviously, a regularizing family $(R_h)_{h>0}$ is a family which should approximate a left inverse of A . Since $A^\dagger$ is an unbounded operator $R_h A$ does not converge to the identity I in the operator norm (by the theorem of Banach–Steinhaus) as h goes to zero. Moreover, the family $(\|R_h\|)_{h>0}$ cannot be bounded.*

Suppose that x^0, y^0, y^ε are given as follows:

$$Ax^0 = y^0 \,,\; \|y^0 - y^\delta\|_Y \le \delta \,. \tag{33}$$

The interpretation is that x^0 is an exact solution of the equation (29) with right hand side y^0 and y^δ is an approximation (measurement, perturbation,...) of y^0 . With a regularizing family $(R_h)_{h>0}$ for A we define

$$x^{\delta,h} := R_h y^\delta \,,\; x^{\delta,0} := R_h y^0 \,,\; h > 0 \,. \tag{34}$$

We want to find the parameter h such that $R_h y^\delta$ deals with the noise δ in an optimal fashion. Since the reconstruction $x^{\delta,h} - x^0$ error can be estimated as follows

$$\begin{aligned} \|x^{\delta,h} - x^0\|_X &\le \|R_h y^\delta - R_h y^0\|_X + \|R_h A x^0 - x^0\|_X \\ &\le \delta \|R_h\| + \|(R_h A - I)x^0\|_X \end{aligned} \tag{35}$$

we observe that two competing effects enter (35). The first one is the ill-posedness effect: as h goes to 0 the norm $\|R_h\|$ tends to ∞ ; so h should not be chosen too

*Any operator norm is denoted by $\|\cdot\|$.

small. The second one is the regularizing effect: as h increases, R_hA becomes a less accurate approximation of the identity; therefore h should not be chosen too large. Thus, we have to balance the two competing terms in (35) by a clever choice of the parameter h (see (27)). Of course, one should try to choose $h = h(\delta)$ such that

$$\delta \| R_{h(\delta)} \| = \| (R_{h(\delta)}A - I)x^0 \|_X \,. \tag{36}$$

This choice is in practice not possible for two reasons. First, the term $\|(R_hA - I)x^0\|_X$ is not known since x^0 is unknown. Second, the equation (36), even when the noice level is known, is not solvable exactly, in general. Usually, we estimate the two competing terms in (35) as follows

$$\delta \| R_{h(\delta)} \| \leq \delta c_1(h(\delta)) \,, \; \| (R_{h(\delta)}A - I)x^0 \|_X \leq c_2(\delta, h(\delta); x^0) \tag{37}$$

and try to find $h = h(\delta)$ such that

$$\delta c_1(h(\delta)) \approx c_2(\delta, h(\delta(h)), x^0) \,. \tag{38}$$

Then

$$\| x^{\delta, h(\delta)} - x^0 \|_X \approx 2\delta c_1(h(\delta)) \,. \tag{39}$$

and we can check whether we may conclude that $\lim_{\delta \downarrow 0} x^{\delta, h(\delta)} = x^0$ holds.

Definition 2.3. Let $(R_h)_{h>0}$ be a regularizing family for A. This family is called ***convergent*** for $y^0 \in \text{ran}(A)$ if there exists a ***parameter choice strategy*** $h = h(\delta)$ such that

$$\lim_{h \downarrow 0} R_{h(\delta)} y^\delta = x^{0,+} \tag{40}$$

where $x^{0,+}$ is the minimal least squares solution of $Ax = y^0$. □

Filtering

A convenient method to construct a regularizing family is given by *filtering* when the operator A is compact. A *compact operator* is a linear continuous operator which admitts a *singular system* $(e^j, f^j, \sigma_j)_{j \in \mathbb{N}}$. This means:

(1) $(e^j)_{j \in \mathbb{N}}, (f^j, \sigma_j)_{j \in \mathbb{N}}$ is an orthonormal basis in X and Y, respectively
(2) $Ae^j = \sigma_j f^j \,, \; A^* f^j = \sigma_j e^j \,, \; j \in \mathbb{N}$
(3) $0 < \sigma_j, j \in \mathbb{N}, \lim_j \sigma_j = 0$

With a singular system we have a *singular value decomposition* of A:

$$Ax = \sum_{j=1}^{\infty} \sigma_j \langle x, e^j \rangle_X f^j \,, \; x \in X \,. \tag{41}$$

The adjoint A^* of A is given by

$$A^* y = \sum_{j=1}^{\infty} \sigma_j \langle y, f^j \rangle_Y e^j \,, \; y \in Y \,. \tag{42}$$

The spectrum of the selfadjoint operators A^*A, AA^* consists of eigenvalues $\sigma_j^2, j \in \mathbb{N}$. The pseudo-inverse is formally given by

$$A^\dagger y = \sum_{j=1}^{\infty} \sigma_j^{-1} \langle y, f^j \rangle_Y e^j \,,\; y \in \mathrm{ran} + \mathrm{ran}(A)^\perp \,. \tag{43}$$

Since the series above does not converge for every $y \in Y$ (due to Picard's lemma) we have to introduce a *filter* for the small singular values. This can be done by choosing a mapping $q : (0, \infty) \times (0, \sigma_1] \longrightarrow \mathbb{R}$ which damps out the contribution of small singular values in the series (43). With such a filter function q we define a potential candidate for a regularizing family in the following way:

$$R_h y := \sum_{j=1}^{\infty} q(h, \sigma) \sigma_j^{-1} \langle y, f^j \rangle_Y e^j \,,\; y \in Y \,. \tag{44}$$

The conditions which the filter q should satisfy can be read off from the following estimates:

$$\begin{aligned} \|R_h y\|_X^2 &= \sum_{j=1}^{\infty} |q(h, \sigma_j)|^2 \sigma_j^{-2} |\langle y, f^j \rangle_Y|^2 \\ &\le \sup_{\sigma \in (0, \sigma_1]} |q(h, \sigma)\sigma^{-1}|^2 \sum_{j=1}^{\infty} |\langle y, f^j \rangle_Y|^2 \\ &= \sup_{\sigma \in (0, \sigma_1]} |q(h, \sigma)\sigma^{-1}|^2 \|y\|_Y^2 \qquad (45) \\ \|R_h A x - x\|_X^2 &= \sum_{j=1}^{\infty} |q(h, \sigma_j) - 1|^2 |\langle x, e^j \rangle_X|^2 \,. \qquad (46) \end{aligned}$$

Theorem 2.1. *Let $A : X \longrightarrow Y$ be an injective compact operator with singular system $(e^j, f^j, \sigma_j)_{j \in \mathbb{N}}$ and let $q : (0, \infty) \times (0, \sigma_1] \longrightarrow \mathbb{R}$ be a filter function which satisfies the following conditions:*

- *F1) $|q(t, \sigma)| \le 1$ for all $t > 0, \sigma \in (0, \sigma_1]$.*
- *F2) For all $t > 0$ there exists a constant $c(t)$ such that for all $\sigma \in (0, \sigma_1]$ we have $|q(t, \sigma)| \le c(t)\sigma$.*
- *F3) $\lim_{t \to 0} q(t, \sigma) = 1$ for every $\sigma \in (0, \sigma_1]$.*

Then the family $(R_h)_{h>0}$ defined in (44) is a regularizing family. Additionally, we have with $x^{\delta,h} := R_h y^\delta$

$$\|x^{\delta,h} - x^0\|_X^2 \le \|x^0\|_X^2 \sup_{\sigma \in (0, \sigma_1]} |q(h, \sigma) - 1|^2 + \delta^2 c(h)^2 \,. \tag{47}$$

Proof:
With the condition F2) we conclude from (45) $\|R_h\| \le c(h), h > 0$. Let $x \in X$ and let $\varepsilon > 0$. Then there exists $N \in \mathbb{N}$ with

$$\sum_{j=N+1}^{\infty} |\langle x, e^j \rangle_X|^2 < \varepsilon^2 \,.$$

According to the condition F3) we can choose a constant $t_0 > 0$ such that

$$|q(t,\sigma_j) - 1|^2 < \varepsilon^2 \text{ for all } j = 1,\ldots,N \text{ and } 0 < t \le t_0 .$$

With the condition F1) we obtain for $h \in (0, t_0]$

$$\begin{aligned}\|R_h Ax - x\|_X^2 &= \sum_{j=1}^{N} |q(h,\sigma_j) - 1|^2 |\langle x, e^j\rangle|_X^2 + \sum_{j=N+1}^{\infty} |q(h,\sigma_j) - 1|^2 |\langle x, e^j\rangle_X|^2 \\ &\le \varepsilon^2 \sum_{j=1}^{N} |\langle x, e^j\rangle_X|^2 + 4\varepsilon^2 \le \varepsilon^2 \|x\|_X^2 + 4\varepsilon^2 .\end{aligned}$$

This shows $\lim_{h\to 0} R_h Ax = x$. The estimate (47) follows immediately from (35), (45), (46). ■

Truncation:
We define

$$q(t,\sigma) := \begin{cases} 1 & , \sigma^2 \ge t \\ 0 & , \sigma^2 < t \end{cases} .$$

This is a filter which truncates the contribution of singular values larger than the threshold parameter $\sqrt{t}$. Here we can verify $c(t) = 1/\sqrt{t}$.

The classical method of Tikhonov

Consider the filtering with

$$q(t,\sigma) := \frac{\sigma^2}{\sigma^2 + t}, \; t \ge 0 .$$

Here we have $c(t) = 1/(2\sqrt{t})$.* The associated regularization family

$$R_h y := \sum_{j=1}^{\infty} \frac{\sigma_j^2}{\sigma_j^2 + h} \sigma_j^{-1} \langle y, f^j\rangle_Y e^j , \; y \in Y , \; h > 0,$$

has – thanks to the assumption **(A1)** – an alternative representation:

$$R_h y = (A^*A + hI)^{-1} A^* y , \; y \in Y , \; h > 0 . \tag{48}$$

This leads the *classical method of Tikhonov* as follows: we consider the *normal equation*

$$A^*Ax = A^*y \quad (x \in X, y \in Y) \tag{49}$$

and regularize this equation in the following way

$$(A^*A + hI)x = A^*y \quad (x \in X, y \in Y, h > 0) \tag{50}$$

The equation (50) is a necessary and sufficient condition for a solution x^{opt} of the optimization problem

*To simplify some expressions we do not always use the best possible constants.

$$\text{Minimize} \quad J_h(x) := \|Ax - y\|_Y^2 + h\|x\|_X^2 \quad \text{subject to } x \in X$$

since the functional J_h is convex. The solution of this optimization problem – J_h is called the *Tikhonov functional* – is given by $R_h y$; see (48). The advantage of the formulation as an optimization problem consists in the fact that now we may omit the assumption that A is injective and compact.

Remark 2.4. In the last section we have considered the regularization of the process of numerical differentiation. The procedure is different from filtering and is called ***regularization by discretization***.

Regularization by iteration

In order to solve the equation (49) we may use the idea of a fixed point iteration applied to

$$x = x - \omega(A^*Ax - A^*y) \quad (x \in X, y \in Y) \tag{51}$$

This leads to the iteration

$$x^{n+1} = x^n - \omega(A^*Ax^n - A^*y)\,,\; n \in \mathbb{N}_0\,. \tag{52}$$

Since

$$x^{n+1} = \sum_{k=0}^{n}(I - \omega A^*A)x + \sum_{j=0}^{n-1}\omega(I - \omega A^*A)A^*Ay\,,\; n \in \mathbb{N}_0\,, \tag{53}$$

ω should be chosen such that $\operatorname{rad}(I - \omega A^*A)$ is smaller than one. This is the case when $0 < \omega < \|A\|^2$. Thus the regularization operator R_h is to be chosen as follows:

$$R_h y = \sum_{k=0}^{\lfloor 1/h \rfloor}(I - \omega A^*A)x + \sum_{j=0}^{\lfloor 1/h \rfloor - 1}\omega(I - \omega A^*A)A^*Ay\,,\; h > 0\,, \tag{54}$$

It is obvious to translate the formula (54) to the case when A is a compact operator.

What concerns linear equations, iterative methods for approximating the generalized inverse (i.e. the least square solution of minimum norm) are based on algorithms for solving fixed point equations related to the normal equation (see [47; 59; 60] for corresponding definitions). The regularization character of these methods (in the case of noisy data) is related to an early termination of the iteration, and a corresponding stopping rule is determined by an *a posteriori* evaluation of the iteration residual.

Standard iterative methods for the solution of ill-posed equations are considered in [14; 13; 10; 68; 90; 122].

Statistical regularization

The philosophy behind the statistical inversion methods is to recast the inverse problem in the form of statistical quest for information. We have directly observable quantities and others that cannot be observed. In inverse problems, some of the unobservable quantities are of primary interest. These quantities depend on each other through models. The objective of statistical inversion theory is to extract information and assess the uncertainty about the variables based on all available knowledge of the measurement process as well as information and models of the unknowns that are available prior to the measurement.

The statistical inversion approach is based on the following principles: All variables included in the model are modelled as random variables, *a priori* information is described by probability distributions, regularization is done by introducing some requirements for these probabilty distributions in the space of solutions.

Statistical regularization is a rather new subject in solving inverse problems. For more information see [89].

2.3. *Calibration*

In this section we describe the aspects of *calibration* and their connection to inverse modelling. Moreover, we describe the least squares method to calibrate models using ideas from elementary optimization methods.

Applied mathematics is full of models which describe phenomena observed in real problems in mathematical terms. As we know, such models contain some parameters which have to be fitted to the reality. *Model calibration* consists of changing values of model input parameters in an attempt to match field conditions within some acceptable criteria.

Introduction

When is calibration needed? In our inverse methodology, this is the case mainly for problems of the type (C). In general, uncertainty of the parameters which govern the model under consideration leads the fact that the predictive validity of the model is in question. Thus, calibration is an approach to *identify/estimate/adjust* the parameters of the model such that behavior of the model is consistent with the real data of the process described by the model. As we shall see, in finance calibration is mainly devoted to the adjustment of the volatility such that pricing is consistent with the market data.

Consider again the problem (C) in Section 1.1:

Given $x \in X$ and $y \in Y$, find $p \in P_{\mathrm{ad}}$ such that $A(p)x = y$.

Here $A(p)$ is assumed to operate linearly on x. Various problems may be extracted from the formulation.

Identifiability Determines the knowledge of y and x the parameter in an unique way or are there are "many solutions"?

Stability Is the problem to determine p well-posed or ill-posed?

Erroroneous data How can erroneous data y, x be handled by a clever solution concept?

Regularization How can a computation scheme be designed to avoid effects of ill-posedness?

As a rule, to obtain identifiability the "dimensionality" of the data x, y have to be of the same size as the "dimensionality" of p. Additionally, some structural assumptions have to be satisfied as we can see from the elementary observation that $x = \theta, y = \theta$ does not allow to determine p. We collect such structural assumptions by saying that the information content of the "experiment" determined by x, y should be rich enough.

In general, even when identifiability holds parameter identification problems are ill-posed when they are considered in the topology which is appropriate to describe the experiment (process of measurement,...). This leads then to the question whether and how the identification of p from erroneous data $\tilde{x}, \tilde{y}$ is possible. We assume that the "input" x is an exact quantity and given. Then we set

$$F(p) := A(p)x \, , \; p \in P_{\mathrm{ad}} \, ,$$

and we assume that the equation $F(p) = y$ for the right hand side $y = y^0$ has a solution:

$$F(p^0) = y^0 \;\; (p^0 \in P_{\mathrm{ad}}) \tag{55}$$

Moreover, we assume that the level of error in the data y^0 is known:

$$\|y^0 - y^\delta\|_Y \le \delta \;\; (\delta \ge 0) \tag{56}$$

Now the solution concept has to be adjusted: instead of the equation $F(p) = y$ we should consider a defect–type formulation. We consider the classical least squares formulation, see below.

Parameter identification are of great importance in applications. According to te type of application different solution concepts and numerical algorithm werde developed: in mechanics and control theory online adaptation schemes for the parameter (see [15; 105] for instance), in biology and reaction kinetics compartmental analysis. Unfortunately, the estimation of the volatility in finance does not fit into these already developed methods. This is why in finance new identifiability results have to be developed.

Least squares formulation

Here is the formulation as a *defect-minimization problem*:

(DMin) Minimize $\|F(p) - y^\delta\|_Y^2$ subject to $p \in P_{\mathrm{ad}}$

Unfortunately, ill-posedness is not avoided and regularization has to be introduced. This may be done by the idea of *Tikhonov-regularization*:

(TMin) Minimize $J_{h,\delta} := \|F(p) - y^\delta\|_Y^2 + h\|p - p^*\|_P^2$ subject to $p \in P_{\text{ad}}$

Here it is assumed that the set P_{ad} of admissible parameters is a subset of a Hilbert space P, $h > 0$ is a *regularization parameter* scaling the penalty term $\|p - p^*\|^2$, p^* should be considered as an initial guess for the solution p^0 of the problem.

Since the mapping F may may be nonlinear the problem (DMin) nor the problem (TMin) can be studied as a convex optimization problem. As a consequence, these optimization problems may have many local minima. In the formulation (TMin) the introduction of the term $h\|p - p^*\|_P^2$ is an attempt to localize the problem to the neighborhood of p^*. Additional, the nonconvexity has the consequence that the existence of a solution is in doubt. As we know, compactness of the level sets and the appropriate continuity of the functional is sufficient for existence. We collect such conditions:*

(L1) The admissible subset P_{ad} is nonempty, closed and convex.
(L2) The level sets $N_c := \{p \in P_{\text{ad}} | \|F(p) - y^0\|_Y^2 \le c\}, c \ge 0$, are weakly compact.

Theorem 2.2. *Under the assumptions (L1), (L2) the problem (TMin) has a solution $p^{h,\delta} \in P_{ad}$ for each $h > 0$.*

Proof:
Consider a positive sequence $(\eta_n)_{n\in\mathbb{N}}$ with $\lim_n \eta_n = 0$. Then there exist for each $n \in \mathbb{N}$ a $p^n \in P_{\text{ad}}$ with $J_{h,\delta}(p^n) \le \inf\{J_{h,\delta}(p) | p \in P_{\text{ad}}\} + \eta_n$. Due to

$$\|F(p^n) - y^\delta\|_Y^2 + h\|p^n - p^*\|^2 \le \inf\{J_{h,\delta}(p) | p \in P_{\text{ad}}\} + \eta_n \,,\; n \in \mathbb{N},$$

the sequences $(\|F(p^n) - y^\delta\|_Y^2)_{n\in\mathbb{N}}, (\|p^n - p^*\|^2)_{n\in\mathbb{N}}$ are bounded in Y and X, respectively. Since in each Hilbert space a bounded sequence has a weakly convergent subsequence we obtain a subsequence $(p^{n_k})_{k\in\mathbb{N}}$ of $(p^n)_{n\in\mathbb{N}}$ and $p^{h,\delta}$ with $p^{h,\delta} =$ weak $- \lim_k p^{n_k}$. Since P_{ad} and the level sets N_c are weakly closed (see (L1),(L2) in connection with (56)) we get $p^{h,\delta} \in P_{\text{ad}}$ and $\|F(p^{h,\delta}) - y^\delta\|_Y^2 + h\|p^{h,\delta} - p^*\|^2 \le \inf\{J_{h,\delta}(p) | p \in P_{\text{ad}}\}$. Here we have used the fact a norm in a Hilbert space is weakly lower continuous. Now, it is clear that $p^{h,\delta}$ is a solution of (TMin). ∎

Now we want to use the optimization problem (TMin) as a *regularization method*, i.e. we ask whether the sequence $(p^{h_n,\delta_n})_{n\in\mathbb{N}}$ is convergent when the sequence of the regularization parameter $(h_n)_{n\in\mathbb{N}}$ and the error level $(\delta_n)_{n\in\mathbb{N}}$ converges to zero in an appropriate way. Since we are not sure that a solution in Theorem 2.2 is unique we have to define an enlarged solution concept.

*A subset C of a Hilbert space U with inner product $\langle \cdot, \cdot \rangle_U$ is weakly closed if every sequence $(u^k)_{k\in\mathbb{N}}$ in C converges weakly to an element $u \in C$, i.e. $\lim_k \langle u^k, v\rangle_U = \langle u, v\rangle_U$ for every $v \in U$.

Definition 2.5. $p^+ \in P_{\mathrm{ad}}$ is a ***p^*-minimum-norm-solution*** of the equation $F(p) = y^0$ if we have

$$\|p^+ - p^*\|_P = \min\{\|p - p^*\|_P | p \in P_{\mathrm{ad}}, F(p) = y^0\}$$

Notice that the set $\{p \in P_{\mathrm{ad}} | F(p) = y^0\}$ is a nonempty set by (55). Therefore, there exist p^*-minimum-norm-solution; the argumentation is similar to the proof of Theorem 2.2.

The next result shows that by the optimization of the Tikhonov-functional in (TMin) leads to a method to construct a p^*-minimum-norm-solution. In this result we become aquainted with the key-idea in regularization: choice of an appropriate regularization parameter.

Theorem 2.3. *Suppose that the assumptions (L1),(L2) are satisfied. Moreover let*

$$(\delta_n)_{n\in\mathbb{N}}, (h_n)_{n\in\mathbb{N}} \text{ be a sequence with } \lim_n \delta_n = \lim_n h_n = \lim_n \frac{\delta_n^2}{h_n} = 0 . \tag{57}$$

Then the associated sequence $(p^n)_{n\in\mathbb{N}}$ with $p^n := p^{\delta_n,\eta_n}$ contains a subsequence which converges to a p^*-minimum-norm-solution. If the p^*-minimum-norm-solution is unique then the sequence $(p^n)_{n\in\mathbb{N}}$ converges to the p^*-minimum-norm-solution.

Proof:
Let p^+ be a p^*-minimum-norm-solution. Let $p^n := p^{h_n,\delta_n}$ be a solution according to Theorem 2.2. Then

$$J_{h_n,\delta_n}(p^n) \le J_{h_n,\delta_n}(p^+) \le \delta_n^2 + h_n\|p^+ - p^*\|_P^2 \tag{58}$$

and we obtain $\lim_n F(p^n) = y^0$, weak $-\lim_n F(p^n) = y^0$. Due to the assumption $\lim_n \frac{\delta_n^2}{h_n} = 0$ (see (57)) we get $\limsup_n \|p^n - p^*\| \le \|p^+ - p^*\|_P$. Now, since $(p^n)_{n\in\mathbb{N}}$ is bounded there exists a weakly convergent subsequence $(p^{n_k})_{k\in\mathbb{N}}$ with $\overline{p} :=$ weak$-\lim_k p^{n_k} \in P_{\mathrm{ad}}$ (see assumption (L1)). Due to (L2) we have $F(\overline{p}) = y^0$. This implies

$$\|\overline{p} - p^*\|_P \le \liminf_k \|p^{n_k} - p^*\|_P \le \limsup_k \|p^{n_k} - p^*\|_P \le \|p^+ - p^*\|_P . \tag{59}$$

Therefore $\overline{p}$ is a p^*-minimum-norm-solution too and we have due to (59) $\lim_k p^{n_k} = \overline{p}$.
The additional assertion follows by a standard argumentation. ∎

Under additional assumptions convergence rates for the sequence $(p^{h_n,\delta_n})_{n\in\mathbb{N}}$ can be proved. but in general, such assumptions are difficult to realize.

A more recent development is devoted to the use of Banach spaces instead of Hilbert spaces. This has in application many nice properties but the analysis is much more difficult; see [85; 119; 48] for the treatment of ill-posed problems in Banach spaces.

2.4. *Data assimilation*

A loose definition of *data assimilation* is:

> Estimation and prediction of an unknown, true state by combining observations and system dynamics (model output).

This definition shows clearly, that algorithms for data assimilation have to consider the problems inherent in inverse problems. The most important application is numerical weather prediction. Other applications are navigation, geosciences, medical imaging. In finance there are just a few papers on this subject, a treatment of problems in finance by applying the following approach would be formidable. For a rather complete overview concerning data assimilation algorithms see the monograph [86].

Mostly, data assimilation algorithms are based on stochastical properties of the fields under consideration and bring together Bayesian analysis and stochastic regularization. We prefer here the deterministic approach to handle data assimilation algorithms. To study data assimilation from a rigorous mathematical point of view one has to combine various themes of ill-posednes and regularization. This is why this subject is placed in this chapter. A rather new sight on these algorithms is to study data assimilation algorithms as iterative or cycled schemes. We sketch this approach and we follow mainly [108]; see also [56; 55; 115]. Notice that we consider the subject in a linear context.

Let's start with an infinite partition

$$0 \le t_0 < t_1 < \cdots < t_k <$$

of the real (time) axis where data are provided. The data are given by a measurement operator

$$H : X \longrightarrow Y \tag{60}$$

where X, Y are infinite dimensional Hilbert spaces endowed with the norm $\|\cdot\|_X$ and $\|\cdot\|_Y$ respectively. The state of the system at time t_k is denoted by x^k. By the measurement operator we obtain data

$$y^k = y^k_{\text{true}} + y^k_\delta = Hx^k_{\text{true}} + y^k_\delta \,,\; k \in \mathbb{N}\,. \tag{61}$$

For the most cases, H is a linear compact operator. Thus, H cannot have a continuous inverse and we need regularization for handling the equation $Hx = y$. Stability is reconstructed by Tikhonov regularization:

$$\text{Minimize } J(x) := \|y^k - Hx^k_b\|^2_Y + \lambda\|x - x^k_b|^2_X \text{ subject to } x \in X \tag{62}$$

Here x^k_b is the background state at time t_k. Let us denote by x^k_a the minimizer of the functional J.

Now, a model for the process comes in. We assume that the transition from t_k to t_{k+1} in the states is modeled by a linear model operator $M_{k+1,k}$:

$$x^{k+1}_b = M_{k+1,k}x^k_a\,. \tag{63}$$

The "iteration steps" (62), (63) define the *cycled Tikhonov regularization.* The analysis of this type of regularization has to consider the interplay of two parts: setup of the Tikhonov regularization and study of the discrete dynamical system $x^{k+1} := M_{k+1,k}x^k$. Spectral theory is a main tool in the analysis.

3. Elementary numerical approaches

In this section we consider methods to find the implied volatility and to complete market data. The common characteristic of these problems is that the reconstructed quantities depend sensitive on the data. The considerations are of elementary nature and the results can be considered more or less as recipes. But nevertheless, already these elementary problems lead to very interesting mathematical considerations. In the following chapters we shall consider some of these problems more on a theoretical basis.

3.1. *Computation of the implied volatility*

The goal

The Black-Scholes model assumes that the volatility is constant across strikes and maturity dates. However as we know, in the world of options this is a very unrealistic assumption: option prices for different maturities change drastically, and option prices for different strikes also experience significant variations. For European options under the Black-Scholes model, calculation of the implied volatility seems to be a straightforward exercise since a closed-form presentation exists for the price. However, this closed-form allows not an analytical computation of the implied volatility. In this section we consider the numerical problem to compute the implied volatilities and the implied volatility surface.

Almost always, the inversion of the Black-Scholes formula to get the implied volatility is done with some sort of solver method, for example, the Newton-Raphson method; see the next subsection. These methods work very well for a single option, usually producing very accurate estimates in negligible computing time. However, frequently one has to invert millions or hundreds of thousands of options at the same time. In these situations, a solver method might prove to be slow, especially for real-time applications. The need to overcome the slow-speed problem in the solver methods have led researchers to consider an alternative to the solver methods, namely, analytical computable approximations. Most of the approximate closed-form inversion methods perform some Taylor expansion to the Black-Scholes formula and then analytically invert the expansion to obtain a formula for the implied volatility. The usual assumption being made to justify the Taylor expansion is that the strike price is close to the forward price. Because of the local nature of the Taylor expansion, these methods work relatively well for very near-the-money options. We sketch one such analytical approximation of the implied volatility.

Results for the analytical inversion in order to get the implied volatility can be found in [31; 35; 88].

Let us first define the normalized call option price $c(\cdot;\cdot)$ by

$$C(S;t;K,T,r,\sigma) = S\,c(\ln(Se^{r(T-t)}/K), \sigma\sqrt{T-t})\,.$$

The expression for the normalized call price $c(x,v)$ is given by

$$c(x;v) = \mathcal{N}\left(\frac{x}{v}+\frac{v}{2}\right) - e^{-x}\mathcal{N}\left(\frac{x}{v}-\frac{v}{2}\right)\,. \tag{64}$$

The variable x is the logarithmic forward-moneyness and we shall call v the *integrated volatility*. x, v are given as follows:

$$x = \ln(Se^{r(T-t)}/K) = \ln(F/K)\,,\; v = \sigma\sqrt{T-t}, \tag{65}$$

where $F = Se^{r(T-t)}$ is the forward price. We call equation (64) the dimensionless Black-Scholes formula because all the three quantities entering the equation are dimensionless. It clearly states that the Black-Scholes formula is essentially a relation between three dimensionless quantities, namely the normalized price c, the integrated volatility v and the moneyness x.

The identity (64) implicitly gives v as a function of c and x and is our starting point for the approximate inversion. Given observed values of S, t, K, T, r and an option price C, we first calculate p according to $p = C/S$ and x according to equation (65) and then use an approximate formula to get v from the equation $p = c(x,v)$. The implied volatility σ can then be obtained by dividing v by $\sqrt{T-t}$. When the option is at-the-money-forward, that is, when $x = 0$, the inversion for $v(x,c) = p$ can be done explicitly as follows:*

$$v(0,p) = \sqrt{8}\,\mathrm{erf}^{-1}(p) = \sqrt{2\pi}\,p + O(p^3) \tag{66}$$

We see that $v(0,p)$ is almost linear in the normalized call price p.†

Newton's method

As a computational method to compute the implied volatility we present the classical Newton-procedure. Let us recall the Black-Scholes-formula for the price of an European call-option.

$$C(S,t;K,\tau,r,\sigma) := S\mathcal{N}(d_+(\sigma)) - Ke^{-r\tau}\mathcal{N}(d_-(\sigma))\,, \tag{67}$$

where

$$d_\pm(\sigma) := d_\pm + (\sigma, S, K, \tau, r) := \frac{\ln(\frac{S}{K}) + (r \pm \frac{\sigma^2}{2})\tau}{\sigma\sqrt{\tau}}\,,$$

and $\mathcal{N}$ is the distribution function of the standard normal distribution. Notice that we have used the remaining time τ as a "new" variable. In the following we use the

*erf^{-1} is the inverse of the so called error-function $\mathcal{N}$.

†We use here and in the following the Landau-symbols $O(\cdot), o(\cdot)$.

identity

$$d_-(\sigma) = d_+(\sigma) - \sigma\sqrt{\tau}\,.$$

The procedure for the "daily" computation of the implied volatility of an underlying is as follows:

> Given at the time t_* the price S_* of the underlying, the constant interest rate r, a european call option with strike price K, maturity T and term to maturity τ.
> Compute the volatility $\sigma := \sigma_{\text{imp}}(K,\tau)$ by solving the equation
>
> $$C(S_*, t_*; K, \tau, r, \sigma) = v \tag{68}$$
>
> where $v := C_{market}$ is the observed market-price of the option.

To abbreviate the notation, we set

$$f(\sigma) := C(S_*, t_*; K, \tau, r, \sigma)\,. \tag{69}$$

As a result, we have to compute a solution of

$$f(\sigma) - v = 0 \tag{70}$$

There is no closed-form solution but we can solve it iteratively. Since f is differentiable, we can apply each variant of the Newton method. The classical Newton-procedure is the following one:

Algorithm 3.1 Computation of the implied volatility by Newton's method

---

IN Price S_* of the underlying at time t_*, strike price K, term to maturity τ, interest rate r, market price v.
Initial guess σ^0 for the implied volatility, accuracy bound $\varepsilon > 0$ for the iterated approximations.

step 0 Set $f(\sigma) := C(S_*, t_*; K, \tau, r, \sigma)$, $f'(\sigma) := \dfrac{\partial C}{\partial \sigma}(S_*, t_*; K, \tau, r, \sigma)$.

step 1 Set $n := 0$.

step 2 Compute

$$d\sigma := -\frac{f(\sigma^n) - v}{f'(\sigma^n)}$$

step 3 If $|d\sigma| \le \varepsilon$ set $L := n$ and return to OUT else

$$\sigma^{n+1} = \sigma^n + d\sigma, n := n+1,$$

and go to step 2 .

OUT Approximate value σ^L for the implied volatility $\sigma_{imp}(K,\tau)$.

Analysis of the computation scheme

Now, we want to analyze whether the conditions are satisfied which guarantee the quadratic convergence of the method. We do that under the assumptions

$$S_* > 0, \tau > 0,$$

which are not restrictive in practice.

Differentiability Obviously, the function f has infintely many derivatives. The first derivative is given as follows:

$$f'(\sigma) = S_* \sqrt{\tau} \mathcal{N}'(d_+(\sigma)).$$

Since f' is positive the Newton iteration is feasible. The derivative f' is the vega, i.e. $f' = \mathcal{V} = \frac{\partial C}{\partial \sigma}$.

Monotonicity f is strictly monotone increasing due to the fact that the first derivative of f is positive. This implies that the solution of (70) is uniquely determined.

Bounds We know from (a) in 1.1 that $(S_* - Ke^{-r\tau})^+ \le f(\sigma) \le S_*$. Obviously, $\lim_{\sigma \to 0} f(\sigma) = (S_* - Ke^{-r\tau})^+$ and $\lim_{\sigma \to \infty} f(\sigma) = S_*$.

Curvature The second derivative of f is given as follows:

$$f''(\sigma) = \frac{\tau}{2\pi} S_* e^{-q\tau} e^{-\frac{1}{2} d_+(\sigma)^2} \frac{d_+(\sigma) d_-(\sigma)}{\sigma}.$$

f'' is called the **Volga**, i.e. Volga $= \frac{\partial^2 C}{\partial \sigma^2}$. Volga* is positive for options away from the money, and initially increases with distance from the money. Specifically, volga is positive where $d_+(\sigma)$ and $d_-(\sigma)$ terms are of the same sign.

Existence of a solution A solution is guaranteed when we can show that

$$r_l := \lim_{\sigma \to 0} f(\sigma) - v \le 0, \; r_u := \lim_{\sigma \to \infty} f(\sigma) - v \ge 0 \tag{71}$$

is satisfied since due to the monotonicity one of the following inequalities $r_l < 0, r_u > 0$ holds true; continuity of f implies the solvability of (70).

Initial guess An initial guess can be determined by the bisection method.

Order of convergence The conditions for the quadratic convergence are given when the initial guess is sufficient close to the solution. The stopping criterion in the algorithm 3.1 is then appropriate.

The solvability of the equation (70) is in doubt when the market prices are not in agreement with the Black-Scholes model, a fact that cannot be excluded; see the smile-observation. Therefore, we cannot be sure that the market price of an option is in the interval $((S - Ke^{-r\tau})^+, S)$. From the numerical point of view, it is already delicate when the observed market price v is in the near of the boundary of this interval. A high instability of a solution of (70) is the consequence since we have

*The interest in the volga is to measure the convexity of an option with respect to volatility.

that the vega vanishes on the boundary of the interval. This corresponds to the fact that $\lim_{\sigma\to 0}\mathcal{V}(\sigma)=\lim_{\sigma\to\infty}\mathcal{V}(\sigma)=0$ holds.

Let us make a remark concerning the problem "inverse crime" in the implementation of the Newton-method; see Section 2.1. The main work in the iteration is done in the evaluation of the functions f, f'. Since this evaluation cannot be done analytically, an approximation procedure has to be used. In a simulation with artificial data the evaluation in obtaining these data and the evaluation in the iteration should be realized by different methods in order to test the robustness of the method.

3.2. *Pricing surface*

Given a set of market prices, we consider the problem to determine a complete surface of implied volatilities and of option prices.

Implied volatility surface

By varying the strike price K and the term to maturity τ we can create a table whose elements represent volatilities for different strikes and maturities. Under this practice, the implied volatility parameters will be different for options with different time-to-expiration τ and strike price K. As we know, this collection of implied volatilities for different strikes and maturities is called the implied volatility surface and the market participants use this table as the option price quotation. For many applications (calibration, pricing of nonliquid or nontraded options,...) we are interested in an implied volatility surface which is complete, i.e. which contains an implied volatility for each pair (K,T) in a reasonable large enough set $[0,K_{\max}]\times[0,T_{\max}]$.

However, in a typical option market, one often observes the prices of a few options with the same time-to-expiration but different strike levels only. To make things worse, some of these option contracts are not liquid at all, i.e. are not traded to an adequate extent. Therefore, we are faced with the problem of how to interpolate/extrapolate the table of implied volatilities. Such methods for completing the table of implied volatilities are well known: polynomials in two variables, linear, quadratic or cubic splines in one or two variables, parametrization of the surface and fitting of the parameters. But it seems to be appropriate to complete the pricing table to a pricing surface instead of completing the volatilities table since the properties of the pricing surface are more deeply related to the assumptions concerning the market.

The construction of the volatility surface is considered in a large number of papers. See for instance
[24; 25; 39; 49; 52; 46; 45; 70; 74; 72; 94; 103; 106; 116].

Interpolation in option pricing

The reality of missing prices requires a good method to interpolate the option price as a continuous function of the strike price (and the remaining life time of the option). Unfortunately, we cannot choose freely an interpolation method. If the pricing table is complete by interpolation we may compute the implied volatilities in order to complete the table of implied volatilities. The no-arbitrage principle determines that a call option price must be a monotonically decreasing and convex function of the strike price; see $(c), (d), (e)$ in 1.1. So, when we use the Black-Scholes formula and the interpolated volatilities to price options, we need to make sure that the final option price function satisfies the decreasingness and convexity restrictions.

We know that for a fixed expire-time convexity with respect the strikes should be realized. This can be done by the requirement, that the second derivative of the price-function is nonnegative. Here we describe a method which is closely related to the usual interpolationg cubic splines. Let us start with a partition

$$0 < a = x_0 < x_1 < \cdots < x_{N+1} = b < \infty\,. \tag{72}$$

We set $h_i := x_i - x_{i-1}, i = 1, \ldots, N+1$. Consider the following constrained interpolation problem:

$$\min_{g \in W_2^2[a,b]} \frac{1}{2}\|g'' - \psi\|_2^2 \text{ such that} \tag{73}$$

$$g(x_i) = y_i\,,\ i = 0, \ldots,\ N+1\,,\ g''(x) \geq 0 \text{ for a.e. } x \in [a,b]\,. \tag{74}$$

Here $\|\cdot\|_2$ is the Lebesgue $L_2[a,b]$-norm. Below, $W_2^k[a,b]$ denotes the *Sobolev space* of functions with absolutely continuous $(k-1)$-th derivatives and k-th derivative in $L_2[a,b], k \in \mathbb{N}$. Therefore, each function g in $W_2^2[a,b]$ can be written as follows:

$$g(x) = g(a) + \int_a^x g'(\eta)d\eta + \int_a^x g''(\xi)d\xi\,,\ x \in [a,b]\,,$$

with $g'' \in L_2[a,b]$. The usual inner product in $W_2^2[a,b]$ is given as follows:

$$\langle g, h\rangle := g(a)h(a) + \int_a^b g''(\xi)h''(\xi)d\xi\,,\ g,h \in W_2^2[a,b]\,.$$

In the context of the option price surface, $\{(x_i, y_i) \in \mathbb{R}^2 | i = 0, 1, 2, \ldots, N+1\}$ are strike levels and the corresponding observed prices of the options with the same underlying S and the same time-to-expiration τ. Concerning the function ψ we assume:

$$\psi \text{ is continuous differentiable} \tag{75}$$

ψ should be interpreted as an approximation of the second derivative of the pricing function.

We want to introduce in the interpolation problem above the second derivative $u := g''$ as the variable which has to be determined. This is possible by using linear splines (hat functions) $M_1, \ldots, M_N$ associated with the grid (72). With these splines the interpolation condition (74) can be written as follows (Peano kernel theorem; see for instance [80]):

$$\langle M_i, u\rangle = \int_a^b M_i(x)g''(x)dx = (Ky)_i\,,\ i = 1, 2, \ldots, N\,.$$

The matrix $K \in \mathbb{R}^{N,N+2} = (k_{i,j})_{i=1,\ldots,N,j=0,\ldots,N+1}$ is defined as follows:

$$\begin{aligned} k_{i,i} &= \frac{1}{(h_{i+1}+h_i)h_{i+1}}\,,\ k_{i,i+1} = \frac{-h_i}{(h_{i+1}+h_i)h_{i+1}} + \frac{-h_{i+1}}{(h_{i+1}+h_i)h_i}\,,\ k_{i,i+2} \\ &= \frac{1}{(h_{i+1}+h_i)h_i} \end{aligned} \tag{76}$$

for $i = 1, \ldots, N$ and $k_{i,j} = 0$ else. Notice that the matrix K is sparse and has full rank.

Now, the interpolation problem above can be rewritten as the following minimization problem:

Problem (I1)

$$\min_{u\in L_2[a,b]} \frac{1}{2}\|u-\psi\|_2{}^2 \tag{77}$$

$$\text{such that } \langle M_i, u\rangle = (Ky)_i\,,\ i = 1, 2, \ldots, N\,. \tag{78}$$

$$u(x) \geq 0 \text{ for a.e. } x \in [a, b]\,. \tag{79}$$

This optimization problem is an infinite-dimensional problem. The constraint condition (78) can be reformulated by

$$u \in V := \{v \in L_2[a,b] | \langle M_i, v\rangle = (Ky)_i, i = 1, \ldots, N\}\,, \tag{80}$$

with the affine subspace V of $L_2[a,b]$.

Theorem 3.1. *Suppose that the condition*

$$U_{ad} \neq \emptyset \tag{81}$$

holds. Then the problem (I1) has a unique solution.

Proof:
The objective function in (141) is lower semicontinuous, strictly convex, and coercive * . Moreover, V is a closed convex subset. Therefore, the feasible set $U_{\text{ad}} := V \cap U_+$ is closed and convex. As a consequence, the problem (I1) is equivalent to find the projection of ψ onto the nonempty, closed and convex set U_{ad} . It is well known that in a Hilbert space such a projection exists and is uniquely determined. ■

*This means: $\lim_{\|u\|_2\to\infty} \frac{1}{2}\|u-\psi\|_2^2 = \infty$

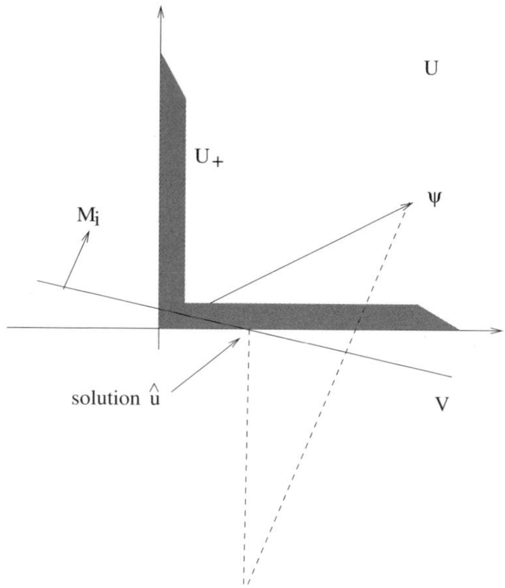

Figure 2. Lagrange multiplier

The constraints in (78), (79) are reformulated by the admissible set $U_{\mathrm{ad}} := V \cap U_+$ where U_+ is the cone $\{u \in L_2[a,b] | u \geq 0 \text{ a.e.}\}$. The main question is, whether the set U_{ad} is nonempty. Since the pricing formula is nonincreasing with respect to the strike variable we may assume that $y_0 \geq y_1 \geq \cdots \geq y_{N+1}$ holds. Moreover, since the pricing formula is convex with respect to the strike variable the second-order divided differences associated with the data, denoted by $(Ky)_i, i = 1, 2, \ldots, N$, are nonnegative. Let us summarize this assumption in

$$(Ky)_i \geq 0, i = 1, \ldots, N\,. \tag{82}$$

Therefore, a necessary condition for $U_{\mathrm{ad}} \neq \emptyset$ is that (82) holds.

In general, the basis for characterizing and computing the solution of an optimization problem is to verify a necessary condition for a solution. The intended necessary condition is a Kuhn-Tucker equation. As it is known, such an equation does not hold without a so called *constrained qualification* condition. Such a constraint qualification is the well known *Slater condition*, which in our case is

$$\mathrm{int}(U_+) \cap V \neq \emptyset. \tag{83}$$

Here $\mathrm{int}(U_+)$ denotes the interior (with respect to the norm-topology) of the cone U_+. Unfortunately, the condition (83) cannot hold since we know $\mathrm{int}(U_+) = \emptyset$, a well known fact in $L_2[a,b]$. Therefore, one has to look for a condition which is sufficient and does not use interior points of the cone U_+. Such a condition (84) is related to the so called *quasi relative interior points*; see [17]. An improvement of (81) is the following condition

$$\text{There exists } \hat{u} \in V \text{ with } \hat{u}(x) > 0 \text{ a.e. in } [a,b]\,. \tag{84}$$

Notice that (84) is not saying that $\hat{u}$ is an interior point of U_+. From this condition we conclude that the assumption (82) has to be changed into

$$(Ky)_i > 0, i = 1, \dots, N\,. \tag{85}$$

Theorem 3.2. *If the assumption* (84) *holds then the problem (I1) has a unique solution in the form*

$$\hat{u}(x) = \left(\sum_{i=1}^{N} \lambda_i M_i(x) + \psi(x)\right)^+ , \, x \in [a,b]\,, \tag{86}$$

where the Lagrange parameter λ *is the solution of the equation*

$$G(\lambda) = d := Ky\,, \tag{87}$$

with

$$G_i(\lambda) = \langle M_i, (\textstyle\sum_{j=1}^{N} \lambda_j M_j + \psi)^+ \rangle\,, \, i = 1, \dots, N\,. \tag{88}$$

Proof:
Set $\rho := \inf_{u \in U_{\mathrm{ad}}} \frac{1}{2}\|u - \psi\|^2$ and let $\hat{u} \in U_{\mathrm{ad}}$ with $\rho = \frac{1}{2}\|\hat{u} - \psi\|^2$; see Theorem 3.1. The main step in the proof of our theorem is the following fact: there exists a parameter $\lambda \in \mathbb{R}^N$ with

$$\rho = \min_{u \in U_+} \left(\frac{1}{2}\|u - \psi\|^2 - \langle Au - Ky, \lambda\rangle\right) = \frac{1}{2}\|\hat{u} - \psi\|^2 - \langle \hat{u}, A^*\lambda\rangle + \langle Ky, \lambda\rangle \tag{89}$$

where A^* is the adjoint of the linear mapping

$$A : U \ni u \longmapsto (M_1(u), \dots, M_N(u)) \in \mathbb{R}^N\,.$$

This fact is the result of a separation theorem which uses the assumption (84) in the form that Ky is a point in the image of the quasi-interior points $\mathrm{qri}(U_+)$ under the mapping A. We have

$$\begin{aligned} &\|\hat{u} - \psi\|^2 - 2\langle \hat{u}, A^*\lambda\rangle + 2\langle Ky, \lambda\rangle \\ &= \|\hat{u} - \psi\|^2 - 2\langle \hat{u} - \psi, A^*\lambda\rangle + 2\langle Ky - A\psi, \lambda\rangle \\ &= \|\hat{u} - \psi - A^*\lambda\|^2 - \|A^*\lambda\|^2 + 2\langle Ky - A\psi, \lambda\rangle. \end{aligned}$$

This implies that $\hat{u}$ is the projection of $\psi + A^*\lambda$ onto the cone U_+. The equation (87) is a consequence of the fact that $\hat{u}$ has to be admissible. ■

For the complete proof of the theorem we refer to [42; 63; 126] in connection with [22].

Interpolation by shape-preserving splines in the presence of noise

The shape-preserving interpolation method considered above presumes that the observed data accurately reflect the behavior of the call option price function to be approximated. Unfortunately, in financial practice, the observed data usually have errors, thus the previous interpolation technique is not appropriate in many cases.

Again let

$$0 < a = x_0 < x_1 < \cdots < x_{N+1} = b < \infty$$

be a partition of the interval $[a, b]$ and let $y_0^\delta, \ldots, y_{N+1}^\delta$ be the noisy data. In the context of finance, this means that $y_0^\delta, \ldots, y_{N+1}^\delta$ are noisy prices of call options: $y_i^\delta \approx C(S_*, t_*; K, \tau)$ for some pair (K, τ). As usual, $\delta \geq 0$ is the size of the noise, i.e. there are data $y_0, \ldots, y_{N+1}$ considered as noiseless, such that

$$|y_i^\delta - y_i| \leq \delta \,.$$

We set $y := (y_0, \ldots, y_{N+1}), y^\delta := (y_0^\delta, \ldots, y_{N+1}^\delta)$.

Here is an approach to regard the presence of noise. The convex interpolation model (I1) is changed to the following minimization problem:

Problem (I2)

$$\min_{u \in L_2[a,b], z=(z_0,\ldots,z_{N+1}) \in \mathbb{R}^N} \quad \frac{1}{2}\|u - \psi\|_2{}^2 + \frac{1}{2\alpha}|Q(z - y^\delta)|^2 \tag{90}$$

$$\text{such that} \quad \langle M_i, u\rangle = (Kz)_i \,,\; i = 1, 2, \ldots, N \,. \tag{91}$$

$$u(x) \geq 0 \text{ for a.e. } x \in [a, b] \,. \tag{92}$$

Here Q is a nonsingular matrix in $\mathbb{R}^{N+2,N+2}$ and $|\cdot|$ denotes the Euclidean norm in $\mathbb{R}^N$. The matrix Q^*Q may be considered as a correlation matrix since $|Q(z - y^\delta)|^2 = \langle z - y^\delta, Q^*Q(z - y^\delta)\rangle$ where $\langle\cdot,\cdot\rangle$ is the euclidean inner product in $\mathbb{R}^{N+2}$.

A solution of (I2) is called a *smoothing spline.* The term $\|u - \psi\|_2{}^2$ is a regularization quantity which is "enhanced" by the positive *regularization parameter* α^{-1}, where α is a parameter which describes the tradeoff between the deviation of the smoothing spline and the a-priori guess and the deviation of the noiseless variable z and the noisy data y^ε.

Again, (I2) is an infinite-dimensional convex optimization problem. It can be reformulated as a problem of the type (I2). We define a Hilbert space W and its inner product $\langle\cdot,\cdot\rangle_W$ by

$$W := U \times \mathbb{R}^{N+2} \,,\; \langle(u, a), (v, b)\rangle_W := \langle u, v\rangle + \alpha^{-1}\langle Qa, Qb\rangle \,.$$

Let $\Psi := (\psi, y^\delta) \in W$, $W_+ := U_+ \times \mathbb{R}^{N+2}$, and define the subspace V of W by

$$V := \{(u, z)|Au - Kz = \theta\} \,,$$

where θ denotes the null vector in $\mathbb{R}^{N+2}$. Then problem (I2) can be written as an interpolation problem in W:

$$\min\{\frac{1}{2}\|w-\Psi\|_W^2 | w \in W_+ \cap V\}.$$

Now, it is clear that the following theorem can be proved similar to Theorem 3.2.

Theorem 3.3. *If assumption* (84) *holds then problem (I2) has a unique solution in the form*

$$\hat{u}(x) = \left(\sum_{i=1}^{N} \hat{\lambda}_i M_i(x) + \psi(x)\right)^+ , \; x \in [a,b] , \tag{93}$$

$$\hat{z} = y^\delta - \alpha(Q^*Q)^{-1}K^*\hat{\lambda} \tag{94}$$

where the Lagrange multiplier $\hat{\lambda}$ *is the solution of the equation*

$$G(\lambda) + \alpha K(Q^*Q)^{-1}K^*\lambda - y^\varepsilon = 0 \tag{95}$$

with

$$G_i(\lambda) = \langle M_i, (\sum\nolimits_{j=1}^{N} \lambda_j M_j + \psi)^+\rangle \, , \, i = 1, \ldots, N \, . \tag{96}$$

Proof:
For the proof of the theorem we refer to [63]. ■

We can see that, as $\alpha \to \infty$, the solution of the minimization problem (I2) tends to the solution of the shape-preserving interpolation problem (I1) when $y = y^\delta$.

Computation of the shape preserving spline by a Newton-type-method

We have to solve the equation

$$G_i(\lambda) = (Ky)_i \, , \, i = 1, \ldots, N \, . \tag{97}$$

Since the mappings

$$\mathbb{R}^n \ni \lambda \longmapsto g_i(\lambda, x) := \sum_{j=1}^{N} (\lambda_j M_j(x) + \psi(x)) M_i(x) \in \mathbb{R} \tag{98}$$

are continuous differentiable with respect to λ for all $x \in [a,b]$ and piecewise continuous differentiable functions with respect to $x \in [a,b]$, $g_i(\cdot,\cdot)$ is Lipschitz continuous with respect to $\lambda \in \mathbb{R}^N$ uniformly in $x \in [a,b]$. Moreover, one can show that the following identity holds:

$$\delta G_i(\lambda) = \int_a^b \delta(g_i(\cdot, x)^+)(\lambda) dx \, ,$$

where δ is a symbol for the generalized Jacobian. Hence, each of the mappings G_i is semismooth in λ, and the composed map G is semismooth at λ too. For the theory of semismooth mappings see for instance [33; 123; 117]. Thus, we have to compute

$$\delta(g_i(\cdot,x))^+(\lambda)\,,\, i=1,\dots,N\,.$$

With

$$\mathbb{M}(x) := \begin{pmatrix} M_1(x) \\ \vdots \\ M_N(x) \end{pmatrix}, \, x \in [a,b]\,,$$

one can show that

$$\delta(g_i(\cdot,x))^+(\lambda) = \begin{cases} M_i(x)\mathbb{M}(x) & \text{if } g_i(\lambda,x) > 0 \\ \{\alpha M_i(x)\mathbb{M}(x) | \alpha \in [0,1]\} & \text{if } g_i(\lambda,x) = 0 \\ 0 & \text{if } g_i(\lambda,x) < 0, \end{cases} \tag{99}$$

where $\lambda \in \mathbb{R}^N, x \in [a,b]\,.$

Let

$$q(\lambda,x) := \sum_{j=1}^{N} \lambda_j M_j(x) + \psi(x)\,,\, \lambda \in \mathbb{R}^N, x \in [a,b],$$

and define

$$T(\lambda) := \{x \in [a,b] | q(\lambda,x) = 0\}\,,\, \overline{T}(\lambda) = [a,b]\backslash T(\lambda)\,,\, \lambda \in \mathbb{R}^N\,.$$

The generalized Jacobian δG of G is now given as

$$\delta G(\lambda) = \delta G^-(\lambda) + DG^+(\lambda)\,,\, \lambda \in \mathbb{R}^N, \tag{100}$$

where, for $j = 1,\dots,N$

$$G_j^-(\lambda) = \int_{T(\lambda)} q(\lambda,x)^+ M_j(x) dt\,,\, G_j^+(\lambda) = \int_{\overline{T}(\lambda)} q(\lambda,x)^+ M_j(x) dt$$

As a consequence,

$$DG^+(\lambda)_{ij} = \int_a^b q(\lambda,x)_+^0 M_i(x) M_j(x) dx\,,\, i,j = 1,\dots,N, \tag{101}$$

where

$$z_+^0 := \begin{cases} 1 & \text{if } z > 0 \\ 0 & \text{if } z \le 0 \end{cases}.$$

Now, one can prove the following result which is the basis for the applicability of Newton's method to the equation (87); see [63].

Suppose that the assumption (84) holds. If λ is a solution of $G(\lambda) = Ky$, then every matrix $Q \in \delta G(\lambda)$ is positive definite.

Algorithm 3.2 Interpolation by smoothing splines using Newton's method

IN Interpolation data $(x_i, y_i), i = 0, \ldots, N+1$.
Initial approximation ψ for the second derivative (of the pricing fuction)
Initial guess λ^0 for the Lagrange multiplier.
Accuracy-bound $\varepsilon > 0$ for the iteration steps.

step 0 Compute $M_1, \ldots, M_N$, $y := (y_0, \ldots, y_N)$, and $v := Ky$.

step 1 For $k = 0, 1, \ldots$ do

- Compute $G(\lambda^k)$
- Choose $Q^k \in \delta G(\lambda^k)$
- Solve the linear equation $Q^k s^k = -G(\lambda^k) + v$
- If $|s^k| \leq \varepsilon$ set $L := k$ and go to OUT
- Set $\lambda^{k+1} := \lambda^k + s^k$

OUT Approximate Lagrange multiplier λ^L and an approximate smoothing spline

$$\hat{u}^L := \left(\sum_{i=1}^{N} \lambda_i^L M_i(x) + \psi(x) \right)^+ , \; x \in [a, b] .$$

Computation of the option price function

The solution to the original problem (73),(74) can be obtained by integrating twice the solution $\hat{u}$ on $[a, b]$ with the result C. Here we give are some hints how to do this; see [63]. $\hat{u}$ may be replaced by the approximation $\hat{u}^L$.

The price function C can be written as

$$C(x) = y_k + (x - x_k)C'(x_k) + \int_{x_k}^{x} \int_{x_k}^{s} \hat{u}(t)dtds \, , \; x \in [x_k, x_{k+1}], k = 0, \ldots, N \, . \quad (102)$$

In this formula, two terms need to be calculated, the first derivative of C at x_k and the integration

$$Q(x_k, x) := \int_{x_k}^{x} \int_{x_k}^{s} \hat{u}(t)dtds \, .$$

If the integration $Q(x_k, x)$ is known, then

$$C'(x_k) = \frac{y_{k+1} - y_k - Q(x_k, x_{k+1})}{x_{k+1} - x_k} \, .$$

We have $\hat{u} = f^+$ with

$$f(x) = \sum_{i=1}^{N} \lambda_i M_i(x) + \psi(x) \, , \; x \in [a, b] \, .$$

Obviously, f is a piecewise differentiable function. Then we can obtain for each $k = 0, \dots, N$ a finite number of node by solving the equation $f(x) = 0$ in $[x_k, x_{k+1}]$. This gives a new enlarged partition

$$a = t_0 < t_1 < \cdots < t_{R+1} = b.$$

In every interval $[t_k, t_{k+1}]$ the solution $\hat{u}$ has a computable representation. By considering several cases, the function $Q(x_k, x)$ can be computed by simple integration (under the assumption that a primitive of ψ is known).

3.3. *Forward rates and discount factors*

In finance, the yield curve is the relation between the (level of) interest rate (or cost of borrowing) and the time to maturity, known as the *term*. In the focus is the financial instrument *zero-coupon-bond.*

Discount factor

Let $B(t,T)$ denote the price at time t of a security making a single payment of 1 unit at time $T, T \geq t$. Such a security is called a *zero coupon bond* and $B(t,T)$ is the *discount factor* over the period $[t,T]$. Discount factors are necessary for the evaluation of financial products. The *yield* of a zero coupon bond may be interpreted as the interest rate implied by the price of the zero coupon bond:

$$B(t,T) = e^{-Y(t,T)(t-T)} \quad \text{or} \quad Y(t;T) = -\frac{\ln(B(t,T))}{T-t}. \tag{103}$$

Suppose that we know the discount factors $B(0,t_1), B(t,t_2)$ with $0 < t_1 < t_2$. Then by an arbitrage-argument we obtain

$$B(0,t_1)B(t_1,t_2) = B(0,t_2) \tag{104}$$

which makes it possible to read off the discount factor $B(t_1,t_2)$ on the interval $[t_1,t_2]$.

Forward rates

A *forward rate* is an interest rate set today for borrowing or lending at some date in the future. Consider the time grid $t < T_1 < T_2$ and let $F(t,T_1,T_2)$ denote the forward rate fixed at time t for borrowing or lending in the period $[T_1,T_2]$. Again, an arbitrage argument shows that forward rates are determined by discount factors, namely:

$$F(t,T_1,T_2) = \frac{1}{T_2 - T_1} \frac{B(t,T_1) - B(t,T_2)}{B(t,T_2)}. \tag{105}$$

With a continously compounded forward rate $f(t,T_1,T_2)$ we obtain the identity

$$F(t,T_1,T_2)(T_2 - T_1) = \exp(f(t,T_1,T_2)(T_2 - T_1)) - 1) - 1 \tag{106}$$

which implies

$$f(t,T_1,T_2)(T_2-T_1) = \frac{\ln(B(t,T_1)) - \ln(B(t,T_2))}{T_2-T_1}. \tag{107}$$

Now define $f(t,T)$ to be the continuously compounded forward rate fixed at t for the maturity T as the limit of $f(t,T,T+h)$ as h approaches 0, whenever it exists:

$$f(t,T) = \lim_{h\downarrow 0} \frac{\ln(B(t,T)) - \ln(B(t,T+h))}{h} = -\frac{\partial}{\partial T}\ln(B(t,\cdot)(T). \tag{108}$$

Using $B(T,T)=1$ we obtain

$$B(t,T) = \exp(-\int_t^T f(t,s)ds). \tag{109}$$

A comparison with (103) implies

$$Y(t,T) = \frac{1}{T-t}\int_t^T f(t,s)ds. \tag{110}$$

i.e., that yields are averages over forward rates.

Forward rates are in the focus when one wants to build a model of *term structure dynamics*. When we introduce the time to maturity τ we obtain from Y the so called *yield curve* y; thus:

$$y(\tau) := Y(T-\tau,T)\,,\ 0\le\tau\le T\,. \tag{111}$$

y is often, but not always, an increasing function of τ.

The market does not contain enough information in order to determine the prices of zero coupon bonds or the yield values for all values of $t\in[0,T]$. Suppose that the discount factor is known for t_1 and t_2. Then the discount factor for $t\in(t_1,t_2)$ may be computed by (local) interpolation:

Linear interpolation $\quad B(t,T) := aB(t_1,T) + (1-a)B(t_2,T)$

Exponential interpolation $\quad B(t,T) := B(t_1,T)^a B(t_2,T)^{1-a}$

where $a := \dfrac{t-t_1}{t_2-t_1}$.

These interpolation procedures may be applied in the case of yield values or forward rates.

Interpolation of market data

It is the goal of the yield-analysis to propose a model for the yield-curve and to fit the outcome of the model to values observed on the financial market. Market data in finance are discrete in time but for many applications we need to construct a function continuous in time which approximates these data in an appropriate way. Various interpolation methods for curve construction are available. In general, an interpolation problem is as follows:

Given real nodes $t_0 < t_1 < \cdots < t_N$ and values $y_0, y_1, \ldots, y_N \in \mathbb{R}$. Construct a continuous function x on an interval (t_l, t_u) containing the nodes with

$$x(t_i) = y_i, i = 0, \ldots, N.$$

Of course, many choices of interpolation function are possible. According to the nature of the problem, one imposes requirements additional to continuity, such as differentiability, twice differentiability, monotonicity, convexity, conditions at the boundary, ...

The typical approach is to require that in each interval the function is described by some low dimensional polynomial, so the requirements of continuity and differentiability reduce to linear equations in the coefficients, which are solved using standard linear algebraic techniques.

Linear interpolation was already described above as local interpolation. Polynomial interpolation is inadequate in the most cases since the interpolating function shows a remarkable oscillatory behavior. Interpolation with splines (piecewise polynomials) is the better method, although problems may occur on the ends of the interval (t_l, t_u). We will not go into the details since this the method of interpolation by splines is described in many textbooks of numerical analysis.

The requirement of convex interpolants leads to the fact that algebraic techniques are not sufficient to solve the interpolation problem. In the last section we studied a method which may be applied in this situation too.

3.4. *Parametric models for the forward rate*

The Nelson-Siegel model

In practice, parametric models for the instantaneous forward rate are used. The most famous is the *Nelson-Siegel model*; see [112]. Here one makes in a domain of definition $[0, T_{\max}]$ the ansatz (τ may be considered as time to maturity)

$$f_{(\beta_0,\beta_1,\beta_2,\lambda)}(\tau) := \beta_0 + \beta_1 e^{-\frac{\tau}{\lambda}} + \beta_2 \frac{\tau}{\lambda} e^{-\frac{\tau}{\lambda}}, \; 0 < \tau \le T. \tag{112}$$

The parameter vector $b := (\beta_0, \beta_1, \beta_2, \lambda)$ determines the shape of the ansatz functions. We see that the instantaneous forward rate $f = f_b = f_{(\beta_0,\beta_1,\beta_2,\lambda)}$ consists of three terms: *long-term, medium-term, short-term.* The long-term component is identified by the asymptotic value β_0. The short-term is $\beta_0 + \beta_1 e^{-\frac{\tau}{\lambda}}$ and it is asymptotical obtained by setting $\tau := 0$ as $\beta_0 + \beta_1$.* The medium-term is identified by $\beta_2 \frac{\tau}{\lambda} e^{-\frac{\tau}{\lambda}}$ and it determines the height of the hump of $f_{(\beta_0,\beta_1,\beta_2,\lambda)}$.

When we plug in the parametric forward rate into the formula for the yield we

*Here we use the rule of de l'Hospital

obtain

$$y(\tau) = \frac{1}{\tau}\int_0^\tau f_{(\beta_0,\beta_1,\beta_2,\lambda)}(s)ds \;=\; \beta_0 + (\beta_1+\beta_2)\frac{1-e^{-\frac{\tau}{\lambda}}}{\frac{\tau}{\lambda}} - \beta_2 e^{-\frac{\tau}{\lambda}}\,,\; 0<\tau\le T\,. \tag{113}$$

Using this expression we try to construct the yield function by estimating the parameter $b := (\beta_0, \beta_1, \beta_2, \lambda)$. Due to the applied situation we have to restrict the parameter b to a set of *admissible parameters* B_{ad}. We set:

$$B_{\mathrm{ad}} := \{b = (\beta_0,\beta_1,\beta_2,\lambda) | \beta_0 \ge 0, \beta_0+\beta_1 \ge 0, \lambda \ge l\} \tag{114}$$

where $l > 0$ has to be chosen appropriately. Once the parameter $b \in B_{\mathrm{ad}}$ is estimated we may compute the yield function by (113) and the prices of the observed bonds. As an optimization criterion to estimate the parameter b we may use the observed prices $\tilde{B}_1, \dots, \tilde{B}_N$ or the observed yield values $\tilde{y}_1, \dots, \tilde{y}_N$ in a least squares functional:

$$\begin{array}{ll} \text{Minimize} & \sum_{i=1}^N (B_i - \tilde{B}_i)^2 \text{ subject to } b \in B_{\mathrm{ad}} \\ \text{Minimize} & \sum_{i=1}^N (y_i - \tilde{y}_i)^2 \text{ subject to } b \in B_{\mathrm{ad}} \end{array}$$

Here, $B_i = B_i(b), y_i = y_i(b), i = 1,\dots,N$, are the prices and yield values computed from the Nelson-Siegel model (see (113) and (115)) for a grid of times $0 \le \tau_1 < \cdots < \tau_n \le T$. For the numerical realization of the minimization problems see below.

After the publication of the Nelson–Siegel model various extensions were proposed that incorporate additional flexibility. Here we mention the four-term model of Svensson (see [125]):

$$f_{(\beta_0,\beta_1,\beta_2,\beta_3,\lambda_1,\lambda_2)}(\tau) := \beta_0 + \beta_1 e^{-\frac{\tau}{\lambda_1}} + \beta_2 \frac{\tau}{\lambda_1} e^{-\frac{\tau}{\lambda_1}} + \beta_3 \frac{\tau}{\lambda_2} e^{-\frac{\tau}{\lambda_2}}\,,\; 0<\tau\le T\,. \tag{115}$$

For this model, five parameters have to be estimated: $\beta_1, \beta_2, \beta_3, \lambda_1, \lambda_2$. It can be done similiar to the considerations below.

Numerical recipes for the calibration

Let us consider the calibration of the Nelson–Siegel model based on the yield-criterion. The vector of the observed data is $\tilde{y} := (\tilde{y}_1, \dots, \tilde{y}_N)$. We set for $\lambda \in \mathbb{R}$

$$\begin{aligned} \phi(\lambda)_i &:= (1, e^{-\frac{\tau_i}{\lambda}}, \frac{\tau_i}{\lambda} e^{-\frac{\tau_i}{\lambda}})\,,\; i = 1,\dots,N\,, \\ \Phi(\lambda)^* &:= (\phi(\lambda)_1, \dots, \phi(\lambda)_N) \in \mathbb{R}^{N,3}\,, \\ e(\lambda,\beta) &:= \Phi(\lambda)\beta - \tilde{y}\,,\; \beta \in B_{\mathrm{ad}} \\ F(\lambda,\beta) &:= \frac{1}{2}\|e(\lambda,\beta)\|_2^2\,,\; \beta \in B_{\mathrm{ad}}\,. \end{aligned}$$

Here are some theoretical and numerical observations:

- The minimization of the objective F is a nonlinear least squares problem.
- Since the objective F is non-convex, local minima for the objective F may exist.

- Non-convex least squares problems require special solvers to ensure convergence.
- The solution quality is heavily depending on the initial guess for the solver.
- The error function e is linear in the parameter β, but "highly nonlinear" in the scaling parameter λ.

The last observation is very important since it shows that our optimization problem is a separable least squares problem; clever algorithms should exploit this fact.

Such problems have been considered very intensively during the last 30 years. For a fixed λ the objective F is minimized with respect to β. This is done formally by the projection of $\tilde{y}$ on the range of the convex set B_{ad} under the mapping $\Phi(\lambda)$. We denote this projection by $\Pi_{B_{\text{ad}}}(\lambda)^{\dagger}\tilde{y}$. Then we complete the optimization by the minimization of $\frac{1}{2}\|\Phi(\lambda)\Pi_{B_{\text{ad}}}(\lambda)^{\dagger}\tilde{y}\|_2^2$ with respect to λ. Thus, the optimization problem is stated as follows:

$$\min \frac{1}{2}\|\Phi(\lambda)\Pi_{B_{\text{ad}}}(\lambda)^{\dagger}\tilde{y}\|_2^2 \quad \text{subject to } \lambda \geq l\,. \tag{116}$$

Algorithm 3.3 Computation of the yield-curve along the Nelson-Siegel model

IN Market data $\tilde{y} := (\tilde{y}_1, \ldots, \tilde{y}_N)$. Accuracy-bound $\varepsilon > 0$.
step 0 Initial guess λ_0.
step 1 For $k := 0, 1, \ldots$ do

- Compute $\Pi_{B_{\text{ad}}}(\lambda^k)^{\dagger}\tilde{y}$
- Compute a descent direction d_k for the given parameter λ^k
- If $|d_k| \leq \varepsilon$ then set $\beta_k := \Pi_{B_{\text{ad}}}(\lambda^k)^{\dagger}\tilde{y}, L := k$ and go to OUT
- Increment $\lambda_{k+1} := \lambda_k + d_k$.

OUT Approximation (β_L, λ_L) for the optimal parameter (β^*, λ^*)

For the computation of $\Pi_{B_{\text{ad}}}(\lambda^k)^{\dagger}\tilde{y}$ we have to solve a quadratic problem with simple linear constraints. Several efficient methods are available. For the computation of a descent direction d_k one should be able to compute the differential of the mapping $\lambda \longmapsto \Pi_{B_{\text{ad}}}(\lambda)^{\dagger}\tilde{y}$. As it is known, this is not possible in a classical sense for two reasons. First, the matrix $\phi(\lambda)$ may be rank-defficient, second, the linear constraint is an inequality constraint and therefore the concept of generalized gradient has to be used. A large amount of literature is devoted to the analysis of the resulting problems.

3.5. *Sensitivities*

In this section we consider simple methods for estimating *sensitivities* of option pricing (greeks).

Parameter dependent expectation values

Consider an expectation value

$$v := \mathbb{E}\left(\Phi(S_T)\right) := \mathbb{E}^Q\left(\Phi(S_T)\right) \tag{117}$$

where $\Phi(S_T)$ is a random variable on a probability space $(\Omega, \mathcal{F}, Q)$ which depends implicitly on a parameter. Its the goal of our considerations to study this dependency. Thus $\Phi(S_T) =: Y(\alpha)$ with a random variable $Y(\alpha)$ dependimg on $\alpha \in A \subset \mathbb{P}$. As a consequence we write

$$u(\alpha) := \mathbb{E}(Y(\alpha)). \tag{118}$$

Especially, we are interested in the derivative $u'(\alpha)$.

Concrete realizations occur in at least two essential different situations: If α is the initial value of a stochastic differential equation then α is a real number (ranging over some interval of the real line). If α stands for the volatility σ, then α is a variable in a function space and the computation of variations of $u(\alpha)$ is much more difficult.

Sensitivities via difference quotients

Let α be a parameter in $\mathbb{R}$, The sensitivity problem consists of finding a way to estimate the derivative $u'(\alpha)$ or an apropriate approximation of it. An obvious approach consists in the approximation of $u'(\alpha)$ by a forward–difference estimator:

$$u'(\alpha) \approx (u(\alpha+h) - u(\alpha))h^{-1} \ \ (h > 0)\,. \tag{119}$$

Going back to the random variable Y the approximation for $u'(\alpha)$ becomes

$$u'(\alpha) \approx \mathbb{E}(Y(\alpha+h) - Y(\alpha))h^{-1} \tag{120}$$

Now, the expectation value is approximated by the average of n replications of $Y(\alpha+h)_1, \ldots, Y_n(\alpha+h)$ and $Y(\alpha)_1, \ldots, Y_n(\alpha)$, respectively. This gives:
In the formulation above it is not required that the random variable Y is realized by independent replications.

If u is twice differentiable at α, then

$$u(\alpha+h) = u(\alpha) + u'(\alpha)h + \frac{1}{2}u''(\alpha)h^2 + o(h^2)\,.$$

and

$$\text{bias}(D_n(\alpha)) := \mathbb{E}(D_n(\alpha) - u'(\alpha)) = \frac{1}{2}u''(\alpha)h + o(h)\,. \tag{121}$$

If we use central differences

$$C_n(\alpha) := \frac{1}{2h}(\hat{Y}(\alpha+h) - \hat{Y}(\alpha-h))$$

Algorithm 3.4 Monte Carlo-simulation of the sensitivity

IN A family of random variables $Y(\alpha))_{\alpha\in A}$, a parameter $\alpha \in A$. Step size $h > 0$ and a number n of replications.

step 1 Compute realizations $Y_1(\alpha), \ldots, Y_n(\alpha)$ and $Y_1(\alpha+h), \ldots, Y_n(\alpha+h)$.

step 2 Compute the averages

$$\hat{Y}(\alpha) := \frac{1}{n}\sum_{i=1}^{n} Y_i(\alpha)\,,\ \hat{Y}(\alpha+h) := \frac{1}{n}\sum_{i=1}^{n} Y_i(\alpha+h)$$

and the number

$$D_n(\alpha) := \frac{\hat{Y}(\alpha+h) - \hat{Y}(\alpha)}{h}\,.$$

OUT Estimator $D_n(\alpha)$ for $u'(\alpha)$.

– the algorithm above may be changed to realize such approximations – we obtain

$$\text{bias}(C_n(\alpha)) := \mathbb{E}(C_n(\alpha) - u'(\alpha)) = \frac{1}{6}u'''(\alpha)h^2 + o(h^2)\,. \tag{122}$$

Clearly, this estimator posesses a higher accuracy, but is more costly.

Let us compute the variance of the estimators $D_n(\alpha)$ and $C_n(\alpha)$:

$$\mathbb{V}(D_n(\alpha)) = h^{-2}\mathbb{V}(\hat{Y}(\alpha+h) - \hat{Y}(\alpha)) \tag{123}$$

$$\mathbb{V}(C_n(\alpha)) = (2h)^{-2}\mathbb{V}(\hat{Y}(\alpha+h) - \hat{Y}(\alpha-h))\,. \tag{124}$$

These identities show the consequences of a small stepsize h in order to obtain a small bias: a disastrous big variance may result. Thus, we can observe the typical situation of an ill-posed problem; see Subsection 2.1.

Suppose that the realizations $Y_1(\alpha), \ldots, Y_n(\alpha)$ and $Y_1(\alpha+h), \ldots, Y_n(\alpha+h)$ are families of identical distributed indepent random variables. Then

$$\mathbb{V}(\hat{Y}(\alpha+h) - \hat{Y}(\alpha)) = \frac{1}{n}\mathbb{V}(Y(\alpha+h) - Y(\alpha))$$

and under the assumption

$$\mathbb{V}(Y(\alpha+h) - Y(\alpha)) = O(h^q) \tag{125}$$

with $q \in \{0, 1, 2\}$ we may analyze the behavior of the estimators more detailed; see [58], pp. 378.

The considerations above may be applied to the computation of greeks. Serious problems arise when the payoff function is not differentiable (in the classical sense). But there are results in the literature which overcome this barrier; see for example [98; 121].

A very deep theory to handle the differentiability of stochastic processes with respect to certain variables is the so called *Malliavan calculus*. This theory presents additional possibilities to compute sensitivities; see [32].

4. Dupire's method: the dual equation

In this section we consider the model of the geometric Brownian motion with a local volatility from the identifiabilty point of view. In the first subsection we sketch the derivation of Dupire's equation. Next we present an identifiability result whose proof is related to this this equation. From the considerations involved in the identifiability problem an integral equation for the local volatility can be extracted. This integral equation may be used to approximate the unknown volatility. In the last subsection we consider the identifiability problem as a problem of numerical differentiation.

4.1. *Dupire's equation*

In Section 5 we will consider two approaches of calibration: one calibration method is based on the volatility-to-solution map of the forward problem of finance. This is a very time consuming method. As first observed by Dupire [43], there is an alternative and a more direct way to solve the inverse problems of finance: the second method is based on this way.

Suppose the option price of a Black-Scholes model with local volatility σ is denoted by

$$C = C(S,t) = C(S,t;K,T,r,\sigma)\,.$$

Given $S_* > 0, t_* > 0$, we introduce a new variable U as follows:

$$U(K,T) = U(K,T;S_*,t_*,r,\sigma) := C(S_*,t_*;K,T,r,\sigma)\,.$$

Then one can show that the function U satisfies the following system:

$$\mathcal{L}_{DU}(U) := \frac{\partial U}{\partial T} - \frac{1}{2}\sigma(K,T)^2K^2\frac{\partial^2 U}{\partial K^2} + rK\frac{\partial U}{\partial K} = 0, \qquad K,T \in (0,\infty)\times(t_*,\infty)\,, \tag{126}$$

$$U(0,T) = S_*\,,\ \lim_{K\to\infty} U(K,T) = 0, T \in (t_*,\infty)\,, \tag{127}$$

$$U(K,t_*) = (S_* - K)^+\,,\ K > 0\,. \tag{128}$$

Several different derivations and proofs of this result are found in the literature; see [3; 24; 43; 92]. In the next section we will sketch the proof in the framework of weak solutions. In [72] a derivation of the system (126), (127), (128) is presented based on a fine analysis of fundamental solutions of parabolic equations as given in [57]. We follow mainly [25; 127].

We set

$$u(K,T) := u(K,T;S,t) := C(S,t;K,T)\,,\ S,K > 0, T > 0, 0 \le t \le T\,.$$

Then we know

$$\lim_{K\to\infty} u(K,T;S,t) = 0 \quad w(K,T;S,t) := \lim_{K\to\infty} \frac{\partial}{\partial K} u(K,T;S,t) = 0$$

$$\phi(K,T;S,t) := \lim_{K\to\infty} \frac{\partial^2}{\partial K^2} u(K,T;S,t) = 0.$$

Here we have already used the fact that these derivatives exist. This can be shown by taking difference quotient with respect to the variable K and using the fact that these difference quotients satisfy an initial boundary value problem. Due to the fact that the final condition of w is of Heavy-side type the final condition of ϕ is the Dirac distribution δ_{S-K}. Moreover, ϕ is a solution of the *dual equation*

$$\frac{\partial v}{\partial T} - \frac{1}{2}\frac{\partial^2}{\partial K^2}\left(\sigma^2(K,T)K^2 v\right) + r\frac{\partial}{\partial K}(Kv) + rv = 0 \ \ (K \in (0,\infty), T \in (0,\infty)). \quad (129)$$

Since

$$\begin{aligned} \int_K^\infty \int_\eta^\infty \frac{\partial}{\partial \xi}\left(\xi\phi(\xi,T)\right) d\xi d\eta &= -\int_K^\infty \eta\phi(\eta,T)d\eta \\ &= \int_K^\infty \eta\frac{\partial}{\partial \xi}\left(\int_\eta^\infty \phi(\xi,T)d\xi\right) d\eta \\ &= -K\int_K^\infty \eta\phi(\xi,T)d\xi - \int_K^\infty \int_\eta^\infty \phi(\xi,T)d\xi d\eta \\ &= K\frac{\partial U}{\partial K}(S,t;\xi,T,\sigma) - U(S,t;\xi,T,\sigma) \end{aligned}$$

we obtain the equation

$$\frac{\partial U}{\partial T} - \frac{1}{2}K^2\sigma^2(K,T)\frac{\partial^2 U}{\partial K^2} - rK\frac{\partial U}{\partial K} = 0 \ \ (K \in (0,\infty), T \in (0,\infty)). \quad (130)$$

By considering boundary values and the final value we obtain the following boundary value problem

$$\frac{\partial U}{\partial T} + \frac{1}{2}\sigma^2(K,\tau)K^2\frac{\partial^2 U}{\partial K^2} + rK\frac{\partial U}{\partial K} + rU = 0 \quad (131)$$
$$(K \in (0,\infty), T \in (t_*,\infty))$$

$$U(0,\tau) = S_*\,, \ \lim_{K\to\infty} U(K,\tau) = 0\,,\ \tau \in (0,t_*)\,. \quad (132)$$

$$U(K,0) = (S_* - K)^+\,,\ K > 0\,. \quad (133)$$

We can compute the same quantity, namely the option price $U(S_*,t_*;K,T,r,q,\sigma)$ as a function of K, with the help of two different systems. First, along the Black-Scholes equation: solve severals Black-Scholes equations for different strikes K. Second, along Dupire's equation: solve Dupire's equation once to obtain the price for several strikes. Theoretically, the values have to be identical, but under numerical methods there may be a difference, especially, since we have to truncate the domain for the variables S, K respectively.

Some attempts are made to derive a dual equation for other models; see [2; 113].

4.2. *The uniqueness theorem for the inverse problem*

Due to the smile effect, the local volatility should depend on both time and the level of the underlying asset. A special case is given when we suppose that the volatility function can be decomposed in the following manner:

$$\sigma(S,t) = \sigma(S)\rho(t)\,,\ (S,t) \in (0,\infty) \times (0,\infty) \tag{134}$$

We sketch here the special case when the volatility is free of time structure, i.e. $\rho(t) = 1$ for all t. The other special case that the volatility depends on the time only which is less complicated is analyzed in [25; 72; 73]. Thus, in our special case the inverse problem is stated as follows:

Identifiability problem in finance

Given $S_*, t_*, r, T \geq t_*$, a nonempty interval $[K_{\min}, K_{\max}]$ of $(0,\infty)$, a continuous function $v : (0,\infty)\backslash[K_{\min}, K_{\max}] \longrightarrow [0,\infty)$ and a continuous function $M_T : [K_{\min}, K_{\max}] \longrightarrow (0,\infty)$.
Prove that there exists a uniquely determined continuous function (volatility function) $\sigma : [0,\infty) \longrightarrow [0,\infty)$ such that

$$U(K,T;S_*,t_*) = M_T(K) \text{ for all } K \in [K_{\min}, K_{\max}],$$
$$\sigma(K) = v(K) \text{ for all } K \in (0,\infty)\backslash[K_{\min}, K_{\max}]\,.$$

Notice that the function M is assumed to be continuous only. In the applied situation when M is assumed to be the result of an observation of the market prices it is important that M is not a function representig exact prices. Then we have to assume that for the exact prices M an approximation M^δ is available only; see Section 2.1.

Theorem 4.1 (Uniqueness of the local volatility). *The volatility function in the identifiability problem in finance is uniquely determined.*

The proof of this theorem was given for the first time in [23]. It is based on a fine study of the fundamental solution of the Black Scholes equation in logarithmic variables combined with the fact that the solution of the Black Scholes equation is an analytic function in the time to maturity. In a more general framework the inverse problem is studied by in [116].

The deviation of the volatility from the constant case may be considered as a perturbation problem. This approach is studied in [26].

4.3. *An integral equation for the unknown local volatility*

We consider again the special case from the last subsection. Two quite different approaches to extract the volatility function from observed option pprices have been proposed in recent years. One is that of Lagnado and Osher ([103]) – we consider that method in the next section – the other approach is that in Bouchouev ([23])

for which we now give a short sketch. This method is method is considered in [37] again and a reduction in complexity is achieved. We follow mainly this treatment. Somehow between these approaches is the investigation via control theory in [106].

Bouchouev ([23]) derives the following integral equation for the volatility function σ :

$$U(\cdot,T;S_*,t_*) = (S_* - \cdot)^+ + I_0(\sigma)(\cdot) + I_1(\sigma)(\cdot) + I_2(\sigma)(\cdot) \tag{135}$$

In the presentation (135) I_0, I_1, I_2 are very complicated integral operators (with singularities in the kernels) which we do not repeat here since the formulae are lengthly. In [37] and [127] proofs of the identity in (135) are given. If we know the values $U(K,T;S_*,t_*), K > 0$, (or their approximations) from an observation of the market we can regard (135) as an integral equation for the volatility σ where S_* is the price of the underlying at the current time t_*. This integral equation can be solved by iteration

$$I_0(\sigma^{k+1}) = U(\cdot,T;S_*,t_*) - (S_* - K)^+ - I_1(\sigma^k) - I_2(\sigma^k)\,,\ k = 0,1\ldots\,, \tag{136}$$

starting with a reasonable approximation σ^0. From the numerical point of view, a huge amount of computational work is necessary to calculate reasonable approximations for σ.

In [37] a reduction of the triple integrals in (135) is achieved by a clever application of the Laplace transform. The main step is to transform I_0 to a functional H of the value $\sigma(K)$ which gives

$$U(\cdot,T;S,t_*) = (S - \cdot)^+ + H(\sigma(\cdot)) + I_1(\sigma) + I_2(\sigma) \tag{137}$$

The iteration for solving (137) becomes

Algorithm 3.5 Computation of the local volatility by an integral equation

---

IN Price S_* of the underlying at time t_*, maturity time T, market prices $U(K,T;S_*,t_*)$ for $K \in [K_{\min}, K_{\max}]$. Initial guess σ^0 for the local volatility. Set $k := 0$.

step 1 Compute the integrals $I_1(\sigma^k), I_2(\sigma^k)$.

step 2 Solve the equation

$$H(\sigma^{k+1}) = U(\cdot,T;S_*,t_*) - (S_* - \cdot)^+ - I_1(\sigma^k)(\cdot) - I_2(\sigma^k)(\cdot)$$

step 3 If the approximation σ^{k+1} is not good enough return to step 2 with $k := k+1$ else set $L := k+1$.

OUT Approximation σ^L for the local volatility.

The initial guess σ^0 can be computed by solving

$$H(\sigma^0) = U(\cdot,T;S_*,t_*) - (S_* - \cdot)^+ .$$

Clearly, some additional considerations are necessary concerning the integration routines, removal of singularities in the integrals and cut off strategies; see [25; 37; 127] for some related details..

4.4. *Computation of the volatility via numerical differentiation*

Now, we come back to a simple exploitation of Dupire's equation. Here we follow mainly [69].

The equation for the local volatility

Given a classical solution of (126),(127),(128) we can solve (126) for the volatility to obtain along

$$\frac{1}{2}\,\sigma(K,T)^2 = \frac{\dfrac{\partial U}{\partial T}(K,T) + rK\dfrac{\partial U}{\partial K}(K,T)}{K^2\dfrac{\partial^2 U}{\partial K^2}(K,T)}\,,\ K \in (0,\infty), T \in (t_*,\infty)\,. \tag{138}$$

the volatility. Despite its simplicity, this approach has severe practical shortcommings which reflects the ill-posedness of the problem. First, financial markets typically allow only a few and prefixed maturity dates, and just a discrete sample of strikes are on sell, too. Second, the Black-Scholes equation is just a model of the real market dynamics, so that (138) is at best an approximate identity.

More on the technical side, in the formula (138) itself some problems are hidden. As we shall see, the denominator in (138) is usually positive when it is evaluated along the model. However, positivity of the nominator may not be an obvious property for real data. Therefore it can easily happen that the computed fractions in (138) change sign, and taking the square root to obtain σ is prohibited. Even if the fraction remains finite and positive the volatility function computed from (138) may exhibit rough oszillations, if the data have not been preprocessed properly; see the examples in [40]. This is not surprising since we have to use more or less implicitly numerical differentiation to evaluate the formula.

The computation of the local volatility via numerical differentiation

We assume that the following market data are available:

- Expiration dates $T_1, \dots, T_N$.
- Traded options with strike prices $K_{i1}, \dots, K_{im}$, for each expiration date T_i .
- Market prices V_{ij} for an option with expiration date T_i and strike price K_{ij} .

To evaluate (138) we need to differentiate these discrete data with respect to the strike variable K and the maturity time T .

Using the interpolation methods in Section 3.2 we may assume that for each maturity time T_i we have a pricing function $u(\cdot, T_i) \in W_2^2[0, K_{\max}]$. Once these functions are determined they can easily be differentiated analytically to compute approximations of $\frac{\partial U}{\partial K}(K,T)$, $\frac{\partial^2 U}{\partial K^2}(K,T)$ wherever needed. Therefore, the assumption that for each i the same number m of option prices V_{ij} are known is no serious restriction.

Differentiation with respect to time is less delicate because of the relatively large time gaps. While this implies that the numerical differentiation cannot be very accurate, it also means that lack of stability is not really an issue here. Therefore we can use simple difference schemes to approximate the time derivatives at the maturity times. More precisely, we may use centered differences at the inner maturities $T_2, \ldots, T_N$, and one-sided differences at the extremal maturities T_1 and T_N. Note that the maturities are not equispaced in general, so that the appropriate central difference quotient is

$$\frac{\partial U}{\partial T}(K_{ij}, T_i)) \approx \frac{1}{h_i + h_{i+1}}\Big(\frac{h_i}{h_{i+1}}u(K_{ij}, T_{i+1}) + (\frac{h_{i+1}}{h_i} - \frac{h_i}{h_{i+1}})u(K_{ij}, T_i) - \frac{h_i}{h_{i+1}}u(K_{ij}, T_{i-1})\Big) \tag{139}$$

with $h_i := T_i - T_{i-1}, i = 2, \ldots, N$.

Let

$$z_{ij} := \sigma(K_{ij}, T_i)^2\,,\ j = 1, \ldots, m_i, i = 1, \ldots, N,$$

the numbers which we want to compute via the formula (138). To use the formula (138) we have to compute the nominator and the denominator. Let

$$b_{ij}, d_{ij} \text{ be these numbers, respectively}\,.$$

We stack all those values in one-dimensional vectors z, d, and $b \in R^N$, $N = (m+1)n$, using a standard lexicographical ordering (with all abscissa for a single maturity in consecutive, increasing, order).The key idea is to consider the resulting expression in (138) as a linear system

$$Dz = b \tag{140}$$

where D is the diagonal matrix whose diagonal coincides with d. Since we cannot exclude tiny entries in the matrix D the linear system (140) may be ill-conditioned in general. Finally, although D is nonnegative by construction the system (140) may not have a nonnegative solution because of possible sign changes in b. For these reasons the linear system (140) needs to be regularized to obtain reasonable approximations of z. We use Tikhonov regularization and consider the problem

Computing the local volatility via numerical differentiation

$$\min_{z \in \mathbb{R}^N} \;\|Dz - b\|_2^2 + \alpha\|Lz\|_2^2 \tag{141}$$

Here, L is a given matrix and $\alpha > 0$ is an appropriate regularization parameter. This minimization problem is equivalent to solving the linear system

$$(D^*D + \alpha L^*L)z = D^*b \tag{142}$$

Notice that $D^* = D$.

For the choice of the regularization parameter α no simple advice can be given. Since we want to solve the equation (140) all recipes for regularizing ill-conditioned systems of linear equations found in the literature may be applied: cross-validation studies ([71]), singular value decomposition ([71]), L-curve analysis ([29; 28]).

Care has to be taken in the choice of L. With L equal to the identity matrix I we cannot guarantee that the solution of the equation (142) is positive. In [69] a choice L different from the identity matrix I is proposed.

5. Calibration/Hilbert space methods

In this section we consider the model of the geometric Brownian motion with a local volatility again. But here we use a modelling in a Hilbert space setting since in the Hilbert space framework calibration methods may benefit from methods considered in the literature of ill-posed problems; see Section 2. Moreover, it is possible to use finite element methods instead of the limited finite differences methods for the computation of solutions. As a consequence, we have to consider weak solutions of the Black Scholes equation and Dupire's equation. At the end, we formulate Kaczmarz's method which is a very powerful method in solving ill-posed problems. Concerning a comprehensive presentation of calibration in finance see [111].

5.1. *Calibration via gradient methods*

In this subsection, we describe a least squares algorithm to calibrate the local volatility by fitting the prices of a set of vanilla european calls available on the market. As we will see later on, for these "simple" derivates a numerically more efficient method is available, namely a least squares method based on Dupire's equation. But nevertheless, for models with a complex structure these methods are not feasible because in this case no replacement for Dupire's equation seems to be possible.

The least squares problem

Suppose market prices of vanilla european calls are given spanning a set of expiration dates $T_1, \dots, T_N$. Assume that for each expiration date T_i options are traded with strike prices $K_{i1}, \dots, K_{im_i}$, $m_i \in \mathbb{N}$. Let

$$c_{ij} := c_{ij}(\sigma) = C(S_*, t_*; K_{ij}, T_i, r, \sigma)\,,\ j = 1, \dots, m_i, i = 1, \dots, N\,, \tag{143}$$

be the prices of the options (based on a model for the dynamics of the underlying with local volatility) $\sigma : [0, \infty) \times [0, T_{\max}] \longrightarrow \mathbb{R}$ where $T_{\max} := \max_{i=1,\dots,N} T_i$.

On the market, the following prices are available

$$v_{ij} \approx c_{ij} \text{ with } V_{ij}^b \le v_{ij} \le V_{ij}^a\,,\ j = 1, \ldots, m_i, i = 1, \ldots, N\,, \tag{144}$$

where V_{ij}^b and V_{ij}^a denote the bid and ask prices, respectively, for an option with expiration date T_i and strike price K_{ij}. v_{ij} may be chosen as the arithmetic mean of the bid and ask prices.

Let Σ denote the space of continuous functions on the domain $[0,\infty) \times [0, T_{\max}]$. The calibration problem to find a volatility function which is in "agreement" with the market prices is transformed into a least squares problem:

$$\text{Minimize } J(\sigma) := \frac{1}{2}\sum_{i=1}^{N}\sum_{j=1}^{m_i} w_{ij}|c_{ij}(\sigma) - v_{ij}|^2 \text{ subject to } \sigma \in \Sigma\,. \tag{145}$$

The nonnegative weights w_{ij} are suitable chosen numbers. These numbers are important in practice because the prices of the options out the money can be very small. A common way to choose these numbers is to relate them to the Vega of the option.

Since the set of data is finite the problem of reconstruction σ from this data set is severely underdetermined. As a rule, there exist several volatility functions σ that match the market option price data. This lack of uniqueness leads to the observation that the optimization problem cannot be solved in a stable way: small perturbations in the price data can result in large changes in the matching function. In general, such an unstable behavior is indicated by oszillations showing up in the matching volatility. To make the problem well-posed we must introduce some type of regularization which surpresses these oszillations. In other words, we have to restrict the set of admissible volatilities to a subset of smooth functions combined with a bound for the derivatives.

Here is a remark concerning gradient methods for minimizing a functional $f : X \longrightarrow \mathbb{R}$ where X is a Hilbert space. A gradient method is the following procedure:

$$x^{k+1} := x^k - \lambda_k \nabla f(x^k)\,,\ k = 0, 1, \ldots, \text{ where } x^0 \text{ is a given initial guess} \tag{146}$$

Here $\nabla f(\cdot)$ is the gradient of the mapping f. For the step size control parameter λ_k a huge amount of proposals can be found in the literature; see for instance [95].

Of course, f has to be differentiable in some sense in order to have a gradient $\nabla f(x)$. The strongest differentiability requirement is that of Frechet-differentiability the "weakest" one is that of subdifferentiability as a lipschitzean mapping; see Subsection 3.2 and [33; 102; 123]. Then, due to the Hilbert space setting, the (generalized) gradient $\nabla f(x)$ is (by the Riesz-Theorem) that element in X with

$$f'(x)(h) = \langle \nabla f(x), h \rangle_X, h \in X\,.$$

In order to apply these comments to the least squares problem above we have to introduce the set Σ of admissibe volatilities as a subset of a Hilbert space; see below.

Regularization of the least squares problem

We change the optimization criterion by adding a regularizing term:

$$\text{Minimize } J(\sigma) + \lambda J_R(\sigma)\,. \tag{147}$$

Now, we have to choose the functional J_R and the parameter λ in an appropriate way. In general, this is done by choosing an appropriate Hilbert space W which contains the admissible volatilities σ. We sketch such a choice in the case when the volatilities are independent of time; see Subsection 4.2.

Again, let $\eta := \eta_\sigma$ denote $\frac{1}{2}\sigma(\cdot)^2$. We consider the vector space

$$W := \left\{ \eta \in H | K \longmapsto (K+1)\frac{\partial \eta}{\partial K} \in H \right\},$$

where $\frac{\partial \eta}{\partial K}$ is the derivative in sense of distributions. Clearly, W is a Hilbert space when it is endowed with the inner product

$$\langle \eta, \rho\rangle_W := \langle \eta, \rho\rangle + \langle (K+1)\frac{\partial \eta}{\partial K}, (K+1)\frac{\partial \rho}{\partial K}\rangle\,,\ \eta, \rho \in W\,.$$

Denote by W' the dual space of W. Then the spaces W, H, W' build a Gelfand triple:

$$W \hookrightarrow H \hookrightarrow W'\,.$$

Additionally, W is continuously imbedded into $L_\infty(\mathbb{R}^+)\cap C(\mathbb{R}_+)$. Now, a candidate for the regularizing functional is given by

$$J_R(\sigma) := \frac{1}{2}\langle \eta_\sigma, \eta_\sigma\rangle_W\,,\ \eta_\sigma \in D\,, \tag{148}$$

where we assume that D is a subset of W satisfying additional conditions in order to allow to apply the existence and uniqueness result 5.1. More details concerning the space W may be found in [1].

Parametrization: a remark

A possible way is to find σ among a family Σ of functions prescribed by a few parameters. Let the local volatility be parametrized in the following way:

$$\Sigma := \{\sigma(a) | a \in A\} \tag{149}$$

where A_{ad} is a set of admissible parameters. With this set of ansatz-functions $\sigma(a)$ the problem of calibration becomes the following optimization problem:

$$\min_{a\in A_{\text{ad}}} \frac{1}{2}\sum_{i=1}^{N}\sum_{j=1}^{m_i} w_{ij}|c_{ij}(\sigma(a)) - v_{ij}|^2\,. \tag{150}$$

The computation of a solution of (150) is possible by applying optimization algorithms, notice however that (150) is not a very structured optimization problem and that each evaluation of the cost function requires N solutions of the model equation; the Black-Scholes formula cannot be used since this formula holds only for constant volatility. Moreover, even when the set A_{ad} of admissible parameters is a subset of a finite dimensional space the associated optimization is ill-conditioned and again regularization is necessary when A_{ad} is not bounded.

The gradient-type method of Lagnado and Osher

In [103] a description of a gradient-type method for the calibration of a model with local volatility. This procedure does not use Dupire's equation. Thus the idea may be applied in a very general setting. We present this method in a slightly changed version.

Let us consider the least squares-functional $J(\sigma)$ used in the last subsection. Regularization is done by the requirement of smoothness: $\nabla\sigma \in X := L_2((0,\infty)\times[0,T_{\max}])$. Consequently

$$J_R(\sigma) := \frac{1}{2}\|\nabla\sigma\|^2 \tag{151}$$

The gradient of the regularizing functional is easy to compute, since it is just an inner product.

To distinguish the informal computation from the rigorous one we denote the linearizations by δ. A formal computation shows

$$\delta J(\sigma) = \sum_{i=1}^{N}\sum_{j=1}^{m_i} w_{ij}(c_{ij}(\sigma) - v_{ij})\delta c_{ij}(\sigma)$$

where $d_{ij} := \delta c_{ij}$ is a solution of the linearized Black Scholes equation:

$$\frac{\partial d_{ij}}{\partial t} + \frac{1}{2}\sigma(S,t)^2 S^2 \frac{\partial^2 d_{ij}}{\partial S^2} + r d_{ij}\frac{\partial d_{ij}}{\partial S} - r d_{ij} = -\delta\left(\frac{1}{2}\sigma(S,t)^2\right)S^2\frac{\partial^2 c_{ij}}{\partial S^2}\,,$$
$$S > 0, t \in (0,T)\,, \tag{152}$$
$$d_{ij}(0,t) = 0\,,\ \lim_{S\to\infty} d_{ij}(S,t) = 0\,,\ t \in (0,T)\,, \tag{153}$$
$$d_{ij}(S,T) = 0\,,\ S > 0\,. \tag{154}$$

As a consequence, to compute the gradient of the least squares-functional is expensive. Therefore the gradient is replaced by a variation defined by simple perturbations of the actual volatility.

$$v_{ij} := \delta c_{ij}(\sigma) := \lim_{\varepsilon\to 0}(c_{ij}(\sigma+\varepsilon h) - c_{ij}(\sigma)) \tag{155}$$

where $h = h_{\xi,\tau}$ is a peak variation:

$$h_{\xi,\tau} := \delta_{\xi,\tau} := \delta_\xi\delta_\tau$$

$\delta_{\xi,\tau}$ is a product of Dirac distributions concerning the points $\xi \in (0,\infty), \tau \in (0,T_{\max})$. Using this peak variation, we obtain that v_{ij} is a solution of the following equation:

$$\frac{\partial v_{ij}}{\partial t} + \frac{1}{2}\sigma(S,t)S^2\frac{\partial^2 v_{ij}}{\partial S^2} + rS\frac{\partial v_{ij}}{\partial S} - rv_{ij} = -\delta_{\xi,\tau}\sigma(S,t)S^2\frac{\partial^2 c_{ij}}{\partial S^2}\,,$$
$$S > 0, t \in (0,T_i)\,, \tag{156}$$
$$v_{ij}(0,t) = 0\,,\ \lim_{S\to\infty} v_{ij}(S,t) = 0\,,\ t \in (0,T_i)\,, \tag{157}$$
$$v_{ij}(S,T_i) = 0\,,\ S > 0\,. \tag{158}$$

For the numerical realization of the method we have to introduce the following initialization:

- The domain $(0,\infty)$ has to truncated, lets say to an interval $(0, S_{\max})$ with $S_{\max}$ large enough; a reasonable proposal is $S_{\max} := 2S_*$.
- We have to choose boundary values in $S = S_{\max}$.
- To apply difference schemes one has to choose a grid $(S_m, t_n), m = 0,\dots,M_S, n = 0,\dots N_t$ in $[0, S_{\max}] \times [0, T_{\max}]$.
- An initial guess for σ in the grid points $(S_m, t_n), m = 0,\dots,M_S, n = 0,\dots N_t$, has to be chosen.
- The regularizing functional has to be replaced by a finite difference approximation using the grid points $(S_m, t_n), m = 0,\dots,M_S, n = 0,\dots N_t$.

Now, the iteration method results in Algorithm 5.1.

Algorithm 3.6 Calibration by the Lagnado-Osher method

IN Given market prices v_{ij}, grid points (S_m, t_n), a discret initial guess σ^0, a regularization parameter λ and a tolerance parameter ε.

step 0 Set $k := 0$.

step 1 Solve the Black Scholes boundary value problems (7),(8),(9) by a finite difference method on the grid $(S_m, t_n), m = 0,\dots,M_S, n = 0,\dots N_t$, to obtain c_{ij}.

step 2 Compute the functional J. If $J(\sigma^k) \le \varepsilon$ then set $L := k$ and go to **OUT**.

step 3 Solve the boundary value problems (156),(157),(158) to obtain v_{ij}

step 4 Evaluate the gradient of the regularizing part of the functional: $g_{ij}, m = 0,\dots,M_S, n = 0,\dots N_t$.

step 5 Iterate as follows:

$$\sigma_{ij}^{k+1} := \sigma_{ij}^k - v_{ij} - \lambda g_{ij}\,,\ m = 0,\dots,M_S, n = 0,\dots N_t, \tag{159}$$

Return with $k := k+1$ to **step 1**.

OUT Discrete approximation σ^L.

The iterative procedure is clearly computational very expensive since a large number of partial differential equations have to be solved. By analyzing the steps, the work can be reduced by exploiting the fact that each solution is done on the same grid with the same differential operator; see [103] for more hints concerning the numerical realization.

In [38] the method of Lagnado and Osher is presented again and an alternative regularization strategy is proposed. Moreover, the resulting method is embedded in the context of Tikhonov regularization. But again, a rigorous analysis of the main steps of the method is not presented. Further extensions of the approach of Lagnado and Osher has been considered in [19; 20; 41; 44; 81].

Here are some additional references ([3; 7; 39; 46; 72; 74]) concerning approaches that are suggested to calibrate the Black Scholes models.

5.2. *Black-Scholes equation: weak solutions*

We consider the boundary value problem (7), (8), (9) now in a variational context. Two problems have to be overcome: first, the equation is not uniformly parabolic, second, the domain "$S \in (0,\infty)$" is not compact. Both problems may be resolved by appropriate function spaces. We follow here mainly the monograph [1].

Appropriate function spaces

As we know, the Black-Scholes-equation governs the model for a put and a call option, the difference is contained in the final condition describing the payoff function. Since the payoff function for the put option has a compact support this case is easier to handle. Later on we discuss then the case of a call option.

The boundary value problem for a put option (with local volatility) is given as follows:

$$\frac{\partial P}{\partial t} + \frac{1}{2}\sigma(S,t)^2 S^2 \frac{\partial^2 P}{\partial S^2} + rS\frac{\partial P}{\partial S} - rP = 0 \quad S \in (0,\infty), t \in (0,T), \tag{160}$$

$$P(0,t) = Ke^{-r(T-t)}, \lim_{S\to\infty} P(S,t) = 0, \, t \in (0,T), \tag{161}$$

$$P(S,T) = (K-S)^+, \, S > 0. \tag{162}$$

Given r, K, T and σ we want to find a sufficiently smooth function $P : (0,\infty) \times (0,T) \longrightarrow \mathbb{R}$, which can be considered as a weak solution of (160), (161), (162).

Let us denote with $H := L_2(\mathbb{R}_+) := L_2(0,\infty)$ the vector space of square-Lebesgue-integrable functions on the domain $[0,\infty)$. H is a Hilbert space endowed with the inner product

$$\langle f, g\rangle := \int_0^\infty f(x)g(x)dx$$

and the associated norm $\|f\| := \langle f, f\rangle^{\frac{1}{2}}, f \in H$. With H we may define a Sobolev-like space of first order by*

$$W := H_0^1(\mathbb{R}_+) :=$$
$$\{f : [0,\infty) \longrightarrow \mathbb{R} \text{ is continuous} | f(0) = 0, f \text{ is absolutely continuous } f' \in H\}.$$

W is a Hilbert space endowed with the inner product

$$\langle f, g\rangle_W := \langle f', g'\rangle \, , \; f, g \in W \, ,$$

and the associated norm $\|f\|_W := \langle f', f'\rangle^{\frac{1}{2}}, f \in W$. We set

$$V := \left\{ v \in H | v : x \longmapsto \frac{1}{x} w(x) \text{ with } w \in W \right\} .$$

and endow V with the inner product†

$$\langle v, u\rangle_V := \langle v, u\rangle + \langle xv, xu\rangle \, , \; v, u \in V \, .$$

We collect some properties of the spaces V and W; for the proof see [1]. V is a Hilbert space and the space $\mathcal{D} := \mathcal{D}_\infty(\mathbb{R}_+)$ of functions with infinite many derivatives with compact support is a dense subspace of W and V. Therefore the space $\mathcal{D}$ may be used as a space of test functions in V . This is important in introducing the weak form of equation (160).

Let us denote by V' the space of continuous functionals on V . Then the triple V, H, V' may be considered as a so called *Gelfand triple*:

$$V \hookrightarrow H \hookrightarrow V'$$

Actually, V' is a Hilbert space where the restriction of the inner product to $H \times V$ is given by the inner product on H . Hence

$$\|\lambda\|_{V'} = \sup \left\{ \frac{\langle \lambda, v\rangle}{\|\lambda\|\|v\|} | v \in V \right\} \, , \; \lambda \in H \, .$$

In the space V we have found the appropriate space for the weak solutions of the Black Scholes equation. The requirement $x \longmapsto xv'(x) \in H$, which is contained in the definition of V takes into account the fact that the term $\frac{1}{2}\sigma(S,t)^2 S^2 \frac{\partial^2 P}{\partial S^2}$ in the equation contains a factor S^2 .

*A continuous function $f : [0,\infty) \longrightarrow \mathbb{R}$ is absolutely continuous if there exist a constant c and an integrable function g such that $f(x) = c + \int_0^x g(t)dt, x \geq 0$.

†Here we use the following abbreviation: xu is the function $x \longmapsto xu(x)$

Weak solutions

We set $\eta := \eta_\sigma, \eta_\sigma(S,t) := \frac{1}{2}\sigma(S,t)^2$. As usual, the weak formulation of the equation is obtained by testing the equation (160) with a function $\phi \in \mathcal{D}$:

$$\begin{aligned}0 = &\int_0^\infty \frac{\partial P}{\partial t}(S,t)\phi(S)dS \\ &- \int_0^\infty \frac{\partial P}{\partial S}(S,t)\Big(\frac{\partial \eta}{\partial S}(S,t)S^2\phi(S) + 2\eta(S,t)S\phi(S) \\ &\qquad\qquad + \eta(S,t)S^2\frac{\partial \phi}{\partial S}(S,t)\Big)dS \\ &+ r\int_0^\infty S\frac{\partial P}{\partial S}(S,t)\phi(S)dS - r\int_0^\infty P(S,t)\phi(S)dS.\end{aligned}$$

This leads to the bilinear form

$$\begin{aligned}a(t;v,\phi;\sigma) := &- \int_0^\infty \Big(\eta_\sigma(S,t)S^2\frac{\partial v}{\partial S}(S)\frac{\partial \phi}{\partial S}(S) - rP(S,t)\phi(S)\Big)\,dS \\ &+ \int_0^\infty \Big(r + 2\eta_\sigma(S,t) + S\frac{\partial \eta_\sigma}{\partial S}(S,t)\Big)\,S\frac{\partial v}{\partial S}(S)\phi(S)ds\,.\end{aligned}$$

Finally, we obtain the weak formulation of the boundary value problem (160), (161), (162):

$$\Big\langle \frac{\partial P}{\partial t}(t), \phi \Big\rangle + a(t;P(t),\phi;\sigma) = 0 \text{ for all } \phi \in V\,. \tag{163}$$

Hereby we used the "evolutionary notation" $P(t) := P(\cdot, t)$.

Definition 5.1. A mapping $P : \mathbb{R}_+ \longrightarrow V$ with $P(T) = (\cdot - K)^+$ is called a **weak solution** of (160), (161), (162) if we have

$$P \in L_2(0,T;V) \cap C(0,T;H), \; \frac{dP}{dt} \in L_2((0,T);V')$$

and if the identity (163) holds for allmost all $t \in (0,T)$.*

For the existence of a solution in the sense of Definition 5.1 we need some assumptions concerning $\eta_\sigma := \frac{1}{2}\sigma(\cdot,\cdot)^2$.

Assumptions 5.2.

(1) There exist numbers $0 < \eta_m \leq \eta_M$ such that

$$\eta_m \leq \eta_\sigma(x,t) \leq \eta_M \text{ for all } t \in [0,T] \text{ and all } x \in \mathbb{R}_+\,. \tag{164}$$

(2) There exists a constant $0 < c$ such that

$$\Big|x\frac{\partial \eta_\sigma}{\partial x}(x,t)\Big| \leq c \text{ for all } t \in [0,T] \text{ and all } x \in \mathbb{R}_+\,. \tag{165}$$

*Here is an illustration of the notation: $P \in L_2(0,T;V)$ means $P(t) \in V$ for allmost all $t \in (0,T)$, $\int_0^T \|P(t)\|_V^2 < \infty$

Under these assumptions we can see that the bilinear form a is continuous (see [1]):

$$|a(t;v,\phi;\sigma)| \leq c_1\|v\|_V\|\phi\|_V \text{ for all } v,\phi \in V, t\in(0,T)\,. \tag{166}$$

In order to apply known results to prove the existence of a weak solution we need a version of *Garding's inequality*:

$$a(t;v,v;\eta)| \geq c_2\|v\|_V^2 - \lambda\|v\|^2 \text{ for all } v,\phi \in V, t\in(0,T)\,. \tag{167}$$

Here $c_1, \lambda \geq 0$ and $c_2 > 0$. For the proof of the inequality (167) see [1].

Theorem 5.1. *Under the assumptions* (164), (164) *we have a uniquely determined weak solution* P *of* (160),(161),(162)*:*

(a) $P \in L_2(0,T;V)\cap C(0,T;H)\,,\ \frac{dP}{dt} \in L_2(0,T;V')\,.$

(b) $\langle \frac{dP}{dt}(t),\phi\rangle + a(t;P(t),\phi;\eta) \;=\; 0\,,\ t\in(0,T)$ *for all* $\phi\in V\,.$

Moreover

$$e^{-2\lambda t}\|P(t)\|^2 + \eta_m\int_0^\infty e^{-2\lambda s}\|P(s)\|_V^2 ds \leq \|P_0\|^2\,,\ t\in(0,T)\,. \tag{168}$$

Proof:
For the proof of $(a),(b)$ see [128], Theorem 26.1. In the proof there we find also the estimate (168). It is more or less a consequence of Gardings inequality. ■

Now, we have an existence result for the put option. In the case of a call option we observe, that the price C_T in the final condition

$$C_T(S) := C(S,T) = (S-K)^+\,,\ S\in(0,\infty)\,,$$

is not a $L_2(\mathbb{R}_+)$–function. In order to overcome this difficulty we use the *Put-Call-Parity*. As we know, under the assumptions concerning the financial market we have

$$C(S,t) := P(S,t) + S - Ke^{-r(t-T)}\,,\ s\in(0,\infty), t\in(0,T)\,. \tag{169}$$

Since the mapping $(S,t) \longmapsto S - Ke^{-r(t-T)}$ is a (classical) solution of (160) we have in (169) a weak solution of (160). Obviously, the boundary condition in $S=0$ is satisfied. The boundary condition in $S=\infty$ is part of the definition of a weak solution P. The final condition in $t=T$ may be extracted from

$$C(S,T) = P(S,T) + S - K = (K-S)^+ + (S-K) = (S-K)^+\,.$$

As a consequence, under the assumption 5.2 we have in (169) a weak solution of the Black Scholes boundary value problem.

In [1] we find a discussion of the stability problem (dependence of the weak solution on the local volatility) in the case when σ is not time dependent.

5.3. *Weak solutions of Dupire's equation*

In this subsection we consider the calibration problem via least sqares using Dupire's equation. Since we want to use a Hilbert space setting we have to introduce weak solution of Dupire's equation. We do this in the case of an approximation of the model: we cut the interval $[0,\infty)$ for each of the variables T and K.

Weak solutions

We consider initial boundary value problem (126), (127), (128) (Dupire's system). We choose $\overline{T} > 0$ and $\overline{K} > 0$. The boundary values in Dupire's system are replaced by

$$U(T,K) = 0\,,\; U(T,\overline{K}) = 0\,,\; T \in (0,\overline{T})\,. \tag{170}$$

Let us fix $\sigma : [0,\overline{K}] \times [0,\overline{T}] \longrightarrow (0,\infty)$ which satisfies assumption 5.2. We take a smooth test function ϕ vaishing in $K = 0$ and $K = \overline{K}$, multiply the equation (126) by ϕ and integrate from $K = 0$ up to $K = \overline{K}$. This gives after a partial integration for $T \in (0,\overline{T})$

$$\begin{aligned}
&-\int_0^{\overline{K}} U(K,T)\frac{\partial\phi}{\partial T}(K)dK \\
&= \int_0^{\overline{K}} \Big(\frac{\partial U}{\partial K}(K)(\frac{\partial\sigma}{\partial K}\sigma(K,T)K^2 + \sigma(K,T)^2K)\phi(K,T) \\
&\qquad\qquad + \frac{1}{2}\sigma(K,T)^2K^2\frac{\partial\phi}{\partial K}\Big)dK \\
&+ \int_0^{\overline{K}} U(K,T)rK\frac{\partial\phi}{\partial K}dK.
\end{aligned}$$

With the bilinear form

$$\begin{aligned}
b(T;K,\phi;\sigma) := -\int_0^{\overline{K}} &\Big(\frac{\partial U}{\partial K}(K)(\frac{\partial\sigma}{\partial K}\sigma(K,T)K^2 + \sigma(K,T)^2K)\phi(K,T) \\
&\qquad + \frac{1}{2}\sigma(K,T)^2K^2\frac{\partial\phi}{\partial K}\Big)dK \\
&- \int_0^{\overline{K}} U(K,T)rK\frac{\partial\phi}{\partial K}dK
\end{aligned}$$

we obtain the following variational formulation of Dupire's equation

$$\Big\langle \frac{\partial U}{\partial T}(T),\phi \Big\rangle + b(T;U(T),\phi;\sigma) = 0 \text{ for all test functions } \phi \in V\,,\; T \in (0,\overline{T}). \tag{171}$$

The Hilbert space setting to make the deriavation rigorous is essentially the same as for the Black Scholes system. We use the Gelfand triple $\overline{V} \hookrightarrow H \hookrightarrow \overline{V}'$ with

$$\overline{H} := L_2[0,\overline{K}]\,,\; \overline{V} := \{v \in L_2(0,\overline{K}) | x\frac{dv}{dx} \in L_2(0,\overline{K}), v(0) = v(\overline{K}) = 0\}\,.$$

Definition 5.3. A mapping $U : (0,\overline{T}) \longrightarrow \overline{V}$ with $U(0) = (S_* - \cdot)^+$ is called a **weak solution** of (126), (127), (128) if we have

$$U \in L_2(0,\overline{T};\overline{V}) \cap C(0,\overline{T};\overline{H}), \frac{\partial U}{\partial T} \in L_2(0,\overline{T};\overline{V}')$$

and if the identity (171) holds for almost all $T \in (0,\overline{T})$

Now, one can prove that under the assumptions 5.2 the system (126), (127), (128) posseses a uniquely determined weak solution. As a consequence, the "parameter-to-solution" mapping is well defined if we choose the domain of definition according to the assumptions 5.2.

Calibration via Dupires equation

We want to apply this method for the calibrating functional $j := J + J_R$. We consider the computation of the functional J in (145) only, the gradient of the regularizing term J_R is easy to compute when it is a a norm in a Hilbert space; see above. For the computation of the gradient J in a volatility σ one has to linearize Dupire's equation with respect to σ. We do that again for a "volatility" only which depends not on time. We fix the volatility σ.

Formally, linearization is no difficult task, the problem consists in the verification of the procedure. Again, to distinguish the formal computation from the rigorous one we denote the linearizations by δ. An informal computation shows

$$\delta J(\sigma) = \sum_{i=1}^{m}\sum_{j=1}^{m_i} w_{ij}(c_{ij}(\sigma) - v_{ij})\delta w(T_i, K_j)$$

where δw is a solution of the linearized Dupire equation:

$$\frac{\partial w}{\partial T}(T) + b(T; w, \cdot; \sigma) = -\frac{\partial b}{\partial \sigma}(T; U, \cdot; \sigma)\delta\sigma \, , \; w(0) \; = \; 0 \tag{172}$$

As it is known from control theory, an adjoint equation is now very helpful to compute the gradient of the optimization criterion. Let $\tilde{b}$ be the bilinear form such that

$$b(T; u, k; \sigma) = \tilde{b}(T; k, u; \sigma) \text{ for all } u, k\,.$$

Then we define the adjoint state p by solving

$$-\frac{\partial p}{\partial T}(T) + \tilde{b}(T; p, \cdot; \sigma) = \sum_{i=1}^{m}\sum_{j=1}^{m_i} w_{ij}(c_{ij}(\sigma) - v_{ij})\delta_{T_i,K_j} \, , \; p(\overline{T}) = 0\,. \tag{173}$$

With this adjoint state the gradient ∇J is given by

$$\nabla J(\sigma) = -\frac{\partial b}{\partial \sigma}(T; U, p; \sigma) \tag{174}$$

Now, we may formulate the gradient method:

Algorithm 3.7 Calibration using Dupire's equation

IN Given market prices v_{ij}, a discret initial guess σ^0, a regularization parameter λ and a tolerance parameter ε.

step 0 Set $k := 0$.

step 1 Compute a weak solution of Dupire's equation using σ^k to obtain $c_{ij} = U(T_i, K_j)$, $j = 1, \ldots, m_i, i = 1, \ldots, N$..

step 2 Compute the functional J. If $J(\sigma^k) \leq \varepsilon$ then set $L := k$ and go to **OUT**.

step 3 Compute a weak solution of the adjoint equation (173) to obtain the adjoint state p.

step 4 Compute the gradient $\nabla J(\sigma^k)$ using (174).

step 5 Evaluate the gradient g of the regularizing part of the functional

step 6 Iterate as follows:

$$\sigma^{k+1} := \sigma^k - \nabla J(\sigma^k) - \lambda g. \tag{175}$$

Return with $k := k + 1$ to **step 1**.

OUT Discrete approximation σ^L.

This algorithm was established in [1] and numerical experiments can be found there.

5.4. *Estimation of volatilities via Kaczmarz method*

This section is devoted to a specific case of an iteration method, namely the so called *algebraic reconstruction technique (ART)* and its generalization to the calibration of options. The ART-algorithm has been developed by Kaczmarz 1936 as method to solve a system of linear equations; see [87] for a reprint. The method was refounded as a method to solve linear systems resulting in computer tomography and in image reconstruction; see [110]. We shall apply the nonlinear realization to the calibration problem in finance. Again, we consider the calibration of the model with a volatility which does not depend on time.

Introduction

Suppose for each expiration time $T_1, \ldots, T_N$ a certain number of strikes is available, and we shall assume that all strike prices are contained in the interval $[K_{\min}, K_{\max}]$ with $K_{\max} \leq \overline{K}, T_i \leq T_{\max},$; see the last subsection. Again, we want to find σ given

$$U(t_*, S_*, K, T_i) = U_i(K)\,,\ K \in [K_{\min}, K_{\max}]\,,\ i = 1, \ldots, N\,.$$

Here t_* is the actual time and S_* is the corresponding value of the underlying. U_j denotes the market price of options given by the Black Scholes model with volatility σ. Since financial markets typically allow only a discrete sample of strikes for each maturity an interpolation procedure has to be applied. After all, the market prices

M_j are known up to some noise level ε only:

$$\|U_i - M_i\|_{\overline{H}} \leq \varepsilon \ , \ i = 1, \ldots, N \tag{176}$$

Thus, the inverse problem can now be formulated as follows: Calibrate the volatility $\sigma = \sigma(K)$ in the system of nonlinear operator equations

$$G_i(\sigma) = M_i, i = 1, \ldots, N, \tag{177}$$

where G_i : again the parameter-to-solution map. Notice that this nonlinear system is inconsistent in general.

Before we continue to analyze the problem we sketch the idea of solving systems of equations by the idea of Kaczmarz

The ART-algorithm

Suppose that a linear system of equations is given by

$$F_i(x) := \langle a^i, x\rangle_2 = y_i, \quad 1 \leq i \leq m, \tag{178}$$

where $a^1, \ldots, a^N \in \mathbb{R}^n$ and $y_1, \ldots, y_N \in \mathbb{R}$; $\langle\cdot,\cdot\rangle_2$ denotes the euclidean inner product in $\mathbb{R}^n$. In practice, the pair (a^i, y_i) may be considered as a representation of an experiment:

y_i is the result of the experiment,
a^i describes the "geometry" of the experiment,
x is the unknown quantity to be estimated from the experiment.

Let $J(x) := \sum_{i=1}^{N} |\langle a^i, x\rangle_2 - y_i|^2, x \in \mathbb{R}^n$. The procedure that we use to find x^* with

$$\alpha := J(x^*) = \min\{J(x) | x \in \mathbb{R}^n\} \tag{179}$$

is iterative and of adaptive type: in any step, an estimation of x at the next iteration step is constructed from that at the preceding step and from a new "observation" given by a pair (a^i, y_i). Adaption is done by moving a certain (small) step in the direction opposite to the current gradient of the objective function J using the i-th defect of the observation only (partial gradient step). This procedure leads to the following form of an iteration:

$$x^{k+1} := x^k - \lambda_k(\langle a^k, x^k\rangle_2 - y_k)a^k, k \in \mathbb{N}_0 \, . \tag{180}$$

Here $(\lambda_k)_{k\in\mathbb{N}}$ is a sequence of relaxation parameters and the data (a^i, y_i) are used in a cyclic order:

$$a^k = a^j, y_k = y_j, \text{ if } j = k \mod m.$$

As usual in Kaczmarz-type algorithms, a group of N subsequent steps (starting at some multiple k of N) shall be called a *cycle*. In general, one uses cyclic relaxation, that is λ_k is constant during a cycle. Then one has to distinguish two cases:

The consistent case Here the linear system has a solution and therefore $\alpha = 0$. Then the sequence of the cyclic relaxation parameters λ_k has to satisfy $0 \le \lambda_k \le 2, \sum_{k=0}^{\infty} \lambda_k(2-\lambda_k) = \infty$, in order to obtain $A^\dagger y = \lim_k x_{kN}$.

The inconsistent case Here the linear system has no solution (due to errors in the data or incomplete modelling) and therefore $\alpha > 0$. Then the sequence of the cyclic relaxation parameters λ_k has to satisfy $0 \le \lambda_k, \sum_{k=0}^{\infty} \lambda_k = \infty, \sum_{k=0}^{\infty} \lambda_k^2 < \infty$. In order to obtain $A^\dagger y = \lim_k x_{kN}$.

For the proof see [12].

The attractivity of the ART-algorithm comes from the following facts:

- the iteration step is easy to implement;
- no extra effort is necessary to add data (a^i, y_i).

The shortcomings of this type of computation scheme are

- that the convergence is slow in general;
- that it is difficult to implement a stopping rule for the iteration, especially when the data are corrupted by noise (leading to the inconsistent case).

A different interpretation of the method above results from the geometric idea of a projection. Each row of the linear system (178) introduces an affine space:

$$V_i := \{x \in \mathbb{R}^n | \langle a^i, x\rangle_2 = y_i\}\,,\ i = 1, \dots, N\,.$$

The projections onto these affine spaces are given as

$$P_i : \mathbb{R}^n \ni x \longmapsto x - (\langle a^i, x\rangle_2 - y_i)a^i\,,\ i = 1, \dots, N\,. \tag{181}$$

Then the method above may be considered as a cyclic projection of the current approximations x^k on the affine spaces. Alternatively, the method may be interpreted as the computation of the common fixed point of $P_1, \cdots, P_N$. Each projection is a nonexpansive mapping but never a contraction. On the other side, the composite mapping $P_N \cdots P_1$ – one cycle in the ART-algorithm – has attractive properties; see for instance [9; 18]. In order to use spectral considerations we should apply the method with the operator $P_1 \cdots P_{N-1} P_N P_N \cdots P_1$ – one cycle forward and one cycle backwards –; see [16]. There is huge amount of literature concerning the fixed point theory of nonexpansive operators, especially in the infinite dimensional setting; see for instance [16; 17; 62].

The Landweber-Kaczmarz method

Here we present a realization of the Kaczmarz-algorithm in the nonlinear case. For recent analysis of Kaczmarz type methods for systems of ill-posed equations, we refer the reader to [13; 14; 27; 66; 65; 104; 90; 67].

The starting point of the approach is the Landweber method which computes the solution of a linear equation $Ax = y$ by the following iteration

$$x^{k+1} := x^k - A^*(Ax^k - y)\,,\ k = 0, 1, \ldots\,,$$

which is actually a fixed point iteration for the mapping $x \longmapsto x - A^*(Ax - y)$.

The problem we are interested in consists of determining an unknown quantity $x \in X$ from the set of data $y := (y_1, \ldots, y_N) \in Y^N$, where X, Y are Hilbert spaces and $N \geq 1$. In practical situations, we do not know the data exactly. Instead, we have approximate measured data $y_i^\delta \in Y, i = 1, \ldots, N$, only, satisfying

$$\|y_i^\delta - y_i\| \leq \delta\,, \quad i = 1, \ldots, N\,, \tag{182}$$

with $\delta \geq 0$ (noise level). The finite set of data is obtained by indirect measurements in a process being described by the model

$$F_i(x) = y_i\,, \quad i = 0, \ldots, N-1\,, \tag{183}$$

where $F_i : D_i \subset X \to Y$ are mappings with corresponding domains of definition D_i. In a combination with the idea of Kaczmarz and using linearization we obtain the following iteration method:

$$x^{k+1} \;=\; x^k - F'_{[k]}(x^k)^*(F_{[k]}(x^k) - y^\delta_{[k]})\,. \tag{184}$$

Here $[k] := k \mod N + 1 \in 1, \ldots, N$, and $x^0 \in X$ is an initial guess, possibly incorporating some *a priori* knowledge about the exact solution. Standard methods for the solution of system (183) use iterative type methods after rewriting (183) as a single equation $F(x) = y$, where

$$F := (F_1, \ldots, F_N) : \bigcap\nolimits_{i=1}^{N} D_i \to Y^N \tag{185}$$

and $y := (y_1, \ldots, y_N)$. However these methods become inefficient if N is large or the evaluations of $F_i(x)$ and $F'_i(x)^*$ are expensive. In such a situation, Kaczmarz type methods which cyclically consider each equation in (183) separately are much faster and are often the method of choice in practice.

The *Landweber-Kaczmarz-iteration* with a "bang-bang" stepsize control consists in

$$x^{k+1} := x^k - \omega_k F'_{[k]}(x^k)^*(F_{[k]}(x^k) - y^\delta_{[k]})\,,\ k = 0, 1, \ldots, \tag{186}$$

where

$$\omega_k := \begin{cases} 1 & \text{if } \|F_{[k]}(x^k) - y^\delta_{[k]}\| > \tau\varepsilon \\ 0 & \text{otherwise} \end{cases} \tag{187}$$

The iteration is stopped when $\omega_k = 0$ during a cycle. This stopping rule is called the *loping-Kaczmarz*-stopping rule. In order to prove convergence results of the method one has to introduce some assumptions:

(A1) The operators F_i are weakly sequentially continuously and Fréchet differentiable and the corresponding domains of definition D_i are weakly closed. Moreover, there exist $x^0 \in X$, $M > 0$, and $\rho > 0$ such that

$$\|F_i'(x)\| \leq M, \quad x \in B_\rho(x^0) \subset \bigcap_{i=0}^{N-1} D_i . \tag{188}$$

where $B_\rho(x^0)$ denotes the ball with radius ρ around x^0 in the space X.

(A2) The *local tangential cone condition*

$$\|F_i(x) - F_i(\bar{x}) - F_i'(\bar{x})(x - \bar{x})\|_Y \leq \eta \|F_i(x) - F_i(\bar{x})\|_Y , \qquad x, \bar{x} \in B_\rho(x_0) \tag{189}$$

holds for some $\eta < 1$.

(A3) There exists and element $x^* \in B_{\rho/4}(x_0)$ such that $F(x^*) = y$, where $y = (y_0, \ldots, y_{N-1})$ are the exact data satisfying (182).

Now, when the parameters α and η are chosen in an appropriate way, the method is a regularizing iteration method. For details see Subsection 2 and [27; 104] and [90].

Recently, Kaczmarz-type methods for systems of ill-posed equations were analyzed. We refer the reader to [65; 66; 104]. The main problem in the application of the method to specific ill-posed problems consits in the verification of the so called *cone-condition* (A1); see [90]. This is an uniform assumption on the nonlinearity of the operators F_i. It is helpful to compare the error in the linearization with the defect in the equation.

The Kaczmarz-method for the computation of local volatilities

Now, we come back to the calibration problem. In [30] it is shown that all the assumptions (A1), (A2) and (A3) can be verified considering Dupire's equation in logarithmic variables. A cycle of the Kaczmarz method consists in N gradient steps. The optimization criteria are

$$J_i(\cdot) := \frac{1}{2}\|c_i - M_i\|_{\overline{H}^2} , \; i = 1, \ldots, m .$$

We formulate the method without regularization term. Regularization is done by stopping the iteration with the bang-bang stopping criterion.

In order to realize the method we use the same ideas as in the last subsection, especially the computation of the adjoint of the derivative via an adjoint equation may be adapted.

$$-\frac{\partial p}{\partial T}(T) + \tilde{b}(T; p, \cdot; \sigma) = f , \; p(\overline{T}) = 0 . \tag{190}$$

A few experiments in [30] show that the method works very well. Notice that no extra effort is necessary to add data T_i.

Recent trends and developments are iterative regularization methods in Banachspaces; see [84; 91]. Up to now, these methods are not applied to inverse problems in finance.

Algorithm 3.8 Calibration via Kaczmarz method

IN Given market prices M_i, an initial guess σ^0, a tolerance parameter ε.

step 0 Set $k := 0$.

step 1 Compute a weak solution of Dupire's equation to obtain $c_i = U_i = U(T_i, \cdot)$, $i = 1, \ldots, N$..

step 2 For i=1...,N do

- Compute $J_i(\sigma)$.
- If $J_i(\sigma^k) \leq \varepsilon$ set $\omega_i := 0$ else $\omega_i := 1$.
- Compute the adjoint state $p_i := P(T_i, \cdot)$ using (190) with right hand side $f = U_i - M_i$.
- Iterate as follows:

$$\sigma^{kN+l} := \sigma^{kN+(l-1)} - \omega_i p_i \,. \tag{191}$$

step 3 If $\omega_1 = \cdots \omega_N = 0$ set $L := kN$ and go to **OUT** else return with $k := k+1$ to **step 2**.

OUT Discrete approximation σ^L.

Bibliography

[1] Achdou, Y. and Pironneau, O.: *Computational methods for option pricing.* SIAM, Philadelphia, 2005.

[2] Amster, P. de Napoli and Zubelli, J.P: Towards a generalization of Dupire's equation for several assets. *J. Math. Anal.*, 355:170–179, 2009.

[3] Andersen, L.B.G. and Brotherton-Ratcliffe, R.: The equity option volatility smile: an implicit finite-difference approach. *Journal Of Computational Finance*, 1:5–37, 1998.

[4] Anderssen, B. and Bloomfield, P.: Numerical differentiation procedures for non-exact data. *Numer. Math.*, 22:157–182, 1973.

[5] Anderssen, B. and De Hoog, F.: Finite difference methods for the numerical differentiation of non-exact data. *Computing*, 33:259–267, 1984.

[6] Atkinson, K. and Han, W.: *Theoretical Numerical Analysis. A Functional Analysis Framework.* Texts in applied mathematics, vol. 39. Springer, New York, 2001.

[7] Avellaneda, M., Friedman, C., Holemes, R. and Samperi, D.: Calibrating volatility surfaces via relative entropy minimization problems. *Applied Mathematical Finance*, 4:37–64, 1997.

[8] Bachelier, L.: Théorie de la spéculation. *Annales Scientifiques de l'Ecole Normale Superiieure*, III-17:21–86, 1900.

[9] Badea, C., Grivaux, S. and Müller, V.: A generalization of the friedrichs angle and the method of alternating projections. *C. R. Acad. Sci. Paris, Ser. I*, 348:53–56, 2010.

[10] Bakushinsky, A.B. and Goncharsky, A.: *Ill-posed problems: theory and applications.* Kluwer, Dordrecht, 1994.

[11] Banks, H.T. and Kunisch, K.: *Estimation Techniques for Distributed Parameter Systems.* Birkhäuser, 1989.

[12] Baumeister, J.: *Stable Solution of Inverse Problems.* Vieweg, Braunschweig, 1987.

[13] Baumeister, J., De Cezaro, A. and Leitao, A.: Modified iterated Tikhonov methods for solving systems of nonlinear equations. *Inverse Problems and Imaging*, 5:1–17, 2011.

[14] Baumeister, J., Kaltenbacher, B. and Leitao, A.: On Levenberg-Marquardt-Kaczmarz iterative methods for solving systems of nonlinear equations. *Inverse Problems and Imaging*, 4:335–350, 2010.

[15] Baumeister, J., Scondo, W., Demetriou, M.A. and Rosen ,I.G.: On–line parameter estimation for infinite dimensional dynamical systems. *SIAM Journal on Control and Optimization*, 35:678 – 713, 1997.

[16] Bauschke, H. H.: The approximation of fixed points of compositions of nonexpansive mappings in Hilbert space. *J. Math. Anal. and Appl.*, 202:150–159, 1996.

[17] Bauschke, H. H. and Borwein, J.M.: On projection algorithms for solving convex feasibility problems. *SIAM Review*, 38:367–426, 1996.

[18] Bauschke, H. H., Deutsch, F. and Hundal, H.: Characterizing arbitrarily slow convergence in the method of alternating projections. *archiv:0710.2387*, 1997.

[19] Berestycki, H., Busca, J. and Florent, I.: An inverse parabolic problem arising in finance. *C. R. Acad. Sci. Paris S r. I Math.*, 331:965–969, 2000.

[20] Berestycki, H., Busca, J. and Florent, I.: Asymptotics and calibration of local volatility models. *Quantitative Finance*, 2:61–69, 2002.

[21] Black, F. and Scholes, M.: The pricing of options and corporate liabilities. *J. of political economy*, 81:637–659, 1973.

[22] Borwein, J.M. and Lewis, A.S.: Partially-finite convex programming, part i: Quasi-relative interiors and duality theory. *Mathematical Programming*, 57:15–48, 1992.

[23] Bouchouev, I.: *Inverse parabolic problems with applications to option pricing.* PhD thesis, Department of Mathematics and Statistics, Wichita State University, 1997.

[24] Bouchouev, I. and Isakov, V.: The inverse problem of option pricing. *Inverse Problems*, 11:L11–L15, 1997.

[25] Bouchouev, I. and Isakov, V.: Uniqueness, stability and numerical methods for the inverse problem that arises in financial markets. *Inverse Problems*, 15:R95–R116, 1999.

[26] Bouchouev, I., Isakov, V. and Valdivia, N.: Recovery of volatility coefficient by linearization. *Quantitative Finance*, 2:257–263, 2002.

[27] Burger, M. and Kaltenbacher, B.: Regularizing Newton-Kaczmarz methods for nonlinear ill-posed problems. *SIAM J. Numer. Anal.*, 44:153–182, 2006.

[28] Calvetti, D., Hansen, P.C. and Reichel, L.: Gmres, l-curves, and discrete ill-posed problems. *BIT*, 42:44–65, 2002.

[29] Calvetti, D., Hansen, P.C. and Reichel, L.: L-curve curvature bounds via lanczos bidiagonalization. *Electronic Trans. Numer. Anal.*, 14:134–149, 2002.

[30] De Cezaro, A.: *On a parabolic inverse problem arising in Finance: convex and iterative regularization.* PhD thesis, Instituto Nacional de Matematica Pura e Aplicada, Rio de Janeiro, 2010.

[31] Chance, D.M.: A generalized simple formula to compute the implied volatility. *Financial Review*, 31:859–867, 1996.

[32] Chen, N. and Glasserman, P.: Mallivian greeks without malliavin calculus. *xxx*, 2006.

[33] Clarke, F.: *Optimization and nonsmooth analysis.* Wiley, 1985.

[34] Colton, D. and Kress, R.: *Integral Equation Methods in Scattering Theory.* Wiley, New York, 1983.

[35] Corrado, C.J. and Miller, T.W.: A note on a simple, accurate formula to compute implied standard deviations. *J. of Banking and Finance*, 20:595–603, 1996.

[36] Cox J.C., Ross, S.A. and Rubinstein, M.: Option pricing: a simplified approach. *J. of Financial Economics*, 7:229–263, 1979.

[37] Chiarella, C., Craddock, M. and El-Hassan, N.: An implementation of Boucheev's method for a short time calibration of option price models. *Computational Economics*, 22:443–438, 2003.

[38] Chiarella, C., Craddock, M. and El-Hassan, n.: The calibration of stock option pricing models using inverse problem methodology. *School of finance and economics university of technology*, 2007.

[39] Crépey, S.: Calibration of the local volatility in a generalized Black-Scoles model using Tikhonov regularization. *SIAM J. Math. Anal.*, 34:1183–1206, 2003.

[40] Crépey, S.: Calibration of the local volatility in a trinomial tree using Tikhonov regularization. *Inverse Problems*, 19:91–127, 2003.

[41] Crépey, S.: *Contribution à des méthodes numériques appliquées la finance et jeux différentiels.* PhD thesis, University of, 2000.

[42] Dontchev, A.L., Qi, H. and Qi, L.: Convergence of Newton's method for convex best interpolation. *Numer. Math.*, 87:435–456, 2001.

[43] Dupire, B.: Pricing with a smile. *Risk Magazine*, 7:18–20, 1994.

[44] Düring, B., Jüngel, A. and Volkwein, S.: A sequential quadratic programming method for volatility estimation in option pricing. *J. of Optim. Theory Appl.*, 139:1–27, 2008.

[45] Egger, H. and Engl, H.W.: Tihonov regularization applied to the inverse problem of option pricing: convergence analysis and rates. *Inverse Problems*, 21:1027–1045, 2007.

[46] Egger, H., Hein, T. and Hofmann, B.: On decoupling of volatility smile and term structure in inverse option pricing. *Inverse Problems*, 22:1247–1259, 2006.

[47] Engl, H.W., Hanke, M. and Neubauer, A.: *Regularization of Inverse Problems.* Kluwer, Dordrecht, 1996.

[48] Louis, A.K., Schoepfer, F. and Schuster, T.: Nonlinear iterative methods for linear ill-posed problems in banach spaces. *Inverse problems*, 22:311–323, 2006.

[49] Fengler, M.R.: *Semiparametric modelling of implied volatility.* Lecture Notes. Springer, 2005.

[50] Föllmer, H.: Alles richtig und trotzdem falsch? Anmerkungen zur Finanzkrise und zur Finanzmathematik.

[51] Föllmer, H. and Schied, A.: *Stochastic Finance: An introduction in Discrete Time.* de Gruyter, 2002.

[52] Foschi, P. and Pascucci, A.: Kolmogorov equations arising in finance: direct and inverse problems. pages 1–14, 2009.

[53] Fournié, E., Lasry, J.M., Lebuchoux, J. and Lions, P.-L.: Applications of Malliavin calculus to Monte Marlo methods in finance II. *Finance and Stochastics*, pages 201–236, 2001.

[54] Fournié, E., Lasry, J.M., Lebuchoux, J., Lions, P.-L. and Touzi, N.: An application of Malliavin calculus to Monte Carlo methods in finance. *Finance and Stochastics*, 3:391–412, 1999.

[55] Freitag, M.: Regularization in variational data assimilation.

[56] Freitag, M.: Data assimilation using 4d-var and links to regularization. Technical report, Dept. of Mathematical Sciences, 2008.

[57] Friedman, A.: *Partial differential equations.* Prentice Hall, 1964.

[58] Glasserman, P.: *Monte Carlo Methods in Financial Engineering.* Springer, Baltimore, 2003.

[59] Groetsch, C.W.: *Generalized Inverses of Linear Operators.* Dekker, New York, 1977.

[60] Groetsch, C.W.: *The theory of Tikhonov regularisation for Fredholm equations of the first kind.* Pittman Publishing, Boston, 1984.

[61] Groetsch, C.W.: *Inverse Problems in Mathematical Sciences.* Vieweg, Braunschweig, 1993.

[62] Gulevich, N.M.: Fixed points of nonexpansive mappings. *Journal of Mathematical Sciences*, 79:755–815, 1996.

[63] Wang, Y., Yin, H. and Qi, L.: Shape-preserving interpolation and smoothing for options market implied volatility. *J. Optim. Theory Appl.*, 142:243–266, 2009.

[64] Hadamard, J.: Sur les problemes aux derivees partielles et leur signification physique. *Princeton University Bulletin*, pages 49–52, 1902.

[65] Haltmeier, M., Leit ao, A. and Scherzer, O.: Kaczmarz methods for regularizing nonlinear ill-posed equations. I. convergence analysis. *Inverse Probl. Imaging*, 1(2):289–298, 2007.

[66] Haltmeier, M., Kowar, R., Leit ao, A. and Scherzer, O.: Kaczmarz methods for regularizing nonlinear ill-posed equations. II. Applications. *Inverse Probl. Imaging*, 1(3):507–523, 2007.

[67] Haltmeier, M., Kowar, R., Leitão, A. and Scherzer, O.: Kacmarz methods for regularizing nonlinear ill-posed equations II: Applications. *Inverse Problems and Imaging*, 1:507–523, 2007.

[68] Hanke, M., Neubauer, A. and Scherzer, O.: A convergence analysis of the Landweber iteration for nonlinear ill-posed problems. *Numerische Mathematik*, 72:21–37, 1995.

[69] Hanke, M. and Rössler, E.: Computation of local volatilities from regularized Dupire equations. *Int. J. Theor. Appl. Finance*, 8:207–222, 2005.

[70] Hanke, M. and Scherzer, O.: Inverse problems light: numerical differentiation. *Amer. Math. Monthly*, 108:512–521, 2001.

[71] Hansen, P.C: Regularization tools: A matlab package for analysis and solution of discrete ill-posed problems. *Numerical Algorithms*, 6:1–35, 1994.

[72] Hein, T.: *Analytische und numerische Studien zu inversen Problemen der Optionspreisbildung.* PhD thesis, Technische Universität Chemnitz, 2003.

[73] Hein, T.: Some analysis of Tikhonov regularization for the inverse problem of option pricing in the price-dependent case. *Zeitschrift für Analysis und Anwendungen*, 24:593–609, 2005.

[74] Hein, T. and Hofmann, B.: On the nature of ill-posedness of an inverse problem in option pricing. *Inverse Problems*, 19:1319–1338, 2003.

[75] Heinz, S.W.: Multilevel monte carlo methoden und deren anwendung. Master's thesis, Goethe University, Frankfurt/Main, 2009.

[76] Heston, S.L.: A closed-form solution for options with stochastic volatility with applications to bond and currency options. *Review of Financial Studies*, 6:327–343, 1993.

[77] Hofmann, B. *Mathematik inverser Probleme.* Teubner, Stuttgart, 1999.

[78] Isakov, V.: *Inverse Source Problems.* American Mathematical Society, Providence, 1990.

[79] Isakov, V.: *Inverse Problems for Partial Differential Equations.* Springer, Berlin, 1998.

[80] Iserles, A.: *A first course in the numerical analysis of differential equations.* Cambridge Texts, Cambridge, 1996.

[81] Jackson, N., Süli, E. and Howison, S.: Computation of deterministic volatility surfaces. *J. of Comput. Finance*, 2:5–32, 1999.

[82] Jeanblanc, M. Yor, M. and Chesnay, M.: *Mathematical Methods for Financial Markets.* Springer, New York, 2009.

[83] Jentzen, A., Kloeden, P. and Neuenkirch, A.: Pathwise approximation of stochastic differential equations on domains: higher order convergence rates without global Lipschitz coefficients. *Num. Math.*, 112:41–64, 2008.

[84] Jin, Q.: Inexact Newton-Landweber iteration for solving nonlinear inverse problems in banach spaces. *Inverse Problems*, 28:15 pp, 2012.

[85] Jin, Q. and Stahls, L.: Nonstationary iterated tikhonov regularization for ill-posed problems in banach spaces. *Inverse Problems*, 28:xx–xx, 2012.

[86] Lakshmivarahan, S.Lewis, J.M. and Dhall, S.: *Dynamic Data Assimilation : A Least Squares Approach.* Cambridge University Press, 2006.

[87] Kaczmarz, S.: Approximate solution of systems of linear equations. *Internat. J. Control*, 57:1269–1271, 1993.

[88] Kagenish, Y.i and Shinohara, Y.: A note on computation of implied volatility. *Asia-Pacific Financial Markets*, 8:361–368, 2004.

[89] Kaipio, J. and Somersalo, E.: *Statistical and Computational Inverse Problems.* Springer, Berlin, 2005.

[90] Kaltenbacher, B., Neubauer, A. and Scherzer, O.: *Iterative Regularization Methods for Nonlinear Ill-Posed problems.* De Gruyter Verlag, Berlin, 2007.

[91] Kaltenbacher, B., Schöpfer, F. and Schuster, T.: Iterative methods for nonlinear ill-posed problems in Banach spaces. *Inverse Problems* 25:19 pp, 2009..

[92] Karatzas, I. and Shreve, S.: *Brownian motion and stochastic calculus.* Springer, New York, 1991.

[93] Karatzas, I. and Shreve, S.: *Methods of mathematical finance.* Springer, New York, 1998.

[94] Kebaier, A.: Statistical Romberg extrapolation: a new variance reduction method and applications to option pricing. *The annals of applied probability*, 15:2681–2705, 2005.

[95] Kelley, C.T.: *Iterative Methods for Optimization.* SIAM, 1999.

[96] Kirchgraber, A., Stoffer, D. and Kirsch, A.: Schlecht gestellte Probleme – oder Wenn das Ungenaue genauer ist. *Mathematische Semesterberichte*, 51:175–206, 2005.

[97] Kirsch, A.: *An Introduction to the Mathematical Theory of Inverse Problems.* Springer, New York, 1996.

[98] Kloeden, P. and Sanz-Chacon, C.: Efficient pricze sensitivity estimation of path-derivatives by weak derivation. *Monte Carlo methods & Applications*, 17:47–73, 2011.

[99] Cyganovski, S., Kloeden, P. and Ombach, J.: *From Elementary Probability to Stochastic Differential Equations with MAPLE.* Springer, Berlin, 2001.

[100] Kloeden, P. and Platen, E.: *Numerical Solution of SDE through Computer Experiments.* Springer, Berlin, 1994.

[101] Korn, R. und Korn, E.: *Optionsbewertung und Portfolio–Optimierung.* Vieweg, Braunschweig, 1999.

[102] Kusraev, A.G. and Kutateladze, S.S.: *Subdifferentials: Theory an applications.* Springer, Berlin, 1995.

[103] Lagnado, R. and Osher, S.: A technique for calibrating derivative security pricing models: numerical solution of an inverse problem. *J. of Comput. Finance*, 1:13–25, 1997.

[104] Leitão, A., De Cezaro, A., Haltmeier, M. and Scherzer, O.: On steepest-descent-kaczmarz methods for regularizing systems of nonlinear ill-posed equations. *Applied Mathematics and Computation*, 202:596–607, 2008.

[105] Lion, P.M.: Rapid identification of linear and nonlinear systems. *AIAA J.*, 5:1835 – 1841, 1987.

[106] Lishang, J. and Youshan, T.: Identifying the volatility of underlying assets from

option prices. *Inverse Problems*, 17:137–155, 2001.

[107] Lu, S. and Pereverzev, S.: Numerical differentiation from a viewpoint of regularization theory. *Mathematics of Computation*, 75:1853–1870, 2006.

[108] Marx, A. and Potthast, W.E.: On instability in data assimilation. Technical report, School of Mathematical and Physical Sciences, 2012.

[109] Merton, R.C.: Theory of rational option pricing. *Bell Journal of Economics and management Science*, 4:141–183, 1973.

[110] Natterer, F.: *The Mathematics of Computerized Tomography*. John Wiley, New York, 1986.

[111] Neftci, S.: *Introduction to Calibration Methods in Finance With Market Applications*. Wiley, London, 2010.

[112] Nelson, C. and Siegel, A.: Parsimonious modeling of yield curves. *Journal of Business*, 60:473–489, 1987.

[113] Pironneau, O.: Dupire identities for complex options. *C.R. Acad. Sci. Paris*, Ser. I 344:127–133, 2007.

[114] Pironneau, O.: Automatic differentiation for financial engineering. *Preprint, University of Paris VI*, pages 1–49, 2008.

[115] Potthast, R., Marx, B., Moodey, A., Lawless, A. and van Leeuwen, P.J.: On instability in data assimilation, 2010.

[116] Pyatkov, S.G.: Inverse problems for the Black-Scholes equation and related problems. *J. Inv. Ill-posed problems*, 15:955–974, 2007.

[117] Qi, L. and Sun, D.: A survey of some nonsmooth equations and smoothing newton methods, 1998.

[118] Ramm, A.G. and Smirnova, B.: On stable numerical differentiation. *Mathem. of Computation*, 70:1131–1153, 1968.

[119] Resmerita, V.: Regularization of ill-posed problems in Banach spaces: convergence rates. *Inverse Problems*, 21:1303–1314, 2005.

[120] Rieder, A.: *Keine Probleme mit Inversen Problemen*. Vieweg, Braunschweig, 2003.

[121] Sandmann, W.: *Simulation seltener Ereignisse mittels importance sampling unter besonderer Berücksichtigung Markovscher Modelle*. PhD thesis, Universität Bonn, 2004.

[122] Scherzer, O.: Convergence criteria of iterative methods based on Landweber iteration for solving nonlinear problems. *J. Math. Anal. Appl.*, 194:911–933, 1995.

[123] Schirotzek, W.: *Nonsmooth analysis*. Springer, Berlin, 2007.

[124] Seydel, R.: *Tools for Computational Finance*. Springer, Berlin, 2002.

[125] Svensson, L.E.O.: Estimating the term structure of interest rates for monetary policy analysis. *Scandinavian Journal of Economics*, 98:163–183, 1996.

[126] Wang, Y., Yin, H. and Qi, L.: No-arbitrage interpolation of the option price function and its reformulation.

[127] Witte, C.: Das inverse Problem der Optionsbewertung. Master's thesis, University of Münster, 1998.

[128] Wloka, J.: *Partielle Differentialgleichungen*. Teubner, Stuttgart, 1982.

[129] Zubelli, J.P.: Inverse problems in finances. In *Conference on Statistical Modelling in Insurance and Finance*, Rio de Janeiro, 2005. IMPA.

Chapter 4

Asymptotic and non asymptotic approximations for option valuation

R. Bompis and E. Gobet*

Centre de Mathématiques Appliquées, Ecole Polytechnique and CNRS, Route de Saclay, 91128 Palaiseau cedex, France

Abstract We give a broad overview of approximation methods to derive analytical formulas for accurate and quick evaluation of option prices. We compare different approaches, from the theoretical point of view regarding the tools they require, and also from the numerical point of view regarding their performances. In the case of local volatility models with general time-dependency, we derive new formulas using the local volatility function at the mid-point between strike and spot: in general, our approximations outperform previous ones by Hagan and Henry-Labordère. We also provide approximations of the option delta.

1. Introduction

In the two last decades, numerous works have been devoted to designing efficient methods in order to give exact or approximative pricing formulas for many financial products in various models. This quest of efficiency originates in the need for more and more accurate methods, when one takes into account an increasing number of sources of risk, while maintaining a competitive computational time. The current interest in real-time tools (for pricing, hedging, calibration) is also very high.

Let us give a brief overview of different computational approaches. While explicit formulas are available in simple models (Black-Scholes model associated to log-normal distribution, or Bachelier model related to normal distributions for instance), in general no closed forms are known and numerical methods have to be used. As a numerical method, it is usual to perform PDE solvers for one or two-dimensional sources of risk (see [1] for instance) or Monte Carlo methods for higher dimensional problems [2]: both approaches are popular, efficiently developed and many improvements have been proposed for years. However, these methods are not intrinsically real-time methods, due to the increasing number of points required in the PDE discretization grid or due to the increasing number of paths needed in the

*This research is part of the Chair Financial Risks of the Risk Foundation, the Chair Derivatives of the Future and the Chair Finance and Sustainable Development.

Monte Carlo procedure. Not being *real-time method* means, for example, that when used for calibration routine based on data consisting of (say) 30 vanilla options, it usually takes more than one minute (in the most favorable situation) to achieve the calibration parameters. The approaches presented below are aimed at reducing this computational time to less than one second.

The class of affine models (such as Heston model, exponential Levy model ...) offers an alternative approach related to Fourier computations: on the one hand, in such models the characteristic function of the marginal distribution of the log-asset is explicitly known; on the other hand, there are general relations between Call/Put prices and the characteristic function of the log-asset. These relations write as follows.

- Following Carr and Madan [3], consider the difference $z_T(k)$ between the Call price in a given model and that price in an arbitrary Black-Scholes model (with volatility σ), both with maturity T and log-strike k. For zero interest rate (to simplify), it is equal to $z_T(k) = \mathbb{E}(e^{X_T} - e^k)_+ - \mathrm{Call}^{\mathrm{BS}}(k)$, where X is the logarithm of the asset. A direct computation gives explicitly the Fourier transform $\widehat{z_T}(v)$ in the log-strike variable:
$$\widehat{z_T}(v) = \int_{\mathbb{R}} e^{ivk} z_T(k)\mathrm{d}k = \frac{\Phi_T^X(1+iv) - \Phi_T^{\mathrm{BS}}(1+iv)}{iv(1+iv)},$$
where $\Phi_T^{X,\mathrm{BS}}(u) := \mathbb{E}(e^{uX_T})$ is either computed in the X model or in the Black-Scholes model. Since $\Phi_T^X(\cdot)$ is required to be known, we get the X-model Call price $z_T(k) + \mathrm{Call}^{\mathrm{BS}}(k)$ simultaneously for any log-strike using a Fast Fourier Transform.
- Alternatively, following the Lewis approach[4] , let $\alpha > 0$ be a damping constant, set $h(y) = (e^y - K)_+ e^{-(1+\alpha)y}$ which belongs to $\mathbf{L}_2(\mathbb{R}, \mathrm{Leb.})$ and assume that $\mathbb{E}(e^{(1+\alpha)X_T}) < +\infty$: from the Parceval-Plancherel identity, assuming that the density p_{X_T} of X_T w.r.t. the Lebesgue measure exists, we obtain
$$\begin{aligned}\mathbb{E}(e^{X_T} - K)_+ &= \int_{\mathbb{R}} h(y) e^{(1+\alpha)y} p_{X_T}(y)\mathrm{d}y\\ &= \frac{1}{2\pi}\int_{\mathbb{R}} \widehat{h}(-\xi)[\widehat{e^{(1+\alpha)\cdot} p_{X_T}(\cdot)}](\xi)\mathrm{d}\xi\\ &= \frac{1}{2\pi}\int_{\mathbb{R}} \frac{e^{-(\alpha+i\xi)\log(K)}}{(i\xi+\alpha)(i\xi+\alpha+1)}\Phi_T^X(1+\alpha+i\xi)\mathrm{d}\xi.\end{aligned}$$
The final identity still holds without assuming the existence of density: this can be proved by adding a small Brownian perturbation (considering $X_T + \varepsilon W_T$ instead of X_T), and taking the limit as the perturbation ε goes to 0. From the above formula, using an extra numerical integration method (to compute the ξ-integral), we recover Call prices. For higher numerical performance, Lewis recommends a variant of the formula above, obtained

through the decomposition $(e^{X_T} - K)_+ = e^{X_T} - \min(e^{X_T}, K)$: it finally comes out as

$$\mathbb{E}(e^{X_T} - K)_+ = S_0 - \frac{1}{2\pi}\int_{\mathbb{R}} \frac{e^{(\frac{1}{2}-i\xi)\log(K)}}{\frac{1}{4}+\xi^2}\Phi_T^X(\frac{1}{2}+i\xi)\mathrm{d}\xi. \quad (1)$$

Regarding computational time, both Fourier-based approaches perform well, since they are essentially reduced to a one-dimensional integration problem. But they can be applied only to specific models for which the characteristic function is given in an explicit and tractable form: in particular, it rules out the local volatility models, the local and stochastic volatility models.

The last approach consists of explicit analytical approximations and this is the main focus of this chapter: it is based on the general principle of expanding the quantity of interest (price, hedge, implied volatility...) with respect to some small-/large parameters (possibly multidimensional). The parameters under consideration may be of very different nature: for instance in the case of Call/Put options of strike K and maturity T, it ranges from the asymptotic behavior as K is small or large, to the case of short or long maturity T, passing through coupled asymptotics, or small/fast volatility variations, and so on... A detailed description with references is presented in Section 2. Due to the plentiful and recent literature on the subject, it is likely that we will not be exhaustive in the references, but we will do our best to give the main trends and to expose whenever possible what are the links between different viewpoints; we will compare the mathematical tools to achieve these approximations (rather PDE techniques or stochastic analysis ones), in order to provide to the reader a clarified presentation of this prolific topic.

The chapter is organized as follows. Section 2 gives an overview of asymptotic and non-asymptotic results: wing formulas, long maturity behavior, large deviations type results, regular and singular perturbation for PDEs, asymptotic expansions of Wiener functionals and other stochastic analysis approaches. The choice of the *small/large* parameter is of course crucial and is usually left to the expertise of the user. In particular, we show that there might be a competition between different small/large parameters and the accuracy order might not be the natural one. This motivates deriving non asymptotic results and this is our emphasize in the next sections. We develop the principle of high order approximations related to an intuitive proxy. In Section 3, we consider the simplest case of second order approximation in local volatility models, using log-normal or normal proxys. We give pedagogic proofs. Section 4 is devoted to a more detailed analysis: we first give arguments based on stochastic analysis (martingales, Malliavin calculus). We compare this derivation with a method mixing stochastic analysis and PDE, and with a pure PDE approach: we show in which respect our methodology is different. In Section 5, we provide various high-order approximation using proxys. In Section 6, approximations of the option delta are provided. Section 7 is gathering numerical experiments illustrating the performance of our formulas compared to those of Hagan etal. [5] and of Henry-Labordère[6]. Some intermediate and complementary results are postponed to Appendix (Section 8).

Notation used throughout the chapter.

▷ **Models.** In all this work, financial products are written w.r.t. a single asset, whose price at time t is denoted by S_t. The dynamics of S is modeled through a filtered probability space $(\Omega, \mathcal{F}, (\mathcal{F}_t)_{t\geq 0}, \mathbb{P})$ where $(\mathcal{F}_t)_{t\geq 0}$ is the natural filtration of a standard linear Brownian motion W, augmented by the $\mathbb{P}$-null sets. The risk-free rate is set* to 0; most of the time and unless stated otherwise, S follows a local volatility model, i.e. it is solution of the stochastic diffusion equation

$$\mathrm{d}S_t = S_t\sigma(t, S_t)\mathrm{d}W_t, \tag{2}$$

where the dynamics is directly under the pricing measure. Assumptions on the local volatility σ are given later. We assume that the complete market framework holds and that an option with payoff $h(S_T)$ paid at maturity T has a fair value at time 0 equal to $\mathbb{E}(h(S_T))$.

For positive S, we define the log-asset $X = \log(S)$ which satisfies

$$\mathrm{d}X_t = a(t, X_t)\mathrm{d}W_t - \frac{1}{2}a^2(t, X_t)\mathrm{d}t, \tag{3}$$

where $a(t, x) = \sigma(t, e^x)$.

▷ **Call options.** Let us denote by $\mathrm{Call}(S_0, T, K)$ the price at time 0 of a Call option with maturity T and strike K, written on the asset S. "Price" usually means the price given by a model on S, that is

$$\mathrm{Call}(S_0, T, K) = \mathbb{E}(S_T - K)_+. \tag{4}$$

This Model Price should equalize the Market Price taken from Market data (calibration step). As usual, ATM (At The Money) Call refers to $S_0 \approx K$, ITM (In The Money) to $S_0 \gg K$, OTM (Out The Money) to $S_0 \ll K$.

▷ **Black-Scholes Call price function.** For convenience of the reader, we give the *Black-Scholes Call price* function depending on log-spot x, total variance y and log-strike z:

$$\mathrm{Call}^{\mathrm{BS}}(x, y, z) = e^x\mathcal{N}(d_1(x, y, z)) - e^k\mathcal{N}(d_2(x, y, z)) \tag{5}$$

where:

$$\mathcal{N}(x) = \int_{-\infty}^{x} \mathcal{N}'(u)\mathrm{d}u, \qquad \mathcal{N}'(u) = \frac{e^{-u^2/2}}{\sqrt{2\pi}},$$

$$d_1(x, y, z) = \frac{x - z}{\sqrt{y}} + \frac{1}{2}\sqrt{y}, \qquad d_2(x, y, z) = d_1(x, y, z) - \sqrt{y}.$$

*for non-zero but deterministic risk-free rate, we are reduced to the previous case by considering the discounted asset; see also the discussion in [7] for stochastic interest rates.

This value $\mathrm{Call}^{\mathrm{BS}}(x,y,z)$ equals $\mathrm{Call}(e^x,T,e^z)$ in (4) when the volatility in (2) is only time-dependent and $y=\int_0^T \sigma^2(t)\mathrm{d}t$. Note that $\mathrm{Call}^{\mathrm{BS}}$ is a smooth function (for $y>0$) and there is in addition a simple relation between its partial derivatives:

$$\partial_y \mathrm{Call}^{\mathrm{BS}}(x,y,z)=\frac{1}{2}(\partial^2_{x^2}-\partial_x)\mathrm{Call}^{\mathrm{BS}}(x,y,z)=\frac{1}{2}(\partial^2_{z^2}-\partial_z)\mathrm{Call}^{\mathrm{BS}}(x,y,z). \quad (6)$$

In the following, $x_0=\log(S_0)$ (which is the initial value of the process X defined in (3)) will represent the log-spot, $k=\log(K)$ the log-strike, $x_{\mathrm{av}}=(x_0+k)/2=\log(\sqrt{S_0K})$ the mid-point between the log-spot and the log-strike, $m=x_0-k=\log(S_0/K)$ the log-moneyness.

The reader can find in Proposition 4.2 the definition of $\mathrm{Vega}^{\mathrm{BS}}$, $\mathrm{Vomma}^{\mathrm{BS}}$ and $\mathrm{Ultima}^{\mathrm{BS}}$ which are the first three derivatives of $\mathrm{Call}^{\mathrm{BS}}$ w.r.t. a volatility parameter.

For (x,T,z) given, the *implied Black-Scholes volatility* of a price $\mathrm{Call}(e^x,T,e^z)$ is the unique non-negative parameter $\sigma_{\mathrm{I}}(x,T,z)$ such that

$$\mathrm{Call}^{\mathrm{BS}}\big(x,\sigma_{\mathrm{I}}^2(x,T,z)T,z\big)=\mathrm{Call}(e^x,T,e^z). \quad (7)$$

$\triangleright$ **Bachelier Call price function.** We now recall the *Bachelier Call price* as a function of spot S, total variance Y and strike Z:

$$\mathrm{Call}^{\mathrm{BA}}(S,Y,Z)=(S-Z)\mathcal{N}\Big(\frac{S-Z}{\sqrt{Y}}\Big)+\sqrt{Y}\mathcal{N}'\Big(\frac{S-Z}{\sqrt{Y}}\Big), \quad (8)$$

which coincides with $\mathrm{Call}(S,T,Z)$ when the volatility in (2) is such that $x\sigma(t,x)=\Sigma(t)$ and $Y=\int_0^T \Sigma^2(t)\mathrm{d}t$. The function $\mathrm{Call}^{\mathrm{BA}}$ is smooth (for $Y>0$) and we have:

$$\partial_Y \mathrm{Call}^{\mathrm{BA}}(S,Y,Z)=\frac{1}{2}\partial^2_{S^2}\mathrm{Call}^{\mathrm{BA}}(S,Y,Z)=\frac{1}{2}\partial^2_{Z^2}\mathrm{Call}^{\mathrm{BA}}(S,Y,Z).$$

We frequently use the notation $S_{\mathrm{av}}=(S_0+K)/2$ and $M=S_0-K$ for the *Bachelier moneyness*. Proposition 4.5 defines the sensibilities of $\mathrm{Call}^{\mathrm{BA}}$ w.r.t. the volatility parameter: $\mathrm{Vega}^{\mathrm{BA}}$, $\mathrm{Vomma}^{\mathrm{BA}}$ and $\mathrm{Ultima}^{\mathrm{BA}}$.

For (S,T,Z) given, the *implied Bachelier volatility* of a price $\mathrm{Call}(S,T,Z)$ is the unique non-negative parameter $\Sigma_{\mathrm{I}}(S,T,Z)$ such that

$$\mathrm{Call}^{\mathrm{BA}}\big(S,\Sigma_{\mathrm{I}}^2(S,T,Z)T,Z\big)=\mathrm{Call}(S,T,Z). \quad (9)$$

Black-Scholes and Bachelier implied volatilities are compared in [8].

2. An overview of approximation results

The increasing need in evaluating financial risks at a very global level and in a context of high-frequency market exchanges is a significant incentive for the computational methods to be efficient in evaluating the exposure of large portfolio to market fluctuations (VaR computations, sensitivity analysis), in quickly calibrating the models to the market data. Hence, in the two last decades, many numerical methods have been developed to meet these objectives: in particular, regarding the option pricing, several approximation results have been derived, following one or another asymptotic point of view. We give a summary of these different approaches, stressing the limits of applicability of the methods.

2.1. *Large and small strikes, at fixed maturity*

The Call price $\mathrm{Call}(S_0, T, K)$ as a function of strike is convex and its left/right derivatives are related to the distribution function of S_T [9] : $\partial_K^- \mathrm{Call}(S_0, T, K) = -\mathbb{P}(S_T \geq K)$ and $\partial_K^+ \mathrm{Call}(S_0, T, K) = -\mathbb{P}(S_T > K)$. Beyond the important fact that the family of Call prices $\{\mathrm{Call}(S_0, T, K) : K \geq 0\}$ completely characterizes the marginal distribution of S_T, this relation also shows that the tails of the law of S_T are intrinsically related to the decay of $\mathrm{Call}(S_0, T, K)$ as $K \to +\infty$. In terms of implicit volatility, the heuristics are the following: the larger the implied volatility of OTM options, the larger the right tail of S_T. This is similar for small strikes K, using Put options. Lee [10] was the first one to quantify these features relating the behavior of implied volatility to the tails of S_T, with an encoding of the tails through the existence of positive/negative moments. These are model-free relations, that can be applied to any model with $\mathbb{E}(S_T) < +\infty$ and not only to local volatility ones like in (2). The well-known Lee moment formulas read as follows, using the log-variables $x_0 = \log(S_0)$ and $m = \log(S_0/K) = x_0 - k$.

Theorem 1. *Define*
- *the maximal finite positive moment order* $p_R := \sup\{p \geq 0 : \mathbb{E}(S_T^{1+p}) < +\infty\}$,
- *the maximal finite negative moment order** $p_L := \sup\{p \geq 0 : \mathbb{E}(S_T^{-p}) < +\infty\}$.

Then, the right tail-wing of the Black-Scholes implied volatility defined in (7) *is such that*

$$\limsup_{m \to -\infty} \frac{T\sigma_{\mathrm{I}}^2(x_0, T, x_0 - m)}{|m|} = \phi(p_R) := \beta_R,$$

while the left tail-wing is such that

$$\limsup_{m \to +\infty} \frac{T\sigma_{\mathrm{I}}^2(x_0, T, x_0 - m)}{m} = \phi(p_L) := \beta_L,$$

where $\phi(x) = 2 - 4(\sqrt{x^2 + x} - x) \in [0, 2]$.

Proof. We refer to [10] for a detailed proof. We only give the two main arguments for proving the right tail-wing, the left one being similar.

• The *first argument* relies on a tight connection between moments and asymptotics of Call/Put as $K \to +\infty$. Indeed, on the one hand, convexity inequalities give $(s - K)_+ \leq \frac{s^{p+1}}{p+1}\left(\frac{p}{p+1}\right)^p \frac{1}{K^p}$ (for $p \geq 0$), and taking the expectation yields

$$\mathrm{Call}(S_0, T, K) \leq \frac{\mathbb{E}(S_T^{p+1})}{p+1} \left(\frac{p}{p+1}\right)^p \frac{1}{K^p}. \tag{10}$$

*$1 + p_R$ and p_L are respectively called right-tail and left-tail indices.

In other words, the more integrability of S_T, the faster the decay of $\mathrm{Call}(S_0, T, K)$ as $K \to +\infty$. Conversely, the Carr formula states that the Call/Put prices form a pricing generating system for any payoff equal to a difference of convex functions: making this principle particular to the *power* payoff, we obtain

$$\mathbb{E}(S_T^{1+p}) = \int_0^\infty p(p+1)K^{p-1}\mathrm{Call}(S_0, T, K)\mathrm{d}K, \tag{11}$$

i.e. the faster the decreasing of $\mathrm{Call}(S_0, T, K)$ as $K \to +\infty$, the higher the integrability of S_T.

• The *second argument* is based on exponential decreasing behaviors of Call/Put in terms of Black-Scholes implied volatility, as the log-moneyness $m \to \pm\infty$. Reparameterizing the implied volatility $\sigma_{\mathrm{I}}(x_0, T, x_0 - m) = \sqrt{\beta|m|/T}$ with $\beta \in (0, 2]$ (β is interpreted as a slope of the total variance per log-moneyness), we obtain

$$\mathrm{Call}^{\mathrm{BS}}(x_0, \beta|m|, x_0 - m) = S_0\mathcal{N}(-\sqrt{f_-(\beta)|m|}) - S_0e^{-m}\mathcal{N}(-\sqrt{f_+(\beta)|m|})$$

where $f_\pm(\beta) = \frac{1}{\beta} + \frac{\beta}{4} \pm 1$. Then, a direct computation shows a dichotomic behavior related to β:

$$\lim_{m\to-\infty} e^{-cm}\mathrm{Call}^{\mathrm{BS}}(x_0, \beta|m|, x_0 - m) = +\infty\mathbf{1}_{c>f_-(\beta)/2}. \tag{12}$$

Comparing (10-11-12) and setting $p_R := f_-(\beta_R)/2$ (or equivalently $\beta_R = \phi(p_R)$) yields the tail-wing formulas. □

Since the original contribution of Lee, several improvements to Theorem 1 have been established. For instance, the lim sup can be removed by a simple limit, under the additional assumptions that S_T has a regularly varying density, see [11]. More recently, Gulisashvili [12] and his co-authors have proved refined expansions of the form

$$\begin{aligned}\sigma_{\mathrm{I}}(x_0, T, k) =& \frac{\sqrt{2}}{\sqrt{T}}\Bigg[\sqrt{\log K + \log\frac{1}{\mathrm{Call}(S_0, T, K)} - \frac{1}{2}\log\log\frac{1}{\mathrm{Call}(S_0, T, K)}}\\ &- \sqrt{\log\frac{1}{\mathrm{Call}(S_0, T, K)} - \frac{1}{2}\log\log\frac{1}{\mathrm{Call}(S_0, T, K)}}\Bigg]\\ &+ O\Big(\Big(\log\frac{1}{\mathrm{Call}(S_0, T, K)}\Big)^{-\frac{1}{2}}\Big)\end{aligned}$$

as K becomes large, which allows precise asymptotics of $\sigma_{\mathrm{I}}(x_0, T, k)$ through those of $\mathrm{Call}(S_0, T, K)$.

These kinds of asymptotics are now well-known for most of the usual models, like CEV models (no right tail-wing), Heston model (tail-wing depending on the maturity)... see [13] for more references. Different models may have the same strike asymptotics. We can use this information on extreme strikes in different manners: first, comparing with the asymptotic market implied volatility smile, it allows for selecting a coherent model. Second, it helps the calibration procedure by setting approximately some parameter values (those having an impact on the tails). Third, it can be used to appropriately extrapolate market data.

In practice, these asymptotic formulas refer to far OTM or ITM options, for which the accuracy of market data is really questionable (large bid-ask spread, low liquidity). Thus, a direct application is usually not straightforward.

2.2. *Long maturities, at fixed strike*

Another asymptotics is large maturity. It has been studied by Rogers and Tehranchi, see [14] and [15], proving the following.

Theorem 2. *Assume that S remains positive with probability 1. Then, for any $\lambda > 0$, $\lim_{T\to+\infty} \sup_{|m|\le\lambda} \left|\sigma_{\mathrm{I}}(x_0, T, x_0 - m) - \sqrt{\frac{8}{T}|\ln(\mathbb{E}(S_T \wedge S_0))|}\right| = 0$.*

As before, the proof is based on the careful derivation of asymptotics of Black-Scholes formula (5). The above limit states that for strikes in a fixed neighborhood of the spot S_0, the implied volatility behaves like $\sqrt{\frac{8}{T}|\ln(\mathbb{E}(S_T \wedge S_0))|}$ for large maturity, and thus it does not depend on the strike. In other words, the implied volatility surface flattens as maturity becomes large, which is coherent with market data. There are also some refined and higher order asymptotics: assuming that the *a.s.* large-time limit of the martingale S is 0, then

$$\begin{aligned}T\sigma_{\mathrm{I}}^2(x_0, T, k) =& 8|\ln(\mathbb{E}(S_T \wedge \frac{K}{S_0}))| - 4\ln(|\ln(\mathbb{E}(S_T \wedge \frac{K}{S_0}))|)\\ &+ 4\ln(\frac{K}{\pi S_0}) + o(1),\end{aligned}$$

where the reminder is locally uniform in the log-moneneyss $m = x_0 - k$.

2.3. *Long maturities, with large/small strikes*

In view of the preceeding results, the asymptotics of the smile for large maturity becomes very simple regarding the strike variable, unless one allows the strike to be large/small together with the maturity. Indeed to recover interesting information at the limit, we should consider strikes of the form $K = S_0 e^{xT}$ with $x \neq 0$, or equivalently $k = x_0 + xT$. From the linearization of the payoff, one obtains

$$\begin{aligned}\mathrm{Call}(S_0, T, S_0 e^{xT}) &= \mathbb{E}\big(S_T \mathbf{1}_{S_T \ge S_0 e^{xT}}\big) - S_0 e^{xT}\mathbb{P}(S_T \ge S_0 e^{xT})\\ &= S_0 \mathbb{P}^S\Big(\frac{1}{T}\log(S_T/S_0) \ge x\Big) - S_0 e^{xT}\mathbb{P}\Big(\frac{1}{T}\log(S_T/S_0) \ge x\Big)\end{aligned}$$

where the new measure $\mathbb{P}^S$ is the one associated to the numéraire S. Under this form, it appears clearly that for x large enough (say larger than the asymptotic $\mathbb{P}$-mean or $\mathbb{P}^S$-mean of $\frac{1}{T}\log(S_T/S_0)$ whenever it exists), both probabilities above correspond to the evaluation of large deviation events. The role of Large Deviation Principle satisfied by the sequence $(\frac{1}{T}\log(S_T/S_0))_{T\ge 0}$ as $T \to +\infty$ has been outlined in [16] in the case of Heston model, and in [17] for more general affine models.

Saddle point arguments combined with Lewis formula (1) have been performed in [18] for the Heston model, to recover the *Stochastic Volatility Inspired* parameterization of Gatheral [19]: the squared implied volatility $\sigma_{\mathrm{I}}^2(x_0, T, x_0 + xT)$ has the simple asymptotic shape

$$\sigma_\infty^2(x) = \frac{\omega_1}{2}\left(1 + \omega_2\rho x + \sqrt{(\omega_2 x + \rho)^2 + 1 - \rho^2}\right). \tag{13}$$

For more general affine models like Heston model, without or with jumps, or Bates model, or Barndorff-Nielsen-Shephard model (see [20] and [17]), it is possible to derive similar limits. Let $\Lambda_t(u) = \log(\mathbb{E}(S_t^u))$ be the exponent of the moment generating function, which is convex in u: in the aforementioned model we can define and compute its asymptotic average $\Lambda(u) = \lim_{t\to\infty} \frac{1}{t}\Lambda_t(u)$, which still satisfies to the convexity feature. We associate its Fenchel-Legendre transform $\Lambda^*(x) = \sup_{u\in\mathbb{R}}(ux - \Lambda(u))$ and it turns out that $(\frac{1}{T}\log(S_T/S_0))_{T\geq 0}$ satisfies an LDP under $\mathbb{P}$ (resp. $\mathbb{P}^S$) with rate function $x \mapsto \Lambda^*(x)$ (resp. $x \mapsto \Lambda^*(x) - x$).

Theorem 3. *Under some assumptions (see [17]), for any $x \in \mathbb{R}$, the asymptotic implied volatility $\sigma_\infty(x)$ is given by*

$$\lim_{T\to\infty} \sigma_{\mathrm{I}}(x_0, T, x_0 + xT) = \sqrt{2}\Big[\mathrm{sgn}(\Lambda'(1) - x)\sqrt{\Lambda^*(x) - x} \\ + \mathrm{sgn}(x - \Lambda'(0))\sqrt{\Lambda^*(x)}\Big].$$

In the Black-Scholes model with constant volatility σ, one has $\Lambda(u) = \frac{\sigma^2}{2}(u^2 - u)$, $\Lambda^*(x) = \frac{1}{2\sigma^2}\left(x + \frac{\sigma^2}{2}\right)^2$, and we get obviously $\sigma_\infty(x) = \sigma$. For Heston model, Λ is explicit as well and we finally recover the SVI parsimonious parameterization (13). Here again, different models may have the same asymptotic smiles, see [17].

2.4. *Non large maturities and non extreme strikes*

To obtain approximation formulas in that situation, the asymptotics should originate from different large/small parameters that are rather related to the model and not to the contract characteristics (maturity and strike). These different asymptotics are generally well intuitively interpreted. For the sake of clarity, we spend time to detail a bit the arguments, in order to make clearer the differences between the further expansion results and the tools to obtain them. To the best of our knowledge, such comparative presentation does not exist in the literature and the reader may find it interesting.

2.4.1. *Small noise expansion*

This is inspired by the Freidlin-Wentzell approach [21] in which the noise in the stochastic differential equation of interest is small. Denote by Y the scalar SDE under study (which can be X or S in our framework), solution of

$$\mathrm{d}Y_t = \mu(t, Y_t)\mathrm{d}t + \nu(t, Y_t)\mathrm{d}W_t, \qquad Y_0 \text{ given.} \tag{14}$$

Assume that ν is small, or equivalently that ν becomes $\varepsilon\nu$ with $\varepsilon \to 0$: after making this small noise parameterization, the model reads

$$\mathrm{d}Y_t^\varepsilon = \mu(t, Y_t^\varepsilon)\mathrm{d}t + \varepsilon\nu(t, Y_t^\varepsilon)\mathrm{d}W_t, \quad Y_0^\epsilon = Y_0.$$

For $\varepsilon = 0$, it reduces to an ODE

$$Y_t^0 = y_{0,t} = Y_0 + \int_0^t \mu(s, y_{0,s})\mathrm{d}s \tag{15}$$

and this deterministic model serves as a zero-order approximation for the further expansion. Under smooth coefficient assumptions [21], we can derive a stochastic expansion of Y^ε in powers of ε:

$$Y_t^\varepsilon = y_{0,t} + \varepsilon Y_{1,t} + \frac{1}{2}\varepsilon^2 Y_{2,t} + o(\varepsilon^2). \tag{16}$$

For instance Y_1 solves a linear Gaussian SDE

$$\begin{aligned}Y_{1,t} &= \int_0^t \partial_x\mu(s, y_{0,s})Y_{1,s}\mathrm{d}s + \int_0^t \nu(s, y_{0,s})\mathrm{d}W_s \\ &= \int_0^t e^{\int_s^t \partial_x\mu(r,y_{0,r})\mathrm{d}r}\nu(s, y_{0,s})\mathrm{d}W_s.\end{aligned}$$

Similarly, Y_2 solves

$$\begin{aligned}Y_{2,t} &= \int_0^t [\partial_x\mu(s, y_{0,s})Y_{2,s} + \partial_x^2\mu(s, y_{0,s})Y_{1,s}^2]\mathrm{d}s + \int_0^t 2\partial_x\nu(s, y_{0,s})Y_{1,s}\mathrm{d}W_s \\ &= \int_0^t e^{\int_s^t \partial_x\mu(r,y_{0,r})dr}\big(\partial_x^2\mu(s, y_{0,s})Y_{1,s}^2\mathrm{d}s + 2\partial_x\nu(s, y_{0,s})Y_{1,s}\mathrm{d}W_s\big).\end{aligned}$$

Higher order expansions are available under higher smoothness assumptions. The notation $o(\varepsilon^2)$ in (16) means that the related error term has a L_p-norm (for any p) that can be neglected compared to ε^2 as $\varepsilon \to 0$. The stochastic expansion (16) becomes a weak expansion result when we compute $\mathbb{E}(h(Y_T))$ for a test function h.

▷ THE CASE OF SMOOTH h. If h is smooth enough, we obviously obtain

$$\begin{aligned}\mathbb{E}(h(Y_T)) =& h(y_{0,T}) + \varepsilon h'(y_{0,T})\mathbb{E}(Y_{1,T}) \\ &+ \varepsilon^2\big(h'(y_{0,T})\mathbb{E}(\frac{Y_{2,T}}{2}) + \frac{1}{2}h''(y_{0,T})\mathbb{E}(Y_{1,T}^2)\big) + o(\varepsilon^2).\end{aligned}$$

Observe that $\mathbb{E}(Y_{1,T}) = 0$ since $Y_{1,T}$ is a Wiener integral. To make the above expansion fully effective in practice, it is necessary to make the coefficients $\mathbb{E}(Y_{2,T})$ and $\mathbb{E}(Y_{1,T}^2)$ explicit: this is quite straightforward thanks to the linear equations solved by $Y_{1,.}$ and $Y_{2,.}$. The L_2-isometry property of the Wiener integral yields $\mathbb{E}(Y_{1,t}^2) = \int_0^t e^{2\int_s^t \partial_x\mu(r,y_{0,r})\mathrm{d}r}\nu^2(s, y_{0,s})\mathrm{d}s$. In addition, we have $\mathbb{E}(Y_{2,t}) = \int_0^t e^{\int_s^t \partial_x\mu(r,y_{0,r})\mathrm{d}r}\partial_x^2\mu(s, y_{0,s})\mathbb{E}(Y_{1,s}^2)\mathrm{d}s$. The coefficients computation is reduced to the evaluation of nested time-integrals which are simple to compute using standard n-points integral discretization, with a computational complexity* of order n. The above expansion analysis is a *regular perturbation* analysis, using a *stochastic analysis* point of view.

To derive this expansion in powers of ε, we could alternatively use a *PDE point of view* based on Feynman-Kac representation, which states that $u^\varepsilon : (t,x) \mapsto u^\varepsilon(t,x) = \mathbb{E}(h(Y_T^\varepsilon)|Y_t^\varepsilon = x)$ solves the perturbed PDE

$$\begin{cases} \partial_t u^\varepsilon(t,x) + \mu(t,x)\partial_x u^\varepsilon(t,x) + \frac{1}{2}\varepsilon^2\nu^2(t,x)\partial_x^2 u^\varepsilon(t,x) = 0 & \text{for } t < T, \\ u^\varepsilon(T,x) = h(x). \end{cases}$$

Setting $\mathcal{L}^\varepsilon = \partial_t + \mu\partial_x + \frac{1}{2}\varepsilon^2\nu^2\partial_x^2 := \mathcal{L}_0 + \varepsilon^2\mathcal{L}_2$, the above PDE writes $\mathcal{L}^\varepsilon u^\varepsilon = 0$ plus boundary conditions at time T. Seeking an expansion of the form $u^\varepsilon = u_0 + \varepsilon u_1 + \frac{1}{2}\varepsilon^2 u_2 + o(\varepsilon^2)$, we obtain

$$\mathcal{L}_0 u_0 + \varepsilon\mathcal{L}_0 u_1 + \varepsilon^2[\frac{1}{2}\mathcal{L}_0 u_2 + \mathcal{L}_2 u_0] + o(\varepsilon^2) = 0.$$

A formal identification of each coefficient of ε^i $(i = 0, 1 \dots)$ to 0, we obtain a system of PDEs:

$$\mathcal{L}_0 u_0 = 0, \qquad \mathcal{L}_0 u_1 = 0, \qquad \frac{1}{2}\mathcal{L}_0 u_2 + \mathcal{L}_2 u_0 = 0,$$

with the boundary conditions $u_0(T,.) = h(.), u_1(T,.) = u_2(T,.) = 0$. The justification of this kind of expansion and its related error analysis can be made under appropriate smoothness assumptions on h, μ and ν; we refer to [22] , [23] or [24] where a similar error analysis is made. The PDE solutions are then given by

$$u_0(t,x) = h(y_T^{t,x}), \qquad u_1 \equiv 0, \qquad u_2(t,x) = \int_t^T 2\mathcal{L}_2 u_0(s, y_s^{t,x})\mathrm{d}s$$

where $(y_s^{t,x})_{s\geq t}$ stands for the solution of the ODE (15) with initial condition (t,x). Under this form of system of PDEs, the derivation of an explicit expression for u_2 is not as easy as within the stochastic analysis approach. However, we can obtain the same expansion (fortunately!), i.e. the same formula for u_2 at $(0, Y_0)$:

$$u_2(0, Y_0) = h'(y_{0,T})\mathbb{E}(Y_{2,T}) + h''(y_{0,T})\mathbb{E}(Y_{1,T}^2) \tag{17}$$

with $\mathbb{E}(Y_{2,T})$ and $\mathbb{E}(Y_{1,T}^2)$ given as before. To see this, start from $\mathcal{L}_2$ and write $u_2(t,x) = \int_t^T \nu^2(s, y_s^{t,x})\partial_y^2 u_0(s, y_s^{t,x})\mathrm{d}s$. We have $\partial_x u_0(t,x) = h'(t, y_T^{t,x})\partial_x y_T^{t,x}$ and $\partial_x^2 u_0(t,x) = h'(t, y_T^{t,x})\partial_x^2 y_T^{t,x} + h''(t, y_T^{t,x})(\partial_x y_T^{t,x})^2$. Then to recover (17), use the notation $y_{0,t} = y_t^{0,Y_0}$, the flow property $y_s^{t,y_{0,t}} = y_{0,s}$ for $s \geq t$, and the explicit expressions for $\partial_x y_s^{t,x}$ and $\partial_x^2 y_s^{t,x}$: for instance $\partial_x y_s^{t,x} = 1 + \int_t^s \partial_x\mu(r, y_r^{t,x})\partial_x y_r^{t,x}\mathrm{d}r = e^{\int_t^s \partial_x\mu(r, y_r^{t,x})\mathrm{d}r}$. We skip further details. This completes the *PDE* approach to derive a *regular perturbation* analysis. Observe that the derivation of explicit formula is

*Observe that although the time integrals are multidimensional, we are reduced to iterative one-dimensional computations since the function to integrate is separable in all its variables.

delicate because of the system of PDEs to solve (more complicated than solving the system of SDEs arising within the stochastic analysis approach).

$\triangleright$ THE CASE OF NON-SMOOTH h. The previous derivation which involves h', h'' and possibly higher derivatives is mathematically incorrect if h is not smooth. This fact is clear using the stochastic analysis approach. It is also clear using PDE arguments: indeed, it would involve the perturbed PDE solution $(t,x) \mapsto \mathbb{E}(h(Y_T^\varepsilon)|Y_t^\varepsilon = x)$, that is not uniformly smooth (because the regularization parameter ε shrinks to 0). If $h(x) = \mathbf{1}_{x\geq K}$ (like digital payoff), i.e. we evaluate $p^\varepsilon = \mathbb{P}(Y_T^\varepsilon \geq K)$, and $y_{0,T} \neq K$, large deviation arguments [25] show that the probability p^ε is exponentially close to 0 or 1 w.r.t. $1/\varepsilon^2$ (i.e. $\log(p^\varepsilon) \approx -c/\varepsilon^2$ if $y_{0,T} < K$), and thus an expansion in power of ε provides zero coefficients at any order. To get a non degenerate and interesting situation, we should consider the case K is close to $y_{0,T}$ in the sense $K = y_{0,T} + \lambda\,\varepsilon$, that is

$$p^\varepsilon = \mathbb{P}(Y_T^\varepsilon \geq y_{0,T} + \lambda\,\varepsilon) = \mathbb{P}(\frac{Y_T^\varepsilon - y_{0,T}}{\varepsilon} \geq \lambda).$$

In other words, to overcome the difficulty of the singularity of h, we have leveraged a homogenization argument (*singular perturbation*), by considering the rescaled variable (usually called fast variable) $Z_t^\varepsilon = \frac{Y_t^\varepsilon - y_{0,t}}{\varepsilon} = Y_{1,t} + \frac{1}{2}\varepsilon Y_{2,t} + o(\varepsilon)$. If the law of $Y_{1,T}$ is non degenerate (for instance Gaussian law with non-zero variance), the latter quantity can be expanded in powers of ε. Actually, for less specific functions h, Watanabe [26] has developed a Malliavin calculus-based machinery to establish a general expansion result of $\mathbb{E}(h(Z_T^\varepsilon))$ in powers of ε, available even for Schwartz distributions h, assuming stochastic expansions in Malliavin sense of

$$Z_T^\varepsilon = Z_{0,T} + \varepsilon Z_{1,T} + \varepsilon^2 Z_{2,T} + \cdots + \varepsilon^n Z_{n,T} + O(\varepsilon^{n+1})$$

for any $n \geq 1$ and asymptotic (in ε) non-degeneracy in Malliavin sense of Z_T^ε:

$$\limsup_{\varepsilon\to 0} \|1/\det(\gamma^{Z_T^\varepsilon})\|_p < +\infty \tag{18}$$

for any $p \geq 1$, where γ^Z is the Malliavin covariance matrix of a random variable Z. The Watanabe result states the existence of random variables $(\pi_k)_{k\geq 1}$ such that for any polynomially bounded function h, we have

$$\mathbb{E}(h(Z_T^\varepsilon)) = \mathbb{E}(h(Z_{0,T})) + \sum_{k=1}^{n} \varepsilon^k \mathbb{E}(h(Z_{0,T})\pi_k) + O(\varepsilon^{n+1}), \qquad \forall n \geq 1. \tag{19}$$

Compared to the non-smooth case, the possibility to get an expansion result is due to the non-degeneracy condition which has a (asymptotic) regularization effect on the non-smooth function h. With our previous notation $Z_T^\varepsilon = \frac{Y_T^\varepsilon - y_{0,T}}{\varepsilon} = Y_{1,T} + \frac{1}{2}\varepsilon Y_{2,T} + o(\varepsilon)$, the asymptotic non-degeneracy (18) implies that the Gaussian random variable $Y_{1,T}$ has a non-zero variance, i.e. $\int_0^T e^{2\int_s^T \partial_x\mu(r,y_{0,r})\mathrm{d}r}\nu^2(s,y_{0,s})\mathrm{d}s > 0$: in the case of time-independent coefficient $\mu(s,y) = \mu(y), \nu(s,y) = \nu(y)$, it reads $\nu(y_{0,T}) \neq 0$. The converse result ($\nu(y_{0,T}) \neq 0$ implies (18)) holds true in the case of time-independent coefficient and in a multidimensional setting, see [26] . Yoshida [27] has weakened the assumption (18) into a localized version allowing degeneracy on a set of polynomially small probability measure. This approach has been successfully applied to different pricing problems in finance, mainly by Yoshida, Takahashi and their co-authors: see [28; 29; 30] or the unpublished work [31]. Their methodology consists in expanding the density of the random variable $Z_T^\varepsilon = \frac{Y_T^\varepsilon - y_{0,T}}{\varepsilon}$ using the Gaussian density as the zero-order term, and then going back to $\mathbb{E}(h(Y_T^\varepsilon))$ by suitable integration computations. The advantage of this approach is that the expansion result (19) holds in a large generality, provided that we assume infinitely differentiable coefficients and uniform non degeneracy. However, observe two difficulties or restrictions:

- within usual financial models like Heston model, the required regularity assumption is not satisfied and we even know that the Malliavin differentiability of high order may fail, see [32].
- the existence of the Malliavin weights $(\pi_k)_k$ does not provide an explicit and numerically computable expansion: very involved additional computations are required to obtain explicit formulas. One might compare these tricky computations to those necessary to solve the aforementioned system of PDEs.

Last, this approach usually leads to normal approximations of financial models (Bachelier prices) whereas log-normal approximations (Black-Scholes prices) might be more accurate (numerical evidences are given in Section 7).

After this presentation of *singular perturbation* using stochastic analysis, we now turn to the PDE approach. It has been developed in the financial context by Hagan and co-authors [5; 33]. To be as close as possible to the quoted work, assume that the drift coefficient is $\mu \equiv 0$. In the case of Call payoff, the original valuation PDE u^ε writes

$$\begin{cases} \partial_t u^\varepsilon(t,x) + \frac{1}{2}\varepsilon^2\nu^2(t,x)\partial_x^2 u^\varepsilon(t,x) = 0 & \text{for } t < T, \\ u^\varepsilon(T,x) = (x-K)_+; \end{cases}$$

now, if we consider ATM strikes ($K - Y_0 = O(\varepsilon)$ similarly to before), we should consider the fast variable $y = (x-K)/\varepsilon$ and the rescaled solution $v^\varepsilon(t,y) = \frac{1}{\varepsilon}u^\varepsilon(t, K+\varepsilon y)$ which solves

$$\begin{cases} \partial_t v^\varepsilon(t,y) + \frac{1}{2}\nu^2(t,K+\varepsilon y)\partial_y^2 v^\varepsilon(t,y) = 0 & \text{for } t < T, \\ v^\varepsilon(T,y) = y_+. \end{cases} \tag{20}$$

At this stage, the analysis follows the routine similar to before, by seeking a solution under the form

$$v^\varepsilon = v_0 + \varepsilon v_1 + o(\varepsilon) \tag{21}$$

solving $\mathcal{L}^\varepsilon v^\varepsilon = 0$ where $\mathcal{L}^\varepsilon = \partial_t + \frac{1}{2}\nu^2(t, K + \varepsilon y)\partial_y^2 = \mathcal{L}_0 + \varepsilon\mathcal{L}_1 + o(\varepsilon)$ with $\mathcal{L}_0 = \partial_t + \frac{1}{2}\nu^2(t,K)\partial_y^2$, $\mathcal{L}_1 = \nu\nu'(t,K)y\partial_y^2$. A formal identification leads to a system of PDEs:

$$\mathcal{L}_0 v_0 = 0,\ v_0(T,y) = y_+ \quad \text{and} \quad \mathcal{L}_0 v_1 + \mathcal{L}_1 v_0 = 0,\ v_1(T,y) = 0.$$

The solution v_0 is obviously given by the Call price in a Bachelier model (8) $\mathrm{d}X_t^{\mathrm{BA}} = \nu(t,K)\mathrm{d}W_t$ with time-dependent diffusion coefficient, and the first correction is given by $v_1(t,y) = \mathbb{E}(\int_t^T \mathcal{L}_1 v_0(s, X_s^{\mathrm{BA}})\mathrm{d}s | X_t^{\mathrm{BA}} = y)$. Although the new terminal function $h(y) = y_+$ is not infinitely smooth, non-zero function ν induces a smoothing effect due to a non-degenerate heat kernel (this feature is analogous to the previous non-degeneracy in the Malliavin sense): hence, v_0 is smooth with derivatives possibly blowing up as time gets close to T and a careful analysis shows that v_1 is well defined too. Here again, the explicit computation of v_1 is not an easy exercise and it requires some tricks. Finally v_1 can be written as the weighted sum of derivatives of v_0 (interpreted as Greeks). In Sections 3 and 5, we provide a more direct and generic way to compute this kind of correction terms using stochastic analysis instead of PDE arguments.
Regarding the careful justification of the above PDE regular expansion with error estimates, quite surprisingly we have not been able to find literature references when the terminal condition is non-smooth (like $y \mapsto y_+$).
Once obtained the expansion of v^ε for a given local volatility function $\sigma(.,.)$ (i.e. $\nu(t,y) = y\,\sigma(t,y)$), one can derive an expansion of the Black-Scholes implied volatility σ_{I} by identifying the previous expansion with that in the case ($\nu_{\mathrm{I}}(t,y) = y\,\sigma_{\mathrm{I}}$): see [5] where the analysis is successfully performed for time-independent volatility $\sigma(t,y) = \sigma(y)$ (or separable function $\sigma(t,y) = \sigma(y)\alpha(t)$ by a simple time-change). It is possible that the case of general time-dependent volatility has been considered out of reach by the authors of [5; 33] using PDE arguments, whereas we will see later how much stochastic analysis tools are suitable even in the case time-dependent coefficients.

2.4.2. *Short maturity*

In this asymptotics, the terminal time T is small. When one has to evaluate $\mathbb{E}(h(Y_T))$ for Y solution of the SDE (14) and for h smooth (say infinitely differentiable with bounded derivatives), iterative applications of Itô's formula give

$$\begin{aligned}\mathbb{E}(h(Y_T)) &= h(Y_0) + \int_0^T \mathbb{E}([\mathcal{L}h](t, Y_t))\mathrm{d}t \\ &= h(Y_0) + T[\mathcal{L}h](0, Y_0) + \int_0^T \int_0^t \mathbb{E}([\mathcal{L}^2 h](s, Y_s))\mathrm{d}s\mathrm{d}t\end{aligned}$$

where $\mathcal{L}$ is the infinitesimal generator associated to Y. Iterating the procedure, we obtain an expansion in powers of T:

$$\mathbb{E}(h(Y_T)) = \sum_{k=0}^{n} \frac{T^k}{k!}[\mathcal{L}^k h](0, Y_0) + O(T^{n+1}), \qquad n \geq 0.$$

The numerical evaluation of such formula is straightforward. We refer the reader to [34] for a more comprehensive exposure of related Itô-Taylor expansions.

As in the case of small noise expansion, the case of non-smooth h requires a different treatment because $\mathcal{L}^k h$ is not defined. For this, we transform the problem of small terminal time with fixed coefficients into a problem of fixed terminal time with small coefficients, by leveraging the scaling property of the Brownian motion. Actually, having T small is equivalent to replace T by $\varepsilon^2 T$ with $\varepsilon \to 0$: then, starting from the SDE (14), we consider the time-rescaled process $(Y_{\varepsilon^2 t})_{0\leq t\leq T}$ which has the same distribution as $(Y_t^\varepsilon)_{0\leq t\leq T}$ defined as the solution of

$$\mathrm{d}Y_t^\varepsilon = \varepsilon^2 \mu(\varepsilon^2 t, Y_t^\varepsilon)\mathrm{d}t + \varepsilon\nu(\varepsilon^2 t, Y_t^\varepsilon)\mathrm{d}W_t, \quad Y_0^\epsilon = Y_0, \tag{22}$$

see [26]. Observe that this leads to a different parameterization compared to the small noise case (in particular, the drift coefficient is multiplied by ε^2). However the expansion methodology is similar: in the case of non-smooth function, it is more appropriate to rescale the process by setting $Z_t^\varepsilon = \frac{Y_t^\varepsilon - y_{0,t}}{\varepsilon} = Y_{1,t} + \frac{1}{2}\varepsilon Y_{2,t} + o(\varepsilon)$, where

$$\begin{aligned} Y_{1,t} &= \partial_\varepsilon Y_t^\varepsilon|_{\varepsilon=0} = \nu(0, Y_0)W_t, \\ Y_{2,t} &= \partial_\varepsilon^2 Y_t^\varepsilon|_{\varepsilon=0} = 2\mu(0, Y_0)t + 2\partial_y\nu(0, Y_0)\int_0^t Y_{1,s}\mathrm{d}W_s \\ &= 2\mu(0, Y_0)t + \partial_y\nu(0, Y_0)\nu(0, Y_0)(W_t^2 - t). \end{aligned}$$

Once the *fast variable* is selected, observe that we are reduced to a regular perturbation problem, that can be handled using stochastic analysis tools (namely Watanabe approach [26]) or using PDE tools. We skip details since it is similar to what have been presented before. See also the book by Henry-Labordère [6], where short-time asymptotics of density functions are derived through geometry considerations.

2.4.3. *Fast volatility*

Since the end of the nineties, another popular approximation approach has been developed by Fouque, Papanicolaou and Sircar, see [35; 36]. It emphasizes that the asset volatility $(\sigma_t)_{t\geq 0}$ has usually slow variations compared to the variations of the asset itself (multiscale modeling). This is achieved in two different ways.

- Either the natural time scale of stochastic volatility is short, which leads to a model of the form (22) for $(\sigma_t)_t$, while the asset dynamics is unchanged. Thus, at order zero, we obtain a Black-Scholes model with a constant volatility equal to the initial stochastic volatility σ_0, see [35] .

- Or the fluctuations of the stochastic volatility $(\sigma_t)_t$ are so fast that they give the appearance of a constant (in time) volatility, when considered at a longer time scale. This second point of view has been much developed by Fouque, Papanicolaou and Sircar and their co-authors, in many respects, and this is presented below.

As an illustration of their methodology, we consider the asset model $\mathrm{d}S_t = S_t\sigma_t\mathrm{d}W_t$ and an Ornstein-Uhlenbeck process for modeling $(\sigma_t = f(\Sigma_t))_t$ with

$$\mathrm{d}\Sigma_t = \frac{1}{\varepsilon}(\overline{\Sigma}_\infty - \Sigma_t)\mathrm{d}t + v\sqrt{\frac{2}{\varepsilon}}\mathrm{d}B_t.$$

For instance in the Scott model [37], $f(x) = e^x$ and (W, B) is a standard bi-dimensional correlated Brownian motion ($\mathrm{d}\langle W, B\rangle_t = \rho\mathrm{d}t$). As time goes to infinity, the random variable Σ_t weakly converges to the stationary Gaussian distribution with mean $\overline{\Sigma}_\infty$ and variance $(v\sqrt{\frac{2}{\varepsilon}})^2/(2\frac{1}{\varepsilon}) = v^2$. In other words, although the fluctuations are fast (the characteristic time being ε), the distribution remains the same (at least for time larger than ε). It allows the application of ergodic theorem to obtain large-time asymptotics of integrals of the realized volatility: for any polynomially bounded function Ψ, we have

$$\lim_{T\to+\infty}\frac{1}{T}\int_0^T \Psi(\sigma_s)\mathrm{d}s = \int_{\mathbb{R}} \Psi(e^y)\frac{1}{\sqrt{2\pi v^2}}e^{-(y-\overline{\Sigma}_\infty)^2/(2v^2)}\mathrm{d}y := \sigma_{BS}^2$$

in the almost sure sense and in the L_1 sense. For $\Psi(y) = y^2$, we obtain a constant large-time approximation of $\frac{1}{T}\int_0^T \sigma_s^2\mathrm{d}s$ to be used as a zero-order approximation in a Black-Scholes formula. To derive correction terms, the authors employ *singular pertubation* PDE techniques: indeed, the price function $u^\varepsilon(t, x, y) = \mathbb{E}(h(S_T)|S_t = x, \Sigma_t = y)$ solves $\mathcal{L}^\varepsilon u^\varepsilon = 0$ with

$$\begin{aligned}\mathcal{L}^\epsilon &= \partial_t + \frac{1}{2}x^2 f^2(y)\partial^2_{xx} + \sqrt{\frac{2}{\varepsilon}}\rho v x f(y)\partial^2_{xy} + \frac{1}{\varepsilon}(v^2\partial^2_y + (\overline{\Sigma}_\infty - y)\partial_y)\\ &:= \mathcal{L}_0 + \frac{1}{\sqrt{\varepsilon}}\mathcal{L}_1 + \frac{1}{\varepsilon}\mathcal{L}_2.\end{aligned}$$

As in the previous approaches, by decomposing u^ε in powers of $\sqrt{\varepsilon}$ and by gathering the contributions of the same order, we obtain a system of PDEs characterizing the main order term and the correction terms. Actually the analysis is quite intricate because one has to take into account the ergodic property of σ (which leads to solving elliptic PDEs of the form of Poisson equation): see [36] where the error analysis is made for smooth payoffs and [38] for the Call option case. The final approximation pricing formula writes

$$\begin{aligned}\mathbb{E}(S_T - K)_+ &= \mathrm{Call}^{\mathrm{BS}}(\log(S_0), T\sigma_{BS}^2, \log(K))\\ &\quad + \sqrt{\varepsilon}\times\text{linear combination of } \partial_S^i\mathrm{Call}^{\mathrm{BS}}(\log(S_0), T\sigma_{BS}^2, \log(K))\\ &\quad \text{for } i = 2, 3 + \ldots\end{aligned}$$

with some explicit coefficients as the factors for the Greeks. Consequently, the approximation formula is straightforward to evaluate on a computer since Black-Scholes price and Greeks are known in closed form and available in any pricing software. In this analysis and similarly to any PDE approaches, assuming time homogeneous coefficients simplifies much the derivation of explicit formula. In the context of fast volatility, some extensions are possible, see [39].

2.4.4. *Proxy expansion*

We complete our overview section by presenting a different point of view, which is going to be developed further in the next sections. As a difference with previous works, this is rather a non-asymptotic approach, relying on the *a priori* knowledge of proxy of the model to handle; for this reason, it may appear as more understandable and more intuitive for practioners. Consider the model (2) on S:

$$\mathrm{d}S_t = S_t\sigma(t, S_t)\mathrm{d}W_t,$$

and assume that by expertise, S behaves closely to a Gaussian model, i.e. the fluctuations of $S_t\sigma(t, S_t)$ are small. Then, it is reasonable to take the Bachelier model S^P with parameter $(\Sigma_t = S_0\sigma(t, S_0))_t$ as a proxy, that is

$$\mathrm{d}S^P_t = \Sigma_t \mathrm{d}W_t, \qquad S^P_0 = S_0. \qquad \text{(NORMAL PROXY)}$$

The Call price in S model should be close to that in the proxy; since this approximation may be too crude, it is recommended to add correction terms.
Alternatively, one could guess that S rather behaves as a log-normal model with parameter $(a_t)_t$, i.e. $X = \log(S)$ may be approximated by

$$\mathrm{d}X^P_t = -\frac{1}{2}a_t^2\mathrm{d}t + a_t dW_t, \qquad X^P_0 = x_0. \qquad \text{(LOG-NORMAL PROXY)}$$

The proxy volatility may be taken to $a_t = a(t, x_0) = \sigma(t, S_0)$ for instance, but another point could be chosen (for instance, the strike K or the mid-point $(K + S_0)/2$). This description does not put an emphasize on a specific asymptotic, but one has to quantify how $S_t\sigma(t, X_t) \approx \Sigma_t$ or $a(t, X_t) \approx a_t$.

To compute correction terms to the relation $\mathbb{E}(S_T - K)_+ \approx \mathbb{E}(S^P_T - K)_+$ or $\mathbb{E}(e^{X^P_T} - K)_+$, it is necessary to derive a convenient representation of the distance to the proxy $S_T - S^P_T$ or $X_T - X^P_T$. The linear interpolation $X^\eta_T = X^P_T + \eta(X_T - X^P_T)$ does not lead to illuminating computations. It is much better to consider the following *interpolation*: for $\eta \in [0, 1]$, define

$$dX^\eta_t = \eta(-\frac{1}{2}a^2(t, X^\eta_t)dt + a(t, X^\eta_t)dW_t), \quad X^\eta_0 = x_0. \tag{23}$$

Note that η is not a small parameter but an *interpolation* parameter. Observe also that this parameterization is different from that in small time or small noise asymptotics.

A direct computation shows that $X^1_t = X_t$, $X^0_t = x_0$ and $\partial_\eta X^\eta_t|_{\eta=0} = \int_0^t a(s, x_0)[\mathrm{d}W_s - \frac{1}{2}a(s, x_0)\mathrm{d}s]$: this shows that $X_t - X^P_t = X^1_t - (X^0_t + \partial_\eta X^\eta_t|_{\eta=0})$

writes as a Taylor formula at order 1. Thus, the natural candidate for the first contribution in $X_t - X_t^P$ is $\frac{1}{2}\partial_\eta^2 X_t^\eta|_{\eta=0}$. The above interpolation is equivalent to

$$d\hat{X}_t^\eta = -\frac{1}{2}a^2(t, x_0 + \eta(\hat{X}_t^\eta - x_0))dt + a(t, x_0 + \eta(\hat{X}_t^\eta - x_0))dW_t, \quad \hat{X}_0^\eta = x_0, \tag{24}$$

which is related to X_t^η by the relation $X_t^\eta = x_0 + \eta(\hat{X}_t^\eta - x_0)$.

2.5. *Asymptotic expansion versus non-asymptotic expansion*

Deriving an asymptotic expansion sheds the light on the crucial role of one model parameter compared to the other ones, to explain and approximate the option prices: for instance, in small noise expansion, we focus only on the volatility by putting an ε in front of the dW-term, and so on. It finally leads to a generic expansion of the form

$$u^\varepsilon = u_0 + \varepsilon u_1 + \frac{1}{2}\varepsilon^2 u_2 + \dots \tag{25}$$

or in powers of $\sqrt{\varepsilon}$ in the *fast volatility* framework. Our previous discussion has shown how this expansion is obtained in a Markovian framework using PDE (regular or singular pertubations), or more generally using Malliavin calculus (Watanabe approach).

Actually, it is important to observe that writing such an expansion implicitly means that apart of the parameter related to ε, the other parameters have no important asymptotics for the problem under consideration. Below we consider a simple toy example to show that there might be a competition between all the model parameters, and moreover there is a necessary trade-off with the payoff regularity. In other words, deriving (25) does not necessarily mean that the first order approximation $u_0 + \varepsilon u_1$ is really accurate and taking more terms do not necessarily improve the accuracy, because of the possible crucial influence of other large or small parameters. Our toy example is the following perturbed Brownian model:

$$X_1^\varepsilon = \sigma W_1 + \sqrt{\varepsilon} B_1$$

where (W, B) is a two-dimensional Brownian motion, and σ is positive. This toy model can be viewed as the simplest way to perturb a volatility model (we could have taken $B = W$ without changing the conclusion of the discussion below) and thus, it is quite realistic compared to the further situations to handle.

(1) *Case* $h(x) = 1 + x^2$. We have $\mathbb{E}[h(X_1^\varepsilon)] = 1 + \sigma^2 + \varepsilon = \mathbb{E}[h(X_1^0)] + \varepsilon$.
(2) *Case* $h(x) = 1 + x_+$. By a scaling argument, we have:

$$\begin{aligned}\mathbb{E}[h(X_1^\varepsilon)] =& 1 + \sqrt{\sigma^2 + \varepsilon}\,\mathbb{E}[(W_1)^+] \\ =& \mathbb{E}[h(X_1^0)] + \frac{1}{2}\frac{\varepsilon}{\sigma}\mathbb{E}[(W_1)^+] + O(\frac{\varepsilon^2}{\sigma^3}).\end{aligned}$$

(3) *Case* $h(x) = 1_{x \le x_0}$. We have:

$$\begin{aligned}\mathbb{E}[h(X_1^\varepsilon)] =& \mathcal{N}\left(\frac{x_0}{\sqrt{\sigma^2+\varepsilon}}\right)\\ =& \mathcal{N}\left(\frac{x_0}{\sigma}\right) - \mathcal{N}'\left(\frac{x_0}{\sigma}\right)\frac{x_0}{\sigma}\frac{\varepsilon}{2\sigma^2} + O(\frac{\varepsilon^2}{\sigma^4}).\end{aligned}$$

These simple computations show that the expansion order depends on the relative magnitude of ε and σ, and also on the regularity of the function h. For instance, if σ is also small, say $\varepsilon = \sigma^3 \to 0$, then the expansion order w.r.t. σ in the case (1), (2), (3) is respectively equal to 3, 2 and 1. These subtleties do not appear in the expansions (19) of Watanabe type or (21) of PDE type, because the focus is made only on a single small parameter ε.

It means that in some situations, asymptotic expansions may be misleading or may not give the best possible approximations; then, we should take into account the influence of all (or many) model parameters. In the context of fast volatility, multi-scale modeling and its related asymptotic analysis are very recently developed in [24]; see also [40].

In the sequel of this work, we consider non asymptotic expansions, mainly for local volatility models, and analyse the approximation error taking into account several parameters at the same time, in order to determine to which extent they play complementary or opposite roles. For instance, it is informative to see the simultaneous influences of maturity, of volatility amplitude or of derivatives of volatility function on the option prices. Their impacts depend on the payoff smoothness: the accuracy is expected to be improved for smooth payoff compared to non-smooth payoffs.

Final considerations. After this (hopefully complete) overview, the reader may wonder what is the best approximation method among those presented. Of course, it depends on the required accuracy and the computational time allowed for the numerical evaluation. From this point of view, all methods are not equivalent. The choice of relevant asymptotics/approximations guarantees to catch the main features of the pricing problem, and as a consequence, it will likely lead to an expansion of low order to achieve a good accuracy (with low computational time or complexity). In these respects, the proxy expansion has immediate advantages: the better or the more intuitive the proxy, the smaller the number of correction terms.

One should also take care of the preservation of some model properties in the approximation.

- One of them is the martingale property of $S = e^X$ (serving as a base for Call/Put parity relation). For instance, a small noise approximation of X defined in (3) does not maintain the martingale property since the volatility coefficient is scaled by ε while the drift remains unchanged: as a result, the final approximation may suffer from numerical arbitrage.

- Another property is positivity of S. Taking a Normal Proxy for S may give wrong results if the values of S close to 0 have a prominent role in the computation of $\mathbb{E}(h(S_T))$ (for instance, Call/Put with small strikes).

These kinds of consideration may help to choose between different methods, with the additional help of comparative numerical tests.

3. Approximation based on proxy

3.1. *Notations and definitions*

The following notations and definitions are repeatedly used in this work.

▷ **Differentiation.** If these derivatives have a meaning, we write $l_t^{(i)}(x) = \partial_{x^i}^i l(t,x)$ for any function l of two variables.

▷ **Integral Operator.** The integral operator ω^T is defined as follows: for any measurable and bounded function l, we set

$$\omega(l)_t^T = \int_t^T l_u du,$$

for $t \in [0,T]$. Its n-times iteration is defined analogously: for any measurable and bounded functions $(l_1, \cdots, l_n)$, we set

$$\omega(l_1, \cdots, l_n)_t^T = \omega(l_1 \omega(l_2, \cdots, l_n)_.^T)_t^T,$$

for $t \in [0,T]$.

▷ **Time reversal.** For any measurable and bounded function l, we denote by $\widetilde{l}$ the function $\widetilde{l}_t = l_{T-t}$ for any $t \in [0,T]$. Notice the relation

$$\omega(\widetilde{l}_1, \widetilde{l}_2, .., \widetilde{l}_n)_0^T = \omega(l_n, l_{n-1}, .., l_1)_0^T \tag{26}$$

available for any measurable and bounded functions $(l_1, \cdots, l_n)$: in other words, reversing the time of integrands is equivalent to change the order of integration.

▷ **Quadratic mean on** $[0,T]$**.** For any measurable function $(l(t,x))_{(t,x)\in[0,T]\times\mathbb{R}}$ of two variables, bounded w.r.t. the time variable for any $x \in \mathbb{R}$, we denote by $\bar{l}_z$ its quadratic mean on $[0,T]$ at the spatial point z defined by:

$$\bar{l}_z = \sqrt{\frac{1}{T}\int_0^T l_t^2(z)\mathrm{d}t}.$$

This notation is frequently used for the function a at the points $z = x_0, k, x_{\text{av}}$ and for the function Σ at $z = S_0, K, S_{\text{av}}$.

▷ **Assumptions on** a **and** Σ**.**

- $(\mathcal{H}^a)$: a is a bounded measurable function of $(t,x) \in [0,T] \times \mathbb{R}$, and five times continuously differentiable in x with bounded* derivatives. Set
$$\mathcal{M}_1(a) = \max_{1\leq i\leq 5} \sup_{(t,x)\in[0,T]\times\mathbb{R}} |\partial_x^i a(t,x)|$$
$$\text{and } \mathcal{M}_0(a) = \max_{0\leq i\leq 5} \sup_{(t,x)\in[0,T]\times\mathbb{R}} |\partial_x^i a(t,x)|.$$
In addition, there exists a constant $c_a > 0$ such that $|a(t,x)| \geq c_a$ for any $(t,x) \in [0,T] \times \mathbb{R}$.
- $(\mathcal{H}_z^a)$: assume $(\mathcal{H}^a)$ by replacing the last uniform ellipticity by the single condition $\int_0^T |a(t,z)|^2 \mathrm{d}t > 0$.

The above hypothesis will be considered at $z = x_0, z = k$ or $z = x_{\mathrm{av}}$.
We define similarly $(\mathcal{H}^\Sigma)$ or $(\mathcal{H}_z^\Sigma)$ by replacing a by Σ in $(\mathcal{H}^a)$ and $(\mathcal{H}_z^a)$. Then the hypothesis will be considered at $z = S_0, z = K$ or $z = S_{\mathrm{av}}$.

▷ **Constants.** Our next error estimates are stated following the notation below.

- "$A = \mathcal{O}(B)$" means that $|A| \leq CB$: here, C stands for a generic constant that is a non-negative increasing function of T, $\mathcal{M}_1(a)$, $\mathcal{M}_0(a)$ and of the oscillation ratio $\frac{\mathcal{M}_0(a)}{c_a}$ (if $(\mathcal{H}^a)$ is fulfilled) or $\frac{[\mathcal{M}_0(a)]^2 T}{\int_0^T |a(t,z)|^2 \mathrm{d}t}$ (if $(\mathcal{H}_z^a)$ is fulfilled).
If $(\mathcal{H}^\Sigma)$ or $(\mathcal{H}_z^\Sigma)$ is satisfied, in the above dependence a has to replaced by Σ.
Usually, a generic constant may depend on S_0, x_0, K and k; nevertheless, it remains uniformly bounded in these variables: it is possible to derive exact dependency but we skip it to keep the analysis short.
- Similarly, if A is positive, $A \leq_c B$ means that $A \leq CB$ for a generic constant C.

3.2. *Proxy approximation: a primer using the local volatility at spot*

▷ **Log-normal proxy.** Assume by expertise that the model (2) behaves closely to a log-normal model, in the sense that a log-normal approximation seems to be reasonable. For instance, in the case of CEV type model
$$S\sigma(t,S) = \nu_t S^{\beta_t}, \tag{27}$$
a log-normal heuristics is associated to β close to 1. Some numerical illustrations are given later.

As a first log-normal approximation, we freeze the volatility in space to the initial spot value: regarding the log-asset X defined in (3), it writes
$$\mathrm{d}X_t^P = -\frac{1}{2}a^2(t,x_0)\mathrm{d}t + a(t,x_0)\mathrm{d}W_t, \qquad X_0^P = x_0.$$

*the boundedness assumption of a and its derivatives could be weakened to L_p-integrability conditions, up to extra work.

We refer to this proxy model as *log-normal proxy with volatility at spot*. The evaluation of the next correction terms requires a suitable representation of the distance between the model and the proxy: for this, we use the interpolated process (23) given by

$$dX^\eta_t = \eta(-\frac{1}{2}a^2(t,X^\eta_t)\mathrm{d}t + a(t,X^\eta_t)\mathrm{d}W_t), \quad X^\eta_0 = x_0.$$

for an interpolation parameter $\eta \in [0,1]$. Under $(\mathcal{H}^a_{x_0})$, the three first derivatives of $\eta \mapsto X^\eta_t$ are well defined (a.s. simultaneously for any t, see [41]). Denote by $X^\eta_{i,t}$ and $X_{i,t}$ the i-th derivative respectively at η and $\eta = 0$. Direct computations yield

$$\begin{aligned} dX^\eta_{1,t} =& -\frac{1}{2}a^2(t,X^\eta_t)dt + a(t,X^\eta_t)dW_t \\ &+ \eta X^\eta_{1,t}(-[a\partial_x a](t,X^\eta_t)\mathrm{d}t + \partial_x a(t,X^\eta_t)\mathrm{d}W_t), \quad X^\eta_{1,0} = 0. \qquad (28) \\ dX^\eta_{2,t} =& 2X^\eta_{1,t}(-[a\partial_x a](t,X^\eta_t)\mathrm{d}t + \partial_x a(t,X^\eta_t)\mathrm{d}W_t) \\ &+ \eta X^\eta_{2,t}(-[a\partial_x a](t,X^\eta_t)\mathrm{d}t + \partial_x a(t,X^\eta_t)\mathrm{d}W_t) \\ &+ \eta [X^\eta_{1,t}]^2(-\partial_x[a\partial_x a](t,X^\eta_t)\mathrm{d}t + \partial^2_x a(t,X^\eta_t)\mathrm{d}W_t), \quad X^\eta_{2,0} = 0. \qquad (29) \\ dX^\eta_{3,t} =& 3X^\eta_{2,t}(-[a\partial_x a](t,X^\eta_t)\mathrm{d}t + \partial_x a(t,X^\eta_t)\mathrm{d}W_t) \\ &+ 3[X^\eta_{1,t}]^2(-\partial_x[a\partial_x a](t,X^\eta_t)\mathrm{d}t + \partial^2_x a(t,X^\eta_t)\mathrm{d}W_t), \\ &+ \eta X^\eta_{3,t}(-[a\partial_x a](t,X^\eta_t)\mathrm{d}t + \partial_x a(t,X^\eta_t)\mathrm{d}W_t) \\ &+ 3\eta[X^\eta_{1,t}][X^\eta_{2,t}](-\partial_x[a\partial_x a](t,X^\eta_t)\mathrm{d}t + \partial^2_x a(t,X^\eta_t)\mathrm{d}W_t) \\ &+ \eta[X^\eta_{1,t}]^3(-\partial^2_x[a\partial_x a](t,X^\eta_t)\mathrm{d}t + \partial^3_x a(t,X^\eta_t)\mathrm{d}W_t), \quad X^\eta_{3,0} = 0. \qquad (30) \end{aligned}$$

Observe that $X^\eta_t|_{\eta=0} = x_0$, thus the derivatives at $\eta = 0$ have simpler expressions:

$$\begin{aligned} dX_{1,t} =& -\frac{1}{2}a^2(t,x_0)\mathrm{d}t + a(t,x_0)\mathrm{d}W_t = dX^P_t, \\ dX_{2,t} =& 2X_{1,t}(-[a\partial_x a](t,x_0)\mathrm{d}t + \partial_x a(t,x_0)\mathrm{d}W_t), \end{aligned}$$

with $X_{i,0} = 0$ for $i \geq 1$. Then notice that $X^P_t = x_0 + X_{1,t}$: hence

$$\begin{aligned} X_T - X^P_T = X^1_T - (x_0 + X_{1,T}) &= \int_0^1 (1-\lambda)X^\lambda_{2,T}\mathrm{d}\lambda \qquad (31) \\ &= \frac{1}{2}X_{2,T} + \int_0^1 \frac{(1-\lambda)^2}{2}X^\lambda_{3,T}\mathrm{d}\lambda \qquad (32) \end{aligned}$$

using the Taylor expansion formula. As a consequence of the above representation, we obtain an approximation of $\mathbb{E}(h(X_T))$ for a smooth function h:

$$\mathbb{E}[h(X_T)] = \mathbb{E}[h(X^P_T + \frac{X_{2,T}}{2} + ...)] = \mathbb{E}[h(X^P_T)] + \mathbb{E}[h^{(1)}(X^P_T)\frac{X_{2,T}}{2}] + ... \qquad (33)$$

The first term is related to a log-normal model and thus, it is expected to be easily computable numerically. The second term is more delicate: actually, we transform it into a weighed sum of sensitivities of $\mathbb{E}[h(X^P_T + \varepsilon)]$ w.r.t. $\varepsilon = 0$. To achieve this transformation, we use a key lemma which proof is given in Subsection 8.4

Lemma 1. *Let φ be a C_b^∞ function and $(\lambda_t)_t$ be a measurable and bounded deterministic function. Let $N \geq 1$ be fixed, and consider measurable and bounded deterministic functions $t \mapsto l_{i,t}$ for $i = 1, \ldots, N$. Then, using the convention $\mathrm{d}W_t^1 = \mathrm{d}W_t$ and $\mathrm{d}W_t^0 = \mathrm{d}t$, for any $(I_1, \ldots, I_N) \in \{0,1\}^N$ we have:*

$$\begin{aligned}
&\mathbb{E}\Big(\varphi(\int_0^T \lambda_t \mathrm{d}W_t) \int_0^T l_{N,t_N} \int_0^{t_N} l_{N-1,t_{N-1}} \cdots \qquad (34)\\
&\qquad \int_0^{t_2} l_{1,t_1} \mathrm{d}W_{t_1}^{I_1} \ldots \mathrm{d}W_{t_{N-1}}^{I_{N-1}} \mathrm{d}W_{t_N}^{I_N}\Big)\\
&= \omega(\widehat{l}_1, \ldots, \widehat{l}_N) \partial_{\varepsilon^{I_1+\cdots+I_N}}^{I_1+\cdots+I_N} \mathbb{E}\Big(\varphi(\int_0^T \lambda_t \mathrm{d}W_t + \varepsilon)\Big)|_{\varepsilon=0}, \qquad (35)
\end{aligned}$$

where $\widehat{l}_{k,t} := l_{k,t}$ if $I_k = 0$ and $\widehat{l}_{k,t} := \lambda_t l_{k,t}$ if $I_k = 1$.

Now, apply the above identity to $\varphi(\cdot) = h^{(1)}(x_0 - \frac{1}{2}\int_0^T a^2(t, x_0)\mathrm{d}t + \cdot)$, $\lambda_t = a(t, x_0)$ and

$$\begin{aligned}
\frac{X_{2,T}}{2} = \int_0^T \Big(\int_0^{t_2} (-\frac{1}{2}a^2(t_1, x_0)\mathrm{d}t_1 + a(t_1, x_0)\mathrm{d}W_{t_1})\Big)&\\
\times (-[a\partial_x a](t_2, x_0)\mathrm{d}t_2 + \partial_x a(t_2, x_0)\mathrm{d}W_{t_2}),&
\end{aligned}$$

to get

$$\mathbb{E}[h^{(1)}(X_T^P)\frac{X_{2,T}}{2}] = C_1(a; x_0)_0^T (\partial_{\varepsilon^3}^3 - \frac{3}{2}\partial_{\varepsilon^2}^2 + \frac{1}{2}\partial_\varepsilon)\mathbb{E}\big(h(X_T^P + \varepsilon)\big)|_{\varepsilon=0}$$

where the operator C_1 is defined by:

$$C_1(l; z)_0^T = \omega(l^2(z), l(z)l^{(1)}(z))_0^T = \int_0^T l_t^2(z) \int_t^T l_s(z) l_s^{(1)}(z) \mathrm{d}s\mathrm{d}t. \qquad (36)$$

Combine this with (33) to obtain that $\mathbb{E}(h(X_T))$ can be approximated by

$$\mathbb{E}[h(X_T^P)] + C_1(a; x_0)_0^T (\partial_{\varepsilon^3}^3 - \frac{3}{2}\partial_{\varepsilon^2}^2 + \frac{1}{2}\partial_\varepsilon)\mathbb{E}\big(h(X_T^P + \varepsilon)\big)|_{\varepsilon=0}.$$

So far, the payoff function h is smooth and this does not fit the Call/Put setting; actually, an extra regularization argument and a careful passing to the limit enables to extend the previous formula to any locally Lipschitz h. Additionally, some error estimates are available (see [42]). All the results are gathered in the following theorem.

Theorem 4. ***(2nd order log-normal approximation with local volatility at spot)**. Assume $(\mathcal{H}_{x_0}^a)$. Assume that h is locally Lipschitz in the following sense: for some constant $C_h \geq 0$,*

$$|h(x)| \leq C_h e^{C_h|x|}, \qquad |\frac{h(y) - h(x)}{y - x}| \leq C_h e^{C_h(|x|+|y|)} \quad (\forall y \neq x).$$

Then

$$\mathbb{E}[h(X_T)] = \mathbb{E}[h(X_T^P)] + C_1(a;x_0)_0^T(\partial^3_{\varepsilon^3} - \frac{3}{2}\partial^2_{\varepsilon^2} + \frac{1}{2}\partial_\varepsilon)\mathbb{E}\big[h(X_T^P+\varepsilon)\big]|_{\varepsilon=0} + \mathcal{O}(\mathcal{M}_1(a)[\mathcal{M}_0(a)]^2T^{\frac{3}{2}}).$$

where the operator C_1 *is defined in* (36) *and* $\mathcal{O}$ *depends notably of the constant* C_h.

This formula is referred to as a second order approximation because the residual term is of order three with respect to the amplitude of the volatility coefficient.

Remark 4.1. The reader should notice that the expansion formulas are exact for the particular payoff function $h(x) = e^x$ (indeed $\mathbb{E}[h(X_T)] = \mathbb{E}[h(X_T^P)] = \partial^i_{\varepsilon^i}\mathbb{E}\big[h(X_T^P+\varepsilon)\big]|_{\varepsilon=0} = e^{x_0}$ and the sum of the corrective terms is equal to zero). This notably implies that the Call/Put parity relationship is preserved within these approximations, which is an essential property. The reader can verify in Section 5 that this martingale property is preserved for higher order approximation formulas.

Under the current assumptions ($\int_0^T a^2(t,x_0)\mathrm{d}t > 0$), the law of X_T^P is a non-degenerate Gaussian r.v. and thus, the above derivatives are meaningful even for non-smooth h. Following [42], the Lipschitz regularity can be weakened to Hölder regularity but error estimates in the case of discontinuous function h are not available so far under the current set of assumptions.

▷ **Normal proxy.** Alternatively to a log-normal proxy, we could prefer the use of normal proxy on the asset S: for CEV-type model described in (27), it can be justified for β close to 0. The same analysis can be done by considering the *normal proxy* with diffusion coefficient computed at spot: it writes

$$\mathrm{d}S_t^P = \Sigma(t,S_0)\mathrm{d}W_t, \qquad S_0^P = S_0.$$

Then, the distance to the proxy is represented through the interpolation process

$$dS_t^\eta = \eta\Sigma(t,S_t^\eta)dW_t, \quad S_0^\eta = S_0.$$

All the previous computations are very similar, and even simpler because there is no $\mathrm{d}t$-term. We skip details and state directly the result (see [42]).

Theorem 5. ***(2nd order normal approximation with local volatility at spot).*** *Assume* $(\mathcal{H}^\Sigma_{S_0})$. *Assume that* h *is locally Lipschitz in the following sense: for some constant* $C_h \geq 0$,

$$|h(x)| \leq C_h(1+|x|^{C_h}), \qquad |\frac{h(y)-h(x)}{y-x}| \leq C_h(1+|x|^{C_h}+|y|^{C_h}) \quad (\forall y \neq x).$$

Then

$$\mathbb{E}[h(S_T)] = \mathbb{E}[h(S_T^P)] + C_1(\Sigma;S_0)_0^T\partial^3_{\varepsilon^3}\mathbb{E}\big[h(S_T^P+\varepsilon)\big]|_{\varepsilon=0} + \mathcal{O}(\mathcal{M}_1(\Sigma)[\mathcal{M}_0(\Sigma)]^2T^{\frac{3}{2}}).$$

Remark 4.2. As for the log-normal proxy (see Remark 4.1), the approximation formulas involving the normal proxy do not suffer from numerical arbitrage when using Call/Put payoffs: indeed they are exact for the particular payoff function $h(x) = x$ (indeed $\mathbb{E}[h(S_T)] = \mathbb{E}[h(S_T^P)] = S_0$ and $\partial^i_{\varepsilon^i}\mathbb{E}\big[h(S_T^P + \varepsilon)\big]|_{\varepsilon=0} = 0$, $\forall i \geq 2$). This property holds again when considering higher order expansions (see Section 5).

Applying two previous results to the pricing of Call option (i.e. $h(x) = (e^x - K)_+$ in the case of log-normal proxy, and $h(x) = (x - K)+$ in the case of normal proxy), we obtain two different expansions using respectively Black-Scholes formula and Bachelier formula.

Theorem 6. ***(2nd order approximations for Call options with local volatility at spot).*** *Assuming* $(\mathcal{H}^a_{x_0})$ *and using the log-normal proxy, one has*

$$\begin{aligned}\mathrm{Call}(e^{x_0}, T, e^k) =& \mathrm{Call}^{\mathrm{BS}}(x_0, \bar{a}^2_{x_0}T, k)\\ &+ C_1(a; x_0)^T_0(\partial^3_{x^3} - \frac{3}{2}\partial^2_{x^2} + \frac{1}{2}\partial_x)\mathrm{Call}^{\mathrm{BS}}(x_0, \bar{a}^2_{x_0}T, k)\\ &+ \mathcal{O}(\mathcal{M}_1(a)[\mathcal{M}_0(a)]^2 T^{\frac{3}{2}}).\end{aligned}$$

Assuming $(\mathcal{H}^\Sigma_{S_0})$ *and using the normal proxy, one has*

$$\begin{aligned}\mathrm{Call}(S_0, T, K) =& \mathrm{Call}^{\mathrm{BA}}(S_0, \bar{\Sigma}^2_{S_0}T, K) + C_1(\Sigma; S_0)^T_0 \partial^3_{S^3}\mathrm{Call}^{\mathrm{BA}}(S_0, \bar{\Sigma}^2_{S_0}T, K)\\ &+ \mathcal{O}(\mathcal{M}_1(\Sigma)[\mathcal{M}_0(\Sigma)]^2 T^{\frac{3}{2}}).\end{aligned}$$

3.3. *Towards Call option approximations with the local volatility at strike and at mid-point*

For general payoff functions, the most natural choice seems to choose a proxy with the local volatility frozen at spot. When we are dealing with Call or Put payoffs, the spot and strike variables play a symmetrical role [43], and there is *a priori* no reason to advantage one or the other one. A first attempt to exploit this duality in proxy expansion is analysed in [44]. In this subsection, we briefly recall the expansion formulas with a local volatility at strike and then we present new expansion formulas with a local volatility at mid-point $x_{\mathrm{av}} = (x_0 + k)/2 = \log\sqrt{S_0 K}$ or $S_{\mathrm{av}} = (S_0 + K)/2$. We detail the analysis only for the log-normal proxy. The proofs for the normal proxy are very similar and are left as an exercise to the reader.

To directly obtain expansions formulas with local volatility frozen at strike, the idea is to follow the Dupire approach [43], using explicitly the PDE satisfied by the Call price function $(T, K) \to \mathrm{Call}(S_0, T, K) = \mathbb{E}[(S_T - K)_+]$. Indeed we have that:

$$\begin{cases}\partial_T \mathrm{Call}(S_0, T, K) = \frac{1}{2}\sigma^2(T, K)K^2\partial^2_{K^2}\mathrm{Call}(S_0, T, K),\\ \mathrm{Call}(S_0, 0, K) = (S_0 - K)_+.\end{cases}$$

Thus we do not consider anymore a PDE in the backward variables (t, S) with a Call payoff as a terminal condition, but we now handle a PDE in the forward variables (T, K), with a put payoff condition. This dual PDE has a probabilistic Feynman-Kac representation:

$$\mathrm{Call}(S_0, T, K) = \mathbb{E}[(S_0 - e^{k_T})_+], \tag{37}$$

where $(k_t)_{t\in[0,T]}$ is the diffusion process defined by:

$$\mathrm{d}k_t = a(T-t, k_t)\mathrm{d}W_t - \frac{1}{2}a^2(T-t, k_t)\mathrm{d}t, \ k_0 = k = \log(K),$$

where we recall that $a(t,z) = \sigma(t, e^z)$. Thus we are in a position to apply Theorem 4 for the Put payoff function $h(z) = (e^{x_0} - e^z)_+$ with log-strike $x_0 = \log(S_0)$, with a log-normal proxy starting from $K = e^k$ and with the local volatility $\widetilde{a}(t,z) = a(T-t, z)$. In the same way, we can apply Theorem 5 with a normal proxy. As a result, we obtain a variant of Theorem 6 where the Greeks w.r.t. the k_T-variable are naturally transformed into Greeks w.r.t. the strike variable. The final statement is the following result.

Theorem 7. ***(2nd order approximations for Call options with local volatility at strike).*** *Assuming* $(\mathcal{H}_k^a)$ *and using the log-normal proxy, one has*

$$\begin{aligned}\mathrm{Call}(e^{x_0}, T, e^k) =& \mathrm{Call}^{\mathrm{BS}}(x_0, \bar{a}_k^2 T, k)\\ &+ C_1(\widetilde{a}; k)_0^T(\partial^3_{z^3} - \frac{3}{2}\partial^2_{z^2} + \frac{1}{2}\partial_z)\mathrm{Call}^{\mathrm{BS}}(x_0, \bar{a}_k^2 T, k)\\ &+ \mathcal{O}(\mathcal{M}_1(a)[\mathcal{M}_0(a)]^2 T^{\frac{3}{2}}).\end{aligned}$$

Assuming $(\mathcal{H}_K^\Sigma)$ *and using the normal proxy, one has*

$$\begin{aligned}\mathrm{Call}(S_0, T, K) =& \mathrm{Call}^{\mathrm{BA}}(S_0, \bar{\Sigma}_K^2 T, K) + C_1(\widetilde{\Sigma}; K)_0^T \partial^3_{Z^3}\mathrm{Call}^{\mathrm{BA}}(S_0, \bar{\Sigma}_K^2 T, K)\\ &+ \mathcal{O}(\mathcal{M}_1(\Sigma)[\mathcal{M}_0(\Sigma)]^2 T^{\frac{3}{2}}).\end{aligned}$$

Now, in order to obtain approximation formulas for the mid-points x_{av} or S_{av}, we perform a Taylor expansion of the local volatility function around these mid-points. We start from the expansions at spot and strike given in Theorems 6 and 7, we consider the average of these expansions and we transform each term to freeze the local volatility function at x_{av} or S_{av}. We only give details for the log-normal proxy. We first analyze the corrective terms.

Lemma 2. *Assume* $(\mathcal{H}_{x_0}^a)$*-*$(\mathcal{H}_k^a)$*-*$(\mathcal{H}_{x_{\mathrm{av}}}^a)$*. We have:*

$$\begin{aligned}&\frac{1}{2}C_1(a; x_0)_0^T(\partial^3_{x^3} - \frac{3}{2}\partial^2_{x^2} + \frac{1}{2}\partial_x)\mathrm{Call}^{\mathrm{BS}}(x_0, \bar{a}_{x_0}^2 T, k)\\ &+ \frac{1}{2}C_1(\widetilde{a}; k)_0^T(\partial^3_{z^3} - \frac{3}{2}\partial^2_{z^2} + \frac{1}{2}\partial_z)\mathrm{Call}^{\mathrm{BS}}(x_0, \bar{a}_k^2 T, k)\\ =& \frac{1}{2}\Big[C_1(a; x_{\mathrm{av}})_0^T - C_1(\widetilde{a}; x_{\mathrm{av}})_0^T\Big](\partial^3_{x^3} - \frac{3}{2}\partial^2_{x^2} + \frac{1}{2}\partial_x)\mathrm{Call}^{\mathrm{BS}}(x_0, \bar{a}_{x_{\mathrm{av}}}^2 T, k)\\ &+ \mathcal{O}(\mathcal{M}_1(a)[\mathcal{M}_0(a)]^2 T^{\frac{3}{2}}).\end{aligned}$$

Proof. We begin with the x_0-Greeks. Perform a zero order Taylor formula for the function $y \to (\partial^3_{x^3} - \frac{3}{2}\partial^2_{x^2} + \frac{1}{2}\partial_x)\mathrm{Call}^{\mathrm{BS}}(x_0, y, k)$ at $y = \bar{a}^2_{x_0}T = \omega(a^2(x_0))^T_0$ around $y = \bar{a}^2_{x_{\mathrm{av}}}T = \omega(a^2(x_{\mathrm{av}}))^T_0$ and $\forall t \in [0, T]$, for the function $x \to a^2_t(x)$ at $x = x_0$ around $x = x_{\mathrm{av}}$ to obtain:

$$\begin{aligned}
&C_1(a; x_0)^T_0(\partial^3_{x^3} - \frac{3}{2}\partial^2_{x^2} + \frac{1}{2}\partial_x)\mathrm{Call}^{\mathrm{BS}}(x_0, \bar{a}^2_{x_0}T, k)\\
=&[C_1(a; x_{\mathrm{av}})^T_0 + R_1][(\partial^3_{x^3} - \frac{3}{2}\partial^2_{x^2} + \frac{1}{2}\partial_x)\mathrm{Call}^{\mathrm{BS}}(x_0, \bar{a}^2_{x_{\mathrm{av}}}T, k) + R_2]\\
=&C_1(a; x_{\mathrm{av}})^T_0(\partial^3_{x^3} - \frac{3}{2}\partial^2_{x^2} + \frac{1}{2}\partial_x)\mathrm{Call}^{\mathrm{BS}}(x_0, \bar{a}^2_{x_{\mathrm{av}}}T, k)\\
&+ (\partial^3_{x^3} - \frac{3}{2}\partial^2_{x^2} + \frac{1}{2}\partial_x)\mathrm{Call}^{\mathrm{BS}}(x_0, \bar{a}^2_{x_{\mathrm{av}}}T, k)R_1 + C_1(a; x_0)^T_0 R_2,
\end{aligned}$$

where:

$$\begin{aligned}
R_1 &= \frac{(x_0 - k)}{2}\int_0^1 (\partial_x C_1(a; x)^T_0)|_{x=\lambda x_0 + (1-\lambda)x_{\mathrm{av}}}\mathrm{d}\lambda,\\
R_2 &= T(\bar{a}^2_{x_0} - \bar{a}^2_{x_{\mathrm{av}}})\int_0^1 (\partial^4_{yx^3} - \frac{3}{2}\partial^3_{yx^2} + \frac{1}{2}\partial^2_{yx})\mathrm{Call}^{\mathrm{BS}}(x_0, y, k)|_{y=T(\lambda\bar{a}^2_{x_0} + (1-\lambda)\bar{a}^2_{x_{\mathrm{av}}})}\mathrm{d}\lambda,\\
T(\bar{a}^2_{x_0} - \bar{a}^2_{x_{\mathrm{av}}}) &= \frac{(x_0 - k)}{2}\int_0^1 (\partial_x \omega(a^2(x))^T_0)|_{x=\lambda x_0 + (1-\lambda)x_{\mathrm{av}}}\mathrm{d}\lambda.
\end{aligned}$$

In view of the definition (36) of C_1, the identity (6), Corollary 4.1 and $(\mathcal{H}^a_{x_0})$-$(\mathcal{H}^a_{x_{\mathrm{av}}})$, we readily obtain

$$\begin{aligned}
&\big|(\partial^3_{x^3} - \frac{3}{2}\partial^2_{x^2} + \frac{1}{2}\partial_x)\mathrm{Call}^{\mathrm{BS}}(x_0, \bar{a}^2_{x_{\mathrm{av}}}T, k)R_1\big|\\
&\le \frac{1}{2}\big|(\partial^3_{x^3} - \frac{3}{2}\partial^2_{x^2} + \frac{1}{2}\partial_x)\mathrm{Call}^{\mathrm{BS}}(x_0, \bar{a}^2_{x_{\mathrm{av}}}T, k)(x_0 - k)\big|\\
&\quad \times \big|\int_0^1 (\partial_x C_1(a; x)^T_0)|_{x=\lambda x_0 + (1-\lambda)x_{\mathrm{av}}}\mathrm{d}\lambda\big|\\
&\le_c [\bar{a}^2_{x_{\mathrm{av}}}T]^{-\frac{1}{2}}\mathcal{M}_1(a)[\mathcal{M}_0(a)]^3T^2 \le_c \mathcal{M}_1(a)[\mathcal{M}_0(a)]^2T^{\frac{3}{2}},
\end{aligned}$$

$$\begin{aligned}
\big|C_1(a; x_0)^T_0 R_2\big| &\le_c [\mathcal{M}_0(a)]^3\mathcal{M}_1(a)T^2\mathcal{M}_0(a)\mathcal{M}_1(a)T\\
&\quad \times \int_0^1 [T(\lambda\bar{a}^2_{x_0} + (1-\lambda)\bar{a}^2_{x_{\mathrm{av}}})]^{-\frac{3}{2}}\mathrm{d}\lambda\ lc\mathcal{M}_1(a)[\mathcal{M}_0(a)]^2T^{\frac{3}{2}}.
\end{aligned}$$

Similarly, using in addition $(\mathcal{H}^a_k)$ we show that:

$$\begin{aligned}
&C_1(\widetilde{a}; k)^T_0(\partial^3_{z^3} - \frac{3}{2}\partial^2_{z^2} + \frac{1}{2}\partial_z)\mathrm{Call}^{\mathrm{BS}}(x_0, \bar{a}^2_k T, k)\\
&= C_1(\widetilde{a}; x_{\mathrm{av}})^T_0(\partial^3_{z^3} - \frac{3}{2}\partial^2_{z^2} + \frac{1}{2}\partial_z)\mathrm{Call}^{\mathrm{BS}}(x_0, \bar{a}^2_{x_{\mathrm{av}}}T, k) + \mathcal{O}(\mathcal{M}_1(a)[\mathcal{M}_0(a)]^2T^{\frac{3}{2}}),\\
&= -C_1(\widetilde{a}; x_{\mathrm{av}})^T_0(\partial^3_{x^3} - \frac{3}{2}\partial^2_{x^2} + \frac{1}{2}\partial_x)\mathrm{Call}^{\mathrm{BS}}(x_0, \bar{a}^2_{x_{\mathrm{av}}}T, k) + \mathcal{O}(\mathcal{M}_1(a)[\mathcal{M}_0(a)]^2T^{\frac{3}{2}}),
\end{aligned}$$

where we have used at the last equality the relation (75) in Proposition 4.3. That completes the proof. □

Second, we analyze the leading order of the formula given in Theorems 6 and 7:

Lemma 3. *Assume* $(\mathcal{H}^a_{x_0})$-$(\mathcal{H}^a_k)$-$(\mathcal{H}^a_{x_{\rm av}})$. *We have:*

$$\frac{1}{2}[\mathrm{Call}^{\mathrm{BS}}(x_0,\bar{a}^2_{x_0}T,k)+\mathrm{Call}^{\mathrm{BS}}(x_0,\bar{a}^2_kT,k)] = \mathrm{Call}^{\mathrm{BS}}(x_0,\bar{a}^2_{x_{\rm av}}T,k) + \mathcal{O}(\mathcal{M}_1(a)[\mathcal{M}_0(a)]^2T^{\frac{3}{2}}).$$

Proof. Apply a first order Taylor formula twice; firstly for the function $y \to \mathrm{Call}^{\mathrm{BS}}(x_0,y,k)$ at $y=\bar{a}^2_{x_0}T$ around $y=\bar{a}^2_{x_{\rm av}}T$ and secondly, for the function $x \to a^2_t(x)$ at $x=x_0$ around $x=x_{\rm av}$, $\forall t\in[0,T]$. It gives

$$\begin{aligned}&\mathrm{Call}^{\mathrm{BS}}(x_0,\bar{a}^2_{x_0}T,k)\\ =&\mathrm{Call}^{\mathrm{BS}}(x_0,\bar{a}^2_{x_{\rm av}}T,k)+\partial_y\mathrm{Call}^{\mathrm{BS}}(x_0,\bar{a}^2_{x_{\rm av}}T,k)T(\bar{a}^2_{x_0}-\bar{a}^2_{x_{\rm av}})+R_1,\\ =&\mathrm{Call}^{\mathrm{BS}}(x_0,\bar{a}^2_{x_{\rm av}}T,k)+\partial_y\mathrm{Call}^{\mathrm{BS}}(x_0,\bar{a}^2_{x_{\rm av}}T,k)\omega(a(x_{\rm av})a^{(1)}(x_{\rm av}))^T_0(x_0-k)\\ &+R_2+R_1.\end{aligned}$$

where:

$$\begin{aligned}R_1=&T^2(\bar{a}^2_{x_0}-\bar{a}^2_{x_{\rm av}})^2\int_0^1(\partial^2_{y^2}\mathrm{Call}^{\mathrm{BS}}(x_0,y,k))|_{y=T(\lambda\bar{a}^2_{x_0}+(1-\lambda)\bar{a}^2_{x_{\rm av}})}(1-\lambda)\mathrm{d}\lambda,\\ R_2=&\partial_y\mathrm{Call}^{\mathrm{BS}}(x_0,\bar{a}^2_{x_{\rm av}}T,k)\frac{(x_0-k)^2}{4}\int_0^1(\partial^2_{x^2}\omega(a^2(x))^T_0)|_{x=\lambda x_0+(1-\lambda)x_{\rm av}}(1-\lambda)\mathrm{d}\lambda.\end{aligned}$$

Similar arguments previously employed in the proof of Lemma 2 easily lead to:

$$\begin{aligned}|R_1|\le_c&[\mathcal{M}_1(a)]^2[\mathcal{M}_0(a)]^2T^2\int_0^1[T(\lambda\bar{a}^2_{x_0}+(1-\lambda)\bar{a}^2_{x_{\rm av}})]^{-\frac{1}{2}}\mathrm{d}\lambda\\ \le_c&\mathcal{M}_1(a)[\mathcal{M}_0(a)]^2T^{\frac{3}{2}},\\ |R_2|\le_c&\mathcal{M}_1(a)[\mathcal{M}_0(a)]^2T^{\frac{3}{2}}.\end{aligned}$$

Similarly we have:

$$\begin{aligned}&\mathrm{Call}^{\mathrm{BS}}(x_0,\bar{a}^2_kT,k)\\ =&\mathrm{Call}^{\mathrm{BS}}(x_0,\bar{a}^2_{x_{\rm av}}T,k)-\partial_y\mathrm{Call}^{\mathrm{BS}}(x_0,\bar{a}^2_{x_{\rm av}}T,k)\omega(a(x_{\rm av})a^{(1)}(x_{\rm av}))^T_0(x_0-k)\\ &+\mathcal{O}(\mathcal{M}_1(a)[\mathcal{M}_0(a)]^2T^{\frac{3}{2}}).\end{aligned}$$

We are finished. □

Lemmas 2 and 3 lead to the following Theorem for the log-normal proxy, while similar arguments apply for the normal proxy.

Theorem 8. ***(2nd order approximations for Call options with local volatility at mid-point).*** *Under* $(\mathcal{H}^a_{x_0})$-$(\mathcal{H}^a_k)$-$(\mathcal{H}^a_{x_{\rm av}})$*, we have*

$$\begin{aligned}\mathrm{Call}(e^{x_0},T,e^k) &= \mathrm{Call}^{\rm BS}(x_0,\bar a^2_{x_{\rm av}}T,k)\\ &+\frac{C_1(a;x_{\rm av})^T_0 - C_1(\widetilde{a};x_{\rm av})^T_0}{2}(\partial^3_{x^3}-\frac{3}{2}\partial^2_{x^2}+\frac{1}{2}\partial_x)\mathrm{Call}^{\rm BS}(x_0,\bar a^2_{x_{\rm av}}T,k)\\ &+\mathcal{O}(\mathcal{M}_1(a)[\mathcal{M}_0(a)]^2T^{\frac{3}{2}}). \end{aligned} \tag{38}$$

Under $(\mathcal{H}^\Sigma_{S_0})$-$(\mathcal{H}^\Sigma_K)$-$(\mathcal{H}^\Sigma_{S_{\rm av}})$*, we have*

$$\begin{aligned}\mathrm{Call}(S_0,T,K) =&\mathrm{Call}^{\rm BA}(S_0,\bar\Sigma^2_{S_{\rm av}}T,K)\\ &+\frac{C_1(\Sigma;S_{\rm av})^T_0 - C_1(\widetilde{\Sigma};S_{\rm av})^T_0}{2}\partial^3_{S^3}\mathrm{Call}^{\rm BA}(S_0,\bar\Sigma^2_{S_{\rm av}}T,K)\\ &+\mathcal{O}(\mathcal{M}_1(\Sigma)[\mathcal{M}_0(\Sigma)]^2T^{\frac{3}{2}}). \end{aligned} \tag{39}$$

Remark 4.3. If a (and consequently Σ) is time-independent or has separable variables, observe that the corrective terms vanish and we obtain remarkably simple formulas: the expansion formulas (38) and (39) reduce to only a Black-Scholes price and a Bachelier price, with the local volatility function frozen at the mid-point.

3.4. *Second order expansion of the implied volatility*

Interestingly, the previous expansions of Call price (Theorems 6, 7 and 8) can be turned into expansions of Black-Scholes and Bachelier implied volatility defined respectively in (7) and (9). To achieve this, we use the relations between Greeks postponed in Propositions 4.3 and 4.6 in order to write the different approximation formulas in terms of the Vega. For example consider the second order log-normal expansion formula based on the ATM local volatility (Theorem 6): thanks to (75) in Proposition 4.3, it becomes:

$$\begin{aligned}\mathrm{Call}(e^{x_0},T,e^k) =&\mathrm{Call}^{\rm BS}(x_0,\bar a^2_{x_0}T,k) - \mathrm{Vega}^{\rm BS}(x_0,\bar a^2_{x_0}T,k)\frac{C_1(a;x_0)^T_0 m}{\bar a^3_{x_0}T^2}\\ &+\mathcal{O}(\mathcal{M}_1(a)[\mathcal{M}_0(a)]^2T^{\frac{3}{2}}),\\ \approx&\mathrm{Call}^{\rm BS}\Big(x_0,(\bar a_{x_0}-\frac{C_1(a;x_0)^T_0}{\bar a^3_{x_0}T^2}m)^2T,k\Big),\end{aligned}$$

where m is the log-moneyness $m = x_0 - k = \log(S_0/K)$. We have paved the way for the following result:

Theorem 9. ***(2nd order expansions of the implied volatility).*** *Assuming*

$(\mathcal{H}^a_{x_0})$-$(\mathcal{H}^a_k)$-$(\mathcal{H}^a_{x_{\mathrm{av}}})$ *and using the log-normal proxy, we have*

$$\sigma_{\mathrm{I}}(x_0,T,k) = \bar{a}_{x_0} - \frac{C_1(a;x_0)_0^T}{\bar{a}^3_{x_0}T^2} m + \mathrm{Error}^{\mathrm{I}}_{2,\mathrm{x_0}}, \tag{40}$$

$$\sigma_{\mathrm{I}}(x_0,T,k) = \bar{a}_k + \frac{C_1(\widetilde{a};k)_0^T}{\bar{a}^3_k T^2} m + \mathrm{Error}^{\mathrm{I}}_{2,\mathrm{k}}, \tag{41}$$

$$\sigma_{\mathrm{I}}(x_0,T,k) = \bar{a}_{x_{\mathrm{av}}} + \frac{(C_1(\widetilde{a};x_{\mathrm{av}})_0^T - C_1(a;x_{\mathrm{av}})_0^T)}{2\bar{a}^3_{x_{\mathrm{av}}}T^2} m + \mathrm{Error}^{\mathrm{I}}_{2,\mathrm{x_{av}}}. \tag{42}$$

Assuming $(\mathcal{H}^\Sigma_{S_0})$-$(\mathcal{H}^\Sigma_K)$-$(\mathcal{H}^\Sigma_{S_{\mathrm{av}}})$ *and using the normal proxy, we have*

$$\Sigma_{\mathrm{I}}(S_0,T,K) = \bar{\Sigma}_{S_0} - \frac{C_1(\Sigma;S_0)_0^T}{\bar{\Sigma}^3_{S_0}T^2} M + \mathrm{Error}^{\mathrm{I}}_{2,\mathrm{S_0}},$$

$$\Sigma_{\mathrm{I}}(S_0,T,K) = \bar{\Sigma}_K + \frac{C_1(\widetilde{\Sigma};K)_0^T}{\bar{\Sigma}^3_K T^2} M + \mathrm{Error}^{\mathrm{I}}_{2,\mathrm{K}},$$

$$\Sigma_{\mathrm{I}}(S_0,T,K) = \bar{\Sigma}_{S_{\mathrm{av}}} + \frac{(C_1(\widetilde{\Sigma};S_{\mathrm{av}})_0^T - C_1(\Sigma;S_{\mathrm{av}})_0^T)}{2\bar{\Sigma}^3_{S_{\mathrm{av}}}T^2} M + \mathrm{Error}^{\mathrm{I}}_{2,\mathrm{S_{av}}},$$

where $S_{\mathrm{av}} = \frac{S_0+K}{2}$ *and* $M = S_0 - K$.

Remark 4.4. We retrieve in our implied volatility approximation formulas the well-known property that at the money (ie $m = 0$) and for short maturity, the value of the implied volatility is equal to the value of the local volatility function and the slope of the local volatility function is twice the slope of the implied volatility. We justify this assertion for the Black-Scholes implied volatility, the work being similar for the Bachelier one. If $T \ll 1$, in view of (40) and the definition (36) of C_1, assuming that $a(t,x_0)$ and $a^{(1)}(t,x_0)$ are continuous at $t=0$, we obtain:

$$\begin{aligned}
[\sigma_{\mathrm{I}}(x_0,T,k)]|_{k=x_0} &\approx a(0,x_0),\\
\partial_k[\sigma_{\mathrm{I}}(x_0,T,k)]|_{k=x_0} &\approx \partial_k[\bar{a}_{x_0}]|_{k=x_0} - \frac{C_1(a;x_0)_0^T}{\bar{a}^3_{x_0}T^2}\partial_k[(x_0-k)]|_{k=x_0}\\
&\approx 0 + \frac{a^3(0,x_0)a^{(1)}(0,x_0)\frac{T^2}{2}}{a^3(0,x_0)T^2} = \frac{a^{(1)}(0,x_0)}{2}.
\end{aligned}$$

We obtain the same estimates starting from (41) and (42), we skip details.

To conclude this paragraph, we estimate the residual terms of the above implied volatility expansions, in terms of $\mathcal{M}_0(a)$, $\mathcal{M}_1(a)$ and so on. Since the Vega is very small for far OTM/ITM Call options, deriving error bounds on implied volatility from Theorems 6, 7 and 8 gives poor estimates for extreme strikes. Actually, in the further numerical experiments, we also observe inaccuracies for extreme strikes. To obtain accurate theoretical error bounds, we restrict to log-moneyness m (resp moneyness M) belonging to a small ball by assuming that $|m| \le \xi\mathcal{M}_0(a)\sqrt{T}$ (resp. $|M| \le \xi\mathcal{M}_0(\Sigma)\sqrt{T}$) for a given $\xi > 0$.

For the sake of brevity, we only analyze the expansion (40), the other approximations being similar. We assume in addition that $\mathcal{M}_0(a)$, $\mathcal{M}_1(a)$ and T are globally small enough to ensure that $\bar{a}_{x_0} - \frac{C_1(a;x_0)_0^T}{\bar{a}_{x_0}^3 T^2} m > 0$. Note that at the money (i.e. $m = 0$), this condition is automatically satisfied. A first order expansion readily gives

$$\begin{aligned}
&\mathrm{Call}^{\mathrm{BS}}(x_0, (\bar{a}_{x_0} - \frac{C_1(a;x_0)_0^T}{\bar{a}_{x_0}^3 T^2} m)^2 T, k) \\
&= \mathrm{Call}^{\mathrm{BS}}(x_0, \bar{a}_{x_0}^2 T, k) - \frac{C_1(a;x_0)_0^T}{\bar{a}_{x_0}^3 T^2} m \mathrm{Vega}^{\mathrm{BS}}(x_0, \bar{a}_{x_0}^2 T, k) \\
&\quad + (\frac{C_1(a;x_0)_0^T}{\bar{a}_{x_0}^3 T^2} m)^2 \int_0^1 \mathrm{Vomma}^{\mathrm{BS}}(x_0, a^2 T, k)|_{a=\bar{a}_{x_0} - \lambda \frac{C_1(a;x_0)_0^T}{\bar{a}_{x_0}^3 T^2} m} (1-\lambda) \mathrm{d}\lambda \\
&= \mathrm{Call}^{\mathrm{BS}}(x_0, \sigma_{\mathrm{I}}^2(x_0, T, k) T, k) + \mathcal{O}(\mathcal{M}_1(a)[\mathcal{M}_0(a)]^2 T^{\frac{3}{2}}) \\
&\quad + (\frac{C_1(a;x_0)_0^T}{\bar{a}_{x_0}^3 T^2} m)^2 \int_0^1 \mathrm{Vomma}^{\mathrm{BS}}(x_0, a^2 T, k)|_{a=\bar{a}_{x_0} - \lambda \frac{C_1(a;x_0)_0^T}{\bar{a}_{x_0}^3 T^2} m} (1-\lambda) \mathrm{d}\lambda,
\end{aligned}$$

applying Theorem 6 and using the definition of the Black-Scholes implied volatility. Expanding $a \to \mathrm{Call}^{\mathrm{BS}}(x_0, a^2 T, k)$ at $a = \sigma_{\mathrm{I}}(x_0, T, k)$ around $a = \bar{a}_{x_0} - \frac{C_1(a;x_0)_0^T}{\bar{a}_{x_0}^3 T^2} m$ gives:

$$\begin{aligned}
&\mathrm{Error}_{2,x_0}^{\mathrm{I}} \int_0^1 \mathrm{Vega}^{\mathrm{BS}}(x_0, a^2 T, k)|_{a=\sigma_{\mathrm{I}}(x_0,T,k) - \lambda\ \mathrm{Error}_{2,x_0}^{\mathrm{I}}} \mathrm{d}\lambda \\
&\quad = \mathcal{O}(\mathcal{M}_1(a)[\mathcal{M}_0(a)]^2 T^{\frac{3}{2}}) - (\frac{C_1(a;x_0)_0^T}{\bar{a}_{x_0}^3 T^2} m)^2 \\
&\quad \times \int_0^1 \mathrm{Vomma}^{\mathrm{BS}}(x_0, a^2 T, k)|_{a=\bar{a}_{x_0} - \lambda \frac{C_1(a;x_0)_0^T}{\bar{a}_{x_0}^3 T^2} m} (1-\lambda) \mathrm{d}\lambda.
\end{aligned}$$

In view of the expression of $\mathrm{Vega}^{\mathrm{BS}}$ (see (68) in Proposition 4.2) and (71) in Corollary 4.2, the hypotheses made on m, $\mathcal{M}_0(a)$, $\mathcal{M}_1(a)$ and T guarantee the existence of a constant $C > 0$ (depending on S_0) such that:

$$\int_0^1 \mathrm{Vega}^{\mathrm{BS}}(x_0, a^2 T, k)|_{a=\sigma_{\mathrm{I}}(x_0,T,k) - \lambda\ \mathrm{Error}_{2,x_0}^{\mathrm{I}}} \mathrm{d}\lambda \geq C\sqrt{T} > 0.$$

In addition (72) and $(\mathcal{H}_{x_0}^a)$ readily yield

$$\begin{aligned}
&|(\frac{C_1(a;x_0)_0^T}{\bar{a}_{x_0}^3 T^2} m)^2 \int_0^1 \mathrm{Vomma}^{\mathrm{BS}}(x_0, a^2 T, k)|_{a=\bar{a}_{x_0} - \lambda \frac{C_1(a;x_0)_0^T}{\bar{a}_{x_0}^3 T^2} m} (1-\lambda) \mathrm{d}\lambda| \\
&\leq_c (\mathcal{M}_1(a)\mathcal{M}_0(a)\sqrt{T})^2 \frac{\sqrt{T}}{\bar{a}_{x_0}} \leq_c \mathcal{M}_1(a)[\mathcal{M}_0(a)]^2 T^{\frac{3}{2}},
\end{aligned}$$

where the generic constant depends in an increasing way on ξ. That finally implies:

$$\mathrm{Error}_{2,x_0}^{\mathrm{I}} = \mathcal{O}(\mathcal{M}_1(a)[\mathcal{M}_0(a)]^2 T).$$

In view of the above upper bound, we interpret our implied volatility formulas as second order expansion ones.

4. Proofs: a comparative discussion between stochastic analysis and PDE techniques

In this section, our aim is to show how three different techniques ranging from stochastic analysis to PDE may lead to the same formulas given in Theorem 4. Although the final result is the same, the derivation is quite different, first regarding the way in which the expansion coefficients are made explicit, second regarding the error estimates and the assumptions used for that.

We admit that our preference is for the stochastic analysis approach, because it is flexible regarding the model and the functionals under consideration, and it is slightly less demanding regarding the assumptions (pointwise ellipticity versus uniform ellipticity for instance). But the reader may argue differently, depending on its own fields of expertise.

As an illustration of flexibility of the stochastic analysis approach, it has been possible to handle Call/Put/digital options in local volatility models with Gaussian jumps [45], Call/Put options in local volatility models with stochastic Gaussian interest rates [7], Call/Put options in time-dependent Heston model [46], general average options (including Asian and Basket options) in local volatility models [42], and more recently local stochastic volatility models [47].

4.1. *A pure stochastic analysis approach*

This is basically the derivation that we have performed in Subsection 3.2.

Smooth payoff h. We first deal with the case of infinitely differentiable function h with exponentially bounded derivatives. Resuming from (31-32-33) and using Taylor's formula, write

$$\begin{aligned}
\mathbb{E}[h(X_T)] =& \mathbb{E}[h(X_T^P)] + \mathbb{E}[h^{(1)}(X_T^P)(X_T - X_T^P)] \\
&+ \int_0^1 \mathbb{E}\big[h^{(2)}(X_T^P + \lambda(X_T - X_T^P))(X_T - X_T^P)^2\big](1-\lambda)\mathrm{d}\lambda \\
=&\mathbb{E}[h(X_T^P)] + \mathbb{E}[h^{(1)}(X_T^P)\frac{X_{2,T}}{2}] + \mathbb{E}\big[h^{(1)}(X_T^P)\int_0^1 \frac{(1-\lambda)^2}{2} X_{3,T}^{\lambda}\big]\mathrm{d}\lambda \\
&+ \int_0^1 \mathbb{E}\big[h^{(2)}(X_T^P + \lambda(X_T - X_T^P))(\int_0^1 X_{2,T}^{\eta}(1-\eta)\mathrm{d}\eta)^2\big](1-\lambda)\mathrm{d}\lambda \\
:=&\mathbb{E}[h(X_T^P)] + \mathbb{E}[h^{(1)}(X_T^P)\frac{X_{2,T}}{2}] + \mathrm{Error}_2(h).
\end{aligned} \tag{43}$$

The first correction term $\mathbb{E}[h^{(1)}(X_T^P)\frac{X_{2,T}}{2}]$ is made explicit using the key Lemma 1, and it is equal to a weighted summation of sensitivities $\partial_\varepsilon^i \mathbb{E}[h(X_T^P + \varepsilon)]|_{\varepsilon=0}$ for $i = 1, 2, 3$ (see the statement of Theorem 4).

The evaluation of $\mathrm{Error}_2(h)$ requires to estimate the L_p-norms of $X_{2,T}^{\lambda}$ and $X_{3,T}^{\lambda}$ (uniformly in $\lambda \in [0,1]$). Direct and standard stochastic calculus inequalities from

(28-29-30) yield

$$|X^{\lambda}_{2,T}|_p \leq_c \mathcal{M}_1(a)\mathcal{M}_0(a)T, \qquad |X^{\lambda}_{3,T}|_p \leq_c \mathcal{M}_1(a)[\mathcal{M}_0(a)]^2T^{3/2} \tag{44}$$

for any $p \geq 1$ and any $\lambda \in [0,1]$. Combining these estimates with Hölder and Minkowski inequalities readily gives $\text{Error}_2(h) = \mathcal{O}(\mathcal{M}_1(a)[\mathcal{M}_0(a)]^2T^{\frac{3}{2}})$, which completes the proof if h is smooth as above. Observe that we have only required the coefficients to be smooth enough, and nothing has been imposed on the non-degeneracy of a.

Locally Lipschitz function h. We now extend the analysis to functions satisfying conditions of Theorem 4 (thus almost everywhere differentiable), assuming additionally $(\mathcal{H}^a_{x_0})$: observe that the pointwise ellipticity condition $\int_0^T a^2(t,x_0)dt > 0$ is necessary to ensure that X^P-Greeks are well defined. The analysis below shows that the condition is also sufficient to obtain the expansion.

The new ingredient consists in *appropriately* smoothing h and in using integration-by-parts formula from Malliavin calculus to get rid of the derivatives of h; this follows the arguments of [42]. Let B be another scalar Brownian motion independent of W and for $\delta > 0$, set

$$h_\delta(x) := \mathbb{E}(h(x+\delta B_{2T})) = \mathbb{E}(h_{\delta/\sqrt{2}}(x+\delta B_T)).$$

For any $\delta > 0$, the function h_δ is smooth and its derivatives are exponentially bounded, so that we can apply the previous expansion to h_δ instead of h in order to obtain:

$$\begin{aligned}\mathbb{E}[h_\delta(X_T)] =& \mathbb{E}[h_\delta(X^P_T)] + C_1(a;x_0)^T_0(\partial^3_{\varepsilon^3} - \frac{3}{2}\partial^2_{\varepsilon^2} + \frac{1}{2}\partial_\varepsilon)\mathbb{E}\big[h_\delta(X^P_T+\varepsilon)\big]|_{\varepsilon=0}\\ &+ \text{Error}_2(h_\delta).\end{aligned}$$

Take $\delta = \mathcal{M}_1(a)[\mathcal{M}_0(a)]^2T$: then replacing $\mathbb{E}(h_\delta(X_T))$ and $\mathbb{E}(h_\delta(X^P_T))$ by $\mathbb{E}(h(X_T))$ and $\mathbb{E}(h(X^P_T))$ readily yields an extra error $\mathcal{O}(\mathcal{M}_1(a)[\mathcal{M}_0(a)]^2T^{\frac{3}{2}})$ which has the right magnitude regarding the expected global error. Moreover using $(\mathcal{H}^a_{x_0})$, we can also prove that computing the sensitivities with respect to h or to h_δ does not deteriorate the global accuracy (see [42]). It remains to prove that $\text{Error}_2(h_\delta) = \mathcal{O}(\mathcal{M}_1(a)[\mathcal{M}_0(a)]^2T^{\frac{3}{2}})$. An inspection of the representation (43) of $\text{Error}_2(h_\delta)$ shows immediately that the first contribution with $h^{(1)}_\delta$ is a $\mathcal{O}(\mathcal{M}_1(a)[\mathcal{M}_0(a)]^2T^{\frac{3}{2}})$, by simply using the exponential growth condition on $h^{(1)}$ and the finiteness of exponential moments of X^η_T .The second contribution with $h^{(2)}_\delta$ is the integral over $(\eta_1,\eta_2,\lambda) \in [0,1]^3$ of $(1-\eta_1)(1-\eta_2)(1-\lambda)$ times

$$\begin{aligned}&\mathbb{E}\big[h^{(2)}_\delta(X^P_T+\lambda(X_T-X^P_T))X^{\eta_1}_{2,T}X^{\eta_2}_{2,T}\big]\\ &= \mathbb{E}\big[h^{(2)}_{\delta/\sqrt{2}}(X^P_T+\lambda(X_T-X^P_T)+\delta B_T)X^{\eta_1}_{2,T}X^{\eta_2}_{2,T}\big]\\ &= \mathbb{E}\big[h^{(1)}_{\delta/\sqrt{2}}(X^P_T+\lambda(X_T-X^P_T)+\delta B_T)H^{\delta,\eta_1,\eta_2,\lambda}_1\big].\end{aligned}$$

The first equality follows from the definition of h_δ, whereas the second one is an integration by parts formula from Malliavin calculus [48]. We do not enter into the derivation details, we only emphasize two points: first, it is allowed since $X_T^P + \lambda(X_T - X_T^P) + \delta B_T$ is a non-degenerate random variable (in Malliavin sense) thanks to the additional perturbation δB_T, and its Malliavin matrix has an inverse of order $(\int_0^T a^2(t,x_0)\mathrm{d}t)^{-1}$ in L_p-norms, owing to the ellipticity assumption in $(\mathcal{H}^a_{x_0})$. Second, the Malliavin norms of $X^\eta_{2,T}$ can be estimated similarly to (44) and it finally gives that $(\mathbb{E}|H_1^{\delta,\eta_1,\eta_2,\lambda}|^2)^{1/2} = \mathcal{O}(\mathcal{M}_1(a)[\mathcal{M}_0(a)]^2 T^{\frac{3}{2}})$. This finishes the proof. Slight modifications in the above arguments would enable to handle functions with local Hölder smoothness.

Arbitrary function h. Here, we do not assume any regularity on h, only exponential growth. The analysis is similar but the regularization step for h is more complex, see [45]: the expansion analysis has been done under the uniform ellipticity condition on $(\mathcal{H}^a)$, and not only under the pointwise ellipticity in $(\mathcal{H}^a_{x_0})$.

As a conclusion to this *stochastic analysis approach*:

- the derivation of expansion coefficients is direct and easy;
- the error analysis relies on delicate Malliavin calculus estimates;
- it applies to general function h under mild non-degeneracy condition.

4.2. *Mixing stochastic analysis and PDE*

Here, we directly prove the expansion result for locally Lipschitz function h. We represent the error $\mathbb{E}[h(X_T)] - \mathbb{E}[h(X_T^P)]$ using the PDE associated to the proxy:

$$u^{P,h}(t,x) = \mathbb{E}[h(X_T^P)|X_t^P = x]$$

To get a smooth solution u^P, assume that $a(t,x_0) \neq 0$ for any $t \in [0,T]$, which is stronger that $\int_0^T a^2(t,x_0)\mathrm{d}t > 0$ considered in $(\mathcal{H}^a_{x_0})$. The generic constants appearing in our next error estimates depend in an increasing way of the oscillation ratio $\frac{\mathcal{M}_0(a)}{\inf_{t\in[0,T]} a(t,x_0)}$. Then,

$$\begin{cases} \partial_t u^{P,h}(t,x) + \frac{1}{2}a^2(t,x_0)(\partial^2_{x^2} - \partial_x)u^{P,h}(t,x) = 0, & \text{for } t < T, \\ u^{P,h}(T,x) = h(x), \end{cases} \tag{45}$$

$$|\partial^n_{x^n} u^{P,h}(t,x)| \leq_c e^{c|x|}\Big(\int_t^T a^2(s,x_0)\mathrm{d}s\Big)^{-\frac{n-1}{2}}. \tag{46}$$

The estimates (46) directly follow from the differentiation of the Gaussian density of X_T^P conditionally to $X_t^P = x$, taking into account the exponential growth of h. Then, apply Itô's formula to $u^{P,h}(t,X_t)$ between $t=0$ and $t=T$, combine this with simplifications coming from the PDE solved by $u^{P,h}$; it gives

$$\begin{aligned}\mathbb{E}[h(X_T)] =& \mathbb{E}[h(X_T^P)] + \frac{1}{2}\mathbb{E}[\int_0^T (a^2(t,X_t) - a^2(t,x_0))(\partial^2_{x^2} - \partial_x)u^{P,h}(t,X_t)\mathrm{d}t] \\ =& \mathbb{E}[h(X_T^P)] + \frac{1}{2}\int_0^T \partial_x[a^2](t,x_0)\mathbb{E}\Big[(X_t - x_0)(\partial^2_{x^2} - \partial_x)u^{P,h}(t,X_t)\Big]\mathrm{d}t \\ &+ \frac{1}{2}\mathbb{E}\Big[\int_0^T [a^2(t,X_t) - a^2(t,x_0) - \partial_x[a^2](t,x_0)(X_t - x_0)] \\ &\qquad \times (\partial^2_{xx} - \partial_x)u^{P,h}(t,X_t)\mathrm{d}t\Big]. \end{aligned} \tag{47}$$

Taking advantage of (46), we can easily bound the last term in (47) by

$$\begin{aligned} &C\int_0^T \mathcal{M}_0(a)\mathcal{M}_1(a)|X_t - x_0|_4^2 (\int_t^T a^2(s,x_0)\mathrm{d}s)^{-\frac{1}{2}}\mathrm{d}t \\ &= \mathcal{O}(\mathcal{M}_1(a)[\mathcal{M}_0(a)]^2 T^{\frac{3}{2}}) \end{aligned}$$

using standard increment estimates and uniform lower and upper bounds on a^2. Observe that the Lipschitz regularity of h gives rise to singular terms of the form $(T-t)^{-\frac{1}{2}}$, which are fortunately integrable at T.
Regarding the second term in (47), we have to approximate $\mathbb{E}\Big[(X_t - x_0)(\partial^2_{x^2} - \partial_x)u^{P,h}(t,X_t)\Big]$ for any $t \in [0,T[$: we apply again the previous decomposition by replacing T by t and $h(x)$ by $\phi_t(x) = (x - x_0)(\partial^2_{xx} - \partial_x)u^{P,h}(t,x)$. We denote by $v_t^{P,\phi}(s,x) = \mathbb{E}[\phi_t(X_t^P)|X_s^P = x]$ the solution of the system (45) on $[0,t[\times\mathbb{R}$ but with terminal condition ϕ_t. The term under study is thus equal to

$$\mathbb{E}[\phi_t(X_t^P)] + \frac{1}{2}\mathbb{E}[\int_0^t (a^2(s,X_s) - a^2(s,x_0))(\partial^2_{x^2} - \partial_x)v_t^{P,\phi}(s,X_s)\mathrm{d}s].$$

It remains to make explicit $v_t^{P,\phi}(s,x)$ in order to compute the first term and to estimate the second. For this, the trick lies in the observation that for any $k \geq 0$, $M_{k,t} = \partial^k_{x^k}u^{P,h}(t,X_t^P)$ is a martingale for $t < T$: this directly follows from the application of Itô's formula, combined with (45) and (46). Hence, successive applications of the equalities $\mathbb{E}[M_{k,t}|X_s^P = x] = \partial^k_{x^k}u^{P,h}(s,x)$ for $s \leq t$ and of the Lemma 1 gives:

$$\begin{aligned} v_t^{P,\phi}(s,x) =& \mathbb{E}\Big[(X_t^P - x_0)(\partial^2_{x^2} - \partial_x)u^{P,h}(t,X_t^P)|X_s^P = x\Big] \\ =& (x - x_0)\mathbb{E}\Big[(\partial^2_{x^2} - \partial_x)u^{P,h}(t,X_t^P)|X_s^P = x\Big] \\ &+ \mathbb{E}\Big[(X_t^P - X_s^P)(\partial^2_{x^2} - \partial_x)u^{P,h}(t,X_t^P)|X_s^P = x\Big] \end{aligned}$$

$$
\begin{aligned}
=&(x-x_0)(\partial^2_{x^2}-\partial_x)u^{P,h}(s,x)\\
&-\frac{1}{2}\int_s^t a^2(\xi,x_0)\mathrm{d}\xi\mathbb{E}\Big[(\partial^2_{x^2}-\partial_x)u^{P,h}(t,X_t^P)|X_s^P=x\Big]\\
&+\mathbb{E}\Big[\Big(\int_s^t a(\xi,x_0)\mathrm{d}W_\xi\Big)(\partial^2_{x^2}-\partial_x)u^{P,h}(t,x\\
&-\frac{1}{2}\int_s^t a^2(\xi,x_0)\mathrm{d}\xi+\int_s^t a(\xi,x_0)\mathrm{d}W_\xi)\Big]\\
=&(x-x_0)(\partial^2_{x^2}-\partial_x)u^{P,h}(s,x)\\
&-\frac{1}{2}\int_s^t a^2(\xi,x_0)\mathrm{d}\xi(\partial^2_{x^2}-\partial_x)u^{P,h}(s,x)\\
&+\int_s^t a^2(\xi,x_0)\mathrm{d}\xi\mathbb{E}\Big[(\partial^3_{x^3}-\partial^2_{x^2})u^{P,h}(t,X_t^P)|X_s^P=x\Big]\\
=&(x-x_0)(\partial^2_{x^2}-\partial_x)u^{P,h}(s,x)+\int_s^t a^2(\xi,x_0)\mathrm{d}\xi(\partial^3_{x^3}-\frac{3}{2}\partial^2_{x^2}\\
&+\frac{1}{2}\partial_x)u^{P,h}(s,x).
\end{aligned}
$$

In particular the above calculus yields $\mathbb{E}[\phi_t(X_t^P)] = v_t^{P,\phi}(0,x_0) = \int_0^t a^2(s,x_0)\mathrm{d}s(\partial^3_{x^3}-\frac{3}{2}\partial^2_{x^2}+\frac{1}{2}\partial_x)u^{P,h}(0,x_0)$ and by multiplying by $\frac{1}{2}\partial_x[a^2](t,x_0)$ and integrating over $t\in[0,T]$ in (47), we recover the correction terms from Theorem 4.

On the other hand, combining (46) and the ellipticity assumption, we easily obtain

$$
\begin{aligned}
&\big|(\partial^2_{x^2}-\partial_x)v_t^{P,\phi}(s,X_s)\big|_p\\
&\le_c |X_s-x_0|_{2p}\big|(\partial^4_{x^4}-2\partial^3_{x^3}+\partial^2_{x^2})u^{P,h}(s,X_s)\big|_{2p}\\
&+\big|(2\partial^3_{x^3}-3\partial^2_{x^2}+\partial_x)u^{P,h}(s,X_s)\big|_p\\
&+\int_s^t a^2(\xi,x_0)\mathrm{d}\xi\big|(\partial^5_{x^5}-\frac{5}{2}\partial^4_{x^4}+2\partial^3_{x^3}-\frac{1}{2}\partial^2_{x^2})u^{P,h}(s,X_s)\big|_p\\
&\le_c \frac{\sqrt{s}}{\inf_{s\in[0,T]} a^2(s,x_0)(T-s)^{\frac{3}{2}}}+\frac{1}{\inf_{s\in[0,T]} a^2(s,x_0)(T-s)},
\end{aligned}
$$

for any $p\ge 1$, $t\in[0,T[$, $s\in[0,t[$. Consequently we obtain for the final error:

$$
\begin{aligned}
&\Big|\int_0^T \partial_x[a^2](t,x_0)\mathbb{E}\Big[\int_0^t (a^2(s,X_s)-a^2(s,x_0))(\partial^2_{x^2}-\partial_x)v_t^{P,\phi}(s,X_s)\mathrm{d}s\Big]\mathrm{d}t\Big|\\
&\le_c \mathcal{M}_1(a)\mathcal{M}_0(a)\int_0^T\int_0^t |X_s-x_0|_2\Big[\frac{\sqrt{s}}{(T-s)^{\frac{3}{2}}}+\frac{1}{(T-s)}\Big]\mathrm{d}s\mathrm{d}t\\
&\le_c \mathcal{M}_1(a)[\mathcal{M}_0(a)]^2T^{\frac{3}{2}}.
\end{aligned}
$$

We have retrieved the error estimate provided in Theorem 4. Once again, we would like to point out that the singular terms $(T-s)^{-3/2}$ and $(T-s)^{-1}$ appearing in the above time iterated integral remain integrable.

As a conclusion to this approach *mixing stochastic analysis and PDE*:

- the error analysis relies on usual estimates of derivatives of heat equations (PDE satisfied by the proxy) and it may be considered easier; however for digital options, the singularities arising in iterated time integrals are not integrable and the current approach seems to be inappropriate.
- this approach requires stronger non-degeneracy assumptions compared to the previous stochastic analysis approach;
- the explicit derivation of expansion coefficients is tricky and relies on appropriate combination of martingale properties and Itô calculus;
- we nevertheless mention that this approach could be potentially used in a framework where the Malliavin calculus fails, e.g. for barrier options.

Actually for higher order expansion, the latter explicit martingale computation is harder to write down, whereas a direct application of Lemma 1 remains direct.

4.3. *A pure PDE approach*

Alternatively, inspired by the interpolation (23-24), consider the solution of the PDE

$$\begin{cases} \partial_t u^\eta(t,x) + \frac{1}{2}a^2(t, x_0 + \eta(x - x_0))(\partial^2_{x^2} - \partial_x)u^\eta(t,x) = 0, & \text{for } t < T, \\ u^\eta(T,x) = h(x). \end{cases}$$

Observe that $u^1(0,x_0)$ coincides with $\mathbb{E}(h(X_T))$ whereas $u^0(0,x_0)$ coincides with $\mathbb{E}(h(X_T^P))$. This PDE has similarities with that of Hagan (20) but it differs here, because the space variable has not been rescaled around the strike. In addition the solution of the principal PDE in the Hagan approach is a Call price in a Bachelier model, whereas $u^0(0,x_0)$ is a Call price in a Black-Scholes model.
To derive the correction terms, we shall apply a regular perturbation analysis by writing $u^\eta = u_0 + \eta u_1 + \dots$, with $u_0 = u^0$, and

$$L^\eta = \partial_t + \frac{1}{2}a^2(t, x_0 + \eta(x - x_0))(\partial^2_{x^2} - \partial_x)$$

$$= L_0 + \eta\frac{1}{2}\partial_x[a^2](t,x_0)(x - x_0)(\partial^2_{x^2} - \partial_x) + \dots .$$

A formal identification of the system to PDEs to solve gives

$$\left[\partial_t + \frac{1}{2}a^2(t,x_0)(\partial^2_{x^2} - \partial_x)\right]u_0(t,x) = 0,$$

$$L_0 u_1 = -\frac{1}{2}\partial_x[a^2](t,x_0)(x - x_0)(\partial^2_{x^2} - \partial_x)u_0$$

with $u_0(T,x) = h(x)$ and $u_1(T,x) = 0$. As mentioned before, in our opinion, an explicit resolution of u_1 is difficult to exhibit without knowing the solution.

However, after tedious calculus involving Gaussian kernels and convolutions, we can retrieve the corrective terms of Theorem 4.

Also, a PDE error analysis (which we have not been able to find in the literature in the case of irregular h) may presumably give error estimates only in powers of η (which equals 1 here!) and not as $\mathcal{O}(\mathcal{M}_1(a)[\mathcal{M}_0(a)]^2T^{\frac{3}{2}})$. Additionally, due to the form of the proxy, our intuition is that the error at $(0,x_0)$ (where we aim at computing the solution) is smaller that the error at arbitrary (t,x). All these reasons indicate that a PDE approach to derive correction terms and error analysis in our proxy setting is probably irrelevant.

5. Higher-order proxy approximation

In this section, we give several expansions formulas with a third order accuracy. First, we recall without proof results obtained in [49] and [44] for expansions based on local volatility at spot and at strike. Second we introduce a new expansion with local volatility frozen at mid-point. Finally new expansions of implied volatility are provided.

5.1. *Third order approximation with the local volatility at spot and at strike.*

We define some integral operators useful to state the next theorems.

Definition 4.1. If the derivatives and the integrals have a meaning, we define for a two variables function l the above operators:

$$\begin{aligned}
C_1(l;z)_0^T =&\omega(l^2(z),l(z)l^{(1)}(z))_0^T,\\
C_2(l;z)_0^T =&\omega(l^2(z),(l^{(1)}(z))^2+l(z)l^{(2)}(z))_0^T,\\
C_3(l;z)_0^T =&\omega(l^2(z),l^2(z),(l^{(1)}(z))^2+l(z)l^{(2)}(z))_0^T,\\
C_4(l;z)_0^T =&\omega(l^2(z),l(z)l^{(1)}(z),l(z)l^{(1)}(z))_0^T.
\end{aligned}$$

We frequently use some linear combinations of these operators:

$$\begin{aligned}
\eta_1(l;z)_0^T =&\frac{C_1(l;z)_0^T}{2}-\frac{C_2(l;z)_0^T}{2}-\frac{C_3(l;z)_0^T}{4}-\frac{C_4(l;z)_0^T}{2},\\
\eta_2(l;z)_0^T =&-\frac{3C_1(l;z)_0^T}{2}+\frac{C_2(l;z)_0^T}{2}+\frac{5C_3(l;z)_0^T}{4}+\frac{7C_4(l;z)_0^T}{2}+\frac{[C_1(l;z)_0^T]^2}{8},\\
\eta_3(l;z)_0^T =&C_1(l;z)_0^T-2C_3(l;z)_0^T-6C_4(l;z)_0^T-\frac{3[C_1(l;z)_0^T]^2}{4},
\end{aligned}$$

$$\eta_4(l;z)_0^T = C_3(l;z)_0^T + 3C_4(l;z)_0^T + \frac{13[C_1(l;z)_0^T]^2}{8},$$
$$\eta_5(l;z)_0^T = -\frac{3[C_1(l;z)_0^T]^2}{2},$$
$$\eta_6(l;z)_0^T = \frac{[C_1(l;z)_0^T]^2}{2},$$
$$\zeta_2(l;z)_0^T = \frac{C_2(l;z)_0^T}{2}, \qquad \zeta_3(l;z)_0^T = C_1(l;z)_0^T,$$
$$\zeta_4(l;z)_0^T = C_3(l;z)_0^T + 3C_4(l;z)_0^T, \qquad \zeta_6(l;z)_0^T = \frac{[C_1(l;z)_0^T]^2}{2}.$$

Theorem 10. ***(3rd order approximations for Call options with the local volatility at spot).*** *Assuming $(\mathcal{H}^a)$ and using the log-normal proxy, one has*

$$\begin{aligned}\mathrm{Call}(e^{x_0}, T, e^k) = &\mathrm{Call}^{\mathrm{BS}}(x_0, \bar{a}_{x_0}^2 T, k) + \sum_{i=1}^{6} \eta_i(a;x_0)_0^T \partial_{x^i}^i \mathrm{Call}^{\mathrm{BS}}(x_0, \bar{a}_{x_0}^2 T, k) \\ &+ \mathcal{O}(\mathcal{M}_1(a)[\mathcal{M}_0(a)]^3 T^2).\end{aligned}$$

Assuming $(\mathcal{H}^\Sigma)$ and using the normal proxy, one has

$$\begin{aligned}\mathrm{Call}(S_0, T, K) = &\ \mathrm{Call}^{\mathrm{BA}}(S_0, \bar{\Sigma}_{S_0}^2 T, K) \\ &+ \sum_{i\in\{2,3,4,6\}} \zeta_i(\Sigma;S_0)_0^T \partial_{S^i}^i \mathrm{Call}^{\mathrm{BA}}(S_0, \bar{\Sigma}_{S_0}^2 T, K) \\ &+ \mathcal{O}(\mathcal{M}_1(\Sigma)[\mathcal{M}_0(\Sigma)]^3 T^2).\end{aligned}$$

The operators ζ_i and η_i in the above expansions are defined in Definition 4.1.

The magnitude of the residual terms in the previous formulas justifies the label of third order approximations.

The above theorem is a straightforward application of Theorems 2.2, 2.3 and 4.2 in [49], taking into account that we slightly modify the notations of the Greek coefficients. Namely, for convenience we merge certain ω operators: for instance the reader can easily check that:

$$\begin{aligned}[C_1(l;z)_0^T]^2 =& [\omega(l(z)^2, l(z)l^{(1)}(z))_0^T]^2 \\ =& \ 4\omega(l(z)^2, l(z)^2, l(z)l^{(1)}(z), l(z)l^{(1)}(z))_0^T \\ &+ 2\omega(l(z)^2, l(z)l^{(1)}(z), l(z)^2, l(z)l^{(1)}(z))_0^T.\end{aligned}$$

We should mention that it seems possible to relax the strong hypothesis $(\mathcal{H}^a)$ which appears in Theorems 2.2 and 4.2 in [49]. As for the second order approximations, $(\mathcal{H}_{x_0}^a)$ may be sufficient.

Using the duality argument introduced in Subsection 3.3 and Theorems 2.2, 2.3 and 4.2 in [49], approximations using the volatility at strike are available too.

Theorem 11. ***(3rd order approximations for Call options with the local volatility at strike).*** *Assuming* $(\mathcal{H}^a)$ *and using the log-normal proxy, one has*

$$\begin{aligned}\mathrm{Call}(e^{x_0}, T, e^k) &= \mathrm{Call}^{\mathrm{BS}}(x_0, \bar{a}_k^2 T, k)\\ &\quad + \sum_{i=1}^{6} \eta_i(\widetilde{a}; k)_0^T \partial^i_{z^i} \mathrm{Call}^{\mathrm{BS}}(x_0, \bar{a}_k^2 T, k)\\ &\quad + \mathcal{O}(\mathcal{M}_1(a)[\mathcal{M}_0(a)]^3 T^2),\end{aligned}$$

Assuming $(\mathcal{H}^\Sigma)$*and using the normal proxy, one has*

$$\begin{aligned}\mathrm{Call}(S_0, T, K) &= \mathrm{Call}^{\mathrm{BA}}(S_0, \bar{\Sigma}_K^2 T, K)\\ &\quad + \sum_{i\in\{2,3,4,6\}} \zeta_i(\widetilde{\Sigma}; K)_0^T \partial^i_{Z^i} \mathrm{Call}^{\mathrm{BA}}(S_0, \bar{\Sigma}_K^2 T, K)\\ &\quad + \mathcal{O}(\mathcal{M}_1(\Sigma)[\mathcal{M}_0(\Sigma)]^3 T^2).\end{aligned}$$

The operators ζ_i *and* η_i *in the above expansions are defined in Definition 4.1.*

5.2. *Third order approximation with the local volatility at mid-point.*

We now state a new result related to third order expansions based on the local volatility at mid-point x_{av} or S_{av}. For a clearer proof, we change the presentation of the corrective terms in comparison with Theorems 10 and 11: instead of gathering them according to the order of the Greeks, we put them together according to the operators C_i introduced in Definition 4.1.

Theorem 12. ***(3rd order approximations for Call options with the local volatility at mid-point).*** *Assuming* $(\mathcal{H}^a)$ *and using the log-normal proxy, one has*

$$\begin{aligned}&\mathrm{Call}(e^{x_0}, T, e^k) = \mathrm{Call}^{\mathrm{BS}}(x_0, \bar{a}_{x_{\mathrm{av}}}^2 T, k)\\ &+ \frac{C_1(a; x_{\mathrm{av}})_0^T - C_1(\widetilde{a}; x_{\mathrm{av}})_0^T}{2}(\partial^3_{x^3} - \frac{3}{2}\partial^2_{x^2} + \frac{1}{2}\partial_x)\mathrm{Call}^{\mathrm{BS}}(x_0, \bar{a}_{x_{\mathrm{av}}}^2 T, k)\\ &+ \frac{C_2(a; x_{\mathrm{av}})_0^T + C_2(\widetilde{a}; x_{\mathrm{av}})_0^T}{2}(\frac{1}{2}\partial^2_{x^2} - \frac{1}{2}\partial_x)\mathrm{Call}^{\mathrm{BS}}(x_0, \bar{a}_{x_{\mathrm{av}}}^2 T, k)\\ &+ \frac{C_3(a; x_{\mathrm{av}})_0^T + C_3(\widetilde{a}; x_{\mathrm{av}})_0^T}{2}(\partial^4_{x^4} - 2\partial^3_{x^3} + \frac{5}{4}\partial^2_{x^2} - \frac{1}{4}\partial_x)\mathrm{Call}^{\mathrm{BS}}(x_0, \bar{a}_{x_{\mathrm{av}}}^2 T, k)\\ &+ \frac{C_4(a; x_{\mathrm{av}})_0^T + C_4(\widetilde{a}; x_{\mathrm{av}})_0^T}{2}(3\partial^4_{x^4} - 6\partial^3_{x^3} + \frac{7}{2}\partial^2_{x^2} - \frac{1}{2}\partial_x)\mathrm{Call}^{\mathrm{BS}}(x_0, \bar{a}_{x_{\mathrm{av}}}^2 T, k)\\ &+ \frac{[C_1(a; x_{\mathrm{av}})_0^T]^2 + [C_1(\widetilde{a}; x_{\mathrm{av}})_0^T]^2}{2}(\frac{1}{2}\partial^6_{x^6} - \frac{3}{2}\partial^5_{x^5} + \frac{13}{8}\partial^4_{x^4} - \frac{3}{4}\partial^3_{x^3} + \frac{1}{8}\partial^2_{x^2})\\ &\times \mathrm{Call}^{\mathrm{BS}}(x_0, \bar{a}_{x_{\mathrm{av}}}^2 T, k) - (x_0 - k)^2 C_5(a; x_{\mathrm{av}})_0^T(\frac{1}{8}\partial^2_{x^2} - \frac{1}{8}\partial_x)\mathrm{Call}^{\mathrm{BS}}(x_0, \bar{a}_{x_{\mathrm{av}}}^2 T, k)\\ &- (x_0 - k)^2 C_6(a; x_{\mathrm{av}})_0^T(\frac{1}{4}\partial^4_{x^4} - \frac{1}{2}\partial^3_{x^3} + \frac{1}{4}\partial^2_{x^2})\mathrm{Call}^{\mathrm{BS}}(x_0, \bar{a}_{x_{\mathrm{av}}}^2 T, k)\\ &+ \mathcal{O}(\mathcal{M}_1(a)[\mathcal{M}_0(a)]^3 T^2),\end{aligned} \tag{48}$$

where the operators C_i for $i = 1..4$ are defined in Definition 4.1 and where the time reversal invariant operators C_5 and C_6 are defined by:*

$$\begin{aligned} C_5(l;z)_0^T =&\omega((l^{(1)}(z))^2 + l(z)l^{(2)}(z))_0^T, \\ C_6(l;z)_0^T =&\omega(l(z)l^{(1)}(z), l(z)l^{(1)}(z))_0^T. \end{aligned}$$

Assuming $(\mathcal{H}^\Sigma)$ and using the normal proxy, one has

$$\begin{aligned} \mathrm{Call}(S_0,T,K) =&\mathrm{Call}^{\mathrm{BA}}(S_0, \bar{\Sigma}^2_{S_{\mathrm{av}}} T, K) \\ &+ \frac{C_1(\Sigma; S_{\mathrm{av}})_0^T - C_1(\widetilde{\Sigma}; S_{\mathrm{av}})_0^T}{2} \partial^3_{S^3} \mathrm{Call}^{\mathrm{BA}}(S_0, \bar{\Sigma}^2_{S_{\mathrm{av}}} T, K) \\ &+ \frac{C_2(\Sigma; S_{\mathrm{av}})_0^T + C_2(\widetilde{\Sigma}; S_{\mathrm{av}})_0^T}{4} \partial^2_{S^2} \mathrm{Call}^{\mathrm{BA}}(S_0, \bar{\Sigma}^2_{S_{\mathrm{av}}} T, K) \\ &+ \frac{C_3(\Sigma; S_{\mathrm{av}})_0^T + C_3(\widetilde{\Sigma}; S_{\mathrm{av}})_0^T}{2} \partial^4_{S^4} (S_0, \bar{\Sigma}^2_{S_{\mathrm{av}}} T, K) \\ &+ 3\frac{C_4(\Sigma; S_{\mathrm{av}})_0^T + C_4(\widetilde{\Sigma}; S_{\mathrm{av}})_0^T}{2} \partial^4_{S^4} (S_0, \bar{\Sigma}^2_{S_{\mathrm{av}}} T, K) \\ &+ \frac{[C_1(\Sigma; S_{\mathrm{av}})_0^T]^2 + [C_1(\widetilde{\Sigma}; S_{\mathrm{av}})_0^T]^2}{4} \partial^6_{S^6} (S_0, \bar{\Sigma}^2_{S_{\mathrm{av}}} T, K) \\ &- (x_0 - k)^2 \frac{C_5(\Sigma; S_{\mathrm{av}})_0^T}{8} \partial^2_{S^2} \mathrm{Call}^{\mathrm{BA}}(S_0, \bar{\Sigma}^2_{S_{\mathrm{av}}} T, K) \\ &- (x_0 - k)^2 \frac{C_6(\Sigma; S_{\mathrm{av}})_0^T}{4} \partial^4_{S^4} (S_0, \bar{\Sigma}^2_{S_{\mathrm{av}}} T, K) \\ &+ \mathcal{O}(\mathcal{M}_1(\Sigma)[\mathcal{M}_0(\Sigma)]^3 T^2). \end{aligned}$$

Proof. We only prove the result for the log-normal proxy. The case of normal proxy is similar, and it is left to the reader as an exercise. The idea is again to consider the average of the third order formulas in spot and strike provided in Theorems 10 and 11 and to perform an expansion around the mid-point.

▷ **Step 1: expansion of the leading term.** Firstly we aim at showing that:

$$\begin{aligned} &(\mathrm{Call}^{\mathrm{BS}}(x_0, \bar{a}^2_{x_0} T, k) + \mathrm{Call}^{\mathrm{BS}}(x_0, \bar{a}^2_k T, k))/2 \\ =&\mathrm{Call}^{\mathrm{BS}}(x_0, \bar{a}^2_{x_{\mathrm{av}}} T, k) + \frac{(x_0 - k)^2}{4} C_5(a; x_{\mathrm{av}})_0^T \partial_y \mathrm{Call}^{\mathrm{BS}}(x_0, \bar{a}^2_{x_{\mathrm{av}}} T, k) \\ &+ (x_0 - k)^2 C_6(a; x_{\mathrm{av}})_0^T \partial^2_{y^2} \mathrm{Call}^{\mathrm{BS}}(x_0, \bar{a}^2_{x_{\mathrm{av}}} T, k) + \mathcal{O}(\mathcal{M}_1(a)[\mathcal{M}_0(a)]^3 T^2), \end{aligned} \quad (49)$$

where the operators C_5 and C_6 are defined in Theorem 12. Perform Taylor expansions to obtain:

*that is $C_5(\tilde{l}; z)_0^T = C_5(l; z)_0^T$ and $C_6(\tilde{l}; z)_0^T = C_6(l; z)_0^T$ using (26).

$$\begin{aligned}
&\mathrm{Call}^{\mathrm{BS}}(x_0, \bar{a}^2_{x_0} T, k)\\
&\quad = \mathrm{Call}^{\mathrm{BS}}(x_0, \bar{a}^2_{x_{\mathrm{av}}} T, k) + \partial_y \mathrm{Call}^{\mathrm{BS}}(x_0, \bar{a}^2_{x_{\mathrm{av}}} T, k) T(\bar{a}^2_{x_0} - \bar{a}^2_{x_{\mathrm{av}}})\\
&\qquad + \frac{1}{2}\partial_{y^2} \mathrm{Call}^{\mathrm{BS}}(x_0, \bar{a}^2_{x_{\mathrm{av}}} T, k) T^2 (\bar{a}^2_{x_0} - \bar{a}^2_{x_{\mathrm{av}}})^2 + R_1\\
&\quad = \mathrm{Call}^{\mathrm{BS}}(x_0, \bar{a}^2_{x_{\mathrm{av}}} T, k) + \partial_y \mathrm{Call}^{\mathrm{BS}}(x_0, \bar{a}^2_{x_{\mathrm{av}}} T, k)\\
&\qquad \times \omega(a(x_{\mathrm{av}}) a^{(1)}(x_{\mathrm{av}}))_0^T (x_0 - k) && (50)\\
&\qquad + \frac{1}{4}\partial_y \mathrm{Call}^{\mathrm{BS}}(x_0, \bar{a}^2_{x_{\mathrm{av}}} T, k) C_5(a; x_{\mathrm{av}})_0^T (x_0 - k)^2\\
&\qquad + \partial^2_{y^2} \mathrm{Call}^{\mathrm{BS}}(x_0, \bar{a}^2_{x_{\mathrm{av}}} T, k) C_6(a; x_{\mathrm{av}})_0^T (x_0 - k)^2 && (51)\\
&\qquad + R_1 + R_2 + R_3, && (52)
\end{aligned}$$

where we have used the relations $\partial_z \omega(l^2(z))_0^T = 2\omega(l(z) l^{(1)}(z))_0^T$, $\partial^2_{z^2}\omega(l^2(z))_0^T = 2C_5(l; z)_0^T$, $[\omega(l(z) l^{(1)}(z))_0^T]^2 = 2C_6(l; z)_0^T$ and where R_1, R_2 and R_3 are defined by:

$$\begin{aligned}
R_1 &= T^3(\bar{a}^2_{x_0} - \bar{a}^2_{x_{\mathrm{av}}})^3\\
&\quad \times \int_0^1 (\partial^3_{y^3} \mathrm{Call}^{\mathrm{BS}}(x_0, y, k))|_{y = T(\lambda \bar{a}^2_{x_0} + (1-\lambda)\bar{a}^2_{x_{\mathrm{av}}})} \frac{(1-\lambda)^2}{2} \mathrm{d}\lambda,\\
R_2 &= \partial_y \mathrm{Call}^{\mathrm{BS}}(x_0, \bar{a}^2_{x_{\mathrm{av}}} T, k) \frac{(x_0 - k)^3}{8}\\
&\quad \times \int_0^1 (\partial^3_x \omega(a^2(x))_0^T)|_{x = \lambda x_0 + (1-\lambda) x_{\mathrm{av}}} \frac{(1-\lambda)^2}{2} \mathrm{d}\lambda,\\
R_3 &= \frac{1}{2}\partial^2_{y^2} \mathrm{Call}^{\mathrm{BS}}(x_0, \bar{a}^2_{x_{\mathrm{av}}} T, k) \frac{(x_0 - k)^2}{4}\\
&\quad \times \int_0^1 (\partial^2_x \omega(a^2(x))_0^T)_{x = \lambda x_0 + (1-\lambda) x_{\mathrm{av}}} (1-\lambda) \mathrm{d}\lambda\\
&\quad \times \Big[\frac{(x_0 - k)^2}{4} \int_0^1 (\partial^2_x \omega(a^2(x))_0^T)_{x = \lambda x_0 + (1-\lambda) x_{\mathrm{av}}} (1-\lambda) \mathrm{d}\lambda\\
&\qquad + 2\omega(a(x_{\mathrm{av}}) a^{(1)}(x_{\mathrm{av}}))_0^T (x_0 - k)\Big].
\end{aligned}$$

Using (6), Corollary 4.1 and $(\mathcal{H}^a)$ we obtain

$$|R_1 + R_2 + R_3| \leq_c \mathcal{M}_1(a)[\mathcal{M}_0(a)]^3 T^2.$$

Similarly, we show that:

$$\begin{aligned}
&\mathrm{Call}^{\mathrm{BS}}(x_0, \bar{a}^2_k T, k)\\
&= \mathrm{Call}^{\mathrm{BS}}(x_0, \bar{a}_{x_{\mathrm{av}}}, k) - \partial_y \mathrm{Call}^{\mathrm{BS}}(x_0, \bar{a}^2_{x_{\mathrm{av}}} T, k) \omega(a(x_{\mathrm{av}}) a^{(1)}(x_{\mathrm{av}}))_0^T (x_0 - k)\\
&+ \frac{1}{4}\partial_y \mathrm{Call}^{\mathrm{BS}}(x_0, \bar{a}^2_{x_{\mathrm{av}}} T, k) C_5(a; x_{\mathrm{av}})_0^T (x_0 - k)^2\\
&+ \partial^2_{y^2} \mathrm{Call}^{\mathrm{BS}}(x_0, \bar{a}^2_{x_{\mathrm{av}}} T, k) C_6(a; x_{\mathrm{av}})_0^T (x_0 - k)^2 + \mathcal{O}(\mathcal{M}_1(a)[\mathcal{M}_0(a)]^3 T^2).
\end{aligned}$$

Combine this with (52) to obtain (49).

$\triangleright$ **Step 2: expansion of the corrective terms.** Firstly we treat the corrective terms with the operators C_2, C_3, C_4 and $[C_1]^2$ in Theorems 10 and 11. We let the reader verify that in the formula with volatility at spot (respectively in strike), we can replace the point x_0 (respectively k) by the point $x_{\rm av}$ in all the corrective terms involving these operators: indeed it induces an extra error of order $\mathcal{M}_1(a)[\mathcal{M}_0(a)]^3T^2$. This is very similar to the proof of Lemma 2 so we skip it. Then we can replace derivatives w.r.t. z with derivatives w.r.t. x in $\mathrm{Call}^{\mathrm{BS}}$ thanks to Proposition 4.3, equations (75)-(77)-(78)-(80). That leads to:

$$
\begin{aligned}
&\frac{1}{2}C_2(a;x_0)_0^T(\frac{1}{2}\partial^2_{x^2}-\frac{1}{2}\partial_x)\mathrm{Call}^{\mathrm{BS}}(x_0,\bar a^2_{x_0}T,k)\\
&+\frac{1}{2}C_2(\widetilde a;k)_0^T(\frac{1}{2}\partial^2_{z^2}-\frac{1}{2}\partial_z)\mathrm{Call}^{\mathrm{BS}}(x_0,\bar a^2_{k}T,k)\\
&+\frac{1}{2}C_3(a;x_0)_0^T(\partial^4_{x^4}-2\partial^3_{x^3}+\frac{5}{4}\partial^2_{x^2}-\frac{1}{4}\partial_x)\mathrm{Call}^{\mathrm{BS}}(x_0,\bar a^2_{x_0}T,k)\\
&+\frac{1}{2}C_3(\widetilde a;k)_0^T(\partial^4_{z^4}-2\partial^3_{z^3}+\frac{5}{4}\partial^2_{z^2}-\frac{1}{4}\partial_z)\mathrm{Call}^{\mathrm{BS}}(x_0,\bar a^2_{k}T,k)\\
&+\frac{1}{2}C_4(a;x_0)_0^T(3\partial^4_{x^4}-6\partial^3_{x^3}+\frac{7}{2}\partial^2_{x^2}-\frac{1}{2}\partial_x)\mathrm{Call}^{\mathrm{BS}}(x_0,\bar a^2_{x_0}T,k)\\
&+\frac{1}{2}C_4(\widetilde a;k)_0^T(3\partial^4_{z^4}-6\partial^3_{z^3}+\frac{7}{2}\partial^2_{z^2}-\frac{1}{2}\partial_z)\mathrm{Call}^{\mathrm{BS}}(x_0,\bar a^2_{k}T,k)\\
&+\frac{1}{2}[C_1(a;x_0)_0^T]^2(\frac{1}{2}\partial^6_{x^6}-\frac{3}{2}\partial^5_{x^5}+\frac{13}{8}\partial^4_{x^4}-\frac{3}{4}\partial^3_{x^3}+\frac{1}{8}\partial^2_{x^2})\mathrm{Call}^{\mathrm{BS}}(x_0,\bar a^2_{x_0}T,k)\\
&+\frac{1}{2}[C_1(\widetilde a;k)_0^T]^2(\frac{1}{2}\partial^6_{z^6}-\frac{3}{2}\partial^5_{z^5}+\frac{13}{8}\partial^4_{z^4}-\frac{3}{4}\partial^3_{z^3}+\frac{1}{8}\partial^2_{z^2})\mathrm{Call}^{\mathrm{BS}}(x_0,\bar a^2_{k}T,k)\\
=&\frac{C_2(a;x_{\rm av})_0^T+C_2(\widetilde a;x_{\rm av})_0^T}{2}(\frac{1}{2}\partial^2_{x^2}-\frac{1}{2}\partial_x)\mathrm{Call}^{\mathrm{BS}}(x_0,\bar a^2_{x_{\rm av}}T,k)\\
&+\frac{C_3(a;x_{\rm av})_0^T+C_3(\widetilde a;x_{\rm av})_0^T}{2}(\partial^4_{x^4}-2\partial^3_{x^3}+\frac{5}{4}\partial^2_{x^2}-\frac{1}{4}\partial_x)\mathrm{Call}^{\mathrm{BS}}(x_0,\bar a^2_{x_{\rm av}}T,k)\\
&+\frac{C_4(a;x_{\rm av})_0^T+C_4(\widetilde a;x_{\rm av})_0^T}{2}(3\partial^4_{x^4}-6\partial^3_{x^3}+\frac{7}{2}\partial^2_{x^2}-\frac{1}{2}\partial_x)\mathrm{Call}^{\mathrm{BS}}(x_0,\bar a^2_{x_{\rm av}}T,k)\\
&+\frac{[C_1(a;x_{\rm av})_0^T]^2+[C_1(\widetilde a;x_{\rm av})_0^T]^2}{2}\\
&\quad\times(\frac{1}{2}\partial^6_{x^6}-\frac{3}{2}\partial^5_{x^5}+\frac{13}{8}\partial^4_{x^4}-\frac{3}{4}\partial^3_{x^3}+\frac{1}{8}\partial^2_{x^2})\mathrm{Call}^{\mathrm{BS}}(x_0,\bar a^2_{x_{\rm av}}T,k)\\
&+\mathcal{O}(\mathcal{M}_1(a)[\mathcal{M}_0(a)]^3T^2).
\end{aligned}\tag{53}
$$

Secondly, we pass to the corrective terms in which appears the operator C_1. For the sake of clarity, we introduce the following notation $\mathcal{A}_x=\partial^3_{x^3}-\frac{3}{2}\partial^2_{x^2}+\frac{1}{2}\partial_x$ and $\mathcal{A}_z=\partial^3_{z^3}-\frac{3}{2}\partial^2_{z^2}+\frac{1}{2}\partial_z$. For example $\partial_y\mathcal{A}_x$ stands for the differential operator $\partial^4_{yx^3}-\frac{3}{2}\partial^3_{yx^2}+\frac{1}{2}\partial^2_{yx}$ and similarly for $\partial^2_{y^2}\mathcal{A}_x$. We recall the following relation $\mathcal{A}_x\mathrm{Call}^{\mathrm{BS}}=-\mathcal{A}_z\mathrm{Call}^{\mathrm{BS}}$ (see (75) in Proposition 4.3). Our purpose is to prove that:

$$\begin{aligned}&\frac{1}{2}C_1(a;x_0)_0^T\mathcal{A}_x\mathrm{Call}^{\mathrm{BS}}(x_0,\bar{a}_{x_0}^2T,k)+\frac{1}{2}C_1(\widetilde{a};k)_0^T\mathcal{A}_z\mathrm{Call}^{\mathrm{BS}}(x_0,\bar{a}_k^2T,k)\\&=\frac{1}{2}(C_1(a;x_{\mathrm{av}})_0^T-C_1(\widetilde{a};x_{\mathrm{av}})_0^T)\mathcal{A}_x\mathrm{Call}^{\mathrm{BS}}(x_0,\bar{a}_{x_{\mathrm{av}}}^2T,k)\\&+\frac{(x_0-k)}{4}[4C_6(a;x_{\mathrm{av}})_0^T\\&+C_2(a;x_{\mathrm{av}})_0^T+C_2(\widetilde{a};x_{\mathrm{av}})_0^T]\mathcal{A}_x\mathrm{Call}^{\mathrm{BS}}(x_0,\bar{a}_{x_{\mathrm{av}}}^2T,k)\\&+[C_4(a;x_{\mathrm{av}})_0^T+C_4(\widetilde{a};x_{\mathrm{av}})_0^T+\omega(a(x_{\mathrm{av}})a^{(1)}(x_{\mathrm{av}}),a^2(x_{\mathrm{av}}),a(x_{\mathrm{av}})a^{(1)}\\&\times(x_{\mathrm{av}}))_0^T](x_0-k)\partial_y\mathcal{A}_x\mathrm{Call}^{\mathrm{BS}}(x_0,\bar{a}_{x_{\mathrm{av}}}^2T,k)\\&+\mathcal{O}(\mathcal{M}_1(a)[\mathcal{M}_0(a)]^3T^2).\end{aligned}\tag{54}$$

Perform a second order Taylor expansion for the function $y\to\mathcal{A}_x\mathrm{Call}^{\mathrm{BS}}(x_0,y,k)$ at $y=\bar{a}_{x_0}^2T=\omega(a^2(x_0))_0^T$ around $y=\bar{a}_{x_{\mathrm{av}}}^2T=\omega(a^2(x_{\mathrm{av}}))_0^T$ and for the function $x\to C_1(a;x)_0^T$ at $x=x_0$ around $x=x_{\mathrm{av}}$:

$$\begin{aligned}&C_1(a;x_0)_0^T\mathcal{A}_x\mathrm{Call}^{\mathrm{BS}}(x_0,\bar{a}_{x_0}^2T,k)\\&=\{C_1(a;x_{\mathrm{av}})_0^T+\partial_x(C_1(a;x)_0^T)|_{x=x_{\mathrm{av}}}\frac{(x_0-k)}{2}+R_1\}\\&\quad\times\{\mathcal{A}_x\mathrm{Call}^{\mathrm{BS}}(x_0,\bar{a}_{x_{\mathrm{av}}}^2T,k)\\&\quad+\partial_y\mathcal{A}_x\mathrm{Call}^{\mathrm{BS}}(x_0,\bar{a}_{x_{\mathrm{av}}}^2T,k)T(\bar{a}_{x_0}^2-\bar{a}_{x_{\mathrm{av}}}^2)+R_2\}\\&=\{C_1(a;x_{\mathrm{av}})_0^T+\partial_x(C_1(a;x)_0^T)|_{x=x_{\mathrm{av}}}\frac{(x_0-k)}{2}+R_1\}\\&\quad\times\{\mathcal{A}_x\mathrm{Call}^{\mathrm{BS}}(x_0,\bar{a}_{x_{\mathrm{av}}}^2T,k)+R_3+R_2\\&\quad+\partial_y\mathcal{A}_x\mathrm{Call}^{\mathrm{BS}}(x_0,\bar{a}_{x_{\mathrm{av}}}^2T,k)\omega(a(x_{\mathrm{av}})a^{(1)}(x_{\mathrm{av}}))_0^T(x_0-k)\}\\&=C_1(a;x_{\mathrm{av}})_0^T\mathcal{A}_x\mathrm{Call}^{\mathrm{BS}}(x_0,\bar{a}_{x_{\mathrm{av}}}^2T,k)\\&\quad+\partial_x(C_1(a;x)_0^T)|_{x=x_{\mathrm{av}}}\frac{(x_0-k)}{2}\mathcal{A}_x\mathrm{Call}^{\mathrm{BS}}(x_0,\bar{a}_{x_{\mathrm{av}}}^2T,k)\\&\quad+C_1(a;x_{\mathrm{av}})_0^T\partial_y\mathcal{A}_x\mathrm{Call}^{\mathrm{BS}}(x_0,\bar{a}_{x_{\mathrm{av}}}^2T,k)\omega(a(x_{\mathrm{av}})a^{(1)}(x_{\mathrm{av}}))_0^T(x_0-k)\\&\quad+R,\end{aligned}\tag{55}$$

where:

$$\begin{aligned}R=&C_1(a;x_0)_0^T[R_3+R_2]+R_1\mathcal{A}_x\mathrm{Call}^{\mathrm{BS}}(x_0,\bar{a}_{x_{\mathrm{av}}}^2T,k)\\&+(x_0-k)\partial_y\mathcal{A}_x\mathrm{Call}^{\mathrm{BS}}(x_0,\bar{a}_{x_{\mathrm{av}}}^2T,k)\omega(a(x_{\mathrm{av}})a^{(1)}(x_{\mathrm{av}}))_0^T\\&\times[R_1+\partial_x(C_1(a;x)_0^T)|_{x=x_{\mathrm{av}}}\frac{(x_0-k)}{2}],\\R_1=&\frac{(x_0-k)^2}{4}\int_0^1(\partial_{x^2}^2(C_1(a;x)_0^T))|_{x=\lambda x_0+(1-\lambda)x_{\mathrm{av}}}(1-\lambda)\mathrm{d}\lambda,\end{aligned}$$

$$
\begin{aligned}
R_2 =& T^2(\bar a^2_{x_0} - \bar a_{x_{\rm av}})^2 \\
&\times \int_0^1 (\partial^2_{y^2}\mathcal{A}_x \mathrm{Call}^{\rm BS}(x_0, y, k))|_{y=T(\lambda \bar a^2_{x_0}+(1-\lambda)\bar a_{x_{\rm av}})}(1-\lambda)\mathrm{d}\lambda, \\
R_3 =& \partial_y \mathcal{A}_x \mathrm{Call}^{\rm BS}(x_0, \bar a^2_{x_{\rm av}}T, k)\frac{(x_0-k)^2}{4} \\
&\times \int_0^1 (\partial^2_{x^2}(\omega(a^2(x))_0^T))|_{x=\lambda x_0+(1-\lambda)x_{\rm av}}(1-\lambda)\mathrm{d}\lambda.
\end{aligned}
$$

On the one hand, we have:

$$
\begin{aligned}
&\partial_z(C_1(l;z)_0^T) = 2C_6(l;z)_0^T + C_2(l;z)_0^T, \\
&C_1(l;z)_0^T\omega(l(z)l^{(1)}(z))_0^T = 2C_4(l;z)_0^T + \omega(l(z)l^{(1)}(z), l^2(z), l(z)l^{(1)}(z))_0^T,
\end{aligned}
$$

and on the other hand, with (6), Corollary 4.1 and $(\mathcal{H}^a)$, it comes:

$$|R| \le_c \mathcal{M}_1(a)[\mathcal{M}_0(a)]^3T^2.$$

We skip further details. Consequently we can write (55) as follows:

$$
\begin{aligned}
&C_1(a;x_0)_0^T\mathcal{A}_x\mathrm{Call}^{\rm BS}(x_0, \bar a^2_{x_0}T, k) \qquad (56)\\
=&C_1(a;x_{\rm av})_0^T\mathcal{A}_x\mathrm{Call}^{\rm BS}(x_0, \bar a^2_{x_{\rm av}}T, k) \\
&+\frac{(x_0-k)}{2}[2C_6(a;x_{\rm av})_0^T + C_2(a;x_{\rm av})_0^T]\mathcal{A}_x\mathrm{Call}^{\rm BS}(x_0, \bar a^2_{x_{\rm av}}T, k) \\
&+[2C_4(a;x_{\rm av})_0^T + \omega(a(x_{\rm av})a^{(1)}(x_{\rm av}), a^2(x_{\rm av}), a(x_{\rm av})a^{(1)}(x_{\rm av}))_0^T](x_0-k) \\
&\quad \times \partial_y\mathcal{A}_x\mathrm{Call}^{\rm BS}(x_0, \bar a^2_{x_{\rm av}}T, k) + \mathcal{O}(\mathcal{M}_1(a)[\mathcal{M}_0(a)]^3T^2).
\end{aligned}
$$

Then using the relation $\mathcal{A}_x\mathrm{Call}^{\rm BS} = -\mathcal{A}_z\mathrm{Call}^{\rm BS}$, the time reversal invariance of $l \to C_6(l,z)_0^T$ and $l \to \omega(l(z)l^{(1)}(z), l^2(z), l(z)l^{(1)}(z))_0^T$ (for any z), one obtains similarly:

$$
\begin{aligned}
&C_1(\widetilde a;k)_0^T\mathcal{A}_z\mathrm{Call}^{\rm BS}(x_0, \bar a^2_{x_0}T, k) \qquad (57)\\
=&-C_1(\widetilde a;x_{\rm av})_0^T\mathcal{A}_x\mathrm{Call}^{\rm BS}(x_0, \bar a^2_{x_{\rm av}}T, k) \\
&+\frac{(x_0-k)}{2}[2C_6(a;x_{\rm av})_0^T + C_2(\widetilde a;x_{\rm av})_0^T]\mathcal{A}_x\mathrm{Call}^{\rm BS}(x_0, \bar a^2_{x_{\rm av}}T, k) \\
&+[2C_4(\widetilde a;x_{\rm av})_0^T + \omega(a(x_{\rm av})a^{(1)}(x_{\rm av}), a^2(x_{\rm av}), a(x_{\rm av})a^{(1)}(x_{\rm av}))_0^T](x_0-k) \\
&\quad \times \partial_y\mathcal{A}_x\mathrm{Call}^{\rm BS}(x_0, \bar a^2_{x_{\rm av}}T, k) + \mathcal{O}(\mathcal{M}_1(a)[\mathcal{M}_0(a)]^3T^2).
\end{aligned}
$$

Compute the average of (56) and (57) to complete the proof of (54).

▷ **Step 3: mathematical reductions.** We gather terms coming from (49) and (54). In view of (6) and equations (74) and (75) in Proposition 4.3, we have:

$$
\begin{aligned}
&\frac{(x_0-k)^2}{4} C_5(a;x_{\rm av})_0^T \partial_y \mathrm{Call}^{\rm BS}(x_0,\bar a^2_{x_{\rm av}}T,k)\\
&+\frac{(x_0-k)}{4}[C_2(a;x_{\rm av})_0^T + C_2(\widetilde a;x_{\rm av})_0^T]\mathcal{A}_x \mathrm{Call}^{\rm BS}(x_0,\bar a^2_{x_{\rm av}}T,k)\\
=&\frac{(x_0-k)^2}{4}\partial_y \mathrm{Call}^{\rm BS}(x_0,\bar a^2_{x_{\rm av}}T,k)\Big(C_5(a;x_{\rm av})_0^T - 2\frac{[C_2(a;x_{\rm av})_0^T + C_2(\widetilde a;x_{\rm av})_0^T]}{\omega(a^2(x_{\rm av}))_0^T}\Big)\\
=&\frac{(x_0-k)^2}{4}\partial_y \mathrm{Call}^{\rm BS}(x_0,\bar a^2_{x_{\rm av}}T,k)(C_5(a;x_{\rm av})_0^T - 2C_5(a;x_{\rm av})_0^T)\\
=&-\frac{(x_0-k)^2}{8} C_5(a;x_{\rm av})_0^T(\partial^2_{x^2}-\partial_x)\mathrm{Call}^{\rm BS}(x_0,\bar a^2_{x_{\rm av}}T,k), \qquad (58)
\end{aligned}
$$

where we have used at the second equality the relation $C_5(l;z)_0^T\omega(l^2(z))_0^T = C_2(l;z)_0^T + C_2(\widetilde l;z)_0^T$ obtained easily with (26). Then (6), (74) and (75) yield

$$
\begin{aligned}
&\partial_y\mathcal{A}_x \mathrm{Call}^{\rm BS}(x_0,\bar a^2_{x_{\rm av}}T,k)\\
=&\partial_y((-\frac{2(x_0-k)}{y}\partial_y)\mathrm{Call}^{\rm BS}(x_0,y,k))|_{y=\bar a^2_{x_{\rm av}}T}\\
=&\frac{2(x_0-k)}{\omega(a^2(x_{\rm av}))_0^T}\Big[\frac{\partial_y \mathrm{Call}^{\rm BS}(x_0,\bar a^2_{x_{\rm av}}T,k)}{\omega(a^2(x_{\rm av}))_0^T} - \partial^2_{y^2}\mathrm{Call}^{\rm BS}(x_0,\bar a^2_{x_{\rm av}}T,k)\Big],
\end{aligned}
$$

and straightforward calculus allows to obtain with (26):

$$
\begin{aligned}
C_6(l;z)_0^T\omega(l^2(z))_0^T =&C_4(l;z)_0^T + C_4(\widetilde l;z)_0^T\\
&+\omega(l(z)l^{(1)}(z),l^2(z),l(z)l^{(1)}(z))_0^T.
\end{aligned}
$$

These two intermediate results give:

$$
\begin{aligned}
&(x_0-k)C_6(a;x_{\rm av})_0^T\,[\mathcal{A}_x \mathrm{Call}^{\rm BS}(x_0,\bar a^2_{x_{\rm av}}T,k)\\
&+(x_0-k)\partial^2_{y^2}\mathrm{Call}^{\rm BS}(x_0,\bar a^2_{x_{\rm av}}T,k)]\\
&+[C_4(a;x_{\rm av})_0^T + C_4(\widetilde a;x_{\rm av})_0^T + \omega(a(x_{\rm av})a^{(1)}(x_{\rm av}),\ldots\\
&\quad a^2(x_{\rm av}),a(x_{\rm av})a^{(1)}(x_{\rm av}))_0^T](x_0-k)\partial_y\mathcal{A}_x\mathrm{Call}^{\rm BS}(x_0,\omega(a^2(x_{\rm av}))_0^T,k)\\
=&-2\frac{(x_0-k)^2}{\omega(a^2(x_{\rm av}))_0^T}C_6(a;x_{\rm av})_0^T\partial_y \mathrm{Call}^{\rm BS}(x_0,\bar a^2_{x_{\rm av}}T,k)\\
&+(x_0-k)^2C_6(a;x_{\rm av})_0^T\partial^2_{y^2}\mathrm{Call}^{\rm BS}(x_0,\bar a^2_{x_{\rm av}}T,k)\\
&+(x_0-k)[C_6(a;x_{\rm av})_0^T\omega(a^2(x_{\rm av}))_0^T]\frac{2(x_0-k)}{\omega(a^2(x_{\rm av}))_0^T}\\
&\times\Big[\frac{\partial_y \mathrm{Call}^{\rm BS}(x_0,\omega(a^2(x_{\rm av}))_0^T,k)}{\omega(a^2(x_{\rm av}))_0^T} - \partial^2_{y^2}\mathrm{Call}^{\rm BS}(x_0,\omega(a^2(x_{\rm av}))_0^T,k)\Big]\\
=&-(x_0-k)^2C_6(a;x_{\rm av})_0^T\partial^2_{y^2}\mathrm{Call}^{\rm BS}(x_0,\bar a^2_{x_{\rm av}}T,k)\\
=&-\frac{(x_0-k)^2}{4}C_6(a;x_{\rm av})_0^T(\partial^4_{x^4}-2\partial^3_{x^3}+\partial^2_{x^2})\mathrm{Call}^{\rm BS}(x_0,\bar a^2_{x_{\rm av}}T,k). \qquad (59)
\end{aligned}
$$

Finally, sum the relations (49-54-53) taking into account the simplifications (58-59) and apply Theorems 10 and 11 to obtain the announced result (48). □

5.3. *Third order expansion of the implied volatility*

We define extra integral operators in order to state a new result about third order expansions of the implied volatility.

Definition 4.2. Provided that the derivatives and the integrals below have a meaning, we define the following operators for a two variables non-negative function l such that $\bar{l}_z > 0$:

$$\begin{aligned}
\gamma_0(l;z)_0^T =& \bar{l}_z + \frac{C_2(l;z)_0^T}{2\bar{l}_z T} - \frac{C_4(l;z)_0^T}{4\bar{l}_z T} - \frac{C_3(l;z)_0^T}{\bar{l}_z^3 T^2} - \frac{3C_4(l;z)_0^T}{\bar{l}_z^3 T^2} \\
&+ \frac{[C_1(l;z)_0^T]^2}{8\bar{l}_z^3 T^2} + \frac{3[C_1(l;z)_0^T]^2}{2\bar{l}_z^5 T^3}, \\
\gamma_1(l;z)_0^T =& \frac{C_1(l;z)_0^T}{\bar{l}_z^3 T^2}, \quad \gamma_2(l;z)_0^T = \frac{C_3(l;z)_0^T}{\bar{l}_z^5 T^3} + 3\frac{C_4(l;z)_0^T}{\bar{l}_z^5 T^3} - \frac{3[C_1(l;z)_0^T]^2}{\bar{l}_z^7 T^4}; \\
\pi_0(l;z)_0^T =& \frac{\gamma_0(l;z)_0^T + \gamma_0(\widetilde{l};z)_0^T}{2}, \qquad \pi_1(l;z)_0^T = \frac{\gamma_1(\widetilde{l};z)_0^T - \gamma_1(l;z)_0^T}{2}, \\
\pi_2(l;z)_0^T =& \frac{\gamma_2(l;z)_0^T + \gamma_2(\widetilde{l};z)_0^T}{2} - \frac{C_5(l;z)_0^T}{8\bar{l}_z T} + \frac{C_6(l;z)_0^T}{4\bar{l}_z^3 T^2}; \\
\chi_0(l;z)_0^T =& \bar{l}_z + \frac{C_2(l;z)_0^T}{2\bar{l}_z T} - \frac{C_3(l;z)_0^T}{\bar{l}_z^3 T^2} - \frac{3C_4(l;z)_0^T}{\bar{l}_z^3 T^2} + \frac{3[C_1(l;z)_0^T]^2}{2\bar{l}_z^5 T^3}, \\
\chi_1(l;z)_0^T =& \gamma_1(l;z)_0^T, \qquad \chi_2(l;z)_0^T = \gamma_2(l;z)_0^T; \\
\Xi_0(l;z)_0^T =& \frac{\chi_0(l;z)_0^T + \chi_0(\widetilde{l};z)_0^T}{2}, \qquad \Xi_1(l;z)_0^T = \pi_1(l;z)_0^T, \\
\Xi_2(l;z)_0^T =& \pi_2(l;z)_0^T.
\end{aligned}$$

Theorem 13. *(3rd order expansions of the implied volatility). Assume* $(\mathcal{H}^a)$. *We have:*

$$\sigma_{\mathrm{I}}(x_0,T,k) = \gamma_0(a;x_0)_0^T - \gamma_1(a;x_0)_0^T m + \gamma_2(a;x_0)_0^T m^2 + \mathrm{Error}^{\mathrm{I}}_{3,\mathrm{x}_0}, \tag{60}$$

$$\sigma_{\mathrm{I}}(x_0,T,k) = \gamma_0(\widetilde{a};k)_0^T + \gamma_1(\widetilde{a};k)_0^T m + \gamma_2(\widetilde{a};k)_0^T m^2 + \mathrm{Error}^{\mathrm{I}}_{3,\mathrm{k}}, \tag{61}$$

$$\sigma_{\mathrm{I}}(x_0,T,k) = \pi_0(a;x_{\mathrm{av}})_0^T + \pi_1(a;x_{\mathrm{av}})_0^T m + \pi_2(a;x_{\mathrm{av}})_0^T m^2 + \mathrm{Error}^{\mathrm{I}}_{3,\mathrm{x}_{\mathrm{av}}}. \tag{62}$$

Under $(\mathcal{H}^{\Sigma})$ *we have*

$$\Sigma_{\rm I}(S_0,T,K) =\chi_0(\Sigma;S_0)_0^T - \chi_1(\Sigma;S_0)_0^T M + \chi_2(\Sigma;S_0)_0^T M^2 + {\rm Error}^{\rm I}_{3,{\rm S_0}}, \tag{63}$$

$$\Sigma_{\rm I}(S_0,T,K) =\chi_0(\widetilde{\Sigma};K)_0^T + \chi_1(\widetilde{\Sigma};K)_0^T M + \chi_2(\widetilde{\Sigma};K)_0^T M^2 + {\rm Error}^{\rm I}_{3,{\rm K}}, \tag{64}$$

$$\Sigma_{\rm I}(S_0,T,K) =\Xi_0(\Sigma;S_{\rm av})_0^T + \Xi_1(\Sigma;S_{\rm av})_0^T M + \Xi_2(\Sigma;S_{\rm av})_0^T M^2 + {\rm Error}^{\rm I}_{3,{\rm S_{av}}}. \tag{65}$$

The operators γ_i, π_i, χ_i *and* Ξ_i *used in the above expansions are defined in Definition 4.2.*

Remark 4.5. We have obtained Black-Scholes (respectively Bachelier) implied volatility approximations which are written as a quadratic function w.r.t. the Black-Scholes log-moneyness (respectively w.r.t. the Bachelier moneyness). At the money, observe that the corresponding approximations are not equal to the local volatility function computed at spot. However, in view of the definition of the operators C_1, C_2, C_3 and C_4 (see Definition 4.1) and the operators γ_0 and χ_0 (see Definition 4.2), we easily obtain the estimate:

$$\begin{aligned}&|\gamma_0(a;x_0)_0^T - \bar{a}_{x_0}| + |\gamma_0(\widetilde{a};x_0)_0^T - \bar{a}_{x_0}|\\&+ |\chi_0(\Sigma;S_0) - \bar{\Sigma}_{S_0}| + |\chi_0(\widetilde{\Sigma};S_0) - \bar{\Sigma}_{S_0}| \leq_c T.\end{aligned}$$

It shows that when the maturity tends to zero, our implied volatility approximations at the money become equal to the local volatility function frozen at spot. We therefore interpret the difference between our implied volatility approximations ATM and the local volatility function frozen at spot as a maturity bias.

Proof. We first focus on the formula (60), the treatment of (61-63-64) being similar. Start from Theorem 10 and apply Proposition 4.3 in order to write the Greeks w.r.t. x (for each operator C_i) in terms of the ${\rm Vega}^{\rm BS}$ and the ${\rm Vomma}^{\rm BS}$. Thus the third order expansion formula based on the ATM local volatility with log-normal proxy can be transformed into:

$$\begin{aligned}&{\rm Call}(e^{x_0},T,e^k)\\
=&{\rm Call}^{\rm BS}(x_0,\bar{a}^2_{x_0}T,k) + {\rm Vega}^{\rm BS}(x_0,\bar{a}^2_{x_0}T,k)\\
&\times\Big[-\frac{C_1(a;x_0)_0^T m}{\bar{a}^3_{x_0}T^2} + \frac{C_2(a;x_0)_0^T}{2\bar{a}_{x_0}T} + \frac{C_3(a;x_0)_0^T m^2}{\bar{a}^5_{x_0}T^3} - \frac{C_3(a;x_0)_0^T}{\bar{a}^3_{x_0}T^2}\\
&+\frac{3C_4(a;x_0)_0^T m^2}{\bar{a}^5_{x_0}T^3} - \frac{3C_4(a;x_0)_0^T}{\bar{a}^3_{x_0}T^2} - \frac{C_4(a;x_0)_0^T}{4\bar{a}_{x_0}T}\\
&+\frac{[C_1(a;x_0)_0^T]^2}{8\bar{a}^3_{x_0}T^2} + \frac{3[C_1(a;x_0)_0^T]^2}{2\bar{a}^5_{x_0}T^3} - \frac{3[C_1(a;x_0)_0^T]^2 m^2}{\bar{a}^7_{x_0}T^4}\Big]\\
&+\frac{1}{2}{\rm Vomma}^{\rm BS}(x_0,\bar{a}^2_{x_0}T,k)\Big(\frac{C_1(a;x_0)_0^T m}{\bar{a}^3_{x_0}T^2}\Big)^2 + \mathcal{O}(\mathcal{M}_1(a)[\mathcal{M}_0(a)]^3T^2)\end{aligned}$$

$$
\begin{aligned}
=&\mathrm{Call}^{\mathrm{BS}}(x_0, \bar{a}^2_{x_0} T, k)\\
&+ \mathrm{Vega}^{\mathrm{BS}}(x_0, \bar{a}^2_{x_0} T, k)\big[\gamma_0(a;x_0)_0^T - \bar{a}_{x_0} - \gamma_1(a;x_0)_0^T m + \gamma_2(a;x_0)_0^T m^2\big]\\
&+ \frac{1}{2}\mathrm{Vomma}^{\mathrm{BS}}(x_0, \bar{a}^2_{x_0} T, k)[\gamma_1(a;x_0)_0^T m]^2 + \mathcal{O}(\mathcal{M}_1(a)[\mathcal{M}_0(a)]^3 T^2)\\
\approx&\mathrm{Call}^{\mathrm{BS}}(x_0, \big[\gamma_0(a;x_0)_0^T - \gamma_1(a;x_0)_0^T m + \gamma_2(a;x_0)_0^T m^2\big]^2 T, k).
\end{aligned}
$$

This reads as an expansion of the implied volatility and achieves the proof of (60).

Now we give the main lines of the derivation of the error estimate in (62), while (65) is left to the reader. Again, we apply Theorem 12 and Proposition 4.3 in order to replace the x_0-Greeks with the $\mathrm{Vega}^{\mathrm{BS}}$ and the $\mathrm{Vomma}^{\mathrm{BS}}$. One obtains similarly:

$$
\begin{aligned}
\mathrm{Call}(e^{x_0}, T, e^k) =& \mathrm{Call}^{\mathrm{BS}}(x_0, \bar{a}^2_{x_{\mathrm{av}}} T, k) \qquad (66)\\
&+ \mathrm{Vega}^{\mathrm{BS}}(x_0, \bar{a}^2_{x_{\mathrm{av}}} T, k)\big[\frac{\gamma_0(a;x_{\mathrm{av}})_0^T + \gamma_0(\widetilde{a};x_{\mathrm{av}})_0^T}{2} - \bar{a}_{x_{\mathrm{av}}}\\
&+ \frac{\gamma_1(\widetilde{a};x_{\mathrm{av}})_0^T - \gamma_1(a;x_{\mathrm{av}})_0^T}{2} m + \frac{\gamma_2(a;x_{\mathrm{av}})_0^T + \gamma_2(\widetilde{a};x_{\mathrm{av}})_0^T}{2} m^2\\
&- \frac{C_5(a;x_{\mathrm{av}})_0^T}{8\bar{a}_{x_{\mathrm{av}}} T} m^2 + \frac{C_6(a;x_{\mathrm{av}})_0^T}{4\bar{a}^3_{x_{\mathrm{av}}} T^2} m^2 + \frac{C_6(a;x_{\mathrm{av}})_0^T}{16\bar{a}_{x_{\mathrm{av}}} T} m^2 - \frac{C_6(a;x_{\mathrm{av}})_0^T}{4\bar{a}^5_{x_{\mathrm{av}}} T^3} m^4\big]\\
&+ \frac{1}{2}\mathrm{Vomma}^{\mathrm{BS}}(x_0, \bar{a}^2_{x_{\mathrm{av}}} T, k) m^2 \Big(\frac{[\gamma_1(a;x_{\mathrm{av}})_0^T]^2 + [\gamma_1(\widetilde{a};x_{\mathrm{av}})_0^T]^2}{2}\Big)\\
&+ \mathcal{O}(\mathcal{M}_1(a)[\mathcal{M}_0(a)]^3 T^2).
\end{aligned}
$$

Then write

$$
\begin{aligned}
&\Big(\frac{[\gamma_1(a;x_{\mathrm{av}})_0^T]^2 + [\gamma_1(\widetilde{a};x_{\mathrm{av}})_0^T]^2}{2}\Big)\\
&= \Big(\frac{\gamma_1(\widetilde{a};x_{\mathrm{av}})_0^T - \gamma_1(a;x_{\mathrm{av}})_0^T}{2}\Big)^2 + \Big(\frac{\gamma_1(\widetilde{a};x_{\mathrm{av}})_0^T + \gamma_1(a;x_{\mathrm{av}})_0^T}{2}\Big)^2,
\end{aligned}
$$

use the fact that (see the definition (69) of $\mathrm{Vomma}^{\mathrm{BS}}$ and the definition of γ_1 in Definition 4.2)

$$
\begin{aligned}
&\frac{1}{2}\mathrm{Vomma}^{\mathrm{BS}}(x_0, \bar{a}^2_{x_{\mathrm{av}}} T, k) m^2 \Big(\frac{\gamma_1(\widetilde{a};x_{\mathrm{av}})_0^T + \gamma_1(a;x_{\mathrm{av}})_0^T}{2}\Big)^2\\
=&\mathrm{Vega}^{\mathrm{BS}}(x_0, \bar{a}^2_{x_{\mathrm{av}}} T, k)(C_1(\widetilde{a};x_{\mathrm{av}})_0^T + C_1(a;x_{\mathrm{av}})_0^T)^2\Big[-\frac{m^2}{32\bar{a}^5_{x_{\mathrm{av}}} T^3} + \frac{m^4}{8\bar{a}^9_{x_{\mathrm{av}}} T^5}\Big],
\end{aligned}
$$

and finally, use the above identity (obtained with the definitions of C_1, C_6 and with the relation (26)):

$$\begin{aligned}
&(C_1(\widetilde{l};x)_0^T + C_1(l;x)_0^T)^2 \\
=&2[\omega(l^2(z))_0^T]^2 C_6(l;z)_0^T = 4\omega(l^2(z),l^2(z))_0^T \omega(l(z)l^{(1)}(z),l(z)l^{(1)}(z))_0^T \\
=&4\big[\omega(l(z)l^{(1)}(z),l(z)l^{(1)}(z),l^2(z),l^2(z))_0^T \\
&+\omega(l^2(z),l^2(z),l(z)l^{(1)}(z),l(z)l^{(1)}(z))_0^T \\
&+\omega(l^2(z),l(z)l^{(1)}(z),l(z)l^{(1)}(z),l^2(z))_0^T \\
&+\omega(l(z)l^{(1)}(z),l^2(z),l^2(z),l(z)l^{(1)}(z))_0^T \\
&+\omega(l^2(z),l(z)l^{(1)}(z),l^2(z),l(z)l^{(1)}(z))_0^T \\
&+\omega(l(z)l^{(1)}(z),l^2(z),l(z)l^{(1)}(z),l^2(z))_0^T\big],
\end{aligned}$$

to cancel the terms $\frac{C_6(a;x_{\rm av})_0^T}{16\bar{a}_{x_{\rm av}}T}m^2$, $-\frac{C_6(a;x_{\rm av})_0^T}{4\bar{a}^5_{x_{\rm av}}T^3}m^4$ and $\frac{1}{2}\text{Vomma}^{\rm BS}(x_0,\bar{a}^2_{x_{\rm av}}T,k)m^2\big(\frac{\gamma_1(\widetilde{a};x_{\rm av})_0^T+\gamma_1(a;x_{\rm av})_0^T}{2}\big)^2$ in (66). That achieves the proof of (62). □

In addition to these implied volatility expansions, one can under additional technical assumptions upper bound the residuals terms. For instance, let us consider (60), for which we can prove

$$\text{Error}^{\rm I}_{3,x_0} = \mathcal{O}(\mathcal{M}_1(a)[\mathcal{M}_0(a)]^3 T^{\frac{3}{2}}), \tag{67}$$

which justifies the label of third order expansion. This is available under the assumptions that $|m| \leq \xi\mathcal{M}_0(a)\sqrt{T}$ (for a given $\xi \geq 0$) and that $\mathcal{M}_0(a)$, $\mathcal{M}_1(a)$ and T are globally small enough to ensure that the implied volatility approximation $\gamma_0(a;x_0)_0^T - \gamma_1(a;x_0)_0^T m + \gamma_2(a;x_0)_0^T m^2$ is bounded away from 0. The method of proof is analogous to that in Subsection 3.4, by performing a third order expansion of BS price w.r.t. the volatility, using the estimate (73) on Ultima$^{\rm BS}$ (see Corollary 4.2), and carefully gathering terms and evaluating their magnitudes.

6. Approximation of the Delta

In this section, we investigate the approximation of the delta of the Call price, i.e. the derivative w.r.t. the spot, by deriving similar expansion formulas. For the sake of brevity we present only results using a log-normal proxy. The results are new. To achieve this goal, we follow again the Dupire approach taking advantage of the symmetry between spot and strike. We start from the Feynman-Kac representation (37) which leads to a nice expression for the delta:

$$\delta(S_0,T,K) = \partial_{S_0}\mathbb{E}[(S_0 - e^{k_T})_+] = \mathbb{P}(e^{k_T} < S_0) = \mathbb{P}(k_T < x_0).$$

Thus we are reduced to compute the price of a binary option on the fictitious asset $(k_t)_t$. This binary payoff is not anymore differentiable, but we can however apply directly [49] to obtain

Theorem 14. ***(1st and 2nd order approximations for delta using local volatility at strike).*** *Assume* $(\mathcal{H}^a)$. *Then we have:*

$$\begin{aligned}\delta(e^{x_0},T,e^k) =&\delta^{\mathrm{BS}}(x_0,\bar{a}_k^2T,k)+C_1(\widetilde{a};k)_0^T(\partial_{z^3}^3-\frac{3}{2}\partial_{z^2}^2+\frac{1}{2}\partial_z)\delta^{\mathrm{BS}}(x_0,\bar{a}_k^2T,k)\\&+\mathcal{O}(\mathcal{M}_1(a)\mathcal{M}_0(a)T),\\ \delta(e^{x_0},T,e^k) =&\delta^{\mathrm{BS}}(x_0,\bar{a}_k^2T,k)+\sum_{i=1}^{6}\eta_i(\widetilde{a};k)_0^T\partial_{k^i}^i\delta^{\mathrm{BS}}(x_0,\bar{a}_k^2T,k)\\&+\mathcal{O}(\mathcal{M}_1(a)[\mathcal{M}_0(a)]^2T^{\frac{3}{2}}),\end{aligned}$$

where δ^{BS} *is Black-Scholes delta function defined by* $\delta^{\mathrm{BS}}(x,y,z)=\mathcal{N}(d_1(x,y,z))$, *with* x *the log-spot,* y *the total variance and* z *the log-strike.*

Remark 4.6. In view of the error estimate, observe that the corresponding second and third order formulas for vanilla payoffs are respectively first and second order approximations for binary payoffs. This is due to the lack of regularity of the payoff (see our discussion in Subsection 2.5).

Like in the previous price approximation formulas, it is possible to perform additional Taylor expansions in order to obtain similar formulas using local volatility function frozen at spot or at mid-point. We announce two Lemmas which proof is very similar to those of Lemmas 2 and 3 and Theorem 12 is left to the reader. Extra technical results are postponed in Appendix, Subsection 8.3.

Lemma 4. *Let* $x\in\{x_0,x_{\mathrm{av}}\}$. *Assume* $(\mathcal{H}^a)$, *then we have*

$$\begin{aligned}&\delta^{\mathrm{BS}}(x_0,\bar{a}_k^2T,k)\\=&\delta^{\mathrm{BS}}(x_0,\bar{a}_x^2T,k)+2\partial_y\delta^{\mathrm{BS}}(x_0,\bar{a}_x^2T,k)(k-x)C_7(a;x)_0^T\\&+\mathcal{O}(\mathcal{M}_1(a)\mathcal{M}_0(a)T),\\=&\delta^{\mathrm{BS}}(x_0,\bar{a}_x^2T,k)+[2(k-x)C_7(a;x)_0^T+(k-x)^2C_5(a;x)_0^T]\\&\times\partial_y\delta^{\mathrm{BS}}(x_0,\bar{a}_x^2T,k)+4\partial_{y^2}^2\delta^{\mathrm{BS}}(x_0,\bar{a}_x^2T,k)(k-x)^2C_6(a;x)_0^T\\&+\mathcal{O}(\mathcal{M}_1(a)[\mathcal{M}_0(a)]^2T^{\frac{3}{2}}),\end{aligned}$$

where $C_7(l;z)_0^T=\omega(l(z)l^{(1)}(z))_0^T$.

Lemma 5. *Let $x \in \{x_0, x_{\rm av}\}$. Assume $(\mathcal{H}^a)$, then we have*

$$
\begin{aligned}
&C_1(\widetilde{a};k)_0^T(\partial^3_{z^3} - \frac{3}{2}\partial^2_{z^2} + \frac{1}{2}\partial_z)\delta^{\rm BS}(x_0, \overline{a}_k^2 T, k)\\
&= C_1(\widetilde{a};x)_0^T(\partial^3_{z^3} - \frac{3}{2}\partial^2_{z^2} + \frac{1}{2}\partial_z)\delta^{\rm BS}(x_0, \overline{a}_x^2 T, k) + \mathcal{O}(\mathcal{M}_1(a)\mathcal{M}_0(a)T),\\
&\sum_{i=1}^{6} \eta_i(\widetilde{a};k)_0^T \partial^i_{z^i}\delta^{\rm BS}(x_0, \overline{a}_k^2 T, k) = \sum_{i=1}^{6} \eta_i(\widetilde{a};x)_0^T \partial^i_{z^i}\delta^{\rm BS}(x_0, \overline{a}_x^2 T, k)\\
&+ [2C_6(\widetilde{a};x)_0^T + C_2(\widetilde{a};x)_0^T](k-x)(\partial^3_{z^3} - \frac{3}{2}\partial^2_{z^2} + \frac{1}{2}\partial_z)\delta^{\rm BS}(x_0, \overline{a}_x^2 T, k)\\
&+ 2(k-x)C_1(\widetilde{a};x)_0^T C_7(\widetilde{a};x)_0^T(\partial^4_{yz^3} - \frac{3}{2}\partial^3_{yz^2} + \frac{1}{2}\partial^2_{yz})\delta^{\rm BS}(x_0, \overline{a}_x^2 T, k)\\
&+ \mathcal{O}(\mathcal{M}_1(a)[\mathcal{M}_0(a)]^2 T^{\frac{3}{2}}).
\end{aligned}
$$

Then remark that:

$$C_1(\widetilde{a};x)_0^T C_7(\widetilde{a};x)_0^T = 2C_4(\widetilde{a};x)_0^T + C_8(\widetilde{a};x)_0^T,$$

where the operator C_8 is defined as follows:

$$C_8(l;z)_0^T = C_8(\widetilde{l};z)_0^T = \omega(l(z)l^{(1)}(z), l^2(z), l(z)l^{(1)}(z))_0^T.$$

An application of Proposition 4.8 finally yields the theorem below.

Theorem 15. ***(1st and 2nd order approximations for delta using local volatility at spot and mid-point).*** *Assume $(\mathcal{H}^a)$ and let $x \in \{x_0, x_{\rm av}\}$. We have:*

$$
\begin{aligned}
\delta(e^{x_0}, T, e^k) &= \delta^{\rm BS}(x_0, \overline{a}_x^2 T, k) + C_1(\widetilde{a};x)_0^T(\partial^3_{z^3} - \frac{3}{2}\partial^2_{z^2} + \frac{1}{2}\partial_z)\delta^{\rm BS}(x_0, \overline{a}_x^2 T, k)\\
&+ (k-x)C_7(a;x)_0^T(\partial^2_{z^2} - \partial_z)\delta^{\rm BS}(x_0, \overline{a}_x^2 T, k) + \mathcal{O}(\mathcal{M}_1(a)\mathcal{M}_0(a)T),
\end{aligned}
$$

$$
\begin{aligned}
\delta(e^{x_0}, T, e^k) &= \delta^{\rm BS}(x_0, \overline{a}_x^2 T, k) + \sum_{i=1}^{6} \eta_i(\widetilde{a};k)_0^T \partial^i_{z^i}\delta^{\rm BS}(x_0, \overline{a}_x^2 T, k)\\
&+ (k-x)[C_7(a;x)_0^T + \frac{(k-x)}{2}C_5(a;x)_0^T](\partial^2_{z^2} - \partial_z)\delta^{\rm BS}(x_0, \overline{a}_x^2 T, k)\\
&+ (k-x)^2 C_6(a;x)_0^T(\partial^4_{z^4} - 2\partial^3_{z^3} + \partial^2_{z^2})\delta^{\rm BS}(x_0, \overline{a}_x^2 T, k)\\
&+ (k-x)[2C_6(\widetilde{a};x)_0^T + C_2(\widetilde{a};x)_0^T](\partial^3_{z^3} - \frac{3}{2}\partial^2_{z^2} + \frac{1}{2}\partial_z)\delta^{\rm BS}(x_0, \overline{a}_x^2 T, k)\\
&+ (k-x)[2C_4(\widetilde{a};x)_0^T + C_8(\widetilde{a};x)_0^T](\partial^5_{z^5} - \frac{5}{2}\partial^4_{z^4} + 2\partial^3_{z^3} - \frac{1}{2}\partial^2_{z^2})\\
&\times \delta^{\rm BS}(x_0, \overline{a}_x^2 T, k) + \mathcal{O}(\mathcal{M}_1(a)[\mathcal{M}_0(a)]^2 T^{\frac{3}{2}}).
\end{aligned}
$$

7. Numerical experiments

7.1. *The set of tests*

For the numerical experiments, we consider a CEV model with constant parameters: $\sigma(t,S) = \nu S^{\beta-1}$. We choose a spot value S_0 equal to 1 and we test two values of ν (a parameter interpreted as a level of volatility): firstly we set $\nu = 0.25$ and we consider either $\beta = 0.8$ (a priori close to the log-normal case) or $\beta = 0.2$ (a priori close to the normal case). Then we investigate the case of a larger volatility with $\nu = 0.4$ and $\beta = 0.5$. For the sake of completeness, we give in Appendix 8.5 the expressions of corrective coefficients allowing the computation of our various approximation formulas proposed throughout the chapter.

We compare the accuracy of different approximations, for various maturities and various strikes gathered in 5 categories. The strikes evolve approximately as

Table 1. Set of maturities and strikes for the numerical experiments

T/K	far ITM		ITM			ATM			OTM			far OTM	
3M	0.70	0.75	0.80	0.85	0.90	0.95	1.00	1.05	1.10	1.15	1.25	1.30	1.35
6M	0.65	0.75	0.80	0.85	0.90	0.95	1.00	1.05	1.10	1.20	1.25	1.35	1.50
1Y	0.55	0.65	0.75	0.80	0.90	0.95	1.00	1.05	1.15	1.25	1.40	1.50	1.80
1.5Y	0.50	0.60	0.70	0.75	0.85	0.95	1.00	1.10	1.15	1.30	1.50	1.65	2.00
2Y	0.45	0.55	0.65	0.75	0.85	0.90	1.00	1.10	1.20	1.35	1.55	1.80	2.30
3Y	0.35	0.50	0.55	0.70	0.80	0.90	1.00	1.10	1.25	1.45	1.75	2.05	2.70
5Y	0.25	0.40	0.50	0.60	0.75	0.85	1.00	1.15	1.35	1.60	2.05	2.50	3.60
10Y	0.15	0.25	0.35	0.50	0.65	0.80	1.00	1.20	1.50	1.95	2.75	3.65	6.30

$S_0 \exp(c\nu\sqrt{T})$ where c takes the value of various quantiles of the standard Gaussian law (1%-5%-10%-20%-30%-40%-50%-60%-70%-80%-90%-95%-99%) which allows to cover far ITM and far OTM options. We report in Tables 2, 3 and 4 the Black-Scholes implied volatilities corresponding to the exact Call prices with constant parameters [50].

Table 2. CEV model ($\beta = 0.8, \nu = 0.25$): BS implied volatilities in %.

3M	25.90	25.73	25.56	25.41	25.26	25.13	25.00	24.88	24.76	24.65	24.45	24.35	24.26
6M	26.09	25.73	25.56	25.41	25.27	25.13	25.00	24.88	24.76	24.55	24.45	24.26	24.00
1Y	26.53	26.10	25.73	25.56	25.27	25.13	25.00	24.88	24.65	24.45	24.17	24.00	23.56
1.5Y	26.78	26.30	25.91	25.73	25.41	25.13	25.00	24.77	24.66	24.35	24.00	23.77	23.31
2Y	27.06	26.53	26.10	25.73	25.41	25.27	25.01	24.77	24.55	24.26	23.92	23.56	22.98
3Y	27.73	26.78	26.53	25.91	25.57	25.27	25.01	24.77	24.45	24.09	23.63	23.25	22.60
5Y	28.64	27.38	26.79	26.31	25.74	25.42	25.01	24.66	24.27	23.85	23.26	22.79	21.94
10Y	30.08	28.66	27.75	26.80	26.12	25.59	25.02	24.57	24.02	23.39	22.57	21.92	20.69

The purpose of the numerical tests is to compare the following approximations:

(1) `ImpVol(AppPriceLN(2,z))` and `ImpVol(AppPriceN(2,z))`: the BS implied volatility of the second order expansions based respectively on the log-normal and normal proxy with local volatility frozen at point z, z being

Table 3. CEV model ($\beta = 0.2, \nu = 0.25$): BS implied volatilities in %.

3M	28.75	28.00	27.31	26.67	26.08	25.53	25.01	24.53	24.07	23.64	22.84	22.48	22.13
6M	29.59	28.02	27.32	26.69	26.09	25.54	25.02	24.53	24.08	23.24	22.85	22.13	21.18
1Y	31.54	29.62	28.05	27.35	26.12	25.56	25.04	24.55	23.66	22.87	21.81	21.19	19.60
1.5Y	32.71	30.57	28.83	28.07	26.74	25.58	25.06	24.11	23.68	22.51	21.20	20.36	18.73
2Y	34.03	31.62	29.69	28.10	26.76	26.16	25.08	24.13	23.29	22.18	20.92	19.62	17.62
3Y	37.34	32.84	31.70	28.92	27.46	26.21	25.12	24.17	22.93	21.55	19.88	18.56	16.40
5Y	42.07	35.80	33.00	30.82	28.27	26.91	25.20	23.80	22.26	20.71	18.59	17.01	14.38
10Y	47.85	41.60	37.46	33.14	30.08	27.76	25.38	23.53	21.41	19.09	16.35	14.32	10.99

Table 4. CEV model ($\beta = 0.5, \nu = 0.4$): BS implied volatilities in %.

3M	43.69	42.97	42.29	41.67	41.08	40.53	40.02	39.53	39.07	38.63	37.82	37.45	37.09
6M	44.51	42.99	42.31	41.68	41.10	40.55	40.03	39.55	39.09	38.23	37.84	37.10	36.11
1Y	46.38	44.55	43.03	42.35	41.13	40.58	40.06	39.58	38.68	37.86	36.78	36.13	34.45
1.5Y	47.49	45.46	43.80	43.06	41.75	40.61	40.10	39.14	38.71	37.51	36.15	35.27	33.52
2Y	48.73	46.47	44.63	43.10	41.79	41.20	40.13	39.17	38.31	37.17	35.87	34.49	32.31
3Y	51.76	47.62	46.55	43.90	42.48	41.26	40.18	39.22	37.97	36.54	34.79	33.36	30.97
5Y	55.94	50.30	47.73	45.69	43.27	41.95	40.28	38.87	37.30	35.69	33.42	31.68	28.66
10Y	60.86	55.20	51.48	47.60	44.80	42.63	40.36	38.55	36.41	33.98	30.97	28.64	24.51

respectively equal to x_0, k or x_{av} and to S_0, K or S_{av}. See Theorems 6-7-8.

(2) `AppImpVolLN(2,z)` and `AppImpVolN(2,z)`: the second order implied volatility expansions (Theorem 9). All the results are converted into Black-Scholes implied volatility. Namely, for the normal proxy, once we have computed Bachelier implied volatility expansions, we first evaluate the price with the Bachelier formula and then compute the related implied Black-Scholes volatility.

(3) `ImpVol(AppPriceLN(3,z))` and `ImpVol(AppPriceN(3,z))`: the implied volatility of the third order expansions (Theorems 10 and 11). In addition for the log-normal proxy, we test the average of approximations based on strike and on spot and we denote it by `Av.ImpVol(AppPriceLN(3,.))`.

(4) `AppImpVolLN(3,z)` and `AppImpVolN(3,z)`: the third order implied volatility expansions (Theorem 13). We use the notation `Av.AppImpVolLN(3,.)` for the average of the expansions in strike and in spot.

(5) Hagan and Henry-Labordère formulas denoted by (`HF`) and (`HLF`) in the following: benchmark implied volatility approximations of Hagan etal. [5] and Henry-Labordère [6] . For the sake of completeness, we recall these

well-known implied volatility approximations in the CEV model:

$$\sigma_{\mathrm{I}}(x_0,T,k) \approx \nu(\frac{S_0+K}{2})^{\beta-1}\Big(1+\frac{(1-\beta)(2+\beta)}{6}(\frac{S_0-K}{S_0+K})^2 + \frac{(\beta-1)^2\nu^2T}{24}(\frac{S_0+K}{2})^{2\beta-2}\Big), \qquad \text{(HF)}$$

$$\sigma_{\mathrm{I}}(x_0,T,k) \approx \frac{\nu(1-\beta)\log(\frac{S_0}{K})}{S_0^{1-\beta}-K^{1-\beta}}\Big(1+\frac{(\beta-1)^2\nu^2T}{24}(\frac{S_0+K}{2})^{2\beta-2}\Big). \qquad \text{(HLF)}$$

We recall that these formulas are essentially available for time-independent volatility, while our formulas allow time dependency.

Our goal is to demonstrate the interest of our approximation formulas in comparison to those of Hagan and Henry-Labordere. We are rather exhaustive with our numerical experiments in order to, on the one hand, select the best approximation formulas among ours, and on the other hand to show that our methods with log-normal proxy involving the mid-point generally outperform Hagan and Henry-Labordère formulas. Full details allow the reader to easily reproduce the results.

In Tables 7 and 9, we report the errors expressed in bps (basis points) on implied volatility for $(\beta,\nu) = (0.8, 0.25)$ using the second and the third order price expansions. Tables 8 and 10 give results for the second and the third order implied volatility expansions. Next in Table 11, we report the errors in bps obtained with the averaged expansions `Av.ImpVol(AppPriceLN(3,.))` and `Av.AppImpVolLN(3,.)` and the benchmarks (HF) and (HLF). Then in Table 12, we compare `Av.ImpVol(AppPriceLN(3,.))`, ImpVol(AppPriceLN(3,x_{av})), `Av.AppImpVolLN(3,.)` and AppImpVolLN(3,x_{av}) with the benchmarks (HF) and (HLF).
After we analyze the case $(\beta,\nu) = (0.2, 0.25)$ and we report in Tables 13 and 14 the errors using ImpVol(AppPriceLN(3,x_{av})), ImpVol(AppPriceN(3,S_{av})), AppImpVolLN(3,x_{av}), AppImpVolN(3,S_{av}) and the benchmarks (HF) and (HLF). Because the other methods in general give globally less accurate results, we just report and compare the best approximations.
Finally in Tables 15 and 16 we establish a comparison between ImpVol(AppPriceLN(3,x_{av})), AppImpVolLN(3,x_{av}) and the benchmarks Hagan and (HLF) for $(\beta,\nu) = (0.5, 0.4)$.

For example, on the first row of Table 7, the value -12 corresponds to the approximation error of ImpVol(AppPriceLN(2,x_0)) for the first strike of the maturity $T = 3M$ (i.e. $K = 0.7$), whereas on the second row, the value -3 corresponds to the approximation error of `ImpVol(AppPriceLN(2,k))` for the third strike of the maturity $T = 6M$ (i.e. $K = 0.8$). If the price approximation does not belong to the non-arbitrage interval for Call options (it may happen for extremes strikes) we just report ND in the tabular.

7.2. *Analysis of results*

▷ **Influence of T and K.** We notice in Tables 7, 8, 9, 10 that errors are increasing w.r.t. T for all the different approximations: this is coherent with the $T^{3/2}$ or T^2-factor of our theoretical error estimates. For ATM options, all the approximations are excellent and errors remain small for a large range of strikes and maturities: with the log-normal proxy, usually smaller than 10 bps up to $10Y$ for strikes corresponding to the Gaussian quantiles range $[10\%, 90\%]$.

▷ **Influence of the proxy.** As expected, approximations based on log-normal proxy perform better than approximations based on normal proxy. On the one hand, we obtain simpler approximation formulas with the normal proxy: on the other hand, the errors become significant when considering slightly OTM or ITM options, even for short maturities and for advanced methods (order 3, local volatility frozen at the mid-point...).

▷ **Influence of the order.** Regarding firstly Tables 7-9 and then Tables 8-10, we notice that as expected, third order approximations are more accurate than second order ones. In addition, for the log-normal proxy case, second order approximations in spot or strike often underestimate the true implied volatility values whereas third order approximations in spot overestimate the true values for OTM options and yield underestimation for ITM options; the converse occurs for the third order approximations in strike. Because the errors have approximately the same magnitude but with opposite signs, approximations are improved by considering the average between the approximations. It is discussed below.

▷ **Influence of the point.** Unquestionably, methods using the local volatility at mid-point systematically give the best results. With `ImpVol(AppPriceLN(2,`x_{av}`))` (Table 7), errors do not exceed 15 bps for the whole set of strikes and maturities, which is already really good, whereas `ImpVol(AppPriceLN(3,`x_{av}`))` and `AppImpVolLN(3,`x_{av}`)` provide errors close to 0 proving an extreme accuracy.

▷ **Price expansions vs implied volatility expansions.** Generally speaking, the implied volatility expansions are more precise and stable. This can be easily observed by comparing on the one hand Tables 7 and 8 and on the other hand Tables 9 and 10. Sometimes, especially for extreme strikes, a simple direct second order second implied volatility expansion is more accurate than the corresponding third order price expansion. Since in addition the formulas are easier to compute, we recommend the use of implied volatility expansions. Moreover, the difference between `ImpVol(AppPriceLN(3,`x_{av}`))` and `AppImpVolLN(3,`x_{av}`)` is not clear, both methods giving similar and excellent results (see Tables 12 or 13) although the direct implied volatility expansion remains more stable especially for $\beta = 0.2$ and/or for large maturities. Last, when the local volatility function is frozen at spot or at strike, there is really an improvement in using implied volatility expansions instead of the corresponding price expansions.

▷ **Comparison with the benchmarks.** In Table 11, we report the performance of the methods `Av.ImpVol(AppPriceLN(3,.))`, `Av.AppImpVolLN(3,.)` and the benchmarks (`HF`) and (`HLF`). Errors on the implied volatility are equal to zero bp for the whole range of maturities and strikes for `Av.AppImpVolLN(3,.)` and the (`HLF`) approximation, whereas `Av.ImpVol(AppPriceLN(3,.))` and (`HF`) provide errors smaller than 45 and 70 bps in absolute value respectively. In Table 12, we compare `Av.ImpVol(AppPriceLN(3,.))`, `ImpVol(AppPriceLN(3,`x_{av}`))`, `Av.AppImpVolLN(3,.)` and `AppImpVolLN(3,`x_{av}`)` with the benchmarks (`HF`) and (`HLF`). In order to observe more clearly the accuracy of the different methods, we partially gather the results and we report the average of errors for different categories of strike (far ITM, ITM, ATM, OTM and far OTM, see Table 1), using a scientific notation for the errors. Computing the average per categories of strikes gives an advantage to methods which errors have non constant sign. These methods may be more reliable than those giving a systematic over/under-estimation.
The best method is clearly `AppImpVolLN(3,`x_{av}`)` which yields errors of 10^{-5} bps for short maturities and 10^{-2} bps for long maturities. The method gives better results than the excellent approximation proposed by Henry-Labordère (errors of 10^{-4} bps for short maturities and 10^{-1} bps for long maturities). `ImpVol(AppPriceLN(3,`x_{av}`))` seems to be slightly better than (`HLF`) but is less robust for extreme strikes than `AppImpVolLN(3,`x_{av}`)`. Significantly better results are obtained by averaging the expansions in spot and strike, thanks to the symmetrical roles played by these two variables. The results are close to those of the corresponding expansions with the mid-point, but they remain less accurate and less robust for extreme strikes. The problem of this averaging method is the risk of huge inaccuracy if one of two approximations in spot and strike fails. (`HF`) is clearly less accurate than all the other approximations.

▷ **Influence of** β**.** In the Table 13, as expected the log-normal proxy provides larger errors than for $\beta = 0.8$. Although the results of the normal proxy are better in comparison with the case $\beta = 0.8$, they remain less accurate and less robust than those obtained with the log-normal proxy. Up to the maturity $5Y$, `AppImpVolLN(3,`x_{av}`)` yields errors in bps smaller than 7 bps which is truly excellent. (`HLF`) gives comparable results. (`HF`) seems less accurate and cruder for extreme strikes. For the maturity $10Y$, we observe that `AppImpVolLN(3,`x_{av}`)` (maximal error close to 159 bps) behaves better than (`HLF`) (maximal error close to 271 bps) for very small strikes, whereas for very large strikes (`HLF`) is slightly better (-5 bps for `AppImpVolLN(3,`x_{av}`)` versus -1 bp for (`HLF`)). Surprisingly (`HF`) yields the smallest maximal error (close to 112 bps) but is more inaccurate for OTM. (`HLF`) and `AppImpVolLN(3,`x_{av}`)` give excellent results with errors of the order of 10^{-3} bps for short maturity ($3M$) and 10^{-1} bps for the maturity $3Y$. We nevertheless notice that ATM, (`HLF`) is better.

▷ **Impact of** ν**.** The level of volatility ν plays a similar role to $\sqrt{T}$, and in Tables 15 and 16, we analyse the impact of a larger volatility on our approximations. We take $\nu = 40\%$ and $\beta = 0.5$. We notice that up to the maturity $5Y$, the errors in bps do not exceed 6 bps for the methods `ImpVol(AppPriceLN(3,`x_{av}`))` or `AppImpVolLN(3,`x_{av}`)` with a maximal error of 92 bps for the maturity $10Y$. Their accuracy is better than those of (`HF`) or (`HLF`) for short and long maturities. (`HLF`) is much more inaccurate ITM for the maturity $10Y$ (maximal error of 286 bps). In Table 16, we aggregate the results per categories of strike up to the maturity $3Y$ and we observe a good accuracy of `ImpVol(AppPriceLN(3,`x_{av}`))` and `AppImpVolLN(3,`x_{av}`)`: 10^{-3} bps for the maturity $3M$ and 10^{-1} for the maturity $3Y$. In particular we notice that ATM, (`HF`) and (`HLF`) are less accurate.

In view of all these tests, we may conclude that `ImpVol (AppPriceLN(3,`x_{av}`))` and particulary `AppImpVolLN(3,`x_{av}`)` give very satisfying results, being at least as good as the Henry-Labordère formula in the worst situations ($\beta = 0.2$ or $\nu = 0.4$) and being often better in the case $\beta = 0.8$. The different current tests prove that our direct implied volatility approximations outperform the corresponding price approximations. In addition, a normal proxy seems not to be the most appropriate for the approximation of a CEV model, in view of the large errors obtained especially for very small strikes. This presumably explains why the Hagan formula is much less accurate than our approximations with log-normal proxy and than that of Henry-Labordère. The Hagan formula is namely close in the spirit to our approximation formulas with normal proxy.
To conclude, our approximations maintain very tight error estimates and allow to deal naturally with general time-dependent local volatility (or with stochastic interest rates, see [7]) which is a significant advantage compared to other approaches.

7.3. *CEV Delta approximations*

Now we test our approximation formulas for the deltas, by choosing again a CEV model with spot value $S_0 = 1$ and constant parameters. We test the values $(\beta, \nu) = (0.8, 0.25)$ and $(\beta, \nu) = (0.2, 0.25)$. We report in Tables 5 and 6 the exact delta values for the set of maturities and strikes defined in Table 1.

Table 5. CEV model ($\beta = 0.8, \nu = 0.25$): deltas in %.

3M	99.75	98.89	96.38	90.83	81.18	67.67	51.99	36.60	23.57	13.91	3.78	1.76	0.76
6M	99.20	95.10	90.44	83.56	74.58	64.05	52.82	41.80	31.77	16.33	11.09	4.64	1.01
1Y	99.09	96.04	88.90	83.53	69.78	61.97	53.98	46.15	31.98	20.75	9.76	5.55	0.82
1.5Y	98.76	95.68	89.46	85.06	74.11	61.40	54.88	42.35	36.62	22.35	10.35	5.44	1.03
2Y	98.75	95.94	90.58	82.55	72.45	66.93	55.63	44.75	34.97	23.04	12.30	5.15	0.75
3Y	99.13	95.36	93.13	83.51	75.22	66.13	56.88	47.99	36.04	23.46	11.44	5.25	0.86
5Y	99.23	95.93	91.89	86.53	76.75	69.63	58.86	48.70	36.85	25.21	12.05	5.54	0.79
10Y	99.13	97.10	93.95	87.68	80.34	72.60	62.46	53.06	40.92	27.19	12.88	5.57	0.54

We test the 6 following approximations:

(1) `AppDeltaLN(1,`x_0`)`, `AppDeltaLN(1,k)` and `AppDeltaLN(1,`x_{av}`)`: first order delta expansions based on the log-normal proxy with local volatility frozen

Table 6. CEV model ($\beta = 0.2, \nu = 0.25$): deltas in %.

3M	99.37	98.08	95.04	89.11	79.43	66.09	50.50	34.98	21.76	12.08	2.60	1.01	0.34
6M	98.13	92.97	88.01	81.08	72.24	61.88	50.71	39.59	29.35	13.70	8.60	2.85	0.35
1Y	97.34	93.15	85.45	80.09	66.66	58.97	51.00	43.08	28.51	17.00	6.39	2.91	0.14
1.5Y	96.23	92.03	85.29	80.89	70.22	57.76	51.23	38.45	32.50	17.73	6.20	2.35	0.13
2Y	95.71	91.77	85.84	77.85	68.06	62.67	51.43	40.30	30.09	17.68	7.29	1.81	0.04
3Y	95.62	90.10	87.57	77.87	69.86	61.01	51.76	42.61	30.02	16.82	5.53	1.36	0.03
5Y	94.70	89.45	84.88	79.47	70.04	63.12	52.32	41.69	28.91	16.52	4.53	0.87	0.00
10Y	94.37	90.36	86.11	79.22	71.81	64.05	53.52	43.32	29.65	14.59	2.76	0.23	0.00

at point x_0, k and x_{av}.

(2) AppDeltaLN(2,x_0), AppDeltaLN(2,k) and AppDeltaLN(2,x_{av}): second order delta expansions based on the log-normal proxy with local volatility frozen at point x_0, k and x_{av}.

Tables 17-18 (respectively 19) give errors on deltas (expressed in bps) using all the approximations with $\beta = 0.8$ (respectively $\beta = 0.2$).
Regarding the results, the accuracy for $\beta = 0.8$ is excellent because, except for AppDeltaLN(1,x_0), we obtain a maximal error (in absolute value) equal to 36 bps. Generally speaking, approximations with local volatility at spot are not as good as related approximations at strike. In addition, for second order formulas, we do not observe any symmetry between the spot and strike approximations (which often overestimate the exact delta), whereas the symmetry slightly appears for the first order expansions (not exactly with the same magnitude but opposite signs). Maybe in this situation, the optimal expansion point is not exactly the convex combination $x_{\mathrm{av}} = (x_0 + k)/2$. However the methods with the mid-point are truly excellent, in particular AppDeltaLN(2,x_{av}) which yields a maximal error (in absolute value) close to 1 bps. From Table 18, we observe that in average, the errors for AppDeltaLN(2,x_{av}) range from 10^{-3} for short maturities to 10^{-1} for long maturities.

In Table 19 ($\beta = 0.2$), without surprise the errors are larger compared to $\beta = 0.8$. The best approximation is still AppDeltaLN(2,x_{av}) which provides errors smaller than 27 bps up to $5Y$ with a global maximal error of 157 bps, which remains quite good. Curiously, for ATM options, the first order approximation may give better estimates even if the related errors quickly for large or small strikes in comparison with the second order approximations.

8. Appendix

8.1. *Computations of derivatives of* Call^{BS} *w.r.t the log spot, the log strike and the total variance*

In the following Proposition, we make explicit the formula for the derivatives at any order of Call^{BS} w.r.t. x and z:

Proposition 4.1. *Let $x, z \in \mathbb{R}$ and $y > 0$. For any integer $n \geq 1$, we have:*

$$\begin{aligned}\partial^n_{x^n} \mathrm{Call}^{\mathrm{BS}}(x,y,z) &= e^x \mathcal{N}(d_1(x,y,z)) \\ &\quad + \mathbb{1}_{n\geq 2}\, e^x \mathcal{N}'(d_1(x,y,z)) \sum_{k=1}^{n-1} \binom{n-1}{k} (-1)^{k-1} \frac{H_{k-1}(d_1(x,y,z))}{y^{\frac{k}{2}}}, \\ \partial^z_{z^n} \mathrm{Call}^{\mathrm{BS}}(x,y,z) &= -e^z \mathcal{N}(d_2(x,y,z)) \\ &\quad + \mathbb{1}_{n\geq 2}\, e^z \mathcal{N}'(d_2(x,y,z)) \sum_{k=1}^{n-1} \binom{n-1}{k} \frac{H_{k-1}(d_2(x,y,z))}{y^{\frac{k}{2}}},\end{aligned}$$

where $(H_k)_{k\in\mathbb{N}}$ are the Hermite polynomials defined for any $n \in \mathbb{N}$ and for any $x \in \mathbb{R}$ by:

$$H_n(x) = (-1)^n e^{x^2/2} \partial^n_{x^n}(e^{-x^2/2})$$

Proof. For $n = 1$ the formulas are easy to obtain. For $n \geq 2$, apply the Leibniz formula to the products $e^x\mathcal{N}(d_1(x,y,z))$ and $e^z\mathcal{N}(d_2(x,y,z))$. □

We deduce a very useful Corollary:

Corollary 4.1. *Let $x, z \in \mathbb{R}$ and $y > 0$. For any integers $n \geq 1$ and $m \geq 1$, we have:*

$$\begin{aligned}|\partial^n_{x^n} \mathrm{Call}^{\mathrm{BS}}(x,y,z)| + |\partial^n_{z^n} \mathrm{Call}^{\mathrm{BS}}(x,y,z)| &\leq_c y^{\frac{1-n}{2}}, \\ |x-z|^m |\partial^n_{x^n} \mathrm{Call}^{\mathrm{BS}}(x,y,z) - e^x \mathcal{N}(d_1(x,y,z))| &\leq_c y^{\frac{1-n+m}{2}}, \\ |x-z|^m |\partial^n_{z^n} \mathrm{Call}^{\mathrm{BS}}(x,y,z) + e^z \mathcal{N}(d_2(x,y,z))| &\leq_c y^{\frac{1-n+m}{2}},\end{aligned}$$

where the generic constants depend polynomially on y.

Remark 4.7. In practice the two last estimates are used when we want to bound $(x-z)^m \sum_{i=1}^n \alpha_i \partial^i_{x^i} \mathrm{Call}^{\mathrm{BS}}(x,y,z)$ or $(x-z)^m \sum_{i=1}^n \alpha_i \partial^i_{z^i} \mathrm{Call}^{\mathrm{BS}}(x,y,z)$ (with $\sum_{i=1}^n \alpha_i = 0$) by a power of y with the highest possible degree.

Proof. We recall that for any polynomial function $\mathcal{P}$, $x \to \mathcal{P}(x)\mathcal{N}'(x)$ is a bounded function. Then the first inequality follows directly from Proposition 4.1. For the second and the third, write $(x-z) = d_1(x,y,z)\sqrt{y} - \frac{1}{2}y = d_2(x,y,z)\sqrt{y} + \frac{1}{2}y$ and conclude similarly. □

In the next Proposition, we provide the formulas of the first, the second and the third derivatives of $\mathrm{Call}^{\mathrm{BS}}$ w.r.t. a positive volatility:

Proposition 4.2. *Let $x, z \in \mathbb{R}$, $\nu > 0$ and $T > 0$. We have:*

$$\begin{aligned}
&\mathrm{Vega}^{\mathrm{BS}}(x, \nu^2 T, z) = \partial_\nu \mathrm{Call}^{\mathrm{BS}}(x, \nu^2 T, z)\\
&= e^x \sqrt{T} \mathcal{N}'(d_1(x, \nu^2 T, z)) = e^z \sqrt{T} \mathcal{N}'(d_2(x, \nu^2 T, z)), && (68)\\
&\mathrm{Vomma}^{\mathrm{BS}}(x, \nu^2 T, z) = \partial_\nu \mathrm{Vega}^{\mathrm{BS}}(x, \nu^2 T, z)\\
&= \frac{\mathrm{Vega}^{\mathrm{BS}}(x, \nu^2 T, z)}{\nu} d_1(x, \nu^2 T, z) d_2(x, \nu^2 T, z)\\
&= \frac{\mathrm{Vega}^{\mathrm{BS}}(x, \nu^2 T, z)}{\nu} [\frac{(x-z)^2}{\nu^2 T} - \frac{\nu^2 T}{4}], && (69)\\
&\mathrm{Ultima}^{\mathrm{BS}}(x, \nu^2 T, z) = \partial_\nu \mathrm{Vomma}^{\mathrm{BS}}(x, \nu^2 T, z)\\
&= -\frac{\mathrm{Vega}^{\mathrm{BS}}(x, \nu^2 T, z)}{\nu^2} [d_1 d_2 (1 - d_1 d_2) + d_1^2 + d_2^2](x, \nu^2 T, z)\\
&= -\frac{\mathrm{Vega}^{\mathrm{BS}}(x, \nu^2 T, z)}{\nu^2} [\frac{(x-z)^2}{2} + \frac{3(x-z)^2}{\nu^2 T} + \frac{\nu^2 T}{4} - \frac{(x-z)^4}{\nu^4 T^2} - \frac{\nu^4 T^2}{16}]. && (70)
\end{aligned}$$

The above Proposition directly implies the following result:

Corollary 4.2. *Let $x, z \in \mathbb{R}$, $\nu > 0$ and $T > 0$. We have the following estimates:*

$$0 < \mathrm{Vega}^{\mathrm{BS}}(x, \nu^2 T, z) \leq_c \sqrt{T}, \tag{71}$$

$$|\mathrm{Vomma}^{\mathrm{BS}}(x, \nu^2 T, z)| \leq_c \frac{\sqrt{T}}{\nu}, \tag{72}$$

$$|\mathrm{Ultima}^{\mathrm{BS}}(x, \nu^2 T, z)| \leq_c \frac{\sqrt{T}}{\nu^2}, \tag{73}$$

where the generic constants depend polynomially of ν.

We finally state relations between the derivatives w.r.t. x or z, the $\mathrm{Vega}^{\mathrm{BS}}$ and the $\mathrm{Vomma}^{\mathrm{BS}}$. These relations allow on the one hand to replace derivatives w.r.t. z with derivatives w.r.t. x and on the other hand to write the differential operators w.r.t. x or z in terms of the $\mathrm{Vega}^{\mathrm{BS}}$ and the $\mathrm{Vomma}^{\mathrm{BS}}$. The verification of these identities is tedious but without mathematical difficulties. For instance, we have used Mathematica to check these relations.

Proposition 4.3. *Let $x, z \in \mathbb{R}$, $\nu > 0$ and $T > 0$. We have:*

$$
\begin{aligned}
(\partial^2_{x^2} - \partial_x)\mathrm{Call}^{\mathrm{BS}}(x, \nu^2 T, z) &= (\partial^2_{z^2} - \partial_z)\mathrm{Call}^{\mathrm{BS}}(x, \nu^2 T, z)\\
&= \frac{e^x}{\nu\sqrt{T}}\mathcal{N}'(d_1(x, \nu^2 T, z))\\
&= \frac{\mathrm{Vega}^{\mathrm{BS}}(x, \nu^2 T, z)}{\nu T}, \qquad (74)\\
(\partial^3_{x^3} - \frac{3}{2}\partial^2_{x^2} + \frac{1}{2}\partial_x)&\mathrm{Call}^{\mathrm{BS}}(x, \nu^2 T, z)\\
&= -(\partial^3_{z^3} - \frac{3}{2}\partial^2_{z^2} + \frac{1}{2}\partial_z)\mathrm{Call}^{\mathrm{BS}}(x, \nu^2 T, z)\\
&= -\frac{e^x(x-z)}{\nu^3 T^{\frac{3}{2}}}\mathcal{N}'(d_1(x, \nu^2 T, z))\\
&= -\mathrm{Vega}^{\mathrm{BS}}(x, \nu^2 T, z)\frac{(x-z)}{\nu^3 T^2}, \qquad (75)\\
(\frac{1}{4}\partial^4_{x^4} - \frac{1}{2}\partial^3_{x^3} + \frac{1}{4}\partial^2_{x^2})&\mathrm{Call}^{\mathrm{BS}}(x, \nu^2 T, z)\\
&= (\frac{1}{4}\partial^4_{z^4} - \frac{1}{2}\partial^3_{z^3} + \frac{1}{4}\partial^2_{z^2})\mathrm{Call}^{\mathrm{BS}}(x, \nu^2 T, z)\\
&= e^x\mathcal{N}'(d_1(x, \nu^2 T, z))\big[\frac{(x-z)^2}{4\nu^5 T^{\frac{5}{2}}} - \frac{1}{16\nu\sqrt{T}} - \frac{1}{4\nu^3 T^{\frac{3}{2}}}\big]\\
&= \mathrm{Vega}^{\mathrm{BS}}(x, \nu^2 T, z)\big[\frac{(x-z)^2}{4\nu^5 T^3} - \frac{1}{16\nu T} - \frac{1}{4\nu^3 T^2}\big], \qquad (76)\\
(\partial^4_{x^4} - 2\partial^3_{x^3} + \frac{5}{4}\partial^2_{x^2} - \frac{1}{4}\partial_x)&\mathrm{Call}^{\mathrm{BS}}(x, \nu^2 T, z)\\
&= (\partial^4_{z^4} - 2\partial^3_{z^3} + \frac{5}{4}\partial^2_{z^2} - \frac{1}{4}\partial_z)\mathrm{Call}^{\mathrm{BS}}(x, \nu^2 T, z)\\
&= e^x\mathcal{N}'(d_1(x, \nu^2 T, z))\big[\frac{(x-z)^2}{\nu^5 T^{\frac{5}{2}}} - \frac{1}{\nu^3 T^{\frac{3}{2}}}\big]\\
&= \mathrm{Vega}^{\mathrm{BS}}(x, \nu^2 T, z)\big[\frac{(x-z)^2}{\nu^5 T^3} - \frac{1}{\nu^3 T^2}\big], \qquad (77)\\
(3\partial^4_{x^4} - 6\partial^3_{x^3} + \frac{7}{2}\partial^2_{x^2} - \frac{1}{2}\partial_x)&\mathrm{Call}^{\mathrm{BS}}(x, \nu^2 T, z)\\
&= (3\partial^4_{z^4} - 6\partial^3_{z^3} + \frac{7}{2}\partial^2_{z^2} - \frac{1}{2}\partial_z)\mathrm{Call}^{\mathrm{BS}}(x, \nu^2 T, z)\\
&= e^x\mathcal{N}'(d_1(x, \nu^2 T, z))\big[3\frac{(x-z)^2}{\nu^5 T^{\frac{5}{2}}} - \frac{3}{\nu^3 T^{\frac{3}{2}}} - \frac{1}{4\nu\sqrt{T}}\big]\\
&= \mathrm{Vega}^{\mathrm{BS}}(x, \nu^2 T, z)\big[3\frac{(x-z)^2}{\nu^5 T^3} - \frac{3}{\nu^3 T^2} - \frac{1}{4\nu T}\big], \qquad (78)
\end{aligned}
$$

$$
\begin{aligned}
&(\frac{1}{2}\partial^6_{x^6} - \frac{3}{2}\partial^5_{x^5} + \frac{13}{8}\partial^4_{x^4} - \frac{3}{4}\partial^3_{x^3} + \frac{1}{8}\partial^2_{x^2})\mathrm{Call}^{\mathrm{BS}}(x,\nu^2T,z)\\
&= (\frac{1}{2}\partial^6_{z^6} - \frac{3}{2}\partial^5_{z^5} + \frac{13}{8}\partial^4_{z^4} - \frac{3}{4}\partial^3_{z^3} + \frac{1}{8}\partial^2_{z^2})\mathrm{Call}^{\mathrm{BS}}(x,\nu^2T,z)\\
&= e^x\mathcal{N}'(d_1(x,\nu^2T,z))\big[\frac{(x-z)^4}{2\nu^9T^{\frac{9}{2}}} - \frac{(x-z)^2}{8\nu^5T^{\frac{5}{2}}} - 3\frac{(x-z)^2}{\nu^7T^{\frac{7}{2}}}\\
&+ \frac{1}{8\nu^3T^{\frac{3}{2}}} + \frac{3}{2\nu^5T^{\frac{5}{2}}}\big] \qquad (79)\\
&= \mathrm{Vega}^{\mathrm{BS}}(x,\nu^2T,z)\big[-3\frac{(x-z)^2}{\nu^7T^4} + \frac{1}{8\nu^3T^2} + \frac{3}{2\nu^5T^3}\big]\\
&+ \frac{1}{2}\mathrm{Vomma}^{\mathrm{BS}}(x,\nu^2T,z)\frac{(x-z)^2}{\nu^6T^4}. \qquad (80)
\end{aligned}
$$

8.2. *Derivatives of* $\mathrm{Call}^{\mathrm{BA}}$ *w.r.t the spot, the strike and the total variance*

Proposition 4.4. *Let $S, Z \in \mathbb{R}$ and $Y > 0$. For any integer $n \geq 1$, we have:*

$$
\begin{aligned}
\partial^n_{S^n}\mathrm{Call}^{\mathrm{BA}}(S,Y,Z) =& \mathbb{1}_{n=1}\mathcal{N}(\frac{S-Z}{\sqrt{Y}}) + \mathbb{1}_{n\geq 2}\mathcal{N}'(\frac{S-Z}{\sqrt{Y}})(-1)^{n-2}\frac{H_{n-2}(\frac{S-Z}{\sqrt{Y}})}{Y^{\frac{n-1}{2}}},\\
\partial^n_{Z^n}\mathrm{Call}^{\mathrm{BA}}(S,Y,Z) =& -\mathbb{1}_{n=1}\mathcal{N}(\frac{S-Z}{\sqrt{Y}}) + \mathbb{1}_{n\geq 2}\mathcal{N}'(\frac{S-Z}{\sqrt{Y}})\frac{H_{n-2}(\frac{S-Z}{\sqrt{Y}})}{Y^{\frac{n-1}{2}}}.
\end{aligned}
$$

Corollary 4.3. *Let $S, Z \in \mathbb{R}$ and $Y > 0$. For any integers $n \geq 2$ and $m \geq 1$, we have:*

$$
|S-Z|^m\big(|\partial^n_{S^n}\mathrm{Call}^{\mathrm{BA}}(S,Y,Z)| + |\partial^n_{Z^n}\mathrm{Call}^{\mathrm{BA}}(S,Y,Z)|\big) \leq_c Y^{\frac{1-n+m}{2}},
$$

where the generic constants depend polynomially on Y.

Proposition 4.5. *Let $S, Z \in \mathbb{R}$, $V > 0$ and $T > 0$. We have:*

$$
\begin{aligned}
\mathrm{Vega}^{\mathrm{BA}}(S,V^2T,Z) =& \partial_V\mathrm{Call}^{\mathrm{BA}}(S,V^2T,Z) = \sqrt{T}\mathcal{N}'(\frac{S-Z}{V\sqrt{T}}),\\
\mathrm{Vomma}^{\mathrm{BA}}(S,V^2T,Z) =& \partial_V\mathrm{Vega}^{\mathrm{BA}}(S,V^2T,Z)\\
=& \frac{\mathrm{Vega}^{\mathrm{BA}}(S,V^2T,Z)}{\nu}\frac{(S-Z)^2}{V^2T},\\
\mathrm{Ultima}^{\mathrm{BA}}(S,V^2T,Z) =& \partial_V\mathrm{Vomma}^{\mathrm{BA}}(S,V^2T,Z)\\
=& -\frac{\mathrm{Vega}^{\mathrm{BA}}(S,V^2T,Z)}{\nu^2}\big[\frac{3(S-Z)^2}{V^2T} - \frac{(S-Z)^4}{V^4T^2}\big].
\end{aligned}
$$

Corollary 4.4. *Let $S, Z \in \mathbb{R}$ $V > 0$ and $T > 0$. We have the following estimates:*

$$0 < \mathrm{Vega}^{\mathrm{BA}}(S, V^2T, Z) \leq_c \sqrt{T},$$

$$|\mathrm{Vomma}^{\mathrm{BA}}(S, V^2T, Z)| \leq_c \frac{\sqrt{T}}{V},$$

$$|\mathrm{Ultima}^{\mathrm{BA}}(S, V^2T, Z)| \leq_c \frac{\sqrt{T}}{V^2},$$

where the generic constants depend polynomially on V.

Proposition 4.6. *Let $S, Z \in \mathbb{R}$ $V > 0$ and $T > 0$. We have:*

$$\begin{aligned}
\partial^2_{S^2}\mathrm{Call}^{\mathrm{BA}}(S, V^2T, Z) &= \partial^2_{Z^2}\mathrm{Call}^{\mathrm{BA}}(x, V^2T, Z) \\
&= \frac{\mathrm{Vega}^{\mathrm{BA}}(S, V^2T, Z)}{VT}, \\
\partial^3_{S^3}\mathrm{Call}^{\mathrm{BA}}(S, V^2T, Z) &= -\partial^3_{Z^3}\mathrm{Call}^{\mathrm{BA}}(S, V^2T, Z) \\
&= -\mathrm{Vega}^{\mathrm{BA}}(S, V^2T, Z)\frac{(S-Z)}{V^3T^2}, \\
\partial^3_{S^4}\mathrm{Call}^{\mathrm{BA}}(S, V^2T, Z) &= \partial^4_{Z^4}\mathrm{Call}^{\mathrm{BA}}(S, V^2T, Z) \\
&= \mathrm{Vega}^{\mathrm{BA}}(S, V^2T, Z)\Big[\frac{(S-Z)^2}{V^5T^3} - \frac{1}{V^3T^2}\Big], \\
\partial^6_{S^6}\mathrm{Call}^{\mathrm{BA}}(S, V^2T, Z) &= \partial^6_{Z^6}\mathrm{Call}^{\mathrm{BA}}(S, V^2T, Z) \\
&= \mathrm{Vega}^{\mathrm{BA}}(S, V, T, Z)\Big[-6\frac{(S-Z)^2}{V^7T^4} + \frac{3}{V^5T^3}\Big] \\
&\quad + \mathrm{Vomma}^{\mathrm{BA}}(S, V^2T, Z)\frac{(S-Z)^2}{V^6T^4}.
\end{aligned}$$

8.3. *Derivatives of δ^{BS} w.r.t the log spot, the log strike and the total variance*

Proposition 4.7. *Let $x, z \in \mathbb{R}$ and $y > 0$. For any integer $n \geq 1$, we have:*

$$\begin{aligned}
\partial^n_{x^n}\delta^{\mathrm{BS}}(x, y, z) &= (-1)^{n-1}\mathcal{N}'(d_1(x, y, z))\frac{H_{n-1}(d_1(x, y, z))}{y^{\frac{n}{2}}}, \\
\partial^n_{z^n}\delta^{\mathrm{BS}}(x, y, z) &= -\mathcal{N}'(d_1(x, y, z))\frac{H_{n-1}(d_1(x, y, z))}{y^{\frac{n}{2}}}.
\end{aligned}$$

Corollary 4.5. *Let $x, z \in \mathbb{R}$ and $y > 0$. For any integers $n \geq 1$ and $m \geq 1$, we have:*

$$|x - z|^m\big(|\partial^n_{x^n}\delta^{\mathrm{BS}}(x, y, z)| + |\partial^n_{z^n}\delta^{\mathrm{BS}}(x, y, z)|\big) \leq_c y^{\frac{m-n}{2}},$$

where the generic constants depend polynomially on y.

Proposition 4.8. *Let $x, z \in \mathbb{R}$ and $y > 0$. We have:*

$$\partial_y\delta^{\mathrm{BS}}(x, y, z) = \frac{1}{2}(\partial^2_{z^2} - \partial_z)\delta^{\mathrm{BS}}(x, y, z) = -\frac{\mathcal{N}'(d_1(x, y, z))}{2y}d_2(x, y, z).$$

8.4. *Proof of Lemma 1*

We proceed by induction. The key is to prove the above technical result:

Lemma 6. *Let $(m_t)_{t\in[0,T]}$ be a square integrable and predictable process, $(\lambda_t)_{t\in[0,T]}$ be a measurable and bounded deterministic function and φ be a C_b^∞ function. Then, we have:*

$$\mathbb{E}\Big(\varphi(\int_0^T \lambda_t \mathrm{d}W_t)\int_0^T m_t \mathrm{d}W_t\Big) = \mathbb{E}\Big(\varphi^{(1)}(\int_0^T \lambda_t \mathrm{d}W_t)\int_0^T \lambda_t m_t \mathrm{d}t\Big).$$

Proof. We propose two proofs: firstly we employ a PDE argument and secondly we show that this is a straightforward application of the Malliavin calculus theory. In the two points of view, we use the common notation for the diffusion process $(Z_t)_{t\in[0,T]} = (\int_0^t \lambda_s \mathrm{d}W_s)_{t\in[0,T]}$ and we recall that $(\mathcal{F}_t)_{t\in[0,T]}$ denotes the augmented filtration of the Brownian motion W.

$\triangleright$ **PDE argument.** We introduce $u(t,x) = \mathbb{E}[\varphi(Z_T)|Z_t = x]$ which solves the following PDE with terminal condition:

$$\begin{cases} \partial_t u(t,x) + \frac{1}{2}\lambda_t^2 \partial_{xx}^2 u(t,x) = 0, & (t,x) \in]0,T[\times\mathbb{R}, \\ u(T,x) = \varphi(x), \quad x \in \mathbb{R}. \end{cases}$$

Thanks to the above PDE and the assumption on φ, $\forall i \in \mathbb{N}$, $\partial_{x^i}^i(u(t,Z_t))_{t\in[0,T]}$ is a martingale and $\forall t \in [0,T]$, we have:

$$\partial_{x^i}^i u(t,Z_t) = \mathbb{E}[\varphi^{(i)}(Z_T)|\mathcal{F}_t] = \mathbb{E}[\varphi^{(i)}(Z_T)] + \int_0^t \partial_{x^{i+1}}^{i+1} u(s,Z_s)\lambda_s \mathrm{d}W_s.$$

Then applying the L_2-isometry for the product $u(T,Z_T)\int_0^T m_t \mathrm{d}W_t = \varphi(Z_T)\int_0^T m_t \mathrm{d}W_t$, it readily comes:

$$\begin{aligned} &\mathbb{E}\Big(\varphi(Z_T)\int_0^T m_t \mathrm{d}W_t\Big) \\ &\quad = \int_0^T \mathbb{E}[\partial_x u(t,Z_t)\lambda_t m_t]\mathrm{d}t = \mathbb{E}\Big(\varphi^{(1)}(Z_T)\int_0^T \lambda_t m_t \mathrm{d}t\Big), \end{aligned}$$

where at the last equality we have used the martingale property of $\partial_x(u(t,Z_t))_{t\in[0,T]}$.

$\triangleright$ **Malliavin calculus approach.** The result directly comes from the duality relationship of Malliavin calculus (see [48]) identifying the Itô integral $\int_0^T m_t \mathrm{d}W_t$ with the Skorohod operator and observing that $(\varphi^{(1)}(Z_T)\lambda_t)_{t\in[0,T]}$ is the first Malliavin derivative of $\varphi(Z_T)$. □

Lemma 6 is a particular case of Lemma 1 for $N = 1$ and $I_N = 1$ noting that $\forall i \in \mathbb{N}$, $\mathbb{E}\Big(\varphi^{(i)}(\int_0^T \lambda_t \mathrm{d}W_t)\Big) = \partial_{\varepsilon^i}^i \mathbb{E}\Big(\varphi(\int_0^T \lambda_t \mathrm{d}W_t + \varepsilon)\Big)|_{\varepsilon=0}$, thanks to the regularity of φ.

For $N = 1$ and $I_N = 0$, there is nothing to prove. Suppose that the formula (35) is true for $N \geq 2$. Then apply Lemma 6 if $I_{N+1} = 1$ to obtain:

$$\begin{aligned}
&\mathbb{E}\Big(\varphi(\int_0^T \lambda_t \mathrm{d}W_t) \int_0^T l_{N+1,t_{N+1}} \int_0^{t_{N+1}} l_{N,t_N} \cdots \\
&\quad \times \int_0^{t_2} l_{1,t_1} \mathrm{d}W_{t_1}^{I_1} \ldots \mathrm{d}W_{t_N}^{I_N} \mathrm{d}W_{t_{N+1}}^{I_{N+1}}\Big) \\
=&\mathbb{E}\Big(\varphi^{(I_{N+1})}(\int_0^T \lambda_t \mathrm{d}W_t) \int_0^T \widehat{l}_{N+1,t_{N+1}} \\
&\quad \times \int_0^{t_{N+1}} l_{N,t_N} \int_0^{t_N} \ldots \int_0^{t_2} l_{1,t_1} \mathrm{d}W_{t_1}^{I_1} \ldots \mathrm{d}W_{t_N}^{I_N} \mathrm{d}t_{N+1}\Big) \\
=&\mathbb{E}\Big(\varphi^{(I_{N+1})}(\int_0^T \lambda_t \mathrm{d}W_t) \int_0^T (l_{N,t_N} \int_{t_N}^T \widehat{l}_{N+1,s} \mathrm{d}s) \int_0^{t_N} \ldots \\
&\quad \times \int_0^{t_2} l_{1,t_1} \mathrm{d}W_{t_1}^{I_1} \ldots \mathrm{d}W_{t_N}^{I_N}\Big),
\end{aligned}$$

where at the last equality we have used the fact that $\int_0^T f_t Z_t \mathrm{d}t = \int_0^T (\int_t^T f_s \mathrm{d}s) dZ_t$ for any continuous semi-martingale Z starting from 0 and any measurable and bounded deterministic function f (apply the Itô formula to the product $(\int_t^T f_s ds) Z_t$). We easily conclude with the induction hypothesis and leave the details to the reader.

8.5. *Applications of the expansions for time-independent CEV model*

We specify in this section the results and the practical calculus of the various expansion coefficients when the volatility has the form:

$$\sigma(S) = \nu S^{\beta - 1},$$

i.e. a CEV-type time-independent volatility with a level ν and a skew $\beta \leq 1$. Although the volatility and its derivatives are not bounded, we expect that our expansions can be generalized to that model. Alternatively, to fit our assumptions, we would need to modify the CEV volatility function σ near 0 and $+\infty$, so that the ellipticity and regularity conditions are met. The impact of such a modification has been studied in the case of Limited CEV model in [51] where the authors show a very small impact on prices. Observe in addition that the correction terms in our expansions do no depend on the modification of σ at 0 and $+\infty$.

To apply our different expansion theorems, we need to give the expressions of the coefficients $(C_i)_{1 \leq i \leq 8}$ defined in Definition 4.1, in Theorem 12 and in Lemmas

4-5. A straightforward calculus leads to:

$$\begin{aligned}
a(x) &= \nu e^{x(\beta-1)}, & a^{(1)}(x) &= (\beta-1)a(x), & a^{(2)}(x) &= (\beta-1)^2 a(x),\\
a(x_0) &= \nu S_0^{\beta-1}, & a(k) &= \nu K^{\beta-1}, & a(x_{\rm av}) &= \nu(S_0K)^{\frac{\beta-1}{2}},\\
\Sigma(S) &= \nu S^{\beta}, & \Sigma^{(1)}(S) &= \beta\frac{\Sigma(S)}{S}, & \Sigma^{(2)}(S) &= \beta(\beta-1)\frac{\Sigma(S)}{S^2},\\
\Sigma(S_0) &= \nu S_0^{\beta}, & \Sigma(K) &= \nu K^{\beta}, & \Sigma(S_{\rm av}) &= \nu(\frac{S_0+K}{2})^{\beta}.
\end{aligned}$$

Thus for $\beta \in [0,1]$, the magnitudes of $\mathcal{M}_0(a)$ and $\mathcal{M}_1(a)$ are mainly linked to those of ν and $\nu(\beta-1)$. At the limit case $\beta = 1$, the model coincides with the log-normal proxy and $\mathcal{M}_1(a) = 0$. In the same spirit, ν and $\nu\beta$ are respectively linked to $\mathcal{M}_0(\Sigma)$ and $\mathcal{M}_1(\Sigma)$. At the limit case $\beta = 0$, the model coincides with the normal proxy and $\mathcal{M}_1(\Sigma) = 0$.

Finally, the expression of the coefficients $(C_i)_{1\le i\le 8}$ are:

$$\begin{aligned}
C_1(a;x)_0^T &= (\beta-1)a^4(x)\frac{T^2}{2}, & C_2(a;x)_0^T &= (\beta-1)^2 a^4(x)T^2,\\
C_3(a;x)_0^T &= (\beta-1)^2 a^6(x)\frac{T^3}{3}, & C_4(a;x)_0^T &= C_8(a;x)_0^T = (\beta-1)^2 a^6(x)\frac{T^3}{6}\\
C_5(a;x)_0^T &= 2(\beta-1)^2 a^2(x)T, & C_6(a;x)_0^T &= (\beta-1)^2 a^4(x)\frac{T^2}{2},
\end{aligned}$$

$$\begin{aligned}
C_7(a;x)_0^T &= (\beta-1)a^2(x)T,\\
C_1(\Sigma;S)_0^T &= \beta\frac{\Sigma^4(S)}{S}\frac{T^2}{2}, & C_2(\Sigma;S)_0^T &= \beta(2\beta-1)\frac{\Sigma^4(S)}{S^2}\frac{T^2}{2},\\
C_3(\Sigma;S)_0^T &= \beta(2\beta-1)\frac{\Sigma^6(S)}{S^2}\frac{T^3}{6}, & C_4(\Sigma;S)_0^T &= C_8(\Sigma;S)_0^T = \beta^2\frac{\Sigma^6(S)}{S^2}\frac{T^3}{6},\\
C_5(\Sigma;S)_0^T &= \beta(2\beta-1)\frac{\Sigma^2(S)}{S^2}T, & C_6(\Sigma;S)_0^T &= \beta^2\frac{\Sigma^4(S)}{S^2}\frac{T^2}{2},\\
C_7(\Sigma;S)_0^T &= \beta\frac{\Sigma^2(S)}{S}T,
\end{aligned}$$

where $x = x_0, k, x_{\rm av}$ and $S = S_0, K, S_{\rm av}$.

We now give the expressions of the coefficients γ_i, π_i, χ_i and Ξ_i defined in

Definition 4.2 useful to compute the implied volatility expansions:

$$\begin{aligned}
\gamma_0(a;x)_0^T &= \frac{(\beta-1)^2}{24} a^3(x) T[1-\frac{a^2(x)T}{4}], & \gamma_1(a;x)_0^T &= \frac{(\beta-1)}{2} a(x),\\
\gamma_2(a;x)_0^T &= \frac{(\beta-1)^2}{12} a(x), & \pi_0(a;x)_0^T &= \gamma_0(a;x)_0^T,\\
\pi_1(a;x)_0^T &= 0, & \pi_2(a;x)_0^T &= -\frac{(\beta-1)^2}{24} a(x),\\
\chi_1(\Sigma;S)_0^T &= \frac{\beta\Sigma(S)}{2S}, & \chi_0(\Sigma;S)_0^T &= \frac{\beta(\beta-2)}{24S^2}\Sigma^3(S)T,\\
\chi_2(\Sigma;S)_0^T &= \frac{\beta(\beta-2)}{12S^2}\Sigma(S), & \Xi_0(\Sigma;S)_0^T &= \chi_0(\Sigma;S)_0^T,\\
\Xi_2(\Sigma;S)_0^T &= -\frac{\beta(\beta+1)}{24S^2}\Sigma(S), & \Xi_1(\Sigma;S)_0^T &= 0.
\end{aligned}$$

For example, the second and third order Black-Scholes and Bachelier implied volatility expansions based on the mid-points are explicitely given by:

$$\begin{aligned}
\sigma_{\mathrm{I}}(x_0,T,k) &\approx \nu(S_0K)^{\frac{\beta-1}{2}},\\
\sigma_{\mathrm{I}}(x_0,T,k) &\approx \nu(S_0K)^{\frac{\beta-1}{2}}\big[1+\frac{(\beta-1)^2\nu^2T}{24}(S_0K)^{\beta-1}(1-\frac{\nu^2T(S_0K)^{\beta-1}}{4})\\
&\qquad -\frac{(\beta-1)^2}{24}\log^2(\frac{S_0}{K})\big],\\
\Sigma_{\mathrm{I}}(S_0,T,K) &\approx \nu(\frac{S_0+K}{2})^{\beta},\\
\Sigma_{\mathrm{I}}(S_0,T,K) &\approx \nu(\frac{S_0+K}{2})^{\beta}\big[1+\frac{\beta(\beta-2)\nu^2T}{24}(\frac{S_0+K}{2})^{2\beta-2}\\
&\qquad -\frac{\beta(\beta+1)}{6}(\frac{S_0-K}{S_0+K})^2\big],
\end{aligned}$$

which are very simple formulas. The last formula coincides with the intermediate equation (A.28b) in [5].

Bibliography

[1] Achdou, Y. and Pironneau, O.: *Computational Methods for Option Pricing.* (SIAM series, Frontiers in Applied Mathematics, Philadelphia, 2005).

[2] Glasserman, P.: *Monte Carlo Methods in Financial Engineering.* (Springer, 2003).

[3] Carr, P. and Madan, D.: Option valuation using the fast Fourier transform, *Journal of Computational Finance.* **2**(4), 61–73, (1998).

[4] Lewis, A.: *Option valuation under stochastic volatility.* (Finance Press, 2000).

[5] Hagan, P. and Woodward, D.: Equivalent Black volatilities, *Applied Mathematical Finance.* **6**, 147–157, (1999).

[6] Henry-Labordère, P.: *Analysis, Geometry, and Modeling in Finance: Advanced Methods in Option Pricing.* (Chapman and Hall, 2008).

[7] Benhamou, E., Gobet, E. and Miri, M.: Analytical formulas for local volatility model with stochastic rates, *Quantitative Finance.* **12**(2), 185–198, (2012).

[8] Schachermayer, W. and Teichmann, J.: How close are the option pricing formulas of Bachelier and Black-Merton-Scholes?, *Math. Finance.* **18**(1), 155–170, (2008).

[9] Musiela, M. and Rutkowski, M.: *Martingale methods in financial modelling.* (Springer Verlag, 2005), second edition.

[10] Lee, R.: Implied volatility: Statics, dynamics, and probabilistic interpretation. In *Recent Advances in Applied Probability.* Springer, (2005).

[11] Benaim, S. and Friz, P.: Regular variation and smile asymptotics, *Mathematical Finance.* **19**(1), 1–12, (2009).

[12] Gulisashvili, A.: Asymptotic formulas with error estimates for call pricing functions and the implied volatility at extreme strikes, *SIAM J. Financial Math.* **1**, 609–641, (2010).

[13] Gulisashvili, A.: Ed., *Analytically Tractable Stochastic Stock Price Models.* (Forthcoming in Springer Finance, 2012).

[14] Tehranchi, M.: Asymptotics of implied volatility far from maturity, *Journal of Applied Probability.* **46**(3), 629–650, (2009).

[15] Rogers, L. and Tehranchi, M.: Can the implied volatility surface move by parallel shifts?, *Finance and Stochastics.* **14**, 235–248, (2010).

[16] Forde, M. and Jacquier, A.: The large-maturity smile for the Heston model, *Finance and Stochastics.* **15**(4), 755–780, (2011).

[17] Jacquier, A., Keller-Ressel, M. and Mijatovic, A.: Large deviations in stochastic volatility with jumps: asymptotic implied volatility for affine models, *preprint, http://arxiv.org/abs/1108.3998.* (2011).

[18] Gatheral, J. and Jacquier, A.: Convergence of Heston to SVI, *Quantitative Finance.* **11**(8), 1129–1132, (2011).

[19] Gatheral, J.: A parsimonious arbitrage-free implied volatility parameterization with application to the valuation of volatility derivatives, *Presentation at Global Derivatives & Risk Management, Madrid, www.math.nyu.edu/fellows_fin_math/gatheral/madrid2004.pdf* (May. 2004).

[20] Duffie, D., Filipovic, D. and Schachermayer, W.: Affine processes and applications in finance, *The Annals of Applied Probability.* **13**(3), 984–1053, (2003).

[21] Freidlin, M. and Wentzell, A.: *Random perturbations of dynamical systems.* (New York, NY: Springer, 1998), second edition. Transl. from the Russian by Joseph Szuecs.

[22] Fleming, W. and Souganidis, P.: Asymptotic series and the method of vanishing viscosity, *Indiana Univ. Math. J.* **35**, 425–447, (1986).

[23] Fournié, E., Lebuchoux, J. and Touzi, N.: Small noise expansion and importance sampling, *Asymptot. Anal.* **14**(4), 361–376, (1997).

[24] Fouque, J., Papanicolaou, G., Sircar, R. and Knut, S.: *Multiscale stochastic volatility for equity, interest rate, and credit derivatives.* (Cambridge University Press, Cambridge, 2011).

[25] Azencott, R.: Densité des diffusions en temps petit: développements asymptotiques. partie i, *Séminaire de Probabilités XVIII - Lecture Notes in Mathematics 1059.* (1982–83).

[26] Watanabe, S.: Analysis of Wiener functionals (Malliavin calculus) and its applications to heat kernels, *Annals of Probability.* **15**(1), 1–39, (1987).

[27] Yoshida, N.: Asymptotic expansions of maximum likelihood estimator for small diffusion via the theory of Malliavin-Watanabe, *Probability Theory and Related Fields.* **92**, 275–311, (1992).

[28] Yoshida, N.: Asymptotic expansion for statistics related to small diffusions, *J. Japan Statistical Society.* **22**(2), 139–159, (1992).

[29] Uchida, M. and Yoshida, N.: Asymptotic expansion for small diffusions applied to

option pricing, *Statistical Inference for Stochastics Processes.* **7**, 189–223, (2004).

[30] Kunitomo, N. and Takahashi, A.: The asymptotic expansion approach to the valuation of interest rate contingent claims, *Mathematical Finance.* **11**(1), 117–151, (2001).

[31] Osajima, Y.: The asymptotic expansion formula of implied volatility for dynamic SABR model and FX hybrid model, *Available at* *http://ssrn.com/abstract=965265*. (2007).

[32] Alòs, E. and Ewald, C.: Malliavin differentiability of the Heston volatility and applications to option pricing, *Advances in Applied Probability.* **40**(1), 144–162, (2008).

[33] Hagan, P., Kumar, D., Lesniewski, A. and Woodward, D.: Managing smile risk, *Willmott Magazine.* pp. 84–108, (2002).

[34] Kloeden, P. and Platen, E.: *Numerical solution of stochastic differential equations.* (Springer Verlag, 1995).

[35] Papanicolaou, G. and Sircar, R.: Stochastic volatility, smile and asymptotics, *Applied Mathematical Finance.* **6**(2), 107–145, (1999).

[36] Fouque, J., Papanicolaou, G. and Sircar, R.: *Derivatives in financial Markets with stochastic volatility.* (Cambridge University Press, 2000).

[37] Scott, L.: Option pricing when the variance changes randomly : Theory, estimation, and an application, *J. Financial and Quantitative Analysis.* **22**, 419–438, (1987).

[38] Fouque, J., Papanicolaou, G., Sircar, R. and Solna, S.: Singular perturbations in option pricing, *SIAM J. Appl. Math.* **63**(5), 1648–1665, (2003).

[39] Fouque, J., Papanicolaou, G., Sircar, R. and Solna, K.: Maturity cycles in implied volatility, *Finance and Stochastics.* **8**(4), 451–477, (2004).

[40] Kevorkian, J. and Cole, J.: *Perturbation Methods in Applied Mathematics.* (Springer-Verlag, 1985).

[41] Kunita, H.: Stochastic flows of diffeomorphisms, *École d'été de Probabilities de St-Flour XII. Lecture notes.* **1097**, (1984).

[42] Gobet, E. and Miri, M.: Weak approximations of averaged diffusion processes, *Forthcoming in Stochastic Processes and their Applications.* (2012).

[43] Dupire, B.: Pricing with a smile, *Risk.* **7**(1), 18–20, (1994).

[44] Gobet, E. and Suleiman, A.: New approximations in local volatility models. In ed. Y. Kabanov, *Musiela Festschrift.* Springer, (2012).

[45] Benhamou, E., Gobet, E. and Miri, M.: Smart expansion and fast calibration for jump diffusion, *Finance and Stochastics.* **13**(4), 563–589, (2009).

[46] Benhamou, E., Gobet, E. and Miri, M.: Time dependent Heston model, *SIAM Journal on Financial Mathematics.* **1**, 289–325, (2010).

[47] Bompis, R. and Gobet, E.: Price expansion formulas for model combining local and stochastic volatility, *Preprint.* (2012).

[48] Nualart, D.: *Malliavin calculus and related topics.* (Springer Verlag, 2006), second edition.

[49] Benhamou, E., Gobet, E. and Miri, M.: Expansion formulas for European options in a local volatility model, *International Journal of Theoretical and Applied Finance.* **13**(4), 603–634, (2010).

[50] Schroder, M.: Computing the constant elasticity of variance option pricing formula, *Journal of Finance.* **44**, 211–219, (1989).

[51] Andersen, L. and Andreasen, J.: Volatility skews and extensions of the Libor market model, *Applied Mathematical Finance.* **7**(1), 1–32, (2000).

Table 7. CEV model ($\beta = 0.8$, $\nu = 0.25$): errors in bps on the BS implied volatility using the 6 second order price approximations ImpVol(AppPriceLN(2,x_0)), ImpVol(AppPriceLN(2,k)), ImpVol(AppPriceLN(2,x_{av})), ImpVol(AppPriceN(2,S_0)), ImpVol(AppPriceN(2,K)) and ImpVol(AppPriceN(2,S_{av})).

3M	−12	−6	−2	−1	0	0	0	0	0	−1	−3	−5	−8
	−17	−7	−3	−1	0	0	0	0	0	−1	−2	−4	−7
	0	0	0	0	0	0	0	0	0	0	0	0	0
	−577	−79	−18	−1	2	2	2	2	2	0	−22	−44	−73
	−124	−59	−20	−2	2	2	2	2	2	1	−21	−53	−125
	21	14	9	6	3	2	2	2	3	4	9	11	15
6M	−13	−3	−1	−1	0	0	0	0	0	−1	−2	−4	−15
	−17	−4	−2	−1	0	0	0	0	0	−1	−2	−4	−11
	1	0	0	0	0	0	0	0	0	0	0	0	1
	−269	−18	0	5	5	4	3	4	4	1	−5	−35	−117
	−138	−26	−4	4	5	4	3	4	4	3	−2	−30	−238
	33	16	11	7	5	4	3	3	4	8	10	16	26
1Y	−23	−8	−2	−1	0	0	0	0	0	−1	−4	−8	−37
	−34	−9	−2	−1	0	0	0	0	0	−1	−4	−7	−23
	1	1	0	0	0	0	0	0	0	0	0	0	1
	−848	−48	6	10	8	7	6	7	9	6	−21	−59	−235
	−240	−69	−1	8	8	7	6	7	9	9	−9	−45	ND
	64	36	20	14	8	7	6	7	9	13	22	29	54
1.5Y	−28	−11	−3	−2	−1	0	0	0	−1	−2	−6	−12	−50
	−41	−12	−4	−2	−1	0	0	0	−1	−2	−5	−10	−30
	2	1	0	0	0	0	0	0	0	0	0	1	2
	−644	−50	10	16	14	10	9	11	12	10	−31	−95	−299
	−291	−90	−4	10	14	10	9	11	12	13	−9	−71	ND
	89	52	30	23	14	10	9	11	12	19	32	44	74
2Y	−36	−14	−5	−2	−1	−1	−1	−1	−1	−2	−6	−18	−91
	−56	−17	−6	−2	−1	−1	−1	−1	−1	−2	−6	−14	−44
	2	1	0	0	0	0	−1	0	0	0	0	1	2
	ND	−65	12	22	18	15	13	14	17	13	−27	−138	−418
	−373	−129	−13	18	18	15	13	14	17	18	−1	−107	ND
	119	72	44	27	17	15	13	14	17	25	39	60	105
3Y	−64	−18	−11	−3	−1	−1	−1	−1	−1	−3	−10	−27	−141
	−122	−21	−13	−3	−1	−1	−1	−1	−1	−3	−9	−21	−57
	4	1	1	0	−1	−1	−1	−1	0	0	1	1	3
	ND	−43	6	35	28	22	19	20	25	17	−62	−208	−534
	−638	−154	−68	27	28	22	19	20	25	27	−8	−159	−2260
	208	102	81	41	28	21	19	20	25	37	61	87	146
5Y	−106	−31	−13	−6	−2	−1	−1	−1	−2	−5	−17	−45	−472
	−256	−38	−14	−6	−2	−1	−1	−1	−2	−5	−15	−32	−88
	7	2	1	0	−1	−1	−1	−1	−1	0	1	2	5
	ND	−41	53	64	49	39	32	35	41	25	−116	−334	−753
	−1000	−295	−55	34	48	39	32	35	43	47	−4	−249	ND
	377	183	119	80	49	38	32	34	43	60	99	140	233
10Y	−172	−69	−30	−10	−4	−3	−2	−2	−4	−9	−35	−103	ND
	ND	−95	−34	−10	−4	−3	−2	−2	−4	−9	−28	−61	−159
	15	6	2	0	−2	−2	−2	−2	−1	0	2	5	10
	ND	40	158	146	109	82	67	70	80	32	−271	−625	−1100
	−1531	−762	−232	75	103	82	67	70	85	95	16	−781	ND
	786	451	289	166	108	80	67	68	84	120	192	267	439

Table 8. CEV model (β = 0.8, ν = 0.25): errors in bps on the BS implied volatility using the 6 second order implied volatility approximations `AppImpVolLN(2,`x_0`)`, `AppImpVolLN(2,k)`, `AppImpVolLN(2,`x_{av}`)`, `AppImpVolN(2,`S_0`)`, `AppImpVolN(2,K)` and `AppImpVolN(2,`S_{av}`)`.

3M	−1	−1	0	0	0	0	0	0	0	0	0	−1	−1
	−1	−1	−1	0	0	0	0	0	0	0	0	−1	−1
	0	0	0	0	0	0	0	0	0	0	0	0	0
	27	18	12	7	4	2	2	2	3	5	11	15	19
	29	19	12	7	4	2	2	2	3	5	11	15	18
	21	14	9	6	3	2	2	2	3	4	9	11	15
6M	−2	−1	−1	0	0	0	0	0	0	0	−1	−1	−1
	−2	−1	−1	0	0	0	0	0	0	0	−1	−1	−1
	1	0	0	0	0	0	0	0	0	0	0	0	1
	41	20	13	9	5	4	3	4	5	10	13	21	36
	44	21	14	9	6	4	3	4	5	9	12	20	33
	33	16	11	7	5	4	3	3	4	8	10	16	26
1Y	−3	−2	−1	−1	0	0	0	0	0	−1	−1	−2	−3
	−4	−2	−1	−1	0	0	0	0	0	−1	−1	−1	−3
	1	1	0	0	0	0	0	0	0	0	0	0	1
	79	44	24	17	9	7	6	7	10	16	28	39	74
	88	48	25	17	9	7	6	7	10	15	27	36	67
	64	36	20	14	8	7	6	7	9	13	22	29	54
1.5Y	−4	−3	−1	−1	−1	0	0	0	−1	−1	−2	−2	−4
	−5	−3	−2	−1	−1	0	0	0	−1	−1	−2	−2	−4
	2	1	0	0	0	0	0	0	0	0	0	1	2
	108	64	36	27	15	10	9	11	13	23	42	59	104
	123	69	38	28	15	10	9	11	13	22	39	54	91
	89	52	30	23	14	10	9	11	12	19	32	44	74
2Y	−6	−4	−2	−1	−1	−1	−1	−1	−1	−1	−2	−3	−6
	−7	−4	−2	−1	−1	−1	−1	−1	−1	−1	−2	−3	−5
	2	1	0	0	0	0	−1	0	0	0	0	1	2
	145	87	52	31	19	15	13	14	19	30	50	80	149
	167	97	55	32	19	15	13	14	18	29	47	72	127
	119	72	44	27	17	15	13	14	17	25	39	60	105
3Y	−10	−5	−4	−2	−1	−1	−1	−1	−1	−2	−3	−5	−9
	−12	−6	−4	−2	−1	−1	−1	−1	−1	−2	−3	−4	−7
	4	1	1	0	−1	−1	−1	−1	0	0	1	1	3
	249	121	96	47	31	22	19	20	28	45	80	120	212
	301	136	105	49	31	22	19	20	28	43	73	106	176
	208	102	81	41	28	21	19	20	25	37	61	87	146
5Y	−18	−9	−5	−4	−2	−1	−1	−1	−2	−3	−5	−8	−14
	−23	−10	−6	−4	−2	−1	−1	−1	−2	−3	−5	−7	−11
	7	2	1	0	−1	−1	−1	−1	−1	0	1	2	5
	443	216	140	92	52	39	32	35	48	74	132	196	352
	571	252	155	98	53	40	32	35	47	69	117	167	277
	377	183	119	80	49	38	32	34	43	60	99	140	233
10Y	−33	−19	−12	−7	−4	−3	−2	−2	−3	−6	−10	−16	−29
	−47	−24	−14	−7	−4	−3	−2	−2	−3	−5	−9	−12	−21
	15	6	2	0	−2	−2	−2	−2	−1	0	2	5	10
	904	522	333	188	117	82	67	70	94	149	264	394	725
	1289	660	390	203	120	83	67	70	91	137	224	313	510
	786	451	289	166	108	80	67	68	84	120	192	267	439

Table 9. CEV model ($\beta = 0.8$, $\nu = 0.25$): errors in bps on the BS implied volatility using the 6 third order price approximations ImpVol(AppPriceLN(3,x_0)), ImpVol(AppPriceLN(3,k)), ImpVol(AppPriceLN(3,x_{av})), ImpVol(AppPriceN(3,S_0)), ImpVol(AppPriceN(3,K)) and ImpVol(AppPriceN(3,S_{av})).

3M	−1	0	0	0	0	0	0	0	0	0	0	0	0
	1	0	0	0	0	0	0	0	0	0	0	0	0
	0	0	0	0	0	0	0	0	0	0	0	0	0
	6	−4	−2	−1	0	0	0	0	0	0	2	0	−6
	−22	−1	2	1	0	0	0	0	0	0	−2	−3	−2
	−1	0	0	0	0	0	0	0	0	0	0	0	0
6M	−1	0	0	0	0	0	0	0	0	0	0	0	1
	1	0	0	0	0	0	0	0	0	0	0	0	−1
	0	0	0	0	0	0	0	0	0	0	0	0	0
	−17	−4	−1	0	0	0	0	0	0	1	2	4	−12
	−13	4	2	0	0	0	0	0	0	−1	−1	−5	−12
	−1	0	0	0	0	0	0	0	0	0	0	0	−1
1Y	1	1	0	0	0	0	0	0	0	0	0	0	1
	−2	0	0	0	0	0	0	0	0	0	0	0	3
	0	0	0	0	0	0	0	0	0	0	0	0	0
	−62	−15	−2	−1	0	0	0	0	0	1	6	9	−56
	−34	13	3	1	0	0	0	0	0	−1	−5	−12	−45
	−3	0	0	0	0	0	0	0	0	0	0	0	−3
1.5Y	−2	0	0	0	0	0	0	0	0	0	0	0	4
	3	0	0	0	0	0	0	0	0	0	0	0	−3
	0	0	0	0	0	0	0	0	0	0	0	0	0
	−100	−23	−4	−2	0	0	0	0	0	2	11	14	−83
	−36	21	7	3	0	0	0	0	0	−1	−8	−23	−98
	−4	−1	0	0	0	0	0	0	0	0	0	−1	−5
2Y	−3	−1	0	0	0	0	0	0	0	0	0	1	10
	4	1	0	0	0	0	0	0	0	0	0	−1	−7
	0	0	0	0	0	0	0	0	0	0	0	0	0
	−163	−36	−8	−1	0	0	0	0	0	3	14	17	−177
	−63	31	14	2	0	0	0	0	0	−2	−9	−38	−243
	−7	−1	0	0	0	0	0	0	0	0	0	−1	−12
3Y	−9	−1	0	0	0	0	0	0	0	0	0	1	16
	11	1	0	0	0	0	0	0	0	0	0	−1	−10
	0	0	0	0	0	0	0	0	0	0	0	0	0
	−818	−48	−23	−2	−1	0	0	0	1	5	29	16	−275
	−250	52	41	4	1	0	0	0	−1	−3	−19	−69	ND
	−24	−2	−1	0	0	0	0	0	0	0	−1	−3	−23
5Y	−18	−1	0	0	0	0	0	0	0	0	1	2	39
	23	1	0	0	0	0	0	0	0	0	−1	−2	−21
	0	0	0	0	0	0	0	0	0	0	0	0	0
	ND	−103	−26	−7	−2	−1	0	0	1	12	57	−10	−522
	−597	90	65	17	1	1	0	−1	−2	−5	−38	−143	ND
	−73	−5	−1	−1	0	0	0	0	0	0	−2	−7	−76
10Y	−34	−5	−1	0	0	0	0	0	0	0	1	7	147
	28	3	1	0	0	0	0	0	0	0	−1	−7	−59
	0	0	0	0	0	0	0	0	0	0	0	0	0
	ND	−340	−83	−16	−6	−4	−1	1	4	42	109	−222	−987
	−1200	−42	230	49	5	1	−1	−3	−4	−13	−94	−588	ND
	−214	−29	−7	−3	−2	−1	−1	−1	−1	−1	−6	−28	ND

Table 10. CEV model ($\beta = 0.8$, $\nu = 0.25$): errors in bps on the BS implied volatility using the 6 third order implied volatility approximations `AppImpVolLN(3,`x_0`)`, `AppImpVolLN(3,k)`, `AppImpVolLN(3,`x_{av}`)`, `AppImpVolN(3,S0)`, `AppImpVolN(3,K)` and `AppImpVolN(3,`S_{av}`)`.

3M	0	0	0	0	0	0	0	0	0	0	0	0	0
	0	0	0	0	0	0	0	0	0	0	0	0	0
	0	0	0	0	0	0	0	0	0	0	0	0	0
	4	2	1	0	0	0	0	0	0	0	−1	−2	−3
	−5	−3	−1	0	0	0	0	0	0	0	1	1	2
	0	0	0	0	0	0	0	0	0	0	0	0	0
6M	0	0	0	0	0	0	0	0	0	0	0	0	0
	0	0	0	0	0	0	0	0	0	0	0	0	0
	0	0	0	0	0	0	0	0	0	0	0	0	0
	7	2	1	0	0	0	0	0	0	−1	−1	−3	−8
	−10	−3	−1	0	0	0	0	0	0	0	1	2	5
	0	0	0	0	0	0	0	0	0	0	0	0	0
1Y	0	0	0	0	0	0	0	0	0	0	0	0	0
	0	0	0	0	0	0	0	0	0	0	0	0	0
	0	0	0	0	0	0	0	0	0	0	0	0	0
	17	6	2	1	0	0	0	0	0	−1	−4	−7	−25
	−30	−10	−2	−1	0	0	0	0	0	1	3	5	14
	1	0	0	0	0	0	0	0	0	0	0	0	1
1.5Y	0	0	0	0	0	0	0	0	0	0	0	0	0
	0	0	0	0	0	0	0	0	0	0	0	0	0
	0	0	0	0	0	0	0	0	0	0	0	0	0
	25	10	3	2	0	0	0	0	0	−2	−7	−15	−43
	−49	−17	−5	−2	0	0	0	0	0	1	5	9	22
	2	1	0	0	0	0	0	0	0	0	0	1	2
2Y	0	0	0	0	0	0	0	0	0	0	0	0	0
	0	0	0	0	0	0	0	0	0	0	0	0	0
	0	0	0	0	0	0	0	0	0	0	0	0	0
	37	16	6	1	0	0	0	0	0	−2	−9	−24	−79
	−80	−29	−9	−2	0	0	0	0	0	2	6	14	35
	4	1	0	0	0	0	0	0	0	0	0	1	4
3Y	0	0	0	0	0	0	0	0	0	0	0	0	0
	0	0	0	0	0	0	0	0	0	0	0	0	0
	0	0	0	0	0	0	0	0	0	0	0	0	0
	77	23	15	2	0	0	0	0	−1	−5	−20	−47	−143
	−214	−48	−28	−4	0	0	0	0	0	3	11	23	55
	12	2	1	0	0	0	0	0	0	0	1	2	8
5Y	−1	0	0	0	0	0	0	0	0	0	0	0	0
	1	0	0	0	0	0	0	0	0	0	0	0	0
	0	0	0	0	0	0	0	0	0	0	0	0	0
	160	50	22	7	0	−1	0	0	−2	−10	−47	−108	−350
	−618	−130	−47	−15	−1	0	0	−1	0	6	22	43	102
	36	7	2	0	0	0	0	0	0	0	2	6	21
10Y	−1	−1	−1	0	0	0	0	0	0	0	1	1	1
	2	1	1	0	0	0	0	0	0	0	0	0	0
	0	0	0	0	0	0	0	0	0	0	0	0	0
	362	155	68	16	0	−3	−1	1	−4	−34	−153	−368	−1307
	−2195	−631	−216	−44	−6	1	−1	−3	0	14	53	102	231
	113	34	10	0	−1	−1	−1	−1	−1	1	8	21	70

Table 11. CEV model ($\beta = 0.8$, $\nu = 0.25$): errors in bps on the BS implied volatility using the 4 approximations `Av.ImpVol(AppPriceLN(3,.))`, `Av.AppImpVolLN(3,.)`, (`HF`), and (`HLF`).

3M	0	0	0	0	0	0	0	0	0	0	0	0	0
	0	0	0	0	0	0	0	0	0	0	0	0	0
	0	0	0	0	0	0	0	0	0	0	0	0	0
	0	0	0	0	0	0	0	0	0	0	0	0	0
6M	0	0	0	0	0	0	0	0	0	0	0	0	0
	0	0	0	0	0	0	0	0	0	0	0	0	0
	0	0	0	0	0	0	0	0	0	0	0	0	0
	0	0	0	0	0	0	0	0	0	0	0	0	0
1Y	0	0	0	0	0	0	0	0	0	0	0	0	0
	0	0	0	0	0	0	0	0	0	0	0	0	0
	−1	0	0	0	0	0	0	0	0	0	0	0	−1
	0	0	0	0	0	0	0	0	0	0	0	0	0
1.5Y	0	0	0	0	0	0	0	0	0	0	0	0	1
	0	0	0	0	0	0	0	0	0	0	0	0	0
	−2	−1	0	0	0	0	0	0	0	0	0	0	−2
	0	0	0	0	0	0	0	0	0	0	0	0	0
2Y	0	0	0	0	0	0	0	0	0	0	0	0	2
	0	0	0	0	0	0	0	0	0	0	0	0	0
	−3	−1	0	0	0	0	0	0	0	0	0	−1	−3
	0	0	0	0	0	0	0	0	0	0	0	0	0
3Y	1	0	0	0	0	0	0	0	0	0	0	0	3
	0	0	0	0	0	0	0	0	0	0	0	0	0
	−9	−2	−1	0	0	0	0	0	0	0	−1	−2	−6
	0	0	0	0	0	0	0	0	0	0	0	0	0
5Y	2	0	0	0	0	0	0	0	0	0	0	0	9
	0	0	0	0	0	0	0	0	0	0	0	0	0
	−24	−5	−2	0	0	0	0	0	0	0	−2	−4	−14
	0	0	0	0	0	0	0	0	0	0	0	0	0
10Y	−3	−1	0	0	0	0	0	0	0	0	0	0	44
	0	0	0	0	0	0	0	0	0	0	0	0	0
	−69	−24	−8	−1	0	0	0	0	0	−1	−6	−14	−44
	0	0	0	0	0	0	0	0	0	0	0	0	0

Table 12. CEV model ($\beta = 0.8$, $\nu = 0.25$): average per categories of strikes of errors in bps on the BS implied volatility using the 6 approximations Av.ImpVol(AppPriceLN(3,.)), ImpVol(AppPriceLN(3,$x_{\rm av}$)), Av.AppImpVolLN(3,.), AppImpVolLN(3,$x_{\rm av}$), (HF), and (HLF).

	far ITM	ITM	ATM	OTM	far OTM
3M	6.7E−2	1.4E−4	−7.9E−6	1.8E−4	1.5E−2
	−2.4E−4	−1.4E−5	−1.3E−5	−1.3E−5	−9.9E−5
	4.3E−4	9.1E−5	−8.2E−6	7.3E−5	2.6E−4
	−3.4E−5	−9.6E−6	−1.3E−5	−8.7E−6	−1.7E−5
	−9.1E−2	−8.6E−3	2.0E−4	−7.3E−3	−4.8E−2
	−6.4E−5	1.7E−4	2.4E−4	1.5E−4	−9.8E−6
6M	3.6E−2	1.6E−4	−4.3E−5	1.5E−4	3.2E−2
	−3.0E−4	−4.3E−5	−5.3E−5	−3.6E−5	−2.4E−4
	1.0E−3	1.5E−4	−4.3E−5	1.4E−4	7.8E−4
	−6.5E−5	−4.3E−5	−5.3E−5	−3.5E−5	−4.9E−5
	−1.6E−1	−7.9E−3	9.2E−4	−8.6E−3	−1.3E−1
	2.5E−4	8.6E−4	9.5E−4	7.1E−4	1.3E−4
1Y	9.6E−2	5.1E−4	−1.9E−4	4.4E−4	1.9E−1
	−1.1E−3	−1.8E−4	−2.1E−4	−1.4E−4	−1.1E−3
	4.2E−3	4.3E−4	−1.9E−4	5.5E−4	3.0E−3
	−2.5E−4	−1.8E−4	−2.1E−4	−1.4E−4	−2.1E−4
	−6.3E−1	−2.2E−2	3.8E−3	−3.6E−2	−5.1E−1
	1.2E−3	3.7E−3	3.8E−3	2.8E−3	4.4E−4
1.5Y	9.2E−2	1.2E−3	−3.9E−4	7.5E−4	2.8E−1
	−1.8E−3	−4.0E−4	−4.5E−4	−3.2E−4	−2.1E−3
	8.5E−3	1.1E−3	−3.9E−4	1.1E−3	6.1E−3
	−4.9E−4	−3.9E−4	−4.5E−4	−3.0E−4	−4.0E−4
	−1.2E+0	−5.5E−2	8.3E−3	−7.3E−2	−9.9E−1
	3.9E−3	8.4E−3	8.6E−3	6.2E−3	1.1E−3
2Y	1.4E−1	1.7E−3	−6.4E−4	1.4E−3	8.7E−1
	−3.1E−3	−7.1E−4	−7.9E−4	−5.3E−4	−4.7E−3
	1.5E−2	1.8E−3	−6.4E−4	1.7E−3	1.1E−2
	−8.9E−4	−7.1E−4	−7.9E−4	−5.1E−4	−8.1E−4
	−2.0E+0	−9.6E−2	1.5E−2	−1.0E−1	−1.9E+0
	7.5E−3	1.5E−2	1.5E−2	1.1E−2	1.2E−3
3Y	5.8E−1	2.5E−3	−1.5E−3	1.9E−3	1.5E+0
	−9.9E−3	−1.6E−3	−1.7E−3	−1.1E−3	−9.4E−3
	3.7E−2	5.3E−3	−1.5E−3	3.8E−3	2.2E−2
	−2.4E−3	−1.5E−3	−1.7E−3	−1.1E−3	−1.6E−3
	−5.2E+0	−3.5E−1	3.4E−2	−2.6E−1	−3.8E+0
	1.8E−2	3.5E−2	3.5E−2	2.4E−2	3.1E−3
5Y	1.2E+0	6.2E−3	−3.6E−3	1.2E−3	4.5E+0
	−3.0E−2	−4.0E−3	−4.2E−3	−2.7E−3	−2.7E−2
	1.0E−1	1.2E−2	−3.5E−3	9.4E−3	5.4E−2
	−7.2E−3	−3.9E−3	−4.3E−3	−2.5E−3	−4.4E−3
	−1.5E+1	−6.9E−1	9.4E−2	−6.7E−1	−9.1E+0
	6.5E−2	1.1E−1	9.7E−2	6.3E−2	7.6E−3
10Y	−2.1E+0	−4.6E−2	−1.1E−2	−3.2E−2	2.2E+1
	−9.7E−2	−1.1E−2	−1.2E−2	−7.5E−3	−1.3E−1
	2.8E−1	3.8E−2	−1.1E−2	2.9E−2	1.7E−1
	−2.3E−2	−1.1E−2	−1.3E−2	−6.9E−3	−1.8E−2
	−4.6E+1	−3.1E+0	3.8E−1	−2.4E+0	−2.9E+1
	4.4E−1	4.8E−1	4.0E−1	2.3E−1	2.2E−2

Table 13. CEV model ($\beta = 0.2$, $\nu = 0.25$): errors in bps on the BS implied volatility using the 6 approximations ImpVol(AppPriceLN(3,x_{av})), AppImpVolLN(3,x_{av}), ImpVol(AppPriceN(3,S_{av})), AppImpVolN(3,S_{av}), (HF) and (HLF).

3M	0	0	0	0	0	0	0	0	0	0	0	0	0
	0	0	0	0	0	0	0	0	0	0	0	0	0
	0	0	0	0	0	0	0	0	0	0	0	0	0
	0	0	0	0	0	0	0	0	0	0	0	0	0
	−1	0	0	0	0	0	0	0	0	0	0	0	0
	0	0	0	0	0	0	0	0	0	0	0	0	0
6M	0	0	0	0	0	0	0	0	0	0	0	0	0
	0	0	0	0	0	0	0	0	0	0	0	0	0
	0	0	0	0	0	0	0	0	0	0	0	0	0
	0	0	0	0	0	0	0	0	0	0	0	0	0
	−1	0	0	0	0	0	0	0	0	0	0	0	−1
	0	0	0	0	0	0	0	0	0	0	0	0	0
1Y	0	0	0	0	0	0	0	0	0	0	0	0	−1
	0	0	0	0	0	0	0	0	0	0	0	0	0
	0	0	0	0	0	0	0	0	0	0	0	0	0
	1	0	0	0	0	0	0	0	0	0	0	0	0
	−5	−1	0	0	0	0	0	0	0	0	0	−1	−2
	0	0	0	0	0	0	0	0	0	0	0	0	0
1.5Y	−1	0	0	0	0	0	0	0	0	0	0	0	−1
	0	0	0	0	0	0	0	0	0	0	0	0	0
	1	0	0	0	0	0	0	0	0	0	0	0	0
	1	1	0	0	0	0	0	0	0	0	0	0	0
	−9	−3	−1	0	0	0	0	0	0	0	−1	−1	−4
	−1	0	0	0	0	0	0	0	0	0	0	0	0
2Y	−1	0	0	0	0	0	0	0	0	0	0	0	−3
	−1	0	0	0	0	0	0	0	0	0	0	0	0
	2	1	0	0	0	0	0	0	0	0	0	0	0
	2	1	1	0	0	0	0	0	0	0	0	0	1
	−15	−5	−1	0	0	0	0	0	0	0	−1	−3	−8
	−1	−1	0	0	0	0	0	0	0	0	0	0	0
3Y	−4	−1	−1	−1	−1	−1	0	0	0	0	0	−1	−8
	−2	−1	−1	−1	−1	−1	0	0	0	0	0	0	−1
	7	2	2	1	0	0	0	0	0	0	0	0	0
	8	3	2	1	0	0	0	0	0	0	0	1	1
	−45	−9	−5	−1	0	0	0	0	0	−1	−2	−5	−15
	−4	−1	−1	0	0	0	0	0	0	0	0	0	0
5Y	2	−1	−2	−2	−2	−2	−1	−1	−1	0	0	−2	−37
	7	−1	−2	−2	−2	−2	−1	−1	−1	0	0	0	−1
	47	13	6	4	2	1	1	1	1	1	1	1	0
	50	13	7	4	2	1	1	1	1	1	1	1	3
	−117	−26	−9	−3	0	0	0	0	0	−1	−6	−12	−31
	4	0	0	0	0	0	0	0	0	0	−1	−1	−1
10Y	148	84	41	12	2	−2	−3	−2	−2	−1	−2	−13	ND
	159	85	41	12	2	−2	−3	−2	−2	−1	−1	−1	−5
	530	221	109	45	21	11	6	3	2	2	2	2	−8
	541	224	111	45	21	11	6	3	2	2	2	3	6
	−112	−6	18	18	12	8	4	2	−1	−5	−17	−33	−73
	271	123	65	29	14	8	4	2	0	−1	−1	−1	−1

Table 14. CEV model ($\beta = 0.2$, $\nu = 0.25$): average per categories of strikes of errors in bps on the BS implied volatility using the 6 approximations `ImpVol(AppPriceLN(3,`x_{av}`))`, `AppImpVolLN(3,`x_{av}`)`, `ImpVol(AppPriceN(3,`S_{av}`))`, `AppImpVolN(3,`S_{av}`)`, (`HF`) and (`HLF`).

	far ITM	ITM	ATM	OTM	far OTM
3M	−5.4E−2	−4.0E−3	−3.4E−3	−3.3E−3	−2.9E−2
	−9.3E−3	−3.1E−3	−3.4E−3	−1.9E−3	−4.1E−3
	2.5E−2	6.3E−3	1.8E−3	3.7E−3	8.9E−3
	3.9E−2	7.2E−3	1.8E−3	4.5E−3	1.6E−2
	−3.8E−1	−3.9E−2	7.9E−5	−2.9E−2	−1.6E−1
	−2.1E−2	−4.6E−3	3.2E−4	−2.6E−3	−7.9E−3
6M	−6.9E−2	−1.5E−2	−1.4E−2	−7.5E−3	−7.4E−2
	−2.1E−2	−1.4E−2	−1.4E−2	−7.2E−3	−1.1E−2
	6.8E−2	1.6E−2	6.9E−3	9.6E−3	2.5E−2
	9.0E−2	1.7E−2	6.9E−3	1.0E−2	4.3E−2
	−7.2E−1	−4.5E−2	1.5E−3	−3.9E−2	−4.3E−1
	−5.6E−2	−7.4E−3	1.8E−3	−5.4E−3	−2.2E−2
1Y	−2.5E−1	−6.6E−2	−5.5E−2	−2.8E−2	−3.8E−1
	−9.9E−2	−6.6E−2	−5.5E−2	−2.6E−2	−4.6E−2
	3.2E−1	6.4E−2	2.8E−2	3.5E−2	7.8E−2
	4.1E−1	6.7E−2	2.8E−2	3.8E−2	1.5E−1
	−3.0E+0	−1.4E−1	8.2E−3	−1.5E−1	−1.6E+0
	−2.6E−1	−2.3E−2	8.8E−3	−1.9E−2	−7.2E−2
1.5Y	−4.4E−1	−1.6E−1	−1.2E−1	−6.2E−2	−7.4E−1
	−2.4E−1	−1.6E−1	−1.2E−1	−5.6E−2	−8.8E−2
	7.4E−1	1.7E−1	6.2E−2	7.1E−2	1.5E−1
	8.8E−1	1.8E−1	6.2E−2	7.8E−2	2.9E−1
	−5.6E+0	−3.5E−1	1.7E−2	−3.0E−1	−3.0E+0
	−5.6E−1	−5.8E−2	1.9E−2	−3.5E−2	−1.3E−1
2Y	−8.0E−1	−3.1E−1	−2.2E−1	−9.6E−2	−1.9E+0
	−4.9E−1	−3.1E−1	−2.2E−1	−9.0E−2	−1.8E−1
	1.5E+0	3.2E−1	1.2E−1	1.1E−1	2.1E−1
	1.7E+0	3.3E−1	1.2E−1	1.2E−1	5.1E−1
	−1.0E+1	−6.1E−1	3.0E−2	−4.3E−1	−5.5E+0
	−1.1E+0	−9.2E−2	3.6E−2	−5.5E−2	−2.1E−1
3Y	−2.3E+0	−8.0E−1	−4.8E−1	−2.0E−1	−4.3E+0
	−1.4E+0	−8.0E−1	−4.8E−1	−1.7E−1	−3.5E−1
	4.8E+0	9.8E−1	2.8E−1	2.4E−1	3.4E−1
	5.4E+0	1.0E+0	2.8E−1	2.6E−1	9.3E−1
	−2.7E+1	−2.1E+0	8.9E−2	−1.0E+0	−9.9E+0
	−2.7E+0	−2.7E−1	9.7E−2	−1.1E−1	−3.7E−1

Table 15. CEV model ($\beta = 0.5$, $\nu = 0.4$): errors in bps on the BS implied volatility using the 4 approximations `ImpVol(AppPriceLN(3,`x_{av}`))`, `AppImpVolLN(3,`x_{av}`)`, (`HF`) and (`HLF`).

3M	0	0	0	0	0	0	0	0	0	0	0	0	0
	0	0	0	0	0	0	0	0	0	0	0	0	0
	−1	0	0	0	0	0	0	0	0	0	0	0	0
	0	0	0	0	0	0	0	0	0	0	0	0	0
6M	0	0	0	0	0	0	0	0	0	0	0	0	0
	0	0	0	0	0	0	0	0	0	0	0	0	0
	−1	0	0	0	0	0	0	0	0	0	0	0	−1
	0	0	0	0	0	0	0	0	0	0	0	0	0
1Y	0	0	0	0	0	0	0	0	0	0	0	0	0
	0	0	0	0	0	0	0	0	0	0	0	0	0
	−4	−1	0	0	0	0	0	0	0	0	0	−1	−3
	0	0	0	0	0	0	0	0	0	0	0	0	0
1.5Y	0	0	0	0	0	0	0	0	0	0	0	0	0
	0	0	0	0	0	0	0	0	0	0	0	0	0
	−8	−2	0	0	0	0	0	0	0	0	−1	−2	−5
	0	0	1	1	0	0	0	0	0	0	0	0	0
2Y	0	0	0	0	0	0	0	0	0	0	0	0	0
	0	0	0	0	0	0	0	0	0	0	0	0	0
	−13	−4	0	1	1	1	1	1	1	0	−1	−3	−10
	1	1	1	1	1	1	1	1	1	0	0	0	0
3Y	−1	−1	−1	−1	−1	−1	−1	−1	0	0	0	0	0
	−1	−1	−1	−1	−1	−1	−1	−1	0	0	0	0	0
	−37	−5	−2	2	2	2	2	2	1	0	−2	−6	−19
	3	3	3	3	2	2	2	2	1	1	0	0	0
5Y	6	1	0	−1	−1	−1	−1	−1	−1	−1	0	0	−1
	6	1	0	−1	−1	−1	−1	−1	−1	−1	0	0	0
	−88	−9	4	7	8	7	6	5	3	1	−5	−14	−43
	25	15	12	10	8	7	6	5	3	2	1	0	−1
10Y	92	61	40	22	13	8	4	2	1	0	0	−1	−8
	91	61	40	22	13	8	4	2	1	0	0	0	−1
	−58	54	76	69	54	42	31	23	15	4	−17	−44	−118
	286	173	120	79	56	42	31	23	16	9	4	1	−2

Table 16. CEV model ($\beta = 0.5$, $\nu = 0.4$): average per categories of strikes of errors in bps on the BS implied volatility using the 4 approximations ImpVol(AppPriceLN(3,x_{av})), AppImpVolLN(3,x_{av}), (HF) and (HLF).

	far ITM	ITM	ATM	OTM	far OTM
3M	−6.9E−3	−5.3E−3	−5.4E−3	−3.6E−3	−3.2E−3
	−4.6E−3	−5.3E−3	−5.4E−3	−3.6E−3	−2.2E−3
	−3.8E−1	−2.9E−2	1.1E−2	−2.4E−2	−1.8E−1
	−5.9E−3	8.0E−3	1.1E−2	5.4E−3	−3.0E−3
6M	−2.3E−2	−2.3E−2	−2.1E−2	−1.5E−2	−1.1E−2
	−2.1E−2	−2.3E−2	−2.1E−2	−1.5E−2	−8.3E−3
	−6.8E−1	4.2E−3	4.4E−2	−1.1E−2	−4.8E−1
	1.1E−2	4.4E−2	4.5E−2	2.8E−2	−3.4E−3
1Y	−9.9E−2	−9.5E−2	−8.2E−2	−5.3E−2	−4.2E−2
	−9.5E−2	−9.4E−2	−8.2E−2	−5.3E−2	−2.8E−2
	−2.6E+0	8.3E−2	1.8E−1	−5.3E−2	−1.8E+0
	8.7E−2	2.0E−1	1.8E−1	1.0E−1	−1.5E−2
1.5Y	−2.3E−1	−2.1E−1	−1.7E−1	−1.1E−1	−7.9E−2
	−2.2E−1	−2.1E−1	−1.7E−1	−1.1E−1	−5.2E−2
	−4.7E+0	2.0E−1	4.1E−1	−7.7E−2	−3.5E+0
	3.4E−1	5.0E−1	4.1E−1	2.4E−1	−2.5E−2
2Y	−4.0E−1	−3.6E−1	−2.9E−1	−1.8E−1	−1.5E−1
	−3.9E−1	−3.6E−1	−2.9E−1	−1.8E−1	−8.1E−2
	−8.2E+0	4.3E−1	7.6E−1	−3.7E−2	−6.7E+0
	8.1E−1	9.7E−1	7.7E−1	4.2E−1	−6.8E−2
3Y	−7.0E−1	−7.2E−1	−5.7E−1	−3.3E−1	−3.0E−1
	−6.6E−1	−7.2E−1	−5.8E−1	−3.3E−1	−1.4E−1
	−2.1E+1	7.9E−1	1.8E+0	−2.2E−1	−1.3E+1
	3.0E+0	2.6E+0	1.8E+0	8.8E−1	−1.2E−1

Table 17. CEV model ($\beta = 0.8$, $\nu = 0.25$): errors in bps on the deltas using the 6 approximations `AppDeltaLN(1,`x_0`)`, `AppDeltaLN(1,k)`, `AppDeltaLN(1,`x_{av}`)`, `AppDeltaLN(2,`x_0`)`, `AppDeltaLN(2,k)` and `AppDeltaLN(2,`x_{av}`)`.

3M	1	2	2	1	1	0	0	0	−1	−1	−2	−2	−2
	1	0	0	−1	−1	0	0	0	0	1	0	0	−1
	0	0	0	0	0	0	0	0	0	0	0	0	0
	0	0	0	0	0	0	0	0	0	0	0	0	0
	0	0	0	0	0	0	0	0	0	0	0	0	0
	0	0	0	0	0	0	0	0	0	0	0	0	0
6M	3	3	2	1	1	1	0	−1	−1	−2	−3	−4	−4
	1	−1	−1	−1	−1	0	0	0	1	1	1	0	−1
	0	0	0	0	0	0	0	0	0	0	0	0	0
	0	0	0	0	0	0	0	0	0	0	0	0	0
	0	0	0	0	0	0	0	0	0	0	0	0	0
	0	0	0	0	0	0	0	0	0	0	0	0	0
1Y	5	5	3	3	1	1	0	−1	−2	−4	−6	−8	−9
	1	−2	−3	−2	−1	0	0	0	1	2	2	1	−3
	0	0	0	1	0	0	0	0	−1	−1	0	0	0
	0	0	0	0	0	0	0	0	0	0	0	0	1
	0	0	0	0	0	0	0	0	0	0	0	0	0
	0	0	0	0	0	0	0	0	0	0	0	0	0
1.5Y	7	7	5	4	2	1	0	−2	−3	−5	−9	−13	−15
	0	−3	−4	−3	−2	0	0	1	1	3	3	1	−4
	0	0	1	1	1	0	0	−1	−1	−1	0	0	0
	1	0	0	0	0	0	0	0	0	0	0	0	1
	0	0	0	0	0	0	0	0	0	0	0	0	0
	0	0	0	0	0	0	0	0	0	0	0	0	0
2Y	9	9	7	5	3	2	0	−2	−4	−7	−12	−18	−20
	0	−4	−6	−4	−2	−1	0	1	2	4	5	0	−6
	0	0	1	1	1	1	0	−1	−1	−1	−1	0	0
	1	0	0	0	0	0	0	0	0	0	0	1	2
	1	0	0	0	0	0	0	0	0	0	0	1	0
	0	0	0	0	0	0	0	0	0	0	0	0	0
3Y	12	12	10	6	4	2	0	−3	−6	−11	−19	−28	−32
	0	−8	−8	−6	−3	−1	0	1	3	6	6	−1	−9
	0	0	1	1	1	1	0	−1	−2	−2	−1	0	1
	1	0	0	0	0	0	0	0	0	0	1	1	4
	1	1	0	0	0	0	0	0	0	0	0	1	0
	0	0	0	0	0	0	0	0	0	0	0	0	0
5Y	16	17	13	10	7	4	0	−5	−10	−18	−34	−51	−61
	−3	−13	−14	−10	−5	−2	0	1	5	10	9	−4	−15
	0	1	2	2	2	1	0	−2	−3	−3	−1	0	1
	3	1	1	0	0	0	0	0	0	1	1	3	10
	3	1	0	0	0	0	0	0	0	0	1	2	−1
	0	0	0	0	0	0	0	0	0	0	0	0	0
10Y	24	26	21	16	12	7	−1	−8	−19	−37	−76	−118	−145
	−16	−26	−26	−17	−9	−4	−1	2	8	20	13	−17	−23
	−1	1	3	4	4	2	−1	−3	−5	−6	−2	1	2
	4	2	1	1	1	1	1	1	1	2	3	8	36
	6	2	0	0	0	1	1	1	0	0	3	5	−6
	0	0	0	0	0	1	1	1	1	0	0	−1	−1

Table 18. CEV model ($\beta = 0.8$, $\nu = 0.25$): average per categories of strikes of errors in bps on the deltas using the 2 approximations `AppDeltaLN(1,`x_{av}`)` and `AppDeltaLN(2,`x_{av}`)`.

	far ITM	ITM	ATM	OTM	far OTM
3M	−3.1E−2	7.1E−2	−1.9E−3	−8.0E−2	4.0E−2
	−1.8E−3	2.4E−4	3.6E−3	4.7E−4	−2.4E−3
6M	−1.1E−2	2.3E−1	−6.0E−3	−2.3E−1	4.8E−2
	−4.6E−3	5.0E−3	1.1E−2	4.4E−3	−6.1E−3
1Y	−4.0E−2	4.5E−1	−1.9E−2	−4.5E−1	7.9E−2
	−1.4E−2	1.7E−2	3.0E−2	1.2E−2	−1.6E−2
1.5Y	−1.9E−2	6.8E−1	−1.2E−1	−6.6E−1	1.4E−1
	−2.4E−2	2.8E−2	5.3E−2	2.4E−2	−3.2E−2
2Y	−2.4E−2	8.7E−1	−4.8E−2	−9.6E−1	2.1E−1
	−3.6E−2	4.4E−2	7.9E−2	3.9E−2	−4.9E−2
3Y	3.9E−2	1.2E+0	−9.4E−2	−1.4E+0	3.6E−1
	−5.7E−2	6.7E−2	1.4E−1	6.2E−2	−9.5E−2
5Y	6.3E−2	2.0E+0	−1.9E−1	−2.3E+0	6.5E−1
	−1.2E−1	1.4E−1	2.8E−1	1.2E−1	−2.1E−1
10Y	8.9E−2	3.4E+0	−5.6E−1	−4.3E+0	1.5E+0
	−3.4E−1	2.5E−1	6.2E−1	2.3E−1	−6.1E−1

Table 19. CEV model ($\beta = 0.2$, $\nu = 0.25$): errors in bps on the deltas using the 6 approximations `AppDeltaLN(1,`x_0`)`, `AppDeltaLN(1,k)`, `AppDeltaLN(1,`x_{av}`)`, `AppDeltaLN(2,`x_0`)`, `AppDeltaLN(2,k)` and `AppDeltaLN(2,`x_{av}`)`.

3M	19	27	25	18	11	6	0	−6	−11	−17	−30	−31	−27
	9	−1	−10	−13	−9	−3	0	3	7	9	−3	−7	−6
	−1	−1	0	2	2	2	0	−2	−2	−1	0	0	0
	5	3	2	1	1	0	0	0	0	1	2	3	5
	3	3	2	0	0	0	0	0	0	0	1	0	−1
	0	0	0	0	0	0	0	0	0	0	0	0	0
6M	51	45	34	24	16	9	0	−8	−15	−31	−41	−59	−58
	−7	−29	−27	−19	−10	−3	0	4	9	17	13	−6	−11
	−2	2	4	5	5	3	0	−2	−4	−3	−2	0	1
	10	3	2	2	1	1	1	1	1	2	2	5	13
	10	3	1	1	1	1	1	1	0	1	2	3	−2
	−1	0	0	0	1	1	1	1	1	0	0	0	0
1Y	99	90	59	46	24	13	1	−10	−32	−53	−93	−117	−113
	−54	−69	−52	−37	−11	−3	1	5	19	32	12	−14	−9
	−3	3	9	11	8	5	1	−3	−7	−7	−2	1	1
	24	10	5	4	3	2	2	2	3	4	7	12	41
	28	9	2	2	2	2	2	2	1	1	6	6	−4
	−2	0	1	1	2	2	2	2	1	1	0	−1	0
1.5Y	146	126	86	70	43	16	2	−26	−38	−77	−140	−183	−180
	−123	−114	−81	−60	−23	−3	2	13	21	45	12	−27	−10
	−2	7	15	17	15	7	2	−7	−9	−10	−3	1	1
	35	16	10	8	5	4	4	4	4	7	13	24	72
	37	12	3	3	4	4	4	3	2	2	11	7	−6
	−3	0	2	3	4	4	4	3	3	1	−1	−1	−1
2Y	189	163	17	81	51	36	3	−29	−58	−102	−173	−254	−224
	−197	−164	−115	−63	−23	−11	3	14	33	58	23	−40	−4
	0	11	20	23	19	14	3	−7	−13	−13	−5	2	1
	50	24	15	11	7	6	5	6	7	11	18	40	125
	45	15	5	5	6	6	5		3	3	14	4	−3
	−4	0	3	5	6	6	5	5	4	2	−1	−2	−1
3Y	262	218	188	118	82	45	6	−33	−86	−157	−286	−393	−354
	−377	−245	−204	−93	−40	−10	6	18	49	82	−3	−54	−3
	0	24	30	36	31	20	6	−7	−19	−18	−4	3	1
	95	36	29	19	14	11	10	10	13	0	35	77	230
	51	13	9	9	11	11	10	8	5	6	26	−4	−2
	−8	2	5	9	10	10	10	9	6	3	−2	−4	−1
5Y	373	323	253	197	131	85	11	−61	−148	−258	−497	−684	−592
	−641	−409	−276	−169	−60	−19	11	34	84	121	−43	−59	0
	0	43	59	64	53	38	11	−13	−30	−28	−3	4	0
	156	62	47	39	28	23	20	20	26	38	73	170	533
	−156	−19	3	15	22	23	20	15	8	13	38	−22	0
	−27	2	13	19	22	21	20	17	12	5	−6	−6	0
10Y	390	381	325	262	06	134	18	−102	−269	−529	−1075	−1477	−1164
	−174	−486	−447	−269	−123	−36	18	56	49	175	−127	−23	0
	−71	−8	39	76	78	58	18	−21	−53	−44	0	4	0
	75	−5	3	25	32	34	38	43	57	88	189	536	1725
	−1252	−598	−237	−43	18	7	38	32	15	47	5	−21	0
	−157	−97	−49	−3	23	34	38	36	26	6	−16	−10	0

PART 2
Algorithms

Chapter 5

Discretization of backward stochastic Volterra integral equations

Christian Bender and Stanislav Pokalyuk

Department for Mathematics, Saarland University, D-66041 Saarbrücken, Germany

Abstract Backward stochastic Volterra integral equations (BSVIEs) can be applied to describe dynamic versions of coherent risk measures, allowing for time inconsistencies, see Yong [26]. We show that under certain regularity conditions the adapted M-solution of a BSVIE can be approximated by a sequence of discrete BSVIEs driven by a binary random walk. The proof relies on a representation formula for BSVIEs via systems of quasilinear partial differential equations of parabolic type.

Keywords: BSVIE; Adapted M-solution; Quasilinear PDEs of parabolic type.

1. Introduction

In the theory of stochastic differential equations (SDEs) numerical methods, allowing to solve these equations, play an important role, because in most cases it is impossible to obtain explicit solutions for SDEs.

In this paper we generalize a numerical method originally designed for backward stochastic differential equations (BSDEs) by Ma, Protter, San Martin, Torres [13], to solve backward stochastic Volterra integral equations (BSVIEs) of the form

$$Y_t = f(W;t) - \int_t^T h(s, Y_s)ds - \int_t^T Z_{t,s}dW_s, \ t \in [0,T], \tag{1}$$

numerically. Here W is a Brownian motion on $[0,T]$, the function h is called the generator, and the free term $f(\cdot;t)$ at time t may depend on the whole path of the Brownian motion up to time T. Similarly to a BSDE a solution of a BSVIE consists of a pair of processes (Y,Z). For a BSVIE Z is a two-parameter process such that $Z_{t,\cdot}$ is adapted to the Brownian filtration for almost every $t \in [0,T]$. Y is also required to be adapted to the Brownian filtration. In order to guarantee uniqueness of the solution of a BSVIE under Lipschitz assumptions, Yong [27] suggested to consider so-called M-solutions only. These solutions determine $Z_{t,s}$ for $t > s$ by a martingale condition. The precise notion of an M-solution for BSDEs will be recalled in Section 2.

The theory of BSDEs and BSVIEs is a relatively modern part of the theory of stochastic differential equations. Intensive research on BSDEs started in the 90s when well-posedness results for nonlinear BSDEs were established (see [16]) and the connection between BSDEs and partial differential equations (PDEs, for short) – in fact a generalization of the well-known Feynman-Kac formula to quasilinear PDEs – was understood (see [17], [18]).

In mathematical finance the theory of BSDEs plays an important role, when questions are focused around pricing and optimal hedging problems for contingent claims in models of financial markets; see [10] for general information. Further, BSDEs can also be used to solve utility maximization problems with backward stochastic dynamics or to describe dynamic risk measures (see [21], [19], [20], [10]).

If one considers a family of BSDEs parametrized in time, one arrives at a special type of backward stochastic Volterra integral equations. In the general form BSVIEs can however not be reduced to BSDEs (see [27]). The first works dedicated to stochastic Volterra integral equations were presented by Berger and Mizel [5] in 1980. In 2002, Lin introduced a class of nonlinear BSVIEs in [12], where he proved existence and uniqueness of the solutions to these BSVIEs under uniform Lipschitz conditions on the generator. Thereafter, Yong [25] formulated BSVIEs in a generalized form. In [27] he proved the well-posedness of adapted M-solutions for these types of BSVIEs. From a financial point of view BSVIEs can be applied to describe time-inconsistent risk measures and preferences, which are known to exist in real world, see e.g. [22] and [11]. A detailed discussion on the modeling of time-inconsistent risk measures and preferences via BSVIEs can be found in Yong [26], Wang and Shi [24] and Wang [23]. BSVIEs also appear as adjoint equations in stochastic control problems for Volterra integral equations, see e.g. Yong [25], [27].

These applications illustrate that numerical algorithms for BSVIEs are called for. There are by now several approaches to provide numerical methods to solve BSDEs, which can basically be divided into two types:

The first type is based on a four step scheme to solve general forward-backward stochastic differential equations via solutions of quasilinear parabolic PDEs proposed in 1994 by Ma, Protter and Yong [14]. In 1996 Douglas, Ma and Protter [7] and in 2006 Milstein and Tretyakov [15] developed two numerical methods using this scheme by approximating numerically the solutions of the corresponding parabolic PDEs.

The second branch of algorithms directly discretizes the BSDE. This approach was initiated by Bally [2] making use of a random time discretization and by Chevance [6] under strong regularity conditions. Convergence of the time dicretized BSDEs under standard Lipschitz assumptions and with arbitrary time grids was shown by Zhang [29]. The time discretized BSDE still requires to compute some nested conditional expectations. Several approaches have been suggested to approximate these conditional expectations, see e.g. Bender and Steiner [3] for a review. One possibility is to replace Brownian motion by a binary random walk, which is the approach followed by Ma, Protter, San Martin, Torres [13], and by Briand, Delyon and Memin [4].

To the best of our knowledge, numerical methods for BSVIEs have not yet been developed. Here, we want to close this gap and present an approximation scheme for the solutions of BSVIEs which generalizes the results from [13]. Actually, we approximate the solution of (1) by the sequence of discrete BSVIEs (DBSVIEs, for short)

$$Y_{t_i}^{(n)} = f(W^{(n)}; t_i) - \sum_{j=i}^{n-1} h(t_j, Y_{t_j}^{(n)})\Delta t_{j+1} - \sum_{j=i}^{n-1} Z_{t_i,t_j}^{(n)} \Delta W_{t_{j+1}}^{(n)} \tag{2}$$

Here, $\Delta W^{(n)}$ denotes the increment of a simple binary random walk (with piecewise linear interpolation) on an equidistant time grid $t_i = Ti/n, 0 \leq i \leq n$, which weakly approximates the Brownian motion. The notion of an M-solution for such DBSVIEs can be introduced analogously to the continuous time case and we refer the reader to Section 2 for the details. We will show that the sequence of discrete solutions $(Y^{(n)})_{n\in\mathbb{N}}$ converges weakly to the continuous process Y in the Skorokhod topology under certain Lipschitz assumptions. As a main argument we use that certain systems of quasilinear PDEs of parabolic type, to which the M-solution of (1) can be connected, are well approached by their discrete analogues.

The rest of the chapter is organized as follows:

In Section 2 we introduce the main spaces and notation. In addition, we recall the well-posedness result of Yong [27] for BSVIEs. Our main result on weak convergence of the solution of a sequence of DBSVIEs to the corresponding continuous time BSVIE is formulated in Section 3. In this section we also discuss well-posedness of DBSVIEs and construct an explicit numerical scheme for solving DBSVIEs. The proof of the main result is provided in Sections 4 and 5. In Section 4 we relate the solution of the continuous time BSVIE to systems of parabolic Cauchy problems under additional smoothness assumptions. These results can be applied to prove the main convergence theorem, when the generator and the free term are smooth and the free term depends on finitely many time points of the Brownian motion only. The proof of the main result in the general case will be given in Section 5. A priori-estimates for BSVIEs are applied to extend convergence of the finite dimensional distributions from the smooth case to the general case. The proof is then completed by a tightness argument. In Section 6 we illustrate the numerical approximation with an example. Here we also obtain empirically the speed of convergence of our algorithm.

2. Preliminaries

In this section we define some spaces needed for the study of BSVIEs, resp. DBSVIEs, and state the well-posedness result for general backward stochastic Volterra integral equations.

We first introduce some general notation.

Let $d, k, k_i, l, p, q, q_i \in \mathbb{N} = \{1, 2, \ldots\}$, for $i = 1, \ldots p$, $T \in (0, \infty)$, $R, S, \tilde{R}, \tilde{S} \in [0, T]$, with $R < S$, $\tilde{R} < \tilde{S}$, $r \in [1, \infty]$, $\alpha \in (0, 1)$ be some variables and $W = (W_t)_{t\in[0,T]}$ be a standard d-dimensional Brownian motion defined on a complete filtered probability space $(\Omega, \mathcal{F}, \mathbb{F}, \mathbb{P})$, where $\mathbb{F} = (\mathcal{F}_t)_{t\in[0,T]}$ is the natural filtration of W augmented by all the $\mathbb{P}$-null sets in $\mathcal{F}$. T is a fixed terminal time.

We use the standard notations for the following spaces:

$\mathbb{R}^k$	k-dimensional real space with the Euclidean norm $\|\|\mathbf{x}\|\|_2 := \sqrt{\mathbf{x}^*\mathbf{x}}$ and inner product $\langle \mathbf{x}, \mathbf{y}\rangle := \mathbf{x}^*\mathbf{y}$, where $[\cdot]^*$ denotes matrix/vector transposition, $\mathbf{x} := (x_1, \ldots, x_k)^*$, $\mathbf{y} := (y_1, \ldots, y_k)^* \in \mathbb{R}^k$.
$\mathbb{R}^{k\times d}$	the Hilbert space of all $(k \times d)$ real matrices with the Euclidean norm $\|\|A\|\|_2 := \sqrt{\mathrm{tr}(AA^*)}$ and inner product $\langle A, B\rangle := \mathrm{tr}(AB^*)$, $A, B \in \mathbb{R}^{k\times d}$.
$C([R, S]; \mathbb{R}^k)$	the space of all continuous functions $\varphi : [R, S] \to \mathbb{R}^k$ with the norm $\|\|\varphi\|\|_\infty := \sup_{t\in[R,S]} \|\|\varphi(t)\|\|_2$.
$C^\infty(X_1 \times \ldots \times X_p; \mathbb{R}^k)$	the space of all infinitely smooth functions $\varphi : X_1 \times \ldots \times X_p \to \mathbb{R}^k$, such that $X_i = \mathbb{R}$ or $X_i = [R, S]$, $i = 1, \ldots, p$.
$C_b^\infty(X_1 \times \ldots \times X_p; \mathbb{R}^k)$	the space of those $\varphi \in C^\infty(X_1 \times \ldots \times X_p; \mathbb{R}^k)$ such that φ and all its derivatives are uniformly bounded.

Further, we define some spaces for random variables and stochastic processes:

$\mathbb{L}^2_{\mathcal{F}_S}(\Omega; \mathbb{R}^k)$	the space of $\mathcal{F}_S$-measurable random variables $X : \Omega \to \mathbb{R}^k$ such that $\mathbb{E}[\|\|X\|\|_2^2] < \infty$.
$L^2_{\mathcal{F}_S}(R, S; \mathbb{R}^k)$	the space of $\mathcal{F}_S \otimes \mathcal{B}([R, S])$-measurable processes $Y : \Omega \times [R, S] \to \mathbb{R}^k$ such that $\mathbb{E}\left[\int_R^S \|\|Y_t\|\|_2^2 dt\right] < \infty$.

$L^2_{\mathbb{F}}(R,S;\mathbb{R}^k)$ the space of $\mathcal{F}_S \otimes \mathcal{B}([R,S])$-measurable processes $Y : \Omega \times [R,S] \to \mathbb{R}^k$ such that Y is $\mathbb{F}$-adapted and $\mathbb{E}\Big[\int_R^S ||Y_t||_2^2 dt\Big] < \infty$. We define $||Y||_{L^2_{\mathbb{F}}(R,S;\mathbb{R}^k)} := \Big(\mathbb{E}\Big[\int_R^S ||Y_t||_2^2 dt\Big]\Big)^{\frac{1}{2}}$.

$L^2(R,S;L^2_{\mathbb{F}}(\tilde{R},\tilde{S};\mathbb{R}^{k\times d}))$ the space of all processes $Z : \Omega \times [R,S] \times [\tilde{R},\tilde{S}] \to \mathbb{R}^{k\times d}$ such that $Z_{t,\cdot} \in L^2_{\mathbb{F}}(\tilde{R},\tilde{S};\mathbb{R}^{k\times d})$ for a.e. $t \in [R,S]$ and $\mathbb{E}\Big[\int_R^S \int_{\tilde{R}}^{\tilde{S}} ||Z_{t,s}||_2^2 ds dt\Big] < \infty$. We define $||Z||_{L^2(R,S;L^2_{\mathbb{F}}(\tilde{R},\tilde{S};\mathbb{R}^{k\times d}))} := \Big(\mathbb{E}\Big[\int_R^S \int_{\tilde{R}}^{\tilde{S}} ||Z_{t,s}||_2^2 ds dt\Big]\Big)^{\frac{1}{2}}$.

$L^{2,\infty}_{\mathbb{F}}(R,S;\mathbb{R}^k)$ the space of $\mathcal{F}_S \otimes \mathcal{B}([R,S])$-measurable processes $Y : \Omega \times [R,S] \to \mathbb{R}^k$ such that Y is $\mathbb{F}$-adapted and $\mathbb{E}[\sup_{t\in[R,S]} ||Y_t||_2^2] < \infty$. We define $||Y||_{L^{2,\infty}_{\mathbb{F}}(R,S;\mathbb{R}^k)} := \Big(\mathbb{E}[\sup_{t\in[R,S]} ||Y_t||_2^2]\Big)^{\frac{1}{2}}$.

$\mathcal{H}^2[R,S]$ $L^2_{\mathbb{F}}(R,S;\mathbb{R}^k) \times L^2(R,S;L^2_{\mathbb{F}}(R,S;\mathbb{R}^{k\times d}))$.

To state the well-posedness of DBSVIEs (2) (see Theorem 3.4 in Section 3) we need to define some discrete time analogues of these spaces. Namely, given a discretization $0 = t_0 < t_1 < \ldots < t_n = T$ of the time interval $[0,T]$ with time-step $\frac{T}{n}$ ($t_i := \frac{iT}{n}$, $i = 0,\ldots,n$, $n \in \mathbb{N}$), replace the d-dimensional Brownian motion W_s in (1) by a simple symmetric random walk $W_s^{(n)}$, whose increments are $\sqrt{T/n}$ and $-\sqrt{T/n}$; i.e. if the number ns/T is an integer

$$W_s^{(n)} := \begin{pmatrix} W_s^{1,(n)} \\ \vdots \\ W_s^{d,(n)} \end{pmatrix} := \sqrt{\frac{T}{n}} \begin{pmatrix} \sum_{j=1}^{ns/T} \varepsilon_j^1 \\ \vdots \\ \sum_{j=1}^{ns/T} \varepsilon_j^d \end{pmatrix}$$

Here, $\{\varepsilon_j^d\}$ is an i.i.d. {-1,1}-symmetric sequence. If ns/T is not an integer then $W_s^{(n)}$ is defined by linear interpolation between the values of $W_{s'}^{(n)}$ and $W_{s''}^{(n)}$ for which ns'/T and ns''/T are integers and the nearest points to the left and right of ns/T.

Letting $\mathbb{F}^{(n)} := (\mathcal{F}_{t_i}^{(n)})_{i=0,\ldots,n}$ be the natural filtration of $W^{(n)}$, we denote the following discrete spaces for $i_0, i_1 \in \{0,\ldots,n\}$:

$L^2_{\mathbb{F}^{(n)}}(t_{i_0}, t_{i_1}; \mathbb{R}^k)$ the space of processes $Y^{(n)} : \Omega \times \{t_{i_0}, \dots, t_{i_1}\} \to \mathbb{R}^k$ such that $Y^{(n)}$ is $\mathbb{F}^{(n)}$-adapted. We define $||Y^{(n)}||_{L^2_{\mathbb{F}^{(n)}}(t_{i_0}, t_{i_1}; \mathbb{R}^k)} := \left(\mathbb{E}\left[\frac{T}{n}\sum_{i=i_0}^{i_1-1} ||Y^{(n)}_{t_i}||_2^2\right]\right)^{\frac{1}{2}}$.

$L^{2,(n)}(t_{i_0}, t_{i_1}; \mathbb{R}^{k\times d})$ the space of all processes $Z^{(n)} : \Omega \times \{t_{i_0}, \dots, t_{i_1}\}^2 \to \mathbb{R}^{k\times d}$ such that $Z^{(n)}_{t_i,\cdot}$ is $(\mathcal{F}^{(n)}_{t_j})_{j=i_0,\dots,i_1}$-adapted for all t_i, $i = i_0, \dots, i_1$. We define $||Z^{(n)}||_{L^{2,(n)}(t_{i_0}, t_{i_1}; \mathbb{R}^{k\times d})} := \left(\mathbb{E}\left[\left(\frac{T}{n}\right)^2 \sum_{i=i_0}^{i_1-1}\sum_{j=i_0}^{i_1-1} ||Z^{(n)}_{t_i,t_j}||_2^2\right]\right)^{\frac{1}{2}}$.

$L^{2,\infty}_{\mathbb{F}^{(n)}}(t_{i_0}, t_{i_1}; \mathbb{R}^k)$ the space of processes $Y^{(n)} : \Omega \times \{t_{i_0}, \dots, t_{i_1}\} \to \mathbb{R}^k$ such that $Y^{(n)}$ is $\mathbb{F}^{(n)}$-adapted. We define $||Y^{(n)}||_{L^{2,\infty}_{\mathbb{F}^{(n)}}(t_{i_0}, t_{i_1}; \mathbb{R}^k)} := \left(\mathbb{E}\left[\sup_{i\in\{i_0,\dots,i_1\}} ||Y^{(n)}_{t_i}||_2^2\right]\right)^{\frac{1}{2}}$.

$\mathcal{H}^{2,(n)}[t_{i_0}, t_{i_1}]$ $L^2_{\mathbb{F}^{(n)}}(t_{i_0}, t_{i_1}; \mathbb{R}^k) \times L^{2,(n)}(t_{i_0}, t_{i_1}; \mathbb{R}^{k\times d})$.

Before we state the well-posedness theorem for solutions of BSVIEs, we define such equations in the general case:

Definition 2.1. For $f : C([0,T];\mathbb{R}^d) \times [0,T] \to \mathbb{R}^k$ and $h : [0,T] \times \mathbb{R}^k \times \mathbb{R}^{k\times d} \times \mathbb{R}^{k\times d} \times [0,T] \to \mathbb{R}^k$ we call a stochastic integral equation of Itô's type of the form

$$Y_t = f(W;t) - \int_t^T h(s, Y_s, Z_{t,s}, Z_{s,t}; t)ds - \int_t^T Z_{t,s}dW_s, \ t \in [0,T], \tag{3}$$

a *backward stochastic Volterra integral equation.* The process f is called the *free term* and the function h the *generator.*

Note, that equation (3) cannot be reduced to a BSDE in general because the free term f or the generator h may depend on t.

Equation (3) has in general infinitely many adapted solutions. To achieve uniqueness, as pointed out in [27], we consider only adapted M-solutions. Such solutions of the BSVIE fulfill a martingale representation property, which determines the process $Z_{t,s}$ for $t \geq s$:

Definition 2.2. Let $S \in [0,T)$. A pair $(Y,Z) \in \mathcal{H}^2[S,T]$ is called an *adapted M-solution* of the BSVIE (3) on $[S,T]$ if (3) holds in the usual Itô's sense for almost all $t \in [S,T]$ and Y_t has the following martingale representation:

$$Y_t = \mathbb{E}[Y_t|\mathcal{F}_S] + \int_S^t Z_{t,s}dW_s, \quad \text{a.e. } t \in [S,T]. \tag{4}$$

Remark 2.3. In general the generator h can also depend on $\omega \in \Omega$ in a non-anticipative way, but throughout this chapter we consider only generators which are independent on ω.

To obtain the well-posedness for the BSVIE (3) we impose the following assumptions for the free term f and generator h:

(V1′) $f : C([0,T];\mathbb{R}^d) \times [0,T] \to \mathbb{R}^k$ is measurable and satisfies

$$\mathbb{E}\left[\int_0^T \|f(W_\cdot;t)\|_2^2 dt\right] < \infty.$$

(V2′) $h : [0,T] \times \mathbb{R}^k \times \mathbb{R}^{k\times d} \times \mathbb{R}^{k\times d} \times [0,T] \to \mathbb{R}^k$ is measurable and satisfies the following Lipschitz condition: There is a constant $L' > 0$ such that

$$\begin{aligned}&||h(s,y_1,z_1,\bar{z}_1;t) - h(s,y_2,z_2,\bar{z}_2;t)||_2 \\ &\leq L'(||y_1 - y_2||_2 + ||z_1 - z_2||_2 + ||\bar{z}_1 - \bar{z}_2||_2)\end{aligned}$$

holds for all $(t,s) \in \{(t,s) \in [0,T]^2 : \ 0 \leq t < s \leq T\}$, $y_i \in \mathbb{R}^k$, $z_i, \bar{z}_i \in \mathbb{R}^{k\times d}$, $i = 1,2$.

The following theorem due to Yong [27] is based on an application of the contraction mapping theorem.

Theorem 2.4. *Let* **(V1′)** *and* **(V2′)** *hold. Then there exists a unique adapted M-solution* $(Y,Z) \in \mathcal{H}^2[0,T]$ *which solves the BSVIE* (3).

3. Main result

We now consider a special case of (3), namely, Equation (1). For this equation, conditions **(V1′)** and **(V2′)** become:

(V1) $f : C([0,T];\mathbb{R}^d) \times [0,T] \to \mathbb{R}^k$ is measurable and satisfies

$$\mathbb{E}\left[\int_0^T \|f(W_\cdot;t)\|_2^2 dt\right] < \infty.$$

(V2) $h : [0,T] \times \mathbb{R}^k \to \mathbb{R}^k$ is measurable and satisfies the following Lipschitz condition: There is a constant $L > 0$ such that

$$||h(s,y_1) - h(s,y_2)||_2 \leq L||y_1 - y_2||_2$$

holds for all $s \in [0,T]$, $y_i \in \mathbb{R}^k$, $i = 1,2$.

We construct a numerical method to approximate the M-solution of (1). In [13] a numerical method is presented to approximate usual BSDEs, where the solutions are related with solutions of quasilinear PDEs of parabolic type. Our aim is to generalize this method to BSVIEs.

The discretization of the BSVIEs (1) is based on replacing the d-dimensional Brownian motion W_s by a simple symmetric random walk $W_s^{(n)}$ (see Section 2).

We introduce the following discretized version of BSVIE (1):

Definition 3.1. For $f : C([0,T];\mathbb{R}^d) \times [0,T] \to \mathbb{R}^k$, $h : [0,T] \times \mathbb{R}^k \to \mathbb{R}^k$ and the discretization $0 = t_0 < t_1 < \ldots < t_n = T$ of the time interval $[0,T]$ with time-step $\frac{T}{n}$ ($t_i := \frac{iT}{n}$, $i = 0, \ldots, n$, $n \in \mathbb{N}$), we call an equation of the form (2) a *discrete backward stochastic Volterra integral equation*, where

$$Y_{t_i}^{(n)} := \begin{pmatrix} Y_{t_i}^{1,(n)} \\ \vdots \\ Y_{t_i}^{k,(n)} \end{pmatrix}, \quad \sum_{j=i}^{n-1} Z_{t_i,t_j}^{(n)} \Delta W_{t_{j+1}}^{(n)} := \begin{pmatrix} \sum_{J=1}^{d} \sum_{j=i}^{n-1} Z_{t_i,t_j}^{1J,(n)} \Delta W_{t_{j+1}}^{J,(n)} \\ \vdots \\ \sum_{J=1}^{d} \sum_{j=i}^{n-1} Z_{t_i,t_j}^{kJ,(n)} \Delta W_{t_{j+1}}^{J,(n)} \end{pmatrix}$$

with $\Delta t_{j+1} := t_{j+1} - t_j = \frac{T}{n}$.

Analogously to (3) also equation (2) can be solved uniquely by a pair $(Y^{(n)}, Z^{(n)})$ only if the pair $(Y^{(n)}, Z^{(n)})$ fulfills a "discrete" martingale representation property:

Definition 3.2. Let $\mathcal{S} \in \{0, 1, \ldots, i-1\}$. A pair $(Y^{(n)}, Z^{(n)}) \in \mathcal{H}^{2,(n)}[0,T]$ is called an *adapted M-solution* of the DBSVIE (2), if $(Y^{(n)}, Z^{(n)})$ solves (2) and $Y_{t_i}^{(n)}$ has the following discrete martingale representation:

$$Y_{t_i}^{(n)} = \mathbb{E}\big[Y_{t_i}^{(n)} \big| \mathcal{F}_{t_\mathcal{S}}^{(n)}\big] + \sum_{j=\mathcal{S}}^{i-1} Z_{t_i,t_j}^{(n)} \Delta W_{t_{j+1}}^{(n)}. \tag{5}$$

The following lemma represents the Z-part of a DBSVIE in terms of the Y-part of the solution. The result is similar to the BSDE case. Therefore the proof is postponed to the appendix.

Lemma 3.3. *Suppose that the DBSVIE has an adapted M-solution $(Y^{(n)}, Z^{(n)})$. Then $Z^{(n)}$ can be expressed in forms of $Y^{(n)}$ as follows:*

$$Z_{t_i,t_\mathcal{S}}^{(n)} = \begin{cases} \dfrac{1}{\Delta t_{\mathcal{S}+1}} \mathbb{E}\big[Y_{t_i}^{(n)} (\Delta W_{t_{\mathcal{S}+1}}^{(n)})^* \big| \mathcal{F}_{t_\mathcal{S}}^{(n)}\big], & \mathcal{S} < i \\ \dfrac{1}{\Delta t_{\mathcal{S}+1}} \mathbb{E}\Big[\Big(f(W^{(n)}; t_i) - \sum_{j=i}^{n-1} h(t_j, Y_{t_j}^{(n)}) \Delta t_{j+1}\Big) (\Delta W_{t_{\mathcal{S}+1}}^{(n)})^* \,\Big|\, \mathcal{F}_{t_\mathcal{S}}^{(n)}\Big], & \mathcal{S} \geq i. \end{cases}$$

We can now state a well-posedness result for BSVIEs.

Theorem 3.4. *Let* (**V1**) and (**V2**) *hold. Then for n large enough (depending on the Lipschitz constant L) there exists a unique adapted M-solution $(Y^{(n)}, Z^{(n)}) \in \mathcal{H}^{2,(n)}[0,T]$ which solves the discrete BSVIE (2).*

Proof. The proof is analogous to the continuous case and we therefore omit the details. It also relies on the contraction mapping theorem. □

The main result of this paper is to show the weak convergence of the sequence $(Y^{(n)})_{n\in\mathbb{N}}$ to Y. For this purpose we need the following additional assumptions:

(V3) The free term f is a Lipschitz function, i.e. there exists a constant $K > 0$ such that

$$||f(g_1;t_1) - f(g_2;t_2)||_2 \leq K(||g_1 - g_2||_\infty + |t_1 - t_2|). \tag{6}$$

holds for all $g_i \in C([0,T];\mathbb{R}^d)$ and $t_i \in [0,T]$, $i = 1, 2$.

(V4) The free term f and the generator h are functions bounded by a constant $D > 0$, and h is continuous.

Remark 3.5. In fact, we do not have to require that h is bounded, since one can show with analogous arguments as in Lemma *3.1* of [*13*] that the processes Y and $Y^{(n)}$ are bounded under the assumptions ***(V1)-(V3)*** with a bounded free term f.

Theorem 3.6. (**Main Convergence Theorem**) *Suppose that the assumptions* (**V1**)-(**V4**) *are fulfilled for BSVIE* (1). *Denote by* $(Y^{(n)}, Z^{(n)})_{n\in\mathbb{N}}$ *the sequence of discrete M-solutions of* (2) *and by* (Y, Z) *the adapted M-solution of* (1). *Then the sequence* $(Y^{(n)})_{n\in\mathbb{N}}$ *converges weakly in the Skorokhod topology to* Y *if it is piecewise constant interpolated between the points* $Y_{t_i}^{(n)}$ *and* $Y_{t_{i+1}}^{(n)}$ *for all* $i = 0, \ldots, n-1$.

The sequence of processes $(Y^{(n)}, Z^{(n)})_{n\in\mathbb{N}}$ can be approximated by a sequence of processes $(\hat{Y}^{(n)}, \hat{Z}^{(n)})_{n\in\mathbb{N}}$ given by explicit expressions. To understand this, note that $Y_{t_i}^{(n)}$ from (2) can be written as

$$\begin{aligned} Y_{t_i}^{(n)} &= Y_{t_{i+1}}^{(n)} + f(W^{(n)};t_i) - f(W^{(n)};t_{i+1}) - h(t_i, Y_{t_i}^{(n)})\Delta t_{i+1} \\ &\quad + \sum_{j=i+1}^{n-1} Z_{t_{i+1},t_j}^{(n)} \Delta W_{t_{j+1}}^{(n)} - \sum_{j=i}^{n-1} Z_{t_i,t_j}^{(n)} \Delta W_{t_{j+1}}^{(n)}. \end{aligned}$$

Taking the conditional expectation of $Y_{t_i}^{(n)}$ given $\mathcal{F}_{t_i}^{(n)}$ one obtains

$$Y_{t_i}^{(n)} = \mathbb{E}\big[Y_{t_{i+1}}^{(n)} + f(W^{(n)};t_i) - f(W^{(n)};t_{i+1})\big|\mathcal{F}_{t_i}^{(n)}\big] - h(t_i, Y_{t_i}^{(n)})\Delta t_{i+1}.$$

In addition, denoting by

$$X_i^0 := \mathbb{E}\big[Y_{t_{i+1}}^{(n)} + f(W^{(n)};t_i) - f(W^{(n)};t_{i+1})\big|\mathcal{F}_{t_i}^{(n)}\big]$$

the map $\Theta_i^{(n)} : \mathbb{L}^2_{\mathcal{F}_{t_i}^{(n)}}(\Omega;\mathbb{R}^k) \to \mathbb{L}^2_{\mathcal{F}_{t_i}^{(n)}}(\Omega;\mathbb{R}^k)$ which maps

$$V \mapsto X_i^0 - h(t_i, V)\Delta t_{i+1}$$

is a contraction for large n with fixed-point $Y_{t_i}^{(n)}$, since for two different random variables V and $\tilde{V}$

$$\begin{aligned} ||\Theta_i^{(n)}(V) - \Theta_i^{(n)}(\tilde{V})||_2 &\leq ||h(t_i, V)\Delta t_{i+1} - h(t_i, \tilde{V})\Delta t_{i+1}||_2 \\ &\leq \frac{LT}{n}||V - \tilde{V}||_2 \quad \mathbb{P}\text{-a.s.}, \end{aligned}$$

holds, where $\frac{LT}{n} < 1$ for large n. Furthermore,

$$\begin{aligned}
&||Y_{t_i}^{(n)} - (X_i^0 - h(t_i, X_i^0)\Delta t_{i+1})||_2 \\
&= ||X_i^0 - h(t_i, Y_{t_i}^{(n)})\Delta t_{i+1} - X_i^0 + h(t_i, X_i^0)\Delta t_{i+1}||_2 \\
&\leq \frac{LT}{n}||Y_{t_i}^{(n)} - X_i^0||_2 \leq \frac{DLT^2}{n^2} \quad \mathbb{P}\text{-a.s.}
\end{aligned}$$

This motivates us to propose the following explicit numerical scheme for the BSVIE (1):

$$\begin{aligned}
\hat{Y}_{t_n}^{(n)} &= f(W^{(n)}; t_n), \quad \hat{Z}_{t_n,t_n}^{(n)} = 0, \\
\hat{X}_{t_i}^{(n)} &= \mathbb{E}\big[\hat{Y}_{t_{i+1}}^{(n)} + f(W^{(n)}; t_i) - f(W^{(n)}; t_{i+1})\big|\mathcal{F}_{t_i}^{(n)}\big], \\
\hat{Y}_{t_i}^{(n)} &= \hat{X}_{t_i}^{(n)} - h(t_i, \hat{X}_{t_i}^{(n)})\Delta t_{i+1}, \\
\hat{Z}_{t_i,t_\mathcal{S}}^{(n)} &= \begin{cases} \dfrac{1}{\Delta t_{\mathcal{S}+1}}\mathbb{E}\big[\hat{Y}_{t_i}^{(n)}(\Delta W_{t_{\mathcal{S}+1}}^{(n)})^*\big|\mathcal{F}_{t_\mathcal{S}}^{(n)}\big], & \mathcal{S} < i \\ \dfrac{1}{\Delta t_{\mathcal{S}+1}}\mathbb{E}\left[\left(f(W^{(n)}; t_i) - \sum_{j=i}^{n-1} h(t_j, \hat{Y}_{t_j}^{(n)})\Delta t_{j+1}\right)(\Delta W_{t_{\mathcal{S}+1}}^{(n)})^* \,\middle|\, \mathcal{F}_{t_\mathcal{S}}^{(n)}\right], & \mathcal{S} \geq i. \end{cases}
\end{aligned}$$

With analogous arguments as in [13], we can estimate the error between $Y^{(n)}$ and $\hat{Y}^{(n)}$ as well as between $Z^{(n)}$ and $\hat{Z}^{(n)}$. Namely,

$$\begin{aligned}
||Y_{t_i}^{(n)} - \hat{Y}_{t_i}^{(n)}||_2 &= ||\mathbb{E}\big[Y_{t_{i+1}}^{(n)} - \hat{Y}_{t_{i+1}}^{(n)}\big|\mathcal{F}_{t_i}^{(n)}\big] - (h(t_i, Y_{t_i}^{(n)}) - h(t_i, \hat{X}_{t_i}^{(n)}))\Delta t_{i+1}||_2 \\
&\leq ||\mathbb{E}\big[Y_{t_{i+1}}^{(n)} - \hat{Y}_{t_{i+1}}^{(n)}\big|\mathcal{F}_{t_i}^{(n)}\big]||_2 + ||h(t_i, Y_{t_i}^{(n)}) - h(t_i, \hat{Y}_{t_i}^{(n)})||_2\Delta t_{i+1} \\
&\quad + ||h(t_i, \hat{Y}_{t_i}^{(n)}) - h(t_i, \hat{X}_{t_i}^{(n)})||_2\Delta t_{i+1} \leq \frac{D(e^{2L}-1)}{n} \quad \mathbb{P}\text{-a.s.}
\end{aligned}$$

and

$$\begin{aligned}
&||Z_{t_i,t_\mathcal{S}}^{(n)} - \hat{Z}_{t_i,t_\mathcal{S}}^{(n)}||_2 \\
&= \begin{cases} \dfrac{1}{\Delta t_{\mathcal{S}+1}}||\mathbb{E}\big[(Y_{t_i}^{(n)} - \hat{Y}_{t_i}^{(n)})(\Delta W_{t_{\mathcal{S}+1}}^{(n)})^*\big|\mathcal{F}_{t_\mathcal{S}}^{(n)}\big]||_2 & ,\mathcal{S} < i \\ \dfrac{1}{\Delta t_{\mathcal{S}+1}}\left|\left|\mathbb{E}\left[\left(\sum_{j=i}^{n-1}(h(t_j, \hat{Y}_{t_j}^{(n)}) - h(t_j, Y_{t_j}^{(n)}))\Delta t_{j+1}\right)(\Delta W_{t_{\mathcal{S}+1}}^{(n)})^* \,\middle|\, \mathcal{F}_{t_\mathcal{S}}^{(n)}\right]\right|\right|_2 & ,\mathcal{S} \geq i \end{cases} \\
&\leq \begin{cases} \dfrac{\frac{1}{\sqrt{T}}D(e^{2L}-1)}{\sqrt{n}} & ,\mathcal{S} < i \\ \dfrac{\sqrt{T}LD(e^{2L}-1)}{\sqrt{n}} & ,\mathcal{S} \geq i. \end{cases}
\end{aligned}$$

Hence, we obtain the following corollary of Theorem 3.6:

Corollary 3.7. *The assertion of Theorem* 3.6 *also holds true if we consider the sequence* $(\hat{Y}^{(n)})_{n\in\mathbb{N}}$ *instead of the sequence* $(Y^{(n)})_{n\in\mathbb{N}}$, .

The following two sections are devoted to the prove of Theorem 3.6. In Section 4 we will prove a representation formula for the solution of the continuous BSVIE in terms of systems of parabolic Cauchy problems. This formula is valid under the assumptions of Theorem 3, when, additionally, the free term f depends on the values of the Brownian motion at finitely many time points only and the coefficient functions f and g are infinitely smooth. This result is the cornerstone to prove Theorem 3.6 in the 'smooth' case. Section 5 contains some a-priori estimates for continuous and discrete BSVIEs which are required to transfer convergence of the finite dimensional distributions from the smooth case to the general case. Moreover, tightness is shown in this Section, which finally completes the of Theorem 3.6.

4. Connection to parabolic Cauchy problems

Generalizing some ideas of [13] from BSDEs to BSVIEs, a method of constructing the M-solutions of (1) via solutions of quasilinear PDEs of parabolic type will be presented in Section 4.1, if the free term depends only on finitely many points of the trajectory of the Brownian motion and the free term and the generator are infinitely smooth. We then discuss the problem of smoothness for the solutions of such PDEs of parabolic type in Section 4.2. Finally, in Section 4.3, we show the weak convergence result of Theorem 3.6 under the additional smoothness assumptions of this section.

Let us first fix some notation. Denote by $\tau := (\tau_0, \dots, \tau_m)$ a partition of $[0,T]$ of length m, i.e. $0 = \tau_0 < \tau_1 < \dots < \tau_m = T$, and consider the BSVIE

$$Y_t^{[m]} = f^{[m]}(W_{\tau_1}, \dots, W_{\tau_m}; t) - \int_t^T h^{[m]}(s, Y_s^{[m]})ds - \int_t^T Z_{t,s}^{[m]} dW_s, \; t \in [0,T], \tag{7}$$

where $f^{[m]} : \mathbb{R}^l \times [0,T] \to \mathbb{R}^k$ with $l := dm$ and $h^{[m]} : [0,T] \times \mathbb{R}^k \to \mathbb{R}^k$ are given maps.

Similarly, for a given partition $\tau^{(n)} := (\tau_0^{(n)}, \dots, \tau_m^{(n)})$ of $[0,T]$ of length m with $\tau_\lambda^{(n)} := \frac{\lfloor \tau_\lambda n \rfloor}{n}$, $\lambda = 0, \dots, m$, we consider the DBSVIE

$$\begin{aligned} Y_{t_i}^{(n),[m]} =& f^{[m]}(W_{\tau_1^{(n)}}^{(n)}, \dots, W_{\tau_m^{(n)}}^{(n)}; t_i) - \frac{T}{n} \sum_{j=i}^{n-1} h^{[m]}(t_j, Y_{t_j}^{(n),[m]}) \\ & - \sum_{j=i}^{n-1} Z_{t_i,t_j}^{(n),[m]} \Delta W_{t_{j+1}}^{(n)}. \end{aligned} \tag{8}$$

For simplicity, in the argument of the free terms we sometimes write $\boldsymbol{W}_{\tau_m} := (W_{\tau_1}, \dots, W_{\tau_m})$ and $\boldsymbol{W}_{\tau_m^{(n)}}^{(n)} := (W_{\tau_1^{(n)}}^{(n)}, \dots, W_{\tau_m^{(n)}}^{(n)})$.

To obtain weak convergence of the solution $Y^{(n),[m]}$ from (8) to the solution $Y^{[m]}$ from (7) we introduce the following assumptions for the free term $f^{[m]}$ and generator $h^{[m]}$:

(M) $f^{[m]} \in C_b^\infty(\mathbb{R}^l \times [0,T]; \mathbb{R}^k)$ and $h^{[m]} \in C_b^\infty([0,T] \times \mathbb{R}^k; \mathbb{R}^k)$.

The following result will be proved in Section 4.3:

Lemma 4.1. *Assume that in* (7) *and* (8) *the assumption* (**M**) *is fulfilled. Let* $m \in \mathbb{N}$, $(Y^{(n),[m]}, Z^{(n),[m]})_{n\in\mathbb{N}}$ *be the sequence of discrete M-solutions of* (8) *and* $(Y^{[m]}, Z^{[m]})$ *be the adapted M-solution of* (7). *Then the sequence* $(Y^{(n),[m]})_{n\in\mathbb{N}}$ *(defined via piecewise constant interpolation) converges weakly to* $Y^{[m]}$ *in the Skorokhod topology.*

4.1. *Construction of the solutions of BSVIEs via PDEs*

For the sake of simplicity we first take $T = d = k = 1$. Further, suppose that the free term $f^{[m]}$ in (7) depends only on two points of the trajectory of a Brownian motion, i.e. $\tau_2 = 1$ (and so $m = 2$), and the assumptions on $f^{[2]}$ and $h^{[2]}$ from (7) for the uniqueness and existence of the adapted M-solution $(Y^{[2]}, Z^{[2]})$ hold.

Our purpose is to show that the M-solution of (7) can be represented via solutions of six systems of PDEs of parabolic type, if $f^{[2]}$ and $h^{[2]}$ are sufficiently smooth. Each system of PDEs is defined on one of the following subsets of $[0,1]^2$ (see Figure 1):

$$\begin{aligned}
S^{2,2} &:= \{(t,s) \in [0,1]^2 |\ t, s \geq \tau_1, t \leq s\},\\
S^{1,2} &:= \{(t,s) \in [0,1]^2 |\ t < \tau_1, s \geq \tau_1\},\\
S^{1,1} &:= \{(t,s) \in [0,1]^2 |\ t < \tau_1, s \leq \tau_1, t \leq s\},\\
S_{2,2} &:= \{(t,s) \in [0,1]^2 |\ t, s \geq \tau_1, t \geq s\},\\
S_{2,1} &:= \{(t,s) \in [0,1]^2 |\ t \geq \tau_1, s \leq \tau_1\},\\
S_{1,1} &:= \{(t,s) \in [0,1]^2 |\ t < \tau_1, s \leq \tau_1, t \geq s\}.
\end{aligned}$$

To construct PDEs for the upper triangle of $[0,1]^2$ we consider (t,s) in all subsets of this domain.

Suppose that $(t,s) \in S^{2,2}$. Recall that according to Theorem 2.4, Equation (7) admits an adapted M-solution $(Y_t^{[2]}, Z_{t,s}^{[2]})$. We consider the ansatz

$$Y_t^{[2]} = u^{2,2}(t, W_{\tau_1}, W_t; t), \ Z_{t,s}^{[2]} = \frac{\partial u^{2,2}}{\partial x_2}(s, W_{\tau_1}, W_s; t)$$

for some deterministic function $u^{2,2}$. If the partial derivatives

$$\frac{\partial u^{2,2}}{\partial s}(s, x_1, x_2; t), \ \frac{\partial^2 u^{2,2}}{\partial x_2^2}(s, x_1, x_2; t)$$

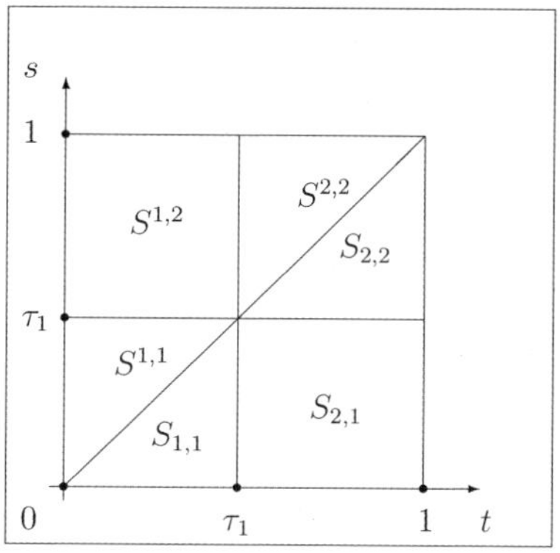

Figure 1. The sets $S^{2,2}$,...,$S_{1,1}$ in $[0,1]^2$.

exist and are continuous we can apply Itô's formula to $u^{2,2}$ and the Brownian motion, and for $r \in [t,1]$ we obtain the following equation:

$$\begin{aligned}u^{2,2}(1, W_{\tau_1}, W_1; t) &= u^{2,2}(r, W_{\tau_1}, W_r; t)\\&+ \int_r^1 \left[\frac{\partial u^{2,2}}{\partial s}(s, W_{\tau_1}, W_s; t) + \frac{1}{2}\frac{\partial^2 u^{2,2}}{\partial x_2^2}(s, W_{\tau_1}, W_s; t)\right] ds\\&+ \int_r^1 \frac{\partial u^{2,2}}{\partial x_2}(s, W_{\tau_1}, W_s; t) dW_s.\end{aligned}$$

Comparing (7) and the previous equation, we see that $u^{2,2}$ is required to solve

$$\begin{cases}\dfrac{\partial u^{2,2}}{\partial s}(s, x_1, x_2; t) + \dfrac{1}{2}\dfrac{\partial^2 u^{2,2}}{\partial x_2^2}(s, x_1, x_2; t) = h^{[2]}(s, u^{2,2}(s, x_1, x_2; s))\\ u^{2,2}(1, x_1, x_2; t) = f^{[2]}(x_1, x_2; t)\end{cases} \tag{9}$$

Theorem 4.2 below states that under condition **(M)** on the free term $f^{[2]}$ and the generator $h^{[2]}$ a unique bounded solution $u^{2,2}$ to (9) exists and is regular enough to guarantee that our application of Itô's formula was correct. Therefore, $Y_t^{[2]}$ and $Z_{t,s}^{[2]}$ can really be written in terms of $u^{2,2}$ and the Brownian motion.

In a similar way we want to find corresponding systems of PDEs for the other subsets on $[0,1]^2$. So in the following we again suppose that $f^{[2]}$ and $h^{[2]}$ fulfill **(M)**. Then the obtained solutions of the PDE systems turn out to be sufficiently differentiable in order to apply Itô's formula thanks to Theorem 4.2 below.

For $t \in [0, \tau_1)$ we write (7) as follows:

$$Y_t^{[2]} = \psi_t^{\tau_1} - \int_t^{\tau_1} h^{[2]}(s, Y_s^{[2]}) ds - \int_t^{\tau_1} Z_{t,s}^{[2]} dW_s, \tag{10}$$

where

$$\psi_t^{\tau_1} = f^{[2]}(W_{\tau_1}, W_1; t) - \int_{\tau_1}^{1} h^{[2]}(s, Y_s^{[2]})ds - \int_{\tau_1}^{1} Z_{t,s}^{[2]} dW_s, \tag{11}$$

which is $\mathcal{F}_{\tau_1}$-measurable for almost all $t \in [0, \tau_1)$.

If $(t, s) \in S^{1,2}$ we set

$$\psi_t^{\tau_1} = u^{1,2}(\tau_1, W_{\tau_1}, W_{\tau_1}; t), \ Z_{t,s}^{[2]} = \frac{\partial u^{1,2}}{\partial x_2}(s, W_{\tau_1}, W_s; t),$$

and, applying Itô's formula, we obtain with equation (11) the system

$$\begin{cases} \dfrac{\partial u^{1,2}}{\partial s}(s, x_1, x_2; t) + \dfrac{1}{2}\dfrac{\partial^2 u^{1,2}}{\partial x_2^2}(s, x_1, x_2; t) = h^{[2]}(s, u^{2,2}(s, x_1, x_2; s)) \\ u^{1,2}(1, x_1, x_2; t) = f^{[2]}(x_1, x_2; t) \end{cases} \tag{12}$$

For $(t, s) \in S^{1,1}$ set,

$$Y_t^{[2]} = u^{1,1}(t, W_t; t), \ Z_{t,s}^{[2]} = \frac{\partial u^{1,1}}{\partial x_1}(s, W_s; t),$$

and from (10) we get the final system

$$\begin{cases} \dfrac{\partial u^{1,1}}{\partial s}(s, x_1; t) + \dfrac{1}{2}\dfrac{\partial^2 u^{1,1}}{\partial x_1^2}(s, x_1; t) = h^{[2]}(s, u^{1,1}(s, x_1; s)) \\ u^{1,1}(\tau_1, x_1; t) = u^{1,2}(\tau_1, x_1, x_1; t) \end{cases} \tag{13}$$

Consequently, the solution of (7) is given by

$$Y_t^{[2]} = \begin{cases} u^{2,2}(t, W_{\tau_1}, W_t; t) & , t \in [\tau_1, 1] \\ u^{1,1}(t, W_t; t) & , t \in [0, \tau_1) \end{cases} \tag{14}$$

$$Z_{t,s}^{[2]} = \begin{cases} \dfrac{\partial u^{2,2}}{\partial x_2}(s, W_{\tau_1}, W_s; t) & , (t,s) \in S^{2,2} \\ \dfrac{\partial u^{1,2}}{\partial x_2}(s, W_{\tau_1}, W_s; t) & , (t,s) \in S^{1,2} \\ \dfrac{\partial u^{1,1}}{\partial x_1}(s, W_s; t) & , (t,s) \in S^{1,1} \end{cases} \tag{15}$$

To obtain systems of PDEs for the lower triangle we have to use the M-condition (4), because this condition determines the process $Z_{t,s}^{[2]}$.

Suppose that $(t, s) \in S_{2,2}$, and that from (4) with $S = \tau_1$ for a.e. $t \in [\tau_1, 1]$ we have

$$Y_t^{[2]} = \mathbb{E}[Y_t^{[2]} | \mathcal{F}_{\tau_1}] + \int_{\tau_1}^{t} Z_{t,s}^{[2]} dW_s. \tag{16}$$

Set

$$Y_t^{[2]} = u_{2,2}(t, W_{\tau_1}, W_t; t), \ \ Z_{t,s}^{[2]} = \frac{\partial u_{2,2}}{\partial x_2}(s, W_{\tau_1}, W_s; t).$$

Applying Itô's formula to $u_{2,2}$ and the Brownian motion we get for $r \in [\tau_1, t]$ the following equation:

$$\begin{aligned} u_{2,2}(r, W_{\tau_1}, W_r; t) &= u_{2,2}(\tau_1, W_{\tau_1}, W_{\tau_1}; t) \\ &\quad + \int_{\tau_1}^{r} \left[\frac{\partial u_{2,2}}{\partial s}(s, W_{\tau_1}, W_s; t) + \frac{1}{2}\frac{\partial^2 u_{2,2}}{\partial x_2^2}(s, W_{\tau_1}, W_s; t) \right] ds \\ &\quad + \int_{\tau_1}^{r} \frac{\partial u_{2,2}}{\partial x_2}(s, W_{\tau_1}, W_s; t) dW_s, \end{aligned}$$

where $u_{2,2}(\tau_1, W_{\tau_1}, W_{\tau_1}; t) = \mathbb{E}[u_{2,2}(r, W_{\tau_1}, W_r; t) | \mathcal{F}_{\tau_1}]$.

Comparing the previous equation and (16), we can write

$$\begin{cases} \dfrac{\partial u_{2,2}}{\partial s}(s, x_1, x_2; t) + \dfrac{1}{2}\dfrac{\partial^2 u_{2,2}}{\partial x_2^2}(s, x_1, x_2; t) = 0 \\ u_{2,2}(t, x_1, x_2; t) = u^{2,2}(t, x_1, x_2; t), \end{cases} \tag{17}$$

from which it follows that

$$\begin{aligned} Y_t^{[2]} &= u^{2,2}(t, W_{\tau_1}, W_t; t) = u_{2,2}(t, W_{\tau_1}, W_t; t) \\ &= u_{2,2}(\tau_1, W_{\tau_1}, W_{\tau_1}; t) + \int_{\tau_1}^{t} \frac{\partial u_{2,2}}{\partial x_2}(s, W_{\tau_1}, W_s; t) dW_s \\ &= \mathbb{E}[Y_t^{[2]} | \mathcal{F}_{\tau_1}] + \int_{\tau_1}^{t} Z_{t,s}^{[2]} dW_s. \end{aligned}$$

Hence, (16) is satisfied.

For $(t, s) \in S_{2,1}$ consider $\varphi_t^{\tau_1} := \mathbb{E}[Y_t^{[2]} | \mathcal{F}_{\tau_1}]$. In order to get (4) with $S = 0$, we require in view of (16) that

$$\varphi_t^{\tau_1} = \mathbb{E}[Y_t^{[2]}] + \int_0^{\tau_1} Z_{t,s}^{[2]} dW_s.$$

Setting

$$\varphi_t^{\tau_1} = u_{2,1}(\tau_1, W_{\tau_1}; t), \ \ Z_{t,s}^{[2]} = \frac{\partial u_{2,1}}{\partial x_1}(s, W_s; t),$$

from Itô's formula we obtain the system of PDE:

$$\begin{cases} \dfrac{\partial u_{2,1}}{\partial s}(s, x_1; t) + \dfrac{1}{2}\dfrac{\partial^2 u_{2,1}}{\partial x_1^2}(s, x_1; t) = 0 \\ u_{2,1}(\tau_1, x_1; t) = u_{2,2}(\tau_1, x_1, x_1; t), \end{cases} \tag{18}$$

since for $r \in [0, \tau_1]$ we have

$$u_{2,1}(r, W_r; t) = u_{2,1}(0, W_0; t) + \int_0^r \frac{\partial u_{2,1}}{\partial x_1}(s, W_s; t) dW_s.$$

For $p \in [\tau_1, 1]$ one obtains

$$\begin{aligned} u_{2,1}(0, W_0; t) &= \mathbb{E}[u_{2,1}(\tau_1, W_{\tau_1}; t)] = \mathbb{E}[u_{2,2}(\tau_1, W_{\tau_1}, W_{\tau_1}; t)] \\ &= \mathbb{E}\left[u^{2,2}(p, W_{\tau_1}, W_p; t) - \int_{\tau_1}^p \frac{\partial u_{2,2}}{\partial x_2}(s, W_{\tau_1}, W_s; t) dW_s\right] \\ &= \mathbb{E}[u^{2,2}(p, W_{\tau_1}, W_p; t)]. \end{aligned}$$

In particular, for $p = t$ we have $u_{2,1}(0, W_0; t) = \mathbb{E}[u^{2,2}(t, W_{\tau_1}, W_t; t)] = \mathbb{E}[Y_t^{[2]}]$, and

$$\varphi_t^{\tau_1} = \mathbb{E}[Y_t^{[2]}] + \int_0^{\tau_1} Z_{t,s}^{[2]} dW_s.$$

Finally, for $(t, s) \in S_{1,1}$, from (4) we obtain

$$Y_t^{[2]} = \mathbb{E}[Y_t^{[2]}] + \int_0^t Z_{t,s}^{[2]} dW_s,$$

for a.e. $t \in [0, \tau_1]$.

Setting

$$Y_t^{[2]} = u_{1,1}(t, W_t; t), \; Z_{t,s}^{[2]} = \frac{\partial u_{1,1}}{\partial x_1}(s, W_s; t),$$

Itô's formula gives

$$\begin{cases} \dfrac{\partial u_{1,1}}{\partial s}(s, x_1; t) + \dfrac{1}{2} \dfrac{\partial^2 u_{1,1}}{\partial x_1^2}(s, x_1; t) = 0 \\ u_{1,1}(t, x_1; t) = u^{1,1}(t, x_1; t) \end{cases} \tag{19}$$

Hence, the solution of (4) is given by

$$Y_t^{[2]} = \begin{cases} u_{2,2}(t, W_{\tau_1}, W_t; t) &, t \in [\tau_1, 1] \\ u_{1,1}(t, W_t; t) &, t \in [0, \tau_1) \end{cases} \tag{20}$$

$$Z_{t,s}^{[2]} = \begin{cases} \dfrac{\partial u_{2,2}}{\partial x_2}(s, W_{\tau_1}, W_s; t) &, (t, s) \in S_{2,2} \\ \dfrac{\partial u_{2,1}}{\partial x_1}(s, W_s; t) &, (t, s) \in S_{2,1} \\ \dfrac{\partial u_{1,1}}{\partial x_1}(s, W_s; t) &, (t, s) \in S_{1,1} \end{cases} \tag{21}$$

Here, of course, (20) coincides with (14).

If the interval $[0,1]$ is divided into m parts at the points $\tau_1,\dots,\tau_m$ $(0<\tau_1<\dots<\tau_m=1)$ then $Y_t^{[m]}$ and $Z_{t,s}^{[m]}$ are computed with the next algorithm:

Consider (7) and a partition of $[0,1]^2$ into $2m-1+\frac{m(m+1)}{2}$ intervals as shown in Figure 2. Thus, one has to solve $2m-1+\frac{m(m+1)}{2}$ PDEs on the following subsets:

$$S^{1,j}:=\{(t,s)\in[0,1]^2|\ 0\le t<\tau_{j-1},\ \tau_{j-1}\le s\le\tau_j\},\ j=2,\dots,m,$$

$$S^{2,j}:=\begin{cases}\{(t,s)\in[0,1]^2|\ \tau_{m-1}\le t\le\tau_m,\ \tau_{m-1}\le s\le\tau_m,\ t\le s\},\ j=m\\ \{(t,s)\in[0,1]^2|\ \tau_{j-1}\le t<\tau_j,\ \tau_{j-1}\le s\le\tau_j,\ t\le s\},\ j=1,\dots,m-1,\end{cases}$$

$$S_{i,j}:=\begin{cases}\{(t,s)\in[0,1]^2|\ \tau_{m-1}\le t\le\tau_m,\ \tau_{j-1}\le s\le\tau_j\},\ i=m,j=1,\dots,m-1,\ i\ne j,\\ \{(t,s)\in[0,1]^2|\ \tau_{i-1}\le t<\tau_i,\ \tau_{j-1}\le s\le\tau_j\},\ i=1,\dots,m-1,\ j=1,\dots,m,\ i\ne j,\\ \{(t,s)\in[0,1]^2|\ \tau_{m-1}\le t\le\tau_m,\ \tau_{m-1}\le s\le\tau_m,\ t\ge s\},\ i=j=m,\\ \{(t,s)\in[0,1]^2|\ \tau_{i-1}\le t<\tau_i,\ \tau_{j-1}\le s\le\tau_j,\ t\ge s\},\ i=j=1,\dots,m-1.\end{cases}$$

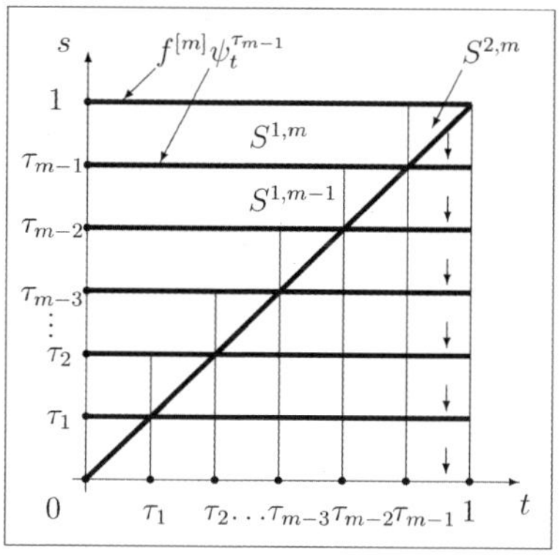

Figure 2. The algorithm of the construction in the general case. The terminal conditions of the systems of PDEs are defined on the thick lines.

We start in the right upper corner $S^{2,m}$. Denoting a vector $(x_1,\dots,x_m)$ by $\boldsymbol{x}_m$, $u^{2,m}$ is given by the system

$$\begin{cases}\dfrac{\partial u^{2,m}}{\partial s}(s,\boldsymbol{x}_m;t)+\dfrac{1}{2}\dfrac{\partial^2 u^{2,m}}{\partial x_m^2}(s,\boldsymbol{x}_m;t)=h^{[m]}(s,u^{2,m}(s,\boldsymbol{x}_m;s))\\ u^{2,m}(1,\boldsymbol{x}_m;t)=f^{[m]}(\boldsymbol{x}_m;t)\end{cases}$$

with $Y_t^{[m]}:=u^{2,m}(t,\boldsymbol{W}_{\tau_{m-1}},W_t;t)$, $Z_{t,s}^{[m]}:=\dfrac{\partial u^{2,m}}{\partial x_m}(s,\boldsymbol{W}_{\tau_{m-1}},W_s;t)$.

With this we can solve on $S^{1,m}$

$$\psi_t^{\tau_{m-1}}=f^{[m]}(\boldsymbol{W}_{\tau_m};t)-\int_{\tau_{m-1}}^{1}h^{[m]}(s,Y_s^{[m]})ds-\int_{\tau_{m-1}}^{1}Z_{t,s}^{[m]}dW_s,$$

which can be translated to a parabolic equation $u^{1,m}$ with terminal condition $f^{[m]}$ and nonhomogeneous term being equal to $h^{[m]}(s, u^{2,m}(s, \boldsymbol{x}_m; s))$.

The most right column of the lower triangle can be solved using the M-condition (4). Namely, we write the solution on $S_{m,m}$ in the following way:

$$Y_t^{[m]} = \mathbb{E}[Y_t^{[m]} | \mathcal{F}_{\tau_{m-1}}] + \int_{\tau_{m-1}}^{t} Z_{t,s}^{[m]} dW_s,$$

which can be translated with Itô's formula into an homogeneous parabolic equation of the form

$$\begin{cases} \dfrac{\partial u_{m,m}}{\partial s}(s, \boldsymbol{x}_m; t) + \dfrac{1}{2} \dfrac{\partial^2 u_{m,m}}{\partial x_m^2}(s, \boldsymbol{x}_m; t) = 0 \\ u_{m,m}(t, \boldsymbol{x}_m; t) = u^{2,m}(t, \boldsymbol{x}_m; t). \end{cases} \tag{22}$$

Iteratively $u_{m,i}$ on $S_{m,i}$ is given by an homogeneous parabolic equation of the form (22) with terminal condition $u_{m,i+1}$.

With this we compute $Y_t^{[m]}$ and $Z_{t,s}^{[m]}$ in the regions $S^{1,m}$, $S^{2,m}$ and $S_{m,j}$, $j = 1, \ldots, m$. The next step is to compute this solution in the regions $S^{1,m-1}$, $S^{2,m-1}$ and $S_{m-1,j}$, $j = 1, \ldots, m-1$.

Observing that

$$Y_t^{[m]} = \psi_t^{\tau_{m-1}} - \int_t^{\tau_{m-1}} h^{[m]}(s, Y_s^{[m]}) ds - \int_t^{\tau_{m-1}} Z_{t,s}^{[m]} dW_s,$$

we can compute $Y_t^{[m]}$ and $Z_{t,s}^{[m]}$ on $S^{1,m-1}$, $S^{2,m-1}$ and $S_{m-1,j}$, $j = 1, \ldots, m-1$, in an analogous way as on $S^{1,m}$, $S^{2,m}$ and $S_{m,j}$, $j = 1, \ldots, m$, with new boundary condition $\psi_t^{\tau_{m-1}}$ instead of $f^{[m]}$. Proceeding in this way we can find $Y_t^{[m]}$, $Z_{t,s}^{[m]}$ on the whole square.

4.2. *Regularity problem for parabolic Cauchy problems*

In this part we study the well-posedness of the PDEs from the previous subsection, i.e. we show that the PDE has a unique bounded classical solution under **(M)**. In addition, we show that even higher order derivatives exist, are continuous and bounded. We will need these derivatives in the proof of Lemma 4.1.

Our regularity result for the solutions of PDEs constructed on the sets from Figure 1 is the following:

Theorem 4.2. *Let condition* **(M)** *on* $f^{[m]}$ *and* $h^{[m]}$ *be fulfilled for the BSVIE (7) and suppose* $l = m = 2$, $k = T = d = 1$. *Then there exist unique bounded classical solutions* u^{i_1,i_2} *and* u_{j_1,j_2}, $j_1 \geq j_2$, *defined on the subsets* S^{1,i_2}, S^{2,i_2}, S_{j_1,j_2} *(see Figure 1),* $i_1, i_2, j_1, j_2 \in \{1, 2\}$, *belonging to* $C_b^\infty([0,1] \times \mathbb{R}^{i_2} \times [0,1]; \mathbb{R})$ *and* $C_b^\infty([0,1] \times \mathbb{R}^{j_2} \times [0,1]; \mathbb{R})$, *respectively.*

Obviously one can extend the result iteratively to the general case:

Corollary 4.3. *Let the conditions* **(M)** *on* $f^{[m]}$ *and* $h^{[m]}$ *be fulfilled for the BSVIE (7). Then there exist unique bounded classical solutions* u^{i_1,i_2} *and* u_{j_1,j_2}, $j_1 \geq j_2$, *defined on the subsets* S^{1,i_2}, S^{2,i_2}, S_{j_1,j_2} *(see Figure 2),* $i_1 \in \{1,2\}, i_2, j_1, j_2 \in \{1,\dots,m\}$, *belonging to* $C_b^\infty([0,T] \times \mathbb{R}^{i_2 \cdot d} \times [0,T]; \mathbb{R}^k)$ *and* $C_b^\infty([0,T] \times \mathbb{R}^{j_2 \cdot d} \times [0,T]; \mathbb{R}^k)$, *respectively.*

Before proving Theorem 4.2 we show the next lemma:

Lemma 4.4. *Let* $g \in C_b^\infty([t,1] \times \mathbb{R}^3; \mathbb{R})$ *depending on the variables* (s, x_2, v, x_1) *be a bounded function,* $t \in [0,1]$. *Then for all* (s, x_2) *the unique bounded classical solution* $v(s, x_2; x_1)$ *of*

$$\begin{cases} \dfrac{\partial v}{\partial s}(s,x_2;x_1) + \dfrac{1}{2}\dfrac{\partial^2 v}{\partial x_2^2}(s,x_2;x_1) = g(s,x_2,v(s,x_2;x_1);x_1), \\ \qquad\qquad\qquad\qquad\qquad\qquad\qquad (s,x_2;x_1) \in [t,1] \times \mathbb{R} \times \mathbb{R} \\ v(1,x_2;x_1) = 0. \end{cases}$$

is C^∞ *in* x_1 *with uniformly bounded derivatives.*

In addition, the mixed partial derivatives $\frac{\partial^{i_1+i_2+i_3} v}{\partial s^{i_3} \partial x_2^{i_2} \partial x_1^{i_1}}$ *exist, are continuous and bounded for* $i_1, i_2, i_3 \in \mathbb{N}_0$.

Proof. We use some ideas from [1] to achieve the claimed differentiability of v in the variable x_1. Namely, using Theorem 4.5, Chapter 7, from [28], the solution of (23) has for x_1 fixed and $(t, x_2) \in [0,1] \times \mathbb{R}$ the following representation:

$$v(t,x_2;x_1) = \mathbb{E}[Y_t^{x_1,x_2,t}] \equiv Y_t^{x_1,x_2,t},$$

where $(Y_s^{x_1,x_2,t}, Z_s^{x_1,x_2,t})$ is the unique adapted solution of the family of BSDEs

$$\begin{cases} Y_r^{x_1,x_2,t} = \displaystyle\int_r^1 g(s, W_s^{t,x_2}, Y_s^{x_1,x_2,t}; x_1) ds + \int_r^1 Z_s^{x_1,x_2,t} dW_s, \quad r \geq t \\ Y_1^{x_1,x_2,t} = 0 \end{cases} \tag{23}$$

with $W_s^{t,x_2} := x_2 + (W_s - W_t)$.

Then one can show that the conditions of Theorem 2.2 in [1] are fulfilled, and hence $Y_s^{x_1,x_2,t}$ is a.s. continuous in s and $Y_s^{x_1,x_2,t}$, $Z_s^{x_1,x_2,t}$ are continuously differentiable in x_1. Therefore, v is differentiable in x_1.

Furthermore, denoting by $(\nabla Y_s^{x_1,x_2,t}, \nabla Z_s^{x_1,x_2,t})$ the derivatives of $Y_s^{x_1,x_2,t}$ and $Z_s^{x_1,x_2,t}$ in x_1, they solve by Theorem 2.1 in [1] the BSDE

$$\begin{cases} \nabla Y_r^{x_1,x_2,t} = \displaystyle\int_r^1 \Big[\frac{\partial g}{\partial x_1}(s, W_s^{t,x_2}, Y_s^{x_1,x_2,t}; x_1) \\ \qquad\qquad + \frac{\partial g}{\partial v}(s, W_s^{t,x_2}, Y_s^{x_1,x_2,t}; x_1) \nabla Y_s^{x_1,x_2,t} \Big] ds - \displaystyle\int_r^1 \nabla Z_s^{x_1,x_2,t} dW_s, \\ \nabla Y_1^{x_1,x_2,t} = 0 \end{cases} \tag{24}$$

for $r \in [t,1]$. In particular, $\tilde{v}(s,x_2;x_1) := \nabla Y_s^{x_1,x_2,s} = \frac{\partial v}{\partial x_1}(s,x_2;x_1)$ solves

$$\begin{cases} \frac{\partial \tilde{v}}{\partial s}(s,x_2;x_1) + \frac{1}{2}\frac{\partial^2 \tilde{v}}{\partial x_2^2}(s,x_2;x_1) \\ \qquad = \frac{\partial g}{\partial x_1}(s,x_2,v(s,x_2;x_1);x_1) + \frac{\partial g}{\partial v}(s,x_2,v(s,x_2;x_1);x_1)\tilde{v}(s,x_2;x_1), \\ \tilde{v}(1,x_2;x_1) = 0. \end{cases} \tag{25}$$

Denoting

$$G_1(s,x_2;x_1) := \frac{\partial g}{\partial x_1}(s,x_2,v(s,x_2;x_1);x_1),$$

$$G_2(s,x_2;x_1) := \frac{\partial g}{\partial v}(s,x_2,v(s,x_2;x_1);x_1)$$

and using Theorem 4.1, Chapter 7, from [28], the unique bounded solution of (25) is given explicitly by

$$\tilde{v}(s,x_2;x_1) = \mathbb{E}\left[-\int_0^{1-s} G_1(s+\theta, W_\theta^{s,x_2};x_1)e^{-\int_0^\theta G_2(\lambda+s,W_\lambda^{s,x_2};x_1)d\lambda}\,d\theta\right]. \tag{26}$$

Note that the solution $\tilde{v}$ is indeed bounded as G_1 and G_2 are bounded. By the smoothness and boundedness condition on g, v can be differentiated a second time with second derivative

$$\frac{\partial^2 v}{\partial x_1^2}(s,x_2;x_1) = \frac{\partial \tilde{v}}{\partial x_1}(s,x_2;x_1)$$

$$= \mathbb{E}\left[-\int_0^{1-s} \frac{\partial}{\partial x_1}\left(G_1(s+\theta, W_\theta^{s,x_2};x_1)e^{-\int_0^\theta G_2(\lambda+s,W_\lambda^{s,x_2};x_1)d\lambda}\right)d\theta\right].$$

Continuing iteratively under the assumptions on the function g we obtain that v is ∞-times differentiable in x_1 with bounded derivatives.

In addition, as $\frac{\partial v}{\partial x_1}$ can be represented as in Equation (26), it follows that also the mixed partial derivatives exist, are continuous and bounded, if the function g fulfills the assumptions stated in the lemma. For $i_1 = 0$, infinite smoothness in (s,x_2) is classical for semilinear parabolic Cauchy problems with smooth data. □

Proof of Theorem *4.2*. We first construct the unique bounded solution $u^{2,2}$ of the system of PDEs in (9) as

$$u^{2,2}(s,x_1,x_2;t) = \tilde{u}^{2,2}(s,x_1,x_2;t) + \bar{u}^{2,2}(s,x_1,x_2), \tag{27}$$

where $\tilde{u}^{2,2}$ and $\bar{u}^{2,2}$ are the unique bounded solutions of the Cauchy problems

$$\begin{cases} \frac{\partial \tilde{u}^{2,2}}{\partial s}(s,x_1,x_2;t) + \frac{1}{2}\frac{\partial^2 \tilde{u}^{2,2}}{\partial x_2^2}(s,x_1,x_2;t) = 0, \\ \tilde{u}^{2,2}(1,x_1,x_2;t) = f^{[2]}(x_1,x_2;t) \end{cases} \tag{28}$$

$$\begin{cases} \dfrac{\partial \bar{u}^{2,2}}{\partial s}(s,x_1,x_2) + \dfrac{1}{2}\dfrac{\partial^2 \bar{u}^{2,2}}{\partial x_2^2}(s,x_1,x_2) = & h^{[2]}(s, \tilde{u}^{2,2}(s,x_1,x_2;s) \\ & +\bar{u}^{2,2}(s,x_1,x_2)). \\ \bar{u}^{2,2}(1,x_1,x_2) = 0 & \end{cases} \tag{29}$$

Using Theorem 4.2, Chapter 4, from [9] for (28) and Lemma 4.4 for (29), we see that $\tilde{u}^{2,2} \in C_b^\infty([0,1] \times \mathbb{R}^2 \times [0,1]; \mathbb{R})$ and $\bar{u}^{2,2} \in C_b^\infty([0,1] \times \mathbb{R}^2; \mathbb{R})$. Then from (27) it follows that $u^{2,2} \in C_b^\infty([0,1] \times \mathbb{R}^2 \times [0,1]; \mathbb{R})$. Consequently, the function $u^{2,2}(s,x,y;t)$ exhibits the required properties.

Due to the differentiability of the functions $f^{[2]}, h^{[2]}$ and $u^{2,2}$, using Theorem 4.1, Chapter 7, from [28], the unique bounded classical solution of $u^{1,2}$ of (12) can be expressed in an explicit form. Using assumption (**M**) from Lemma 4.1, it follows that $u^{1,2} \in C_b^\infty([0,1] \times \mathbb{R}^2 \times [0,1]; \mathbb{R})$. In the same manner we show that $u^{1,1} \in C_b^\infty([0,1] \times \mathbb{R} \times [0,1]; \mathbb{R})$ writing $u^{1,1}$ as a sum of two functions solving easier PDEs similar to $\tilde{u}^{2,2}$ and $\bar{u}^{2,2}$.

Notice that the solutions of the PDEs (17), (18), (19) represent Cauchy problems which are of the same type as the problem (28). We have the same regularity properties because the terminal condition of u_{j_1,j_2} is u^{i_1,i_2} which lies in $C_b^\infty([0,1] \times \mathbb{R}^{i_2} \times [0,1]; \mathbb{R})$. Hence, $u_{j_1,j_2} \in C_b^\infty([0,1] \times \mathbb{R}^{j_2} \times [0,1]; \mathbb{R})$. □

4.3. *Proof of Lemma 4.1*

In this subsection we give the proof of Lemma *4.1*. For notational simplicity, we show this lemma for $T = d = k = 1$ and $m = 2$, since the proof can be easily generalized.

In order to compute a discretization of u^{i_1,i_2}, $i_1 \le i_2$ $(i_1, i_2 = 1,2)$, we consider $(s_\eta, x_{1,i}, x_{2,j}; t_\kappa) = (\eta\Delta s, i\delta, j\delta; \kappa\Delta t)$, where $\delta = \delta(n) = 1/\sqrt{n}$, $\Delta s = \Delta t = 1/n$.

From a Taylor expansion of $u^{2,2}$ we get the following recurrence equation:

$$\begin{aligned}
&\frac{1}{2}\Big(u^{2,2}(s_{\eta+1}, x_{1,i}, x_{2,j}+\delta; t_\kappa) + u^{2,2}(s_{\eta+1}, x_{1,i}, x_{2,j}-\delta; t_\kappa)\Big) \\
&= u^{2,2}(s_\eta, x_{1,i}, x_{2,j}; t_\kappa) + \frac{\partial u^{2,2}}{\partial s}(s_\eta, x_{1,i}, x_{2,j}; t_\kappa)\delta^2 \\
&\quad + \frac{1}{2}\frac{\partial^2 u^{2,2}}{\partial x_2^2}(s_\eta, x_{1,i}, x_{2,j}; t_\kappa)\delta^2 \\
&\quad + \frac{1}{4}\Big(\frac{\partial^2 u^{2,2}}{\partial x_2^2}(s_\eta^+, x_{1,i}, x_{2,j}^+; t_\kappa)\delta^2 - \frac{\partial^2 u^{2,2}}{\partial x_2^2}(s_\eta, x_{1,i}, x_{2,j}; t_\kappa)\delta^2\Big) \\
&\quad + \frac{1}{2}\frac{\partial^2 u^{2,2}}{\partial x_2 \partial s}(s_\eta^+, x_{1,i}, x_{2,j}^+; t_\kappa)\delta^3 + \frac{1}{4}\frac{\partial^2 u^{2,2}}{\partial s^2}(s_\eta^+, x_{1,i}, x_{2,j}^+; t_\kappa)\delta^4 \\
&\quad + \frac{1}{4}\Big(\frac{\partial^2 u^{2,2}}{\partial x_2^2}(s_\eta^-, x_{1,i}, x_{2,j}^-; t_\kappa)\delta^2 - \frac{\partial^2 u^{2,2}}{\partial x_2^2}(s_\eta, x_{1,i}, x_{2,j}; t_\kappa)\delta^2\Big)
\end{aligned}$$

$$
\begin{aligned}
&+\frac{1}{2}\frac{\partial^2 u^{2,2}}{\partial x_2 \partial s}(s_\eta^-, x_{1,i}, x_{2,j}^-; t_\kappa)\delta^3 + \frac{1}{4}\frac{\partial^2 u^{2,2}}{\partial s^2}(s_\eta^-, x_{1,i}, x_{2,j}^-; t_\kappa)\delta^4 \\
=&u^{2,2}(s_\eta, x_{1,i}, x_{2,j}; t_\kappa) + h^{[2]}(s_\eta, u^{2,2}(s_\eta, x_{1,i}, x_{2,j}; s_\eta))\Delta s + O(\delta^3), \qquad (30)
\end{aligned}
$$

for appropriate values $(s_\eta^\pm, x_{2,j}^\pm) \in [s_\eta, s_\eta \pm \delta^2] \times [x_{2,j}, x_{2,j} \pm \delta]$ and with $u^{2,2}(n, x_{1,i}, x_{2,j}; t_\kappa) = f^{[2]}(x_{1,i}, x_{2,j}; t_\kappa)$, since

$$
\begin{aligned}
&\left|\frac{\partial^2 u^{2,2}}{\partial x_2^2}(s_\eta^\pm, x_{1,i}, x_{2,j}^\pm; t_\kappa)\delta^2 - \frac{\partial^2 u^{2,2}}{\partial x_2^2}(s_\eta, x_{1,i}, x_{2,j}; t_\kappa)\delta^2\right| \\
&\leq \left\|\frac{\partial^3 u^{2,2}}{\partial x_2^3}\right\|_\infty \delta^3 + \left\|\frac{\partial^3 u^{2,2}}{\partial x_2^2 \partial s}\right\|_\infty \delta^4
\end{aligned}
$$

and the involved partial derivatives exist and are uniformly bounded in accordance with the assumptions, see Theorem 4.2.

This motivates to define $U_n^{2,2}(\eta, i, j; \kappa)$ for $k_1 = \lfloor \tau_1 n \rfloor$ and $\tau_1^{(n)} = k_1/n$ by

$$
\begin{cases}
\frac{1}{2}\big(U_n^{2,2}(\eta+1, i, j+1; \kappa) + U_n^{2,2}(\eta+1, i, j-1; \kappa)\big) \\
= U_n^{2,2}(\eta, i, j; \kappa) + h^{[2]}(\eta, U_n^{2,2}(\eta, i, j; \eta))\frac{1}{n} \\
U_n^{2,2}(n, i, j; \kappa) = f^{[2]}(x_{1,i}, x_{2,j}; t_\kappa),
\end{cases} \qquad (31)
$$

where $\kappa = n, \ldots, k_1,\ \eta = n-1, \ldots, \kappa,\ i, j \in \mathbb{Z}$.

Using Lipschitz condition on h, the previous difference equation has a unique solution for large n.

Similarly, we approximate (12) and (13):

$$
\begin{cases}
\frac{1}{2}\big(U_n^{1,2}(\eta+1, i, j+1; \kappa) + U_n^{1,2}(\eta+1, i, j-1; \kappa)\big) \\
= U_n^{1,2}(\eta, i, j; \kappa) + h^{[2]}(\eta, U_n^{2,2}(\eta, i, j; \eta))\frac{1}{n} \\
U_n^{1,2}(n, i, j; \kappa) = f^{[2]}(x_{1,i}, x_{2,j}; t_\kappa),\ \kappa = k_1, \ldots, 0, \\
\qquad\qquad\qquad \eta = n-1, \ldots, k_1,\ i, j \in \mathbb{Z}
\end{cases}
$$

$$
\begin{cases}
\frac{1}{2}\big(U_n^{1,1}(\eta+1, j+1; \kappa) + U_n^{1,1}(\eta+1, j-1; \kappa)\big) \\
= U_n^{1,1}(\eta, j; \kappa) + h^{[2]}(\eta, U_n^{1,1}(\eta, j; \eta))\frac{1}{n} \\
U_n^{1,1}(k_1, j; \kappa) = U_n^{1,2}(k_1, j, j; \kappa),\ \kappa = k_1, \ldots, 0,\ \eta = k_1 - 1, \ldots, \kappa,\ j \in \mathbb{Z}
\end{cases}
$$

One easily shows that for

$$
Y_{t_\eta, t_\kappa}^{(n),[2]} = \begin{cases}
U_n^{2,2}(\eta, W_{\tau_1^{(n)}}^{(n)}, W_{t_\eta}^{(n)}; \kappa)\ , \kappa = n, \ldots, k_1,\ \eta = n, \ldots, \kappa, \\
U_n^{1,2}(\eta, W_{\tau_1^{(n)}}^{(n)}, W_{t_\eta}^{(n)}; \kappa)\ , \kappa = k_1, \ldots, 0,\ \eta = n, \ldots, k_1, \\
U_n^{1,1}(\eta, W_{t_\eta}^{(n)}; \kappa) \qquad\quad , \kappa = k_1, \ldots, 0,\ \eta = k_1, \ldots, \kappa,
\end{cases}
$$

we have

$$
Y_{t_\eta, t_\eta}^{(n),[2]} = \mathbb{E}\left[f^{[2]}(W_{\tau_1^{(n)}}^{(n)}, W_1^{(n)}; t_\eta) - \frac{1}{n}\sum_{\lambda=\eta}^{n-1} h^{[2]}(t_\lambda, Y_{t_\lambda, t_\lambda}^{(n),[2]})\middle|\mathcal{F}_{t_\eta}^{(n)}\right]
$$

and, hence,

$$Y_{t_\eta}^{(n),[2]} = \begin{cases} U_n^{2,2}(\eta, W_{\tau_1^{(n)}}^{(n)}, W_{t_\eta}^{(n)}; \eta) \ , t_\eta \in [\tau_1^{(n)}, 1], \\ U_n^{1,1}(\eta, W_{t_\eta}^{(n)}; \eta) \qquad\qquad , t_\eta \in [0, \tau_1^{(n)}). \end{cases} \tag{32}$$

Now we want to obtain an upper bound for

$$\theta_n^{2,2}(\eta) := \sup_{i,j\in\mathbb{Z}} \big| u^{2,2}(\eta\Delta s, x_{1,i}, x_{2,j}; \kappa\Delta s) - U_n^{2,2}(\eta, i, j; \kappa) \big|,$$

$\kappa = n, \ldots, k_1, \ \eta = n-1, \ldots, \kappa$. From (30) and the definition of $U^{2,2}$, see (31), we can find a constant C_1 such that

$$\theta_n^{2,2}(\eta) \le \gamma\big[\theta_n^{2,2}(\eta+1) + \frac{C_1}{n\sqrt{n}}\big], \ \gamma = \frac{1}{1 - L\Delta s},$$

which yields the inequality

$$\theta_n^{2,2}(\eta) \le \gamma^{n-\eta}\theta_n^{2,2}(n) + \frac{C_1}{n\sqrt{n}} \sum_{p=1}^{n-\eta} \gamma^p = \gamma^{n-\eta}\theta_n^{2,2}(n) + \frac{C_1}{n\sqrt{n}}\gamma\frac{\gamma^{n-\eta}-1}{\gamma - 1}$$

and, therefore,

$$\max_{k_1 \le \kappa \le n} \max_{\kappa \le \eta \le n-1} \theta_n^{2,2}(\eta) \le e^{2L}\theta_n^{2,2}(n) + \frac{C_1}{n\sqrt{n}}\gamma\frac{e^{2L}-1}{\gamma - 1} = \frac{C_1(e^{2L}-1)}{L\sqrt{n}},$$

$$\theta_n^{2,2}(n) = 0.$$

Analogously, we can find a constant C_2 such that

$$\begin{aligned} \theta_n^{1,2}(\eta) &:= \sup_{i,j\in\mathbb{Z}} \big| u^{1,2}(\eta\Delta s, x_{1,i}, x_{2,j}; \kappa\Delta s) - U_n^{1,2}(\eta, i, j; \kappa) \big| \\ &\le \frac{C_1(e^{2L}-1) + C_2}{\sqrt{n}}, \end{aligned}$$

where $\kappa = k_1, \ldots, 0, \ \eta = n-1, \ldots, k_1$, and

$$\max_{0 \le \kappa \le k_1} \max_{n-1 \le \eta \le k_1} \theta_n^{1,2}(\eta) \le \frac{C_1(e^{2L}-1) + C_2}{\sqrt{n}}.$$

Finally we consider

$$\theta_n^{1,1}(\eta) := \sup_{j\in\mathbb{Z}} \big| u^{1,1}(\eta\Delta s, x_{2,j}; \kappa\Delta s) - U_n^{1,1}(\eta, j; \kappa) \big|,$$

$\kappa = k_1, \ldots, 0, \ \eta = k_1 - 1, \ldots, \kappa$. Applying the bound for $\theta_n^{1,2}(\eta)$ and denoting by C a bound for the partial derivatives of $u^{1,1}$ and $u^{1,2}$ we obtain,

$$\begin{aligned} \theta_n^{1,1}(k_1) &= \sup_{j\in\mathbb{Z}} \big| u^{1,1}(k_1\Delta s, x_{2,j}; \kappa\Delta s) - U_n^{1,2}(k_1, j, j; \kappa) \big| \\ &\le \theta_n^{1,2}(k_1) + 2C(\tau_1 - k_1\Delta s) \le \frac{C_1(e^{2L}-1) + C_2 + 2C}{\sqrt{n}}. \end{aligned}$$

For $k_1 - 1 \leq \eta \leq \kappa$ we can find a constant C_3 such that

$$\theta_n^{1,1}(\eta) \leq \gamma\big[\theta_n^{1,1}(\eta+1) + \frac{C_3}{n\sqrt{n}}\big],$$

which yields the upper bound

$$\begin{aligned}\max_{0\leq\kappa\leq k_1}\max_{k_1-1\leq\eta\leq\kappa}\theta_n^{1,1}(\eta) &\leq e^{2L}\theta_n^{1,1}(k_1) + \frac{C_3(e^{2L}-1)}{L\sqrt{n}} \\ &\leq \left(e^{2L}(C_2+2C) + \frac{(e^{2L}-1)(Le^{2L}C_1+C_3)}{L}\right)\frac{1}{\sqrt{n}}.\end{aligned}$$

Hence, for a suitable constant A, we obtain the following estimate:

$$\max_{\eta}\left\{\theta_n^{2,2}(\eta), \theta_n^{1,1}(\eta)\right\} \leq \frac{A}{\sqrt{n}}. \tag{33}$$

We now define

$$\tilde{Y}_t^{(n),[2]} = \begin{cases} u^{2,2}(t, \tilde{W}_t^{(n)}; t)\ , t \in [\tau_1, 1], \\ u^{1,1}(t, \tilde{W}_t^{(n)}; t)\ , t \in [0, \tau_1), \end{cases}$$

where $\tilde{W}^{(n)}$ is the piecewise constant interpolation of the binary random walk $W_{t_\eta}^{(n)}$. By the representation result (14) for the solution Y_t of the continuous BSVIE and the smoothness of $u^{1,1}$ and $u^{2,2}$ we obtain by Donsker's theorem and the continuous mapping theorem that $\tilde{Y}^{(n),[2]}$ converges weakly to $Y^{[2]}$ in the Skorokhod topology. We can finally deduce from (32), (33), and Slutsky's theorem that the same convergence result also holds true for $Y^{(n),[2]}$.

5. Proof of Theorem 3.6

In this section we provide the proof of Theorem 3.6 in the general case. To this end we first show how to extend convergence of the finite dimensional distributions from the smooth case to the general case. Then, as the final step we prove tightness.

5.1. *Convergence of the finite dimensional distributions*

As a preparation for the proof of convergence in finite dimensional distributions we require some *a-priori* estimates for continuous and discrete BSVIEs.

The next lemma estimates the difference of solutions Y and $\bar{Y}$ of Equation (1), if Y and $\bar{Y}$ have the free term f and $\bar{f}$ and the generator h and $\bar{h}$, respectively.

Lemma 5.1. *Let $f, \bar{f}$ be two free terms and $h, \bar{h}$ be two generators satisfying* (**V1**) *and* (**V2**). *Let (Y, Z) and $(\bar{Y}, \bar{Z})$ be two adapted M-solutions of* (1) *with f and h replaced by $\bar{f}$ and $\bar{h}$, respectively. Then*

$$||Y - \bar{Y}||_{L_{\mathbb{F}}^{2,\infty}(0,T;\mathbb{R}^k)} \leq \sqrt{C\big(\mathbb{E}\big[||f-\bar{f}||_\infty^2\big] + ||h-\bar{h}||_\infty^2\big)}$$

holds for some constant $C < \infty$ depending only on the Lipschitz constants of h and $\bar{h}$, where

$$||f - \bar{f}||_\infty := \sup_{s\in[0,T]} ||f(W;s) - \bar{f}(W;s)||_2$$

and

$$||h - \bar{h}||_\infty := \sup_{(s,y)\in[0,T]\times\mathbb{R}^k} ||h(s,y) - \bar{h}(s,y)||_2.$$

Proof. The proof of this lemma is analogous to the proof from [13], which is based on Gronwall's lemma and Doob's inequality. □

The analogous result in the discrete case is stated in the following lemma.

Lemma 5.2. *Let $f, \bar{f}$ be two free terms and $h, \bar{h}$ be two generators satisfying* (**V1**) *and* (**V2**). *Let $(Y^{(n)}, Z^{(n)})_{n\in\mathbb{N}}$ and $(\bar{Y}^{(n)}, \bar{Z}^{(n)})_{n\in\mathbb{N}}$ be two discrete M-solutions of* (2) *with f and h replaced by $\bar{f}$ and $\bar{h}$, respectively. Then*

$$\limsup_{n\to\infty} ||Y^{(n)} - \bar{Y}^{(n)}||_{L^{2,\infty}_{\mathbb{F}^{(n)}}(0,T;\mathbb{R}^k)} \leq \sqrt{C\big(\mathbb{E}\big[||f - \bar{f}||^2_\infty\big] + ||h - \bar{h}||^2_\infty\big)}$$

holds for some constant $C < \infty$ depending only on the Lipschitz constants of h and $\bar{h}$, where

$$||f - \bar{f}||_\infty := \sup_{s\in[0,T]} ||f(W;s) - \bar{f}(W;s)||_2$$

and

$$||h - \bar{h}||_\infty := \sup_{(s,y)\in[0,T]\times\mathbb{R}^k} ||h(s,y) - \bar{h}(s,y)||_2.$$

Proof. The proof is the discrete analogue to the proof of Lemma 5.1 with the application of the discrete Gronwall lemma and Doob's inequality together with the fact that $W^{(n)}$ converges weakly to W by Donsker's Theorem. □

With these *a-priori* estimates to our disposal we can prove the following lemma.

Lemma 5.3. *Under the assumptions of Theorem 3.6, convergence of the finite dimensional distributions holds true.*

Proof. Let $m \in \mathbb{N}$ and $\tau^{\langle m\rangle} := (\tau_0^{\langle m\rangle}, \ldots, \tau_m^{\langle m\rangle})$ be a partition of $[0,T]$ $(0 = \tau_0^{\langle m\rangle} < \tau_1^{\langle m\rangle} < \ldots < \tau_m^{\langle m\rangle} = T)$.

For a point $\boldsymbol{x} := (\boldsymbol{x}_1, \ldots, \boldsymbol{x}_m) \in \mathbb{R}^l$, where $\boldsymbol{x}_i \in \mathbb{R}^d$, $i = 1, \ldots, m$, $m \in \mathbb{N}$, $l = dm$, define the linear interpolation $\mathbb{I}_{\tau^{\langle m \rangle}, \boldsymbol{x}} : [0,T] \to \mathbb{R}^d$ of $\tau^{\langle m \rangle}$ and $\boldsymbol{x}$ as

$$\mathbb{I}_{\tau^{\langle m \rangle}, \boldsymbol{x}}(t) := \begin{cases} 0 & , t = 0 \\ \dfrac{t}{\tau_1^{\langle m \rangle}} \boldsymbol{x}_1 & , t \in [0, \tau_1^{\langle m \rangle}] \\ \dfrac{\tau_2^{\langle m \rangle} - t}{\tau_2^{\langle m \rangle} - \tau_1^{\langle m \rangle}} \boldsymbol{x}_1 + \dfrac{t - \tau_1^{\langle m \rangle}}{\tau_2^{\langle m \rangle} - \tau_1^{\langle m \rangle}} \boldsymbol{x}_2 & , t \in [\tau_1^{\langle m \rangle}, \tau_2^{\langle m \rangle}] \\ \vdots & \\ \dfrac{\tau_m^{\langle m \rangle} - t}{\tau_m^{\langle m \rangle} - \tau_{m-1}^{\langle m \rangle}} \boldsymbol{x}_{m-1} + \dfrac{t - \tau_{m-1}^{\langle m \rangle}}{\tau_m^{\langle m \rangle} - \tau_{m-1}^{\langle m \rangle}} \boldsymbol{x}_m & , t \in [\tau_{m-1}^{\langle m \rangle}, \tau_m^{\langle m \rangle}] \end{cases}.$$

Define $f^{\langle m \rangle} : \mathbb{R}^l \times [0,T] \to \mathbb{R}^k$ by $f^{\langle m \rangle}(\boldsymbol{x}; t) := f(\mathbb{I}_{\tau^{\langle m \rangle}, \boldsymbol{x}}; t)$ and consider

$$Y_t^{\langle m \rangle} = f^{\langle m \rangle}(\boldsymbol{W}_{\tau_m^{\langle m \rangle}}; t) - \int_t^T h(s, Y_s^{\langle m \rangle}) ds - \int_t^T Z_{t,s}^{\langle m \rangle} dW_s, \ t \in [0,T]. \tag{34}$$

By Theorem 2.4 Equation (34) has a unique adapted M-solution $(Y^{\langle m \rangle}, Z^{\langle m \rangle})$. Furthermore, by Theorem 3.4 there exists a unique M-solution $(Y^{(n),\langle m \rangle}, Z^{(n),\langle m \rangle})$ for

$$Y_{t_i}^{(n),\langle m \rangle} = f^{\langle m \rangle}(\boldsymbol{W}^{(n)}_{\tau_m^{(n),\langle m \rangle}}; t_i) - \frac{T}{n} \sum_{j=i}^{n-1} h(t_j, Y_{t_j}^{(n),\langle m \rangle}) - \sum_{j=i}^{n-1} Z_{t_i,t_j}^{(n),\langle m \rangle} \Delta W_{t_{j+1}}^{(n)} \tag{35}$$

Assume, that $(\tau^{\langle m \rangle})_{m \in \mathbb{N}}$ is a sequence of partitions, such that

$$\lim_{m \to \infty} \max_{i \in \{0, \ldots, m-1\}} |\tau_{i+1}^{\langle m \rangle} - \tau_i^{\langle m \rangle}| = 0.$$

Then from Lemma 5.1 it follows that $Y^{\langle m \rangle}$ converges to Y in $L_{\mathbb{F}}^{2,\infty}$ for $m \to \infty$ as $f(\mathbb{I}_{\tau^{\langle m \rangle}, \boldsymbol{W}_{\tau_m^{\langle m \rangle}}}; t)$ converges to $f(W; t)$ in $L_{\mathbb{F}}^{2,\infty}$. The analogous result for $Y^{(n),\langle m \rangle}$ is stated in Lemma 5.2.

In addition, as f is bounded, $f^{\langle m \rangle}$ is also bounded. Moreover, $f^{\langle m \rangle}$ is Lipschitz with Lipschitz constant independent of m. Hence, we can approximate $f^{\langle m \rangle}$ and h by a sequence of smooth functions $f^{\langle m \rangle, \{p\}}$ and $h^{\{p\}}$, such that

$$\lim_{p \to \infty} \sup_m \mathbb{E}[\sup_{s \in [0,T]} ||f^{\langle m \rangle, \{p\}}(\boldsymbol{W}_{\tau_m^{\langle m \rangle}}; s) - f^{\langle m \rangle}(\boldsymbol{W}_{\tau_m^{\langle m \rangle}}; s)||_2^2] = 0,$$

and

$$\lim_{p \to \infty} ||h - h^{\{p\}}||_\infty = 0,$$

Actually, we only need uniform convergence of $h^{\{p\}}$ to h on a sufficiently large compact set, because $Y^{\langle m \rangle}$, Y, $Y^{(n),\langle m \rangle}$ and $Y^{(n)}$ are uniformly bounded by the boundedness assumption on f and h.

With this we consider

$$Y_t^{\langle m\rangle,\{p\}} = f^{\langle m\rangle,\{p\}}(\boldsymbol{W}_{\tau_m^{\langle m\rangle}};t) - \int_t^T h^{\{p\}}(s, Y_s^{\langle m\rangle,\{p\}})ds - \int_t^T Z_{t,s}^{\langle m\rangle,\{p\}} dW_s,$$

$$t \in [0,T]. \tag{36}$$

The free term $f^{\langle m\rangle,\{p\}}$ and the generator $h^{\{p\}}$ fulfill assumption (**M**) of Lemma 4.1. Therefore we define $\tau^{(n),\langle m\rangle} := (\tau_0^{(n),\langle m\rangle}, \dots, \tau_m^{(n),\langle m\rangle})$ with $\tau_\lambda^{(n),\langle m\rangle} := \frac{\lfloor \tau_\lambda^{\langle m\rangle} n\rfloor}{n}$ for $\lambda = 0, \dots, m$, and consider

$$Y_{t_i}^{(n),\langle m\rangle,\{p\}} = f^{\langle m\rangle,\{p\}}(\boldsymbol{W}^{(n)}_{\tau_m^{(n),\langle m\rangle}};t_i) \tag{37}$$
$$-\frac{T}{n}\sum_{j=i}^{n-1} h^{\{p\}}(t_j, Y_{t_j}^{(n),\langle m\rangle,\{p\}}) - \sum_{j=i}^{n-1} Z_{t_i,t_j}^{(n),\langle m\rangle,\{p\}} \Delta W_{t_{j+1}}^{(n)}.$$

Due to Theorem 3.4 there exists a unique M-solution $(Y^{(n),\langle m\rangle,\{p\}}, Z^{(n),\langle m\rangle,\{p\}})$ of Equation (38) and Lemma 4.1 implies weak convergence of $Y^{(n),\langle m\rangle,\{p\}}$ to $Y^{\langle m\rangle,\{p\}}$ when n tends to infinity.

In order to show convergence of the finite dimensional distributions of $Y^{(n)}$ to those of Y, let $G : \mathbb{R}^{rk} \to \mathbb{R}$ be a bounded continuous function, $r \geq 1$, and $0 \leq \tilde{t}_1 < \ldots < \tilde{t}_r \leq T$ some partition of $[0,T]$. Then we obtain

$$\begin{aligned}
&\left|\mathbb{E}[G(Y_{\tilde{t}_1}, \dots, Y_{\tilde{t}_r})] - \mathbb{E}[G(Y_{\tilde{t}_1}^{(n)}, \dots, Y_{\tilde{t}_r}^{(n)})]\right| \\
&\leq \mathbb{E}\big[|G(Y_{\tilde{t}_1}, \dots, Y_{\tilde{t}_r}) - G(Y_{\tilde{t}_1}^{\langle m\rangle}, \dots, Y_{\tilde{t}_r}^{\langle m\rangle})|\big] \\
&\quad + \mathbb{E}\big[|G(Y_{\tilde{t}_1}^{\langle m\rangle}, \dots, Y_{\tilde{t}_r}^{\langle m\rangle}) - G(Y_{\tilde{t}_1}^{\langle m\rangle,\{p\}}, \dots, Y_{\tilde{t}_r}^{\langle m\rangle,\{p\}})|\big] \\
&\quad + \left|\mathbb{E}[G(Y_{\tilde{t}_1}^{\langle m\rangle,\{p\}}, \dots, Y_{\tilde{t}_r}^{\langle m\rangle,\{p\}})] - \mathbb{E}[G(Y_{\tilde{t}_1}^{(n),\langle m\rangle,\{p\}}, \dots, Y_{\tilde{t}_r}^{(n),\langle m\rangle,\{p\}})]\right| \\
&\quad + \mathbb{E}\big[|G(Y_{\tilde{t}_1}^{(n),\langle m\rangle,\{p\}}, \dots, Y_{\tilde{t}_r}^{(n),\langle m\rangle,\{p\}}) - G(Y_{\tilde{t}_1}^{(n),\langle m\rangle}, \dots, Y_{\tilde{t}_r}^{(n),\langle m\rangle})|\big] \\
&\quad + \mathbb{E}\big[|G(Y_{\tilde{t}_1}^{(n),\langle m\rangle}, \dots, Y_{\tilde{t}_r}^{(n),\langle m\rangle}) - G(Y_{\tilde{t}_1}^{(n)}, \dots, Y_{\tilde{t}_r}^{(n)})|\big].
\end{aligned}$$

Now letting first n tend to infinity, the third summand goes to zero by Lemma 4.1. Then, for $p \to \infty$, the second and the fourth summand tend to zero by Lemmas 5.1 and 5.2. Finally, the same lemmas yield convergence of the first and fifth term to zero, as finally $m \to \infty$. □

5.2. *Tightness*

As the final step of the proof of Theorem 3.6, we will now proof tightness of the sequence $(Y^{(n)})_{n\in\mathbb{N}}$.

Lemma 5.4. *Assume that the assumptions* (**V1**)-(**V4**) *are fulfilled. Let* $(Y^{(n)}, Z^{(n)})_{n\in\mathbb{N}}$ *be the sequence of discrete M-solutions from* (2). *Then the sequence* $(Y^{(n)})_{n\in\mathbb{N}}$ *is tight in the Skorokhod topology.*

For the proof we need the next lemma.

Lemma 5.5. *The jumps of $Y^{(n)}$ in (2) converge uniformly to zero, more precisely,*

$$||Y^{(n)}_{t_{i+1}} - Y^{(n)}_{t_i}||_2 \leq \frac{Ke^{2L}}{\sqrt{n}} + \frac{KT+D}{n} \quad \mathbb{P}\text{-a.s.}$$

Proof. Note that from [27], we can interpret (1) as a parameterized BSDE in the following sense:

$$Y_{t,r} = f(W;t) - \int_r^T h(s, Y_{s,s})ds - \int_r^T Z_{t,s}dW_s, \; r \in [t,T],$$

with $Y_{t,\cdot}$, $Z_{t,\cdot}$ being $\mathbb{F}$-adapted for almost all $t \in [0,T]$ and $Y_s := Y_{s,s}$.

From the above we can write the following discretization of the solution $Y_{t,r}$:

$$Y^{(n)}_{t_i,t_k} = f(W^{(n)};t_i) - \frac{T}{n}\sum_{j=k}^{n-1} h(t_j, Y^{(n)}_{t_j,t_j}) - \sum_{j=k}^{n-1} Z^{(n)}_{t_i,t_j}\Delta W^{(n)}_{t_{j+1}}.$$

In particular, we notice that $Y^{(n)}_{t_i,t_i}$ coincides with $Y^{(n)}_{t_i}$ in (2).

Thus, we obtain

$$\begin{aligned}
||Y^{(n)}_{t_{i+1}} - Y^{(n)}_{t_i}||_2 &= ||Y^{(n)}_{t_{i+1},t_{i+1}} - Y^{(n)}_{t_i,t_{i+1}} + Y^{(n)}_{t_i,t_{i+1}} - Y^{(n)}_{t_i,t_i}||_2 \\
&\leq ||Y^{(n)}_{t_{i+1},t_{i+1}} - Y^{(n)}_{t_i,t_{i+1}}||_2 + ||Y^{(n)}_{t_i,t_{i+1}} - Y^{(n)}_{t_i,t_i}||_2 =: A + B,
\end{aligned}$$

where due to adaptedness of $Y^{(n)}$, $Z^{(n)}_{t_{i+1},\cdot}$, $Z^{(n)}_{t_i,\cdot}$

$$\begin{aligned}
A &= \left|\left|\mathbb{E}\left[Y^{(n)}_{t_{i+1},t_{i+1}} - Y^{(n)}_{t_i,t_{i+1}} \middle| \mathcal{F}^{(n)}_{t_{i+1}}\right]\right|\right|_2 \\
&= \Bigg|\Bigg|\mathbb{E}\Bigg[f(W^{(n)};t_{i+1}) - \frac{T}{n}\sum_{j=i+1}^{n-1} h(t_j, Y^{(n)}_{t_j}) - \sum_{j=i+1}^{n-1} Z^{(n)}_{t_{i+1},t_j}\Delta W^{(n)}_{t_{j+1}} \\
&\quad - f(W^{(n)};t_i) + \frac{T}{n}\sum_{j=i+1}^{n-1} h(t_j, Y^{(n)}_{t_j}) + \sum_{j=i+1}^{n-1} Z^{(n)}_{t_i,t_j}\Delta W^{(n)}_{t_{j+1}} \Bigg| \mathcal{F}^{(n)}_{t_{i+1}}\Bigg]\Bigg|\Bigg|_2 \\
&\leq \left|\left|\mathbb{E}\left[f(W^{(n)};t_{i+1}) - f(W^{(n)};t_i) \middle| \mathcal{F}^{(n)}_{t_{i+1}}\right]\right|\right|_2 \\
&\leq K|t_{i+1} - t_i| = \frac{KT}{n}.
\end{aligned}$$

Moreover, using Lemma 3.2 in [13] we obtain,

$$B \leq \frac{Ke^{2L}}{\sqrt{n}} + \frac{D}{n}.$$

Hence, for all $i = 0, \ldots, n$,

$$||Y^{(n)}_{t_{i+1}} - Y^{(n)}_{t_i}||_2 \leq \frac{Ke^{2L}}{\sqrt{n}} + \frac{KT+D}{n},$$

which proves the lemma. □

Proof of Lemma *5.4*. Consider the Doob decomposition for the discretized solution $Y^{(n)}_{t_i}$, $i = 0, \dots, n$:

$$Y^{(n)}_{t_i} = U^{(n)}_{t_i} + A^{(n)}_{t_i},$$

where

$$U^{(n)}_{t_i} := \sum_{j=0}^{i-1} \left(Y^{(n)}_{t_{j+1}} - \mathbb{E}\left[Y^{(n)}_{t_{j+1}} \middle| \mathcal{F}^{(n)}_{t_j}\right]\right)$$

and

$$A^{(n)}_{t_i} := Y^{(n)}_{t_i} - U^{(n)}_{t_i}.$$

Then $A^{(n)}$ is a predictable process and $U^{(n)}$ is a square integrable martingale. Thus, $Y^{(n)}_{t_i}$ is a square integrable semimartingale and, as pointed out in [13], we can use Theorem 2.3 (with condition C2) in [8] for the process

$$G^{(n)} := \sum_{l=1}^{k} [U^{(n),l}, U^{(n),l}] + V(A^{(n),l})$$

with $[U^{(n),l}, U^{(n),l}]$ and $V(A^{(n),l})$ being the quadratic, resp., total variation of the lth component $U^{(n),l}$ of $U^{(n)}$ and $A^{(n),l}$ of $A^{(n)}$. Namely, using Lemma 5.5 one can see that $G^{(n)}$ is bounded by some constant independent of n. Indeed,

$$\begin{aligned}
&\sum_{i=0}^{n-1} ||A^{(n)}_{t_{i+1}} - A^{(n)}_{t_i}||_2 \\
&= \sum_{i=0}^{n-1} \left|\left| \mathbb{E}\left[f(W^{(n)}; t_{i+1}) \middle| \mathcal{F}^{(n)}_{t_i}\right] - \mathbb{E}\left[f(W^{(n)}; t_i) \middle| \mathcal{F}^{(n)}_{t_i}\right] + \frac{T}{n} h(t_i, Y^{(n)}_{t_i}) \right|\right|_2 \\
&\le \sum_{i=0}^{n-1} \left(\frac{KT}{n} + \frac{DT}{n}\right) = T(K + D),
\end{aligned}$$

and

$$\begin{aligned}
&\sum_{J=0}^{n-1} ||U^{(n)}_{t_{J+1}} - U^{(n)}_{t_J}||_2^2 = \sum_{J=0}^{n-1} ||Y^{(n)}_{t_{J+1}} - \mathbb{E}\left[Y^{(n)}_{t_{J+1}} \middle| \mathcal{F}^{(n)}_{t_J}\right]||_2^2 \\
&= \sum_{J=0}^{n-1} \left|\left| Y^{(n)}_{t_{J+1}} - Y^{(n)}_{t_J} - \frac{T}{n} h(t_J, Y^{(n)}_{t_J}) \right.\right. \\
&\qquad \left.\left. + \mathbb{E}\left[f(W^{(n)}; t_J) - f(W^{(n)}; t_{J+1}) \middle| \mathcal{F}^{(n)}_{t_J}\right] \right|\right|_2^2 \\
&\le \sum_{J=0}^{n-1} \left(\frac{Ke^{2L}}{\sqrt{n}} + \frac{KT + D}{n} + \frac{T(K + D)}{n}\right)^2 \le C
\end{aligned}$$

for some constant C. Hence, from [8] the sequence $(Y^{(n)})$ is relatively compact in the Skorokhod topology and, by Prokhorov's theorem, it is tight. □

6. Numerical example

Consider for (1) the BSVIE:

$$Y_t = te^{\frac{1}{2}}\sin(W_1) + \int_t^1 Y_s ds - \int_t^1 Z_{t,s} dW_s, \ t \in [0,1], \tag{38}$$

with $T = d = k = 1$.

Since conditions (**V1**) and (**V2**) hold, Equation (38) admits a unique solution (Y, Z).

Note that $Y_t = e^{\frac{t}{2}}\sin(W_t)$, since $e^{\frac{t}{2}}\sin(W_t)$ is a martingale and hence

$$\mathbb{E}\left[te^{\frac{1}{2}}\sin(W_1) + \int_t^1 Y_s ds \middle| \mathcal{F}_t\right] = tY_t + (1-t)Y_t = Y_t.$$

Using (2) we find the discretization $Y_{t_i}^{(n)}$. If we interpolate piecewise constantly between values $Y_{t_i}^{(n)}$ and $Y_{t_{i+1}}^{(n)}$, from Theorem 3.6, we obtain that $Y^{(n)}$ converges weakly to Y. In particular, for every bounded continuous function $g : \mathbb{D}([0,1];\mathbb{R}) \to \mathbb{R}$, we have

$$\lim_{n\to\infty} \left(\mathbb{E}\left[g(Y^{(n)})\right] - \mathbb{E}\left[g(Y)\right]\right) = 0. \tag{39}$$

For our numerical test example we choose the function $g(y) := \int_0^1 \min\{y(s)^2, M\}ds$ for some constant $M \geq e$. The truncation guarantees that g is bounded. By the choice of M we have $Y_t^2 = \min\{Y_t^2, M\}$ for every $t \in [0,1]$. Hence, by Itô's formula

$$\begin{aligned}\mathbb{E}\left[\int_0^1 \min\{Y_s^2, M\}ds\right] &= \mathbb{E}\left[\int_0^1 Y_s^2 ds\right] = \int_0^1 e^s \mathbb{E}\left[\sin^2(W_s)\right]ds \\ &= \frac{1}{2}\int_0^1 e^s(1-e^{-2s})ds = \frac{(e-1)^2}{2e}.\end{aligned}$$

We examine the convergence behavior by plotting the weak approximation error

$$n \mapsto \left(\mathbb{E}\left[\int_0^1 \min\{Y_s^2, M\}ds\right] - \mathbb{E}\left[\frac{1}{n}\sum_{i=0}^{n-1} \min\{(Y_{t_i}^{(n)})^2, M\}\right]\right) =: X^{(n)}, \quad M \geq e,$$

see Figure 3. For the computation of

$$\frac{1}{n}\sum_{i=0}^{n-1} \mathbb{E}\left[\min\{(Y_{t_i}^{(n)})^2, M\}\right],$$

we exploit that the solution of the DBSVIE $Y_{t_i}^{(n)}$ can be written as

$$Y_{t_i}^{(n)} = \frac{n}{n-1}\mathbb{E}\left[t_i e^{\frac{1}{2}} \sin(W_1^{(n)}) + \frac{1}{n}\sum_{j=i+1}^{n-1} Y_{t_j}^{(n)} \middle| \mathcal{F}_{t_i}^{(n)}\right].$$

We plot the weak approximation error $X^{(n)}$ in logarithmic scales on both axes for $n = 10, 20, 40, \ldots, 400$ with $M = 10$. The slope in Figure 3 is about -0.928. This indicates that the rate of convergence is close to order 1, which is the weak convergence rate that one may expect in this type of example.

We also note that, by weak convergence, $Y_0^{(n)} \to Y_0 = 0$. Table 1 illustrates that even for a moderate grid size (e.g. $n = 100$) the numerical approximation almost perfectly (up to rounding errors) matches the true value.

Table 1. Values of $X^{(n)}$ and $Y_0^{(n)}$.

n	$X^{(n)}$	$Y_0^{(n)}$
100	0.0339	0.0039×10^{-16}
200	0.0172	-0.0078×10^{-16}
300	0.0116	0.1067×10^{-16}
400	0.0087	0.0503×10^{-16}

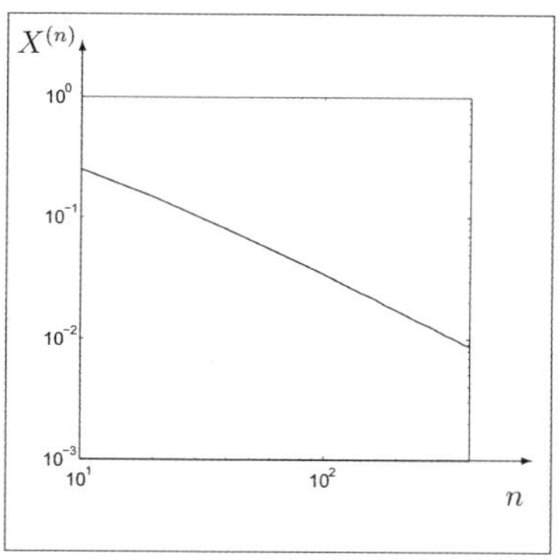

Figure 3. Speed of convergence of the numerical algorithm: The weak approximation error $n \mapsto X^{(n)}$ is plotted on logarithmic scales on both the axes for $n = 10, 20, 40, \ldots, 400$; the slope is about -0.928.

7. Proof of Lemma 3.3

Proof of Lemma *3.3*. In the case $\mathcal{S} < i$, multiplying Equation (5) by $\Delta W^{(n)}_{t_{\mathcal{S}+1}}$ and taking the conditional expectation of this expression given $\mathcal{F}^{(n)}_{t_\mathcal{S}}$ one obtains

$$\mathbb{E}\big[(Y^{(n)}_{t_i} - \mathbb{E}\big[Y^{(n)}_{t_i}\big|\mathcal{F}^{(n)}_{t_\mathcal{S}}\big])(\Delta W^{(n)}_{t_{\mathcal{S}+1}})^*\big|\mathcal{F}^{(n)}_{t_\mathcal{S}}\big] \\ = \sum_{j=\mathcal{S}}^{i-1} \mathbb{E}\big[(Z^{(n)}_{t_i,t_j}\Delta W^{(n)}_{t_{j+1}})(\Delta W^{(n)}_{t_{\mathcal{S}+1}})^*\big|\mathcal{F}^{(n)}_{t_\mathcal{S}}\big].$$

On the other hand, using the martingale property of the random walk and the tower property of conditional expectation, it follows that

$$\mathbb{E}\big[(Y^{(n)}_{t_i} - \mathbb{E}\big[Y^{(n)}_{t_i}\big|\mathcal{F}^{(n)}_{t_\mathcal{S}}\big])(\Delta W^{(n)}_{t_{\mathcal{S}+1}})^*\big|\mathcal{F}^{(n)}_{t_\mathcal{S}}\big] = \mathbb{E}\big[Y^{(n)}_{t_i}(\Delta W^{(n)}_{t_{\mathcal{S}+1}})^*\big|\mathcal{F}^{(n)}_{t_\mathcal{S}}\big]$$

and

$$\mathbb{E}\big[(Z^{(n)}_{t_i,t_j}\Delta W^{(n)}_{t_{j+1}})(\Delta W^{(n)}_{t_{\mathcal{S}+1}})^*\big|\mathcal{F}^{(n)}_{t_\mathcal{S}}\big] = \begin{cases} Z^{(n)}_{t_i,t_\mathcal{S}}\Delta t_{\mathcal{S}+1} & , j = \mathcal{S} \\ 0 & , j > \mathcal{S}. \end{cases}$$

Hence,

$$Z^{(n)}_{t_i,t_\mathcal{S}} = \frac{1}{\Delta t_{\mathcal{S}+1}}\mathbb{E}\big[Y^{(n)}_{t_i}(\Delta W^{(n)}_{t_{\mathcal{S}+1}})^*\big|\mathcal{F}^{(n)}_{t_\mathcal{S}}\big].$$

In the same spirit, for the case $\mathcal{S} \geq i$ we multiply Equation (2) by $\Delta W^{(n)}_{t_{\mathcal{S}+1}}$ and take the conditional expectation given $\mathcal{F}^{(n)}_{t_\mathcal{S}}$. One obtains

$$0 = \mathbb{E}\big[Y^{(n)}_{t_i}(\Delta W^{(n)}_{t_{\mathcal{S}+1}})^*\big|\mathcal{F}^{(n)}_{t_\mathcal{S}}\big] \\ = \mathbb{E}\Bigg[\Bigg(f(W^{(n)};t_i) - \sum_{j=i}^{n-1} h(t_j, Y^{(n)}_{t_j})\Delta t_{j+1}\Bigg)(\Delta W^{(n)}_{t_{\mathcal{S}+1}})^*\Bigg|\mathcal{F}^{(n)}_{t_\mathcal{S}}\Bigg] \\ - \sum_{j=i}^{n-1} \mathbb{E}\big[(Z^{(n)}_{t_i,t_j}\Delta W^{(n)}_{t_{j+1}})(\Delta W^{(n)}_{t_{\mathcal{S}+1}})^*\big|\mathcal{F}^{(n)}_{t_\mathcal{S}}\big]$$

where

$$\mathbb{E}\big[(Z^{(n)}_{t_i,t_j}\Delta W^{(n)}_{t_{j+1}})(\Delta W^{(n)}_{t_{\mathcal{S}+1}})^*\big|\mathcal{F}^{(n)}_{t_\mathcal{S}}\big] = \begin{cases} Z^{(n)}_{t_i,t_\mathcal{S}}\Delta t_{\mathcal{S}+1} & , j = \mathcal{S} \\ 0 & , j \neq \mathcal{S}. \end{cases}$$

Thus,

$$Z^{(n)}_{t_i,t_\mathcal{S}} = \frac{1}{\Delta t_{\mathcal{S}+1}}\mathbb{E}\Bigg[\Bigg(f(W^{(n)};t_i) - \sum_{j=i}^{n-1} h(t_j, Y^{(n)}_{t_j})\Delta t_{j+1}\Bigg)(\Delta W^{(n)}_{t_{\mathcal{S}+1}})^*\Bigg|\mathcal{F}^{(n)}_{t_\mathcal{S}}\Bigg].$$

□

Bibliography

[1] Ankirchner, S., Imkeller, P., Gonçalo, D.R., Classical and variational differentiability of BSDEs with quadratic growth. *Electr. Jour. of Probab.* **12**, 1418-1453 (2007).

[2] Bally, V., Approximation scheme for solutions of BSDE. *El Karoui., N., et al. (Edts), Backward stochastic differential equations* **364**, 177-191 (1997).

[3] Bender, C., Steiner, J., Least-squares Monte-Carlo for BSDEs. In: Carmona et al. (eds.), *Numerical Methods in Finance*, Springer, 257-290 (2012).

[4] Briand, P., Delyon, B., Memin, J., Donsker-type theorem of BSDEs. *Elect. Comm. in Probab.* **6**, 1-14 (2001).

[5] Berger, M., Mizel, V., Volterra equations with Itô integrals. *Int. Eq.* **2**, 187-245, 319-337 (1980).

[6] Chevance, D., Numerical methods for backward stochastic differential equations. *L.C.G. Rogers, et al. (Edts), Numerical methods in finance, University Press, Cambridge*, 232-244 (1997).

[7] Douglas, J., Ma, J., Protter, P., Numerical methods for forward-backward stochastic differential equations. *Ann. Appl. Probab.* **6**, 940-968 (1996).

[8] Jacod, J., Memin, J., Metivier, M., On tightness and stopping times. *Stoch. Proc. Appl.* **14**, 109-146 (1983).

[9] Karatzas, I., Shreve, S.E., Brownian motion and stochastic calculus. *Springer-Verlag*, New York (1988).

[10] El Karoui, N., Peng, S., Quenez, M. C., Backward stochastic differential equations in finance. *Math. Finance* **7**, 1-71 (1997).

[11] Laibson, D., Golden eggs and hyperbolic discounting. *Q. J. Econ.* **42**, 443-477 (1997).

[12] Lin, J., Adapted solution of backward stochastic nonlinear Volterra integral equations. *Stoch. Anal. Appl.* **20**, 165-183 (2002).

[13] Ma, J., Protter, P., San Martin, J., Torres, S., Numerical method for backward stochastic differential equations. *Ann. Appl. Probab.* **12**, 302-316 (2002).

[14] Ma, J., Protter, P., Yong, J., Solving forward-backward stochastic differential equations explicitly - a four step scheme. *Probab. Theory Relat. Fields* **98**, 339-359 (1994).

[15] Milstein, G. N., Tretyakov, M.V., Numerical algorithms for forward-backward stochastic differential equations. *SIAM J. Sci. Comput.* **28**, 561-582 (2006).

[16] Pardoux, E., Peng, S., Adapted solution of a backward stochastic differential equation. *Systems & Control Letters* **14**, 55-61 (1990).

[17] Pardoux, E., Peng, S., Backward stochastic differential equations and quasilinear parabolic partial differential equations. *Lecture Notes in Control and Information Sciences, Springer* **176**, 200-217 (1992).

[18] Peng, S., A nonlinear Feynman-Kac formula and applications. *Proc. Symposium of System Sciences and Control Theory, Chen & Yong eds.*, 173-184, World Scientific, Singapore (1992).

[19] Peng, S., Nonlinear expectations, nonlinear evaluations and risk measures. *Stoch. Methods in Fin., Lecture Notes in Math., Springer* **1856**, 165-253 (2004).

[20] Pham, H., Continuous-time stochastic control and optimization with financial applications. *Springer-Verlag*, Berlin (2009).

[21] Rouge, R., El Karoui, N., Pricing via utility maximization and entropy. *Math. Finance* **10(2)**, 259-276 (2000).

[22] Strotz, R. H., Myopia and inconsistency in dynamic utility maximization. *Rev.*

Econ. Stud. **23**, 165-180 (1956).

[23] Wang, T., BSVIEs with stochastic Lipschitz coefficients and applications in finance. *Preprint*, (2010).

[24] Wang, T., Shi, Y., Symmetrical solutions of backward stochastic Volterra integral equations and their applications. *Preprint*, (2009).

[25] Yong, J., Backward stochastic Volterra integral equations and some related problems. *Stoch. Proc. Appl.* **116**, 779-795 (2006).

[26] Yong, J., Continuous-time dynamic risk measures by backward stochastic Volterra integral equations. *Appl. Ann.* **86**, 1429-1442, (2007).

[27] Yong, J., Well-posedness and regularity of backward stochastic Volterra integral equations. *Probab. Theory Relat. Fields* **142**, 21-77, (2008).

[28] Yong, J., Zhou, X.Y., Stochastic controls: Hamiltonian systems and HJB equations. *Springer-Verlag*, New York (1999).

[29] Zhang, J., A numerical scheme for BSDEs, *Ann. Appl. Probab.* **14**, 459–488 (2004).

Chapter 6

Semi-Lagrangian schemes for parabolic equations

Kristian Debrabant and Espen Robstad Jakobsen

University of Southern Denmark, Department of Mathematics and Computer Science, Campusvej 55, 5230 Odense M, Denmark

Norwegian University of Science and Technology, NO–7491, Trondheim, Norway

1. Introduction

In this chapter we discuss a class of numerical schemes for linear and fully non-linear parabolic equations of second order. These schemes are monotone and based on differencing and interpolation. Moreover, they are very robust in the sense that they preserve monotonicity and convergence even for arbitrary degenerate problems and low regularity solutions. E. g. we can approximate diffusion equations with arbitrary positive semidefinite diffusion matrices, while the standard assumption for finite difference methods is that the matrix is diagonally dominant. We also consider fractional or nonlocal versions of the equations of order less than or equal to two. These are equations with (singular) integral terms, and in this case we also add numerical quadrature.

In linear form the equation (the initial value problem) we will consider takes the form

$$u_t - L[u](t,x) - J[u](t,x) - c(t,x)u - f(t,x) = 0 \quad \text{in } Q_T, \tag{1}$$

$$u(0,x) = g(x) \quad \text{in } \mathbb{R}^N, \tag{2}$$

where $Q_T := (0,T] \times \mathbb{R}^N$,

$$L[\phi](t,x) := \operatorname{tr}\big[a(t,x)D^2\phi\big] + b(t,x)D\phi,$$

$$J[\phi](t,x) := \int_{\mathbb{R}^M\setminus\{0\}} \Big(\phi(t,x+j(t,x,z)) - \phi - \mathbf{1}_{|z|\leq 1}j(t,x,z)D\phi\Big)\nu(dz),$$

for some positive definite matrix $a \in \mathbb{R}^{N,N}$ and vectors $b, c, f, j \in \mathbb{R}^N$. The Lévy measure $\nu(dz)$ is a positive Radon measure on $\mathbb{R}^M\setminus\{0\}$ satisfying

$$\int_{|z|>0} |z|^2 \wedge 1 \; \nu(dz) < \infty.$$

This equation models e. g. the value of options in finance with a fixed realization date.

Under suitable assumptions the solution u has a stochastic representation,

$$u(T-t,x) = E\Big(g(X_T) + \int_t^T f(s,X_s)ds\Big) \tag{3}$$

(here we assumed for simplicity $c \equiv 0$), where the stochastic process X_t (e. g. stock prices) is the solution of the following SDE:

$$dX_s = \sigma(s,X_s)dW_s + b(s,X_s)ds + \int_{|z|>0} j(s,X_s,z)\tilde{N}(dz,ds), \tag{4}$$

$$s > t, \quad X_t = x,$$

and $a = \frac{1}{2}\sigma\sigma^T$ for some $\sigma \in \mathbb{R}^{N,P}$ and W and $\tilde{N}$ denote Brownian motion and (compensated) Poisson random measure respectively. When $j \equiv 0$ this could e. g. be the Black-Scholes model or local/stochastic volatility models in finance. When $j \not\equiv 0$, the Poisson random measure produces jumps (discontinuities) in the paths of X_t, infinitely many on each time-interval if $\int_{|z|>0} \nu(dz) = \infty$. Many such (Lévy) models have been used in finance to remedy the shortcomings of the Black-Scholes model, we refer to the books [12; 31].

In general the equations we consider may be fully non-linear, degenerate parabolic integro partial differential equations (IPDEs) of Bellman type:

$$u_t + \sup_{\alpha\in\mathcal{A}} \Big\{ - L^\alpha[u](t,x) - J^\alpha[u](t,x) - c^\alpha(t,x)u - f^\alpha(t,x)\Big\} = 0 \quad \text{in } Q_T, \tag{5}$$

where $\mathcal{A}$ is the value set of the controls and the coefficients in L^α and J^α now depend also on α. Note that (1) is a special case of (5) by taking $\mathcal{A}$ equal to a point. Equation (5) is convex and non-local and arises as the dynamic programming or Bellman equation in optimal control of SDEs like the one above. In finance it is, starting with Merton's pioneering work [28; 29], applied to portfolio optimization type of problems.

In typical applications, one is faced with diffusions that may degenerate (at least on the boundary) and value functions of the optimal control problem that often are not smooth enough to be interpreted as classical solutions. However, under quite general assumptions, the value function is the unique viscosity solution [13; 21] of the IPDE (5), which is also the correct framework for typical finance problems, see [19]. Therefore, in the following we consider (5) in the case that a unique bounded continuous viscosity solution exists, see Section 5 for some relevant well-posedness and regularity results. Then, classical finite difference methods do typically not necessarily converge - stability and consistency are no longer sufficient for convergence. However, Barles and Souganidis [4; 21] showed in the case $J^\alpha \equiv 0$ that if the numerical approximation method is in addition monotone, then it will converge to the viscosity solution. This result has later found many applications in finance, see, e. g. [1; 32; 20; 33].

Classical finite difference approximations (FDMs) of (10) are not monotone unless the matrix a^α satisfies additional assumptions like e. g. being diagonally dominant [26]. More general assumptions are given in e. g. [4; 17] but at the cost of increased stencil length. In fact, Dong and Krylov [17] proved that *no fixed stencil FDM* can approximate equations with a second derivative term involving a general positive semi-definite matrix function a^α. Note that this type of result has been known for a long time, see e. g. [30; 14]. Some very simple examples of such "bad" matrices are given by

$$\begin{pmatrix} x_1^2 & \frac{1}{2}x_1x_2 \\ \frac{1}{2}x_1x_2 & x_2^2 \end{pmatrix} \text{ in } [0,1]^2, \quad \begin{pmatrix} \alpha^2 & \alpha_1\alpha_2 \\ \alpha_1\alpha_2 & \alpha_2^2 \end{pmatrix} \text{ for } \begin{pmatrix} \alpha_1 \\ \alpha_2 \end{pmatrix} \in \mathcal{A} = [0,1]^2.$$

In the following we will consider monotone approximation schemes relying on monotone interpolation. We do a motivating derivation of the schemes in the linear, local case in Section 2. In Section 3 we consider a generalized family of SL schemes for local PDEs of Bellman type and give results on consistency, stability, monotonicity and convergence in case the interpolation used is monotone. Moreover, we give robust error estimates which are valid also for degenerate equations and non-smooth solutions. The method is applied to a super-replication problem under gamma constraints. In Section 4 we derive SL schemes applicable also to IPDEs (5) of Bellman type.

Note that we will not discuss American type of option pricing problems, even though it could be done. The numerical and analytical challenges of this case fall between the linear and fully-nonlinear cases, and it should be easy to adapt the ideas herein in this direction.

2. Some motivation for the schemes

We now formally derive a numerical scheme for the PDE (1) by a two step procedure: (i) Discretize (in time) the associated representation formula (3) and SDE (4), and (ii) introduce a spatial grid via interpolation in space.

This idea has been explored in the general case for control problems in [16; 11; 18; 27; 9]. We illustrate this approach by deriving an explicit scheme in the linear case and when the coefficients $c \equiv 0$ and $j \equiv 0$. Note that in this case the equation is local and $J \equiv 0$. Let $s_n = n\Delta t$ for $n = 0, \dots, M$ be discrete times and let $T = M\Delta t$. We consider a simple quadrature approximation of (3) combined with a *suitably chosen* weak Euler approximation of (4):

$$\tilde{u}(T - s_m, x) = E\Big[\sum_{k=m}^{M-1} f(\tilde{X}_k)\,\Delta t + g(\tilde{X}_M)\Big], \tag{6}$$

$$\tilde{X}_m = x, \quad \tilde{X}_n = \tilde{X}_{n-1} + \sigma(\tilde{X}_{n-1})\,k\,\xi_n + b(\tilde{X}_{n-1})\,k^2\,\eta_n, \; n > m, \tag{7}$$

where $k = \sqrt{(P+1)\Delta t}$ and $\xi_n = (\xi_{n,1}, \dots, \xi_{n,P})^\top$ and η_n are sequences of i. i. d.

random variables satisfying

$$P\Big((\xi_{n,1},\dots,\xi_{n,P},\eta_n)=\pm e_j\Big)=\frac{1}{2(P+1)}\quad\text{if } j\in\{1,\dots,P\},$$
$$P\Big((\xi_{n,1},\dots,\xi_{n,P},\eta_n)=e_{P+1}\Big)=\frac{1}{P+1},$$

(e_j denotes the j-th unit vector) and all other values of $(\xi_{n,1},\dots,\xi_{n,P},\eta_n)$ have probability zero. By the definition of $\tilde{u}$ and properties of the conditional expectation (or the dynamic programming principle in the control setting) one sees that

$$\tilde{u}(T-s_m,x)=E\Big[f(\tilde{X}_m)\,\Delta t+\tilde{u}(T-s_{m+1},\tilde{X}_{m+1})\Big].$$

By the definition of $\tilde{X}$ (7), we find that

$$\tilde{u}(T-s_m,x)=E\Big[f(x)\,\Delta t+\tilde{u}(T-s_{m+1},x+\sigma(x)\,k\,\xi_n+b(x)\,k^2\,\eta_n)\Big],$$

and after evaluating the expectation that

$$\begin{aligned}\tilde{u}(T-s_m,x)=\;& f(x)\Delta t+\frac{1}{P+1}\tilde{u}(T-s_{m+1},x+b(x)k^2)\\ &+\frac{1}{P+1}\sum_{j=1}^{P}\frac{1}{2}\Big(\tilde{u}(T-s_{m+1},x+\sigma_j(x)k)+\tilde{u}(T-s_{m+1},x-\sigma_j(x)k)\Big).\end{aligned}$$

Subtract $\tilde{u}(T-s_{m+1},x)$ from both sides and divide by $\Delta t=\frac{k^2}{P+1}$. If we let $m=M-n$ and $t_n=T-s_{M-n}$ (and $t_{n-1}=T-s_{(M-n)+1}$), we then find that

$$\frac{\tilde{u}(t_n,x)-\tilde{u}(t_{n-1},x)}{\Delta t}=f(x)+\tilde{L}[\tilde{u}](t_{n-1},x),\tag{8}$$

where

$$\begin{aligned}\tilde{L}[\phi](t,x)=&\sum_{j=1}^{P}\frac{\phi(t,x+\sigma_j(x)k)-2\phi(t,x)+\phi(t,x-\sigma_j(x)k)}{k^2}\\ &+\frac{\phi(t,x+b(x)k^2)-\phi(t,x)}{k^2}.\end{aligned}$$

Note that $-\tilde{L}$ is a non-negative operator, and that hence the explicit scheme is monotone.

We have obtained a derivative free discrete in time scheme. It remains to discretize also in space. The idea we will use is to replace $\tilde{u}$ in (8) by some interpolant $\mathcal{I}u$ associated to a non-degenerate spatial grid $X_{\Delta x}=\{x_\alpha\}_{\alpha\in\mathbb{Z}^N}$ where the distance between the nodes is of the order Δx. Note that this means that

$$\mathcal{I}\phi(x_\alpha)=\phi(x_\alpha)\qquad\text{and}\qquad\mathcal{I}\phi(x)\to\phi(x)\quad\text{as}\quad\Delta x\to 0.$$

A fully discrete scheme can then be obtained by requiring this new scheme to hold in all grid points:

$$\frac{U(t_n,x_\alpha)-U(t_{n-1},x)}{\Delta t}=f(x_\alpha)+\tilde{L}[\mathcal{I}U](t_{n-1},x_\alpha)\quad\text{for}\quad n\geq 0,\alpha\in\mathbb{Z}^N.\tag{9}$$

If $\mathcal{I}$ represents a monotone interpolation in the sense that $U \leq V$ implies that $\mathcal{I}U \leq \mathcal{I}V$ (e. g. piecewise linear interpolation), then it is easy to see that $-\tilde{L} \circ \mathcal{I}$ is a non-negative operator and again the explicit scheme is monotone.

Remark 6.1. We did not require a CFL condition for the above explicit scheme to be monotone. The reason is essentially that the discretization of space derivatives here depends directly on Δt through k but not on Δx.

The best monotone interpolation is piecewise linear interpolation with quadratic $O(\Delta x^2)$ convergence. The local truncation error of the scheme has contributions from discretization in time, approximation of derivatives, and interpolation error. In the case of linear interpolation it takes the form:

$$O\Big(\Delta t + k^2 + \frac{\Delta x^2}{k^2}\Big),$$

and the optimal choices are $\Delta t \sim k^2 \sim \Delta x$, i. e. a rather good linear "CFL" condition but with a linear rate of $O(\Delta x)$.

Remark 6.2. By construction, the arguments of Φ occurring in the definition of $\tilde{L}[\Phi]$ can be seen as a short time approximation of (7). Hence the scheme (9) tracks particle paths approximately. In view of the discussion above we might say that the scheme follows particles in the mean because of the expectation. For first order PDEs, schemes defined in this way are called SL schemes by e. g. Falcone, so we will also denote the generalized schemes considered in the following as SL schemes.

3. Local PDEs of Bellman type

In this section we restrict to local PDEs of Bellman type

$$u_t - \inf_{\alpha \in \mathcal{A}} \Big\{L^\alpha[u](t,x) + c^\alpha(t,x)u + f^\alpha(t,x)\Big\} = 0 \qquad \text{in} \quad Q_T, \tag{10}$$

$$u(0,x) = g(x) \qquad \text{in} \quad \mathbb{R}^N, \tag{11}$$

where

$$L^\alpha[u](t,x) = \mathrm{tr}[a^\alpha(t,x)D^2u(t,x)] + b^\alpha(t,x)Du(t,x),$$

i. e. $J^\alpha \equiv 0$, and generalize the above construction. Consider general finite difference approximations of the differential operator $L^\alpha[\phi]$ in (10) defined as

$$L_k^\alpha[\phi](t,x) := \sum_{i=1}^{M} \frac{\phi(t, x + y_{k,i}^{\alpha,+}(t,x)) - 2\phi(t,x) + \phi(t, x + y_{k,i}^{\alpha,-}(t,x))}{2k^2}, \tag{12}$$

for $k > 0$ and some $M \geq 1$, where for all smooth functions ϕ,

$$|L_k^\alpha[\phi] - L^\alpha[\phi]| \leq C(|D\phi|_0 + \cdots + |D^4\phi|_0)k^2. \tag{13}$$

This approximation and interpolation yield a semi-discrete approximation of (10),

$$U_t - \inf_{\alpha \in \mathcal{A}} \Big\{L_k^\alpha[\mathcal{I}U](t,x) + c^\alpha(t,x)U + f^\alpha(t,x)\Big\} = 0 \quad \text{in} \quad (0,T) \times X_{\Delta x},$$

and the final scheme can then be found after discretizing in time using a parameter $\theta \in [0,1]$,

$$\delta_{\Delta t_n} U_i^n = \inf_{\alpha \in \mathcal{A}} \left\{ L_k^\alpha [\mathcal{I}\bar{U}_\cdot^{\theta,n}]_i^{n-1+\theta} + c_i^{\alpha,n-1+\theta} \bar{U}_i^{\theta,n} + f_i^{\alpha,n-1+\theta} \right\} \tag{14}$$

in G, where $U_i^n = U(t_n, x_i)$, $f_i^{\alpha,n-1+\theta} = f^\alpha(t_{n-1} + \theta \Delta t_n, x_i)$, ... for $(t_n, x_i) \in G$,

$$\delta_{\Delta t}\phi(t,x) = \frac{\phi(t,x) - \phi(t - \Delta t, x)}{\Delta t}, \qquad \text{and} \qquad \bar{\phi}_\cdot^{\theta,n} = (1-\theta)\phi_\cdot^{n-1} + \theta \phi_\cdot^n.$$

As initial conditions we take

$$U_i^0 = g(x_i) \quad \text{in} \quad X_{\Delta x}. \tag{15}$$

For the choices $\theta = 0, 1$, and $1/2$ the time discretization corresponds to respectively explicit Euler, implicit Euler and midpoint rule. For $\theta = 1/2$, the full scheme can be seen as generalized Crank-Nicolson type discretization.

For the approximation L_k^α and interpolation $\mathcal{I}$ we assume

$$\text{(Y1)} \begin{cases} \sum_{i=1}^M [y_{k,i}^{\alpha,+} + y_{k,i}^{\alpha,-}] = 2k^2 b^\alpha + \mathcal{O}(k^4), \\ \sum_{i=1}^M [y_{k,i}^{\alpha,+} y_{k,i}^{\alpha,+\top} + y_{k,i}^{\alpha,-} y_{k,i}^{\alpha,-\top}] = 2k^2 \sigma^\alpha \sigma^{\alpha\top} + \mathcal{O}(k^4), \\ \sum_{i=1}^M [y_{k,i,j_1}^{\alpha,+} y_{k,i,j_2}^{\alpha,+} y_{k,i,j_3}^{\alpha,+} + y_{k,i,j_1}^{\alpha,-} y_{k,i,j_2}^{\alpha,-} y_{k,i,j_3}^{\alpha,-}] = \mathcal{O}(k^4), \\ \sum_{i=1}^M [y_{k,i,j_1}^{\alpha,+} y_{k,i,j_2}^{\alpha,+} y_{k,i,j_3}^{\alpha,+} y_{k,i,j_4}^{\alpha,+} + y_{k,i,j_1}^{\alpha,-} y_{k,i,j_2}^{\alpha,-} y_{k,i,j_3}^{\alpha,-} y_{k,i,j_4}^{\alpha,-}] = \mathcal{O}(k^4), \end{cases}$$

for all $j_1, j_2, j_3, j_4 = 1, 2, \ldots, N$ indicating components of the y-vectors.

(I.1) There are $K \geq 0, r \in \mathbb{N}$ such that for all smooth functions ϕ

$$|(\mathcal{I}\phi) - \phi|_0 \leq K |D^r \phi|_0 \Delta x^r.$$

(I.2) There is a set of non-negative functions $\{w_j(x)\}_j$ such that

$$(\mathcal{I}\phi)(x) = \sum_j \phi(x_j) w_j(x),$$

and

$$w_j(x) \geq 0, \qquad w_i(x_j) = \delta_{ij}$$

for all $i, j \in \mathbb{N}$.

Under assumption (Y1), a Taylor expansion shows that L_k^α is a second order consistent approximation satisfying (13). If we assume also (I.1), it then follows that $L_k^\alpha[\mathcal{I}\phi]$ is a consistent approximation of $L^\alpha[\phi]$ if $\frac{\Delta x^r}{k^2} \to 0$. Indeed

$$|L_k^\alpha[\mathcal{I}\phi] - L^\alpha[\phi]| \leq |L_k^\alpha[\mathcal{I}\phi] - L_k^\alpha[\phi]| + |L_k^\alpha[\phi] - L^\alpha[\phi]|,$$

where $|L_k^\alpha[\phi] - L^\alpha[\phi]|$ is estimated in (13), and by (I.1),

$$|L_k^\alpha[\mathcal{I}\phi] - L_k^\alpha[\phi]| \leq C|D^r\phi|_0 \frac{\Delta x^r}{k^2}.$$

Remark 6.3. Assumption (Y1) is similar to the local consistency conditions used in [26]. The $O(k^4)$ terms insure that the method is second order accurate as $k \to 0$. Convergence will still be achieved if we relax $O(k^4)$ to $o(k^2)$ as $k \to 0$.

Remark 6.4. An interpolation satisfying (I.2) is said to be *positive* and is *monotone* in the sense explained just before Remark 6.1. Such an interpolation $\mathcal{I}\phi$ does not use (exact) derivatives to reconstruct the function ϕ. Typically $\mathcal{I}$ will be constant, linear, or multi-linear interpolation (i. e. $r \leq 2$ in (I.1)), because higher order interpolation is not monotone in general.

3.1. *Examples of approximations L_k^α*

We present several examples of approximations of the term $L^\alpha[\phi]$ of the form $L_k^\alpha[\phi]$, including previous approximations that have appeared in [18; 27; 9; 14] plus some new approximations to show the flexibility of our framework.

(1) The approximation of Falcone [18] (see also [11]),

$$b^\alpha D\phi \approx \frac{\mathcal{I}\phi(x + hb^\alpha) - \mathcal{I}\phi(x)}{h},$$

corresponds to our L_k^α if $k = \sqrt{h}$, $y_k^{\alpha,\pm} = k^2 b^\alpha$.

(2) The approximation of Crandall-Lions [14],

$$\frac{1}{2}\mathrm{tr}[\sigma^\alpha \sigma^{\alpha\top} D^2\phi] \approx \sum_{j=1}^{P} \frac{\mathcal{I}\phi(x + k\sigma_j^\alpha) - 2\mathcal{I}\phi(x) + \mathcal{I}\phi(x - k\sigma_j^\alpha)}{2k^2},$$

corresponds to our L_k^α if $y_{k,j}^{\alpha,\pm} = \pm k\sigma_j^\alpha$ and $M = P$.

(3) The corrected version of the approximation of Camilli-Falcone [9] (see also [27]),

$$\frac{1}{2}\mathrm{tr}[\sigma^\alpha \sigma^{\alpha\top} D^2\phi] + b^\alpha D\phi$$
$$\approx \sum_{j=1}^{P} \frac{\mathcal{I}\phi(x + \sqrt{h}\sigma_j^\alpha + \frac{h}{P}b^\alpha) - 2\mathcal{I}\phi(x) + \mathcal{I}\phi(x - \sqrt{h}\sigma_j^\alpha + \frac{h}{P}b^\alpha)}{2h},$$

corresponds to our L_k^α if $k = \sqrt{h}$, $y_{k,j}^{\alpha,\pm} = \pm k\sigma_j^\alpha + \frac{k^2}{P}b^\alpha$ and $M = P$.

(4) The new approximation obtained by combining approximations 1 and 2,

$$\frac{1}{2}\mathrm{tr}[\sigma^\alpha\sigma^{\alpha\top}D^2\phi]+b^\alpha D\phi\approx\frac{\mathcal{I}\phi(x+k^2b^\alpha)-\mathcal{I}\phi(x)}{k^2}+\sum_{j=1}^{P}\frac{\mathcal{I}\phi(x+k\sigma_j^\alpha)-2\mathcal{I}\phi(x)+\mathcal{I}\phi(x-k\sigma_j^\alpha)}{2k^2},$$

corresponds to our L_k^α if $y_{k,j}^{\alpha,\pm}=\pm k\sigma_j^\alpha$ for $j\leq P$, $y_{k,P+1}^{\alpha,\pm}=k^2b^\alpha$ and $M=P+1$.

(5) Yet another new approximation,

$$\begin{aligned}&\frac{1}{2}\mathrm{tr}[\sigma^\alpha\sigma^{\alpha\top}D^2\phi]+b^\alpha D\phi\\&\approx\sum_{j=1}^{P-1}\frac{\mathcal{I}\phi(x+k\sigma_j^\alpha)-2\mathcal{I}\phi(x)+\mathcal{I}\phi(x-k\sigma_j^\alpha)}{2k^2}\\&+\frac{\mathcal{I}\phi(x+k\sigma_P^\alpha+k^2b^\alpha)-2\mathcal{I}\phi(x)+\mathcal{I}\phi(x-k\sigma_P^\alpha+k^2b^\alpha)}{2k^2},\end{aligned}$$

corresponds to our L_k^α if $y_{k,j}^{\alpha,\pm}=\pm k\sigma_j^\alpha$ for $j<P$, $y_{k,P}^{\alpha,\pm}=\pm k\sigma_P^\alpha+k^2b^\alpha$ and $M=P$.

When σ^α does not depend on α but b^α does, approximations (4) and (5) are much more efficient than approximation (3).

3.2. *Linear interpolation SL scheme (LISL)*

To keep the scheme (14) monotone, linear or multi-linear interpolation is the most accurate interpolation one can use in general. In this typical case we call the full scheme (14)–(15) the LISL scheme. In the following, we denote by $c^{\alpha,+}$ the positive part of c^α, and we use the technical assumptions (A.1) and (A.2) from the appendix. Then we have the following result by [16]:

Theorem 1. *Assume that (A.1), (A.2), (I.1), (I.2) and (Y1) hold.*

(a) The LISL scheme is monotone if the CFL conditions

$$(1-\theta)\Delta t\Big[\frac{M}{k^2}-c_i^{\alpha,n-1+\theta}\Big]\leq 1\ \text{ and }\ \theta\Delta t\, c_i^{\alpha,n-1+\theta}\leq 1 \text{ for all } \alpha,n,i \tag{16}$$

hold.

(b) The truncation error of the LISL scheme is $O(|1-2\theta|\Delta t+\Delta t^2+k^2+\frac{\Delta x^2}{k^2})$, so it is first order accurate when $k=O(\Delta x^{1/2})$ and $\Delta t=O(\Delta x)$, or if $\theta=\frac{1}{2}$, $\Delta t=O(\Delta x^{1/2})$.

(c) If $2\theta\Delta t\sup_\alpha|c^{\alpha,+}|_0\leq 1$ and (16) hold, then there exists a unique bounded and L^∞-stable solution U of the LISL scheme converging uniformly to the solution u of (10)–(11) as $\Delta t, k, \frac{\Delta x}{k}\to 0$.

From this result it follows that the scheme is at most *first order accurate*, has *wide and increasing stencil* and a *good CFL condition*. From the truncation error and the definition of L_k^α the stencil is wide since the scheme is consistent only if $\Delta x/k \to 0$ as $\Delta x \to 0$ and has stencil length proportional to

$$l := \frac{\max_{t,x,\alpha,i} |y_{k,i}^{\alpha,-}| \vee |y_{k,i}^{\alpha,+}|}{\Delta x} \sim \frac{k}{\Delta x} \to \infty \quad \text{as} \quad \Delta x \to 0.$$

Here we have used that if (Y1) holds and $\sigma \not\equiv 0$, then typically $y_{k,i}^{\alpha,\pm} \sim k$. Note that if $k = \Delta x^{1/2}$, then $l \sim \Delta x^{-1/2}$. Finally, in the case $\theta \neq 1$ the CFL condition for (14) is $\Delta t \leq Ck^2 \sim \Delta x$ when $k = O(\Delta x^{1/2})$, and it is much less restrictive than the usual parabolic CFL condition, $\Delta t = O(\Delta x^2)$.

Remark 6.5. The LISL scheme is consistent and monotone for arbitrary degenerating diffusions, without requiring that a^α is diagonally dominant or similar conditions. In comparison to other schemes applicable in this situation, like the ones of Bonnans-Zidani [4; 7], it is much easier to analyze and to implement and faster in the sense that the computational cost for approximating the diffusion matrix is for fixed x, t, α independent of the stencil size.

3.3. *Error estimates*

In the following we take a uniform time-grid, $G = \Delta t\{0, 1, \ldots, N_T\} \times X_{\Delta x}$, for simplicity. Let $Q_{\Delta t,T} := \Delta t\{0, 1, \ldots, N_T\} \times \mathbb{R}^N$. To apply the regularization method of Krylov [25] we need a continuity and continuous dependence result for the scheme that relies on the following additional (covariance-type) assumptions: Whenever two sets of data σ, b and $\tilde\sigma, \tilde b$ are given, the corresponding approximations $L_k^\alpha, y_{k,i}^{\alpha,\pm}$ and $\tilde L_k^\alpha, \tilde y_{k,i}^{\alpha,\pm}$ in (12) satisfy

$$\text{(Y2)} \qquad \begin{cases} \displaystyle\sum_{i=1}^M [y_{k,i}^{\alpha,+} + y_{k,i}^{\alpha,-}] - [\tilde y_{k,i}^{\alpha,+} + \tilde y_{k,i}^{\alpha,-}] \leq 2k^2(b^\alpha - \tilde b^\alpha), \\ \displaystyle\sum_{i=1}^M [y_{k,i}^{\alpha,+} y_{k,i}^{\alpha,+\top} + y_{k,i}^{\alpha,-} y_{k,i}^{\alpha,-\top}] + [\tilde y_{k,i}^{\alpha,+} \tilde y_{k,i}^{\alpha,+\top} + \tilde y_{k,i}^{\alpha,-} \tilde y_{k,i}^{\alpha,-\top}] \\ \qquad -[y_{k,i}^{\alpha,+} \tilde y_{k,i}^{\alpha,+\top} + \tilde y_{k,i}^{\alpha,+} y_{k,i}^{\alpha,+\top} + y_{k,i}^{\alpha,-} \tilde y_{k,i}^{\alpha,-\top} + \tilde y_{k,i}^{\alpha,-} y_{k,i}^{\alpha,-\top}] \\ \leq 2k^2(\sigma^\alpha - \tilde\sigma^\alpha)(\sigma^\alpha - \tilde\sigma^\alpha)^\top + 2k^4(b^\alpha - \tilde b^\alpha)(b^\alpha - \tilde b^\alpha)^\top, \end{cases}$$

when $\sigma, b, y_k^\pm$ are evaluated at (t, x) and $\tilde\sigma, \tilde b, \tilde y_k^\pm$ are evaluated at (t, y) for all t, x, y.

Then one can prove the following error estimate [16]:

Theorem 2 (Error Bound). *Assume that (A.1), (A.2), (I.1), (I.2), (Y1), (Y2), and the CFL conditions* (16) *hold, and that* $k \in (0, 1)$ *and* $\Delta t \leq (2k_0 \wedge 2k_1)^{-1}$. *If* u *solves* (10)–(11) *and* U *solves* (14)–(15), *then*

$$|u - U| \leq C(|1 - 2\theta|\Delta t^{1/4} + \Delta t^{1/3} + k^{1/2} + \frac{\Delta x}{k^2}) \quad \textit{in} \quad G.$$

This error bound holds also for unstructured grids. For more regular solutions it is possible to obtain better error estimates, but general and optimal results are not available. The best estimate in our case is $O(\Delta x^{1/5})$ which is achieved when $k = O(\Delta x^{2/5})$ and $\Delta t = O(k^2)$. Note that the CFL conditions (16) already imply that $\Delta t = O(k^2)$ if $\theta < 1$. Also note that the above bound does not show convergence when k is optimal for the LISL scheme ($k = O(\Delta x^{1/2})$).

Remark 6.6. These results are consistent with results for LISL type schemes for stationary Bellman equations. In fact if all coefficients are independent of time and $c^\alpha(x) < -c < 0$, then by combining the results of [9] and [3], exactly the same error estimate is obtained for the solution of a particular stationary LISL scheme and the unique *stationary* Lipschitz solution of (10).

3.4. *Boundary conditions*

When solving PDEs on bounded domains, the SL (like other) schemes may exceed the domain if they are not modified near the boundary. The reason is of course the wide stencil. This may or may not be a problem depending on the equation and the type of boundary condition: (i) For Dirichlet conditions the scheme needs to be modified near the boundary or boundary conditions must be extrapolated. This may result in a loss of accuracy or monotonicity near the boundary. (ii) Homogeneous Neumann conditions can be implemented exactly by extending in the normal direction the values of the solution on the boundary to the exterior. (iii) If the boundary has no regular points, no boundary conditions can be imposed. In this case the SL schemes will not leave the domain if the normal diffusion tends to zero fast enough when the boundary is approached. Typical examples are equations of Black-Scholes type.

3.5. *Convergence test for a super-replication problem*

We consider a test problem from [6] which was used to test convergence rates for numerical approximations of a super-replication problem from finance. The corresponding PDE is

$$\inf_{\alpha_1^2+\alpha_2^2=1} \left\{ \alpha_1^2 u_t(t,x) - \frac{1}{2}\mathrm{tr}\left(\sigma^\alpha(t,x)\sigma^{\alpha\top}(t,x)D^2u(t,x)\right)\right\} = f(t,x),$$
$$0 \le x_1, x_2 \le 3 \tag{17}$$

with $\sigma^\alpha(t,x) = \begin{pmatrix} \alpha_1 x_1 \sqrt{x_2} \\ \alpha_2 \eta(x_2) \end{pmatrix}$ and $\eta(x) = x(3-x)$. We take $u(t,x) = 1+t^2-e^{-x_1^2-x_2^2}$ as exact solution as in [6], and then f is forced to be

$$f(t,x) = \frac{1}{2}\left(u_t - \frac{1}{2}x_1^2 x_2 u_{x_1x_1} - \frac{1}{2}x_2^2(3-x_2)^2 u_{x_2x_2}\right.$$
$$\left. -\sqrt{\left(-u_t + \frac{1}{2}x_1^2 x_2 u_{x_1x_1} - \frac{1}{2}x_2^2(3-x_2)^2 u_{x_2x_2}\right)^2 + \left(x_1\sqrt{x_2}^3(3-x_2)u_{x_1x_2}\right)^2}\right).$$

In [6] $\eta(x) = x$, while we take $\eta(x) = x(3-x)$ to prevent the LISL scheme from overstepping the boundaries. Note that changing η does *not* change the solutions as long as $\eta > 0$ in the interior of the domain, see [6], and hence the above equation is equivalent to the equation used in [6]. The initial values and Dirichlet boundary values at $x_1 = 0$ and $x_2 = 0$ are taken from the exact solution. As in [6], at $x = 3$ and $y = 3$ homogeneous Neumann boundary conditions are implemented. To approximate the values of α_1, α_2, the Howard algorithm is used (see [6]), which requires an implicit time discretization, so we choose $\theta = 1$. As stop criterion of the iterations we require that the change of the maximal component and the sum over all components of the residual in Howard's algorithm are both smaller than 0.01. The minimization is done over $\alpha_{1,k} + i\alpha_{2,k} = e^{2\pi i k/2N_{\Delta x}}$, $k = 1,\ldots,N_{\Delta x}$, where $N_{\Delta x} = 3/\Delta x$ is the number of space grid points in one dimension. The calculations are done in MATLAB, on an INTEL(R) Core(TM)2 Duo P8700, 2.54 Ghz Laptop. The linear systems involved are solved by the standard MATLAB back slash operator, using internally UMFPACK [15]. We choose $k = \sqrt{\Delta x}$ and a regular triangular grid. The numbers of time steps are chosen as $\frac{1}{\Delta x}$.

The results at $t = 1$ are given in Table 1. The numerical order of convergence is approximately one. The CPU times are better than expected: They get multiplied roughly by 10 when Δx is divided by 2, a property which can also be observed in [6]. The reason is that the Howard algorithm needs fewer iterations when the time step becomes smaller.

Table 1. Results for the convergence test for the super-replication problem at $t = 1$

Δx	error	rate	time in s
1.50e-1	2.01e-1		0.71
7.50e-2	9.49e-2	1.08	5.52
3.75e-2	4.29e-2	1.15	59.32
1.87e-2	1.94e-2	1.15	803.26

Remark 6.7. Equation (17) can not be written in a form (10) satisfying the assumptions of this section, so the results of this section do not apply to this problem. However, it seems possible to extend them to cover this problem using comparison results from [6] along with L^∞-bounds on the numerical solution that follow from the maximum principle.

Remark 6.8. If we compare naively these results to the results of [6], we find that the LISL scheme is about 10 times faster than the method of [6]. Of course, this comparison is not fair, e. g. it could be that a less efficient linear solver is used in [6].

3.6. *A super-replication problem*

We apply our method to solve a problem from finance, the super-replication problem under gamma constraints considered in [6]. It consists of solving equation (17) with $f \equiv 0$, Neumann boundary conditions and σ^α as in Subsection 3.5, and initial and Dirichlet conditions given by

$$u(t,x) = \max(0, 1 - x_1), \quad t = 0 \quad or \quad x_1 = 0 \quad or \quad x_2 = 0.$$

The solution obtained with the LISL scheme is given in Figure 1 and coincides with the solution found in [6]. It gives the price of a put option of strike and maturity 1, and x_1 and x_2 are respectively the price of the underlying and the price of the forward variance swap on the underlying.

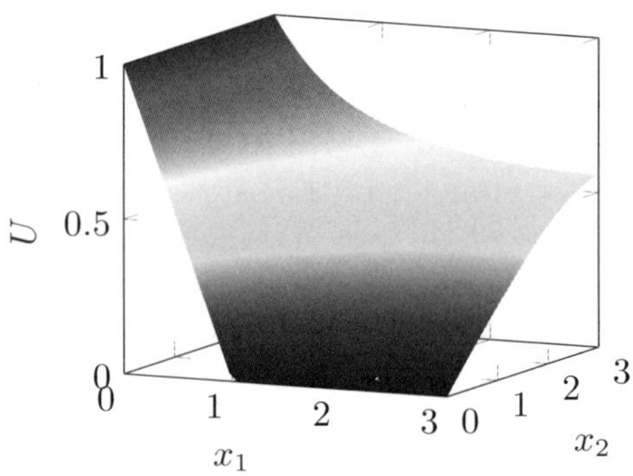

Figure 1. Numerical solution of super-replication problem at $t = 1$

4. Nonlocal PDEs of Bellman type

In this section we derive a Semi-Lagrangian scheme for equation (5) in the non-local case. We also discuss very briefly how this scheme can be shown to converge and how error estimates can be obtained. We repeat that the non-local equations arise in non-Gaussian models of (financial) markets, e. g. if you want to incorporate heavy tail distributions or skew distributions. Such models have been extensively treated in the books by Cont-Tankov [12] and Schoutens [31].

4.1. *Derivation of the scheme for linear problems*

Now we return to the non-local scheme which will be a time-dependent version of the stationary scheme derived and extensively analyzed in Camilli-Jakobsen [10]. In the following we will derive the scheme in the linear case and when $c \equiv 0$, modifying the derivation of the local scheme in Section 3 to treat the nonlocal case when $j \not\equiv 0$. Here we also assume that

$$\bar{\nu} = \int_{|z|>0} \nu(dz) < \infty, \tag{18}$$

and we remark that this means that the jump part of the process is a compound Poisson process with intensity $\bar{\nu}$ and increment distribution (jump length and direction) $\frac{\nu}{\bar{\nu}}$. We will come back to this assumption. Compared to the local construction, we only need to change the approximation of the SDE. We take e. g. an interlacing weak Euler approximation of (4) from [10]:

$$\begin{cases} \tilde{X}_0 = x, \\ \tilde{X}_n = \tilde{X}_{n-1} + \Delta t b(\tilde{X}_{n-1}, t_{n-1}) + \sqrt{P\Delta t}\sum_{m=1}^{P} \sigma_m(X_{n-1}, t_{n-1})\xi_{n-1}^m, \\ \qquad\qquad\qquad\qquad \text{for } n = N_i + 1, N_i + 2, \ldots, N_{i+1} - 1, \\ X_{N_{i+1}} = X_{N_{i+1}-1} + j(t_{N_{i+1}-1}, X_{N_{i+1}-1}, Z_i), \end{cases}$$

where ξ_n^m, $m = 1, \ldots, d$ are random variables taking values in $\{-1, 0, 1\}$ such that

$$P[\{\xi_n^i = \pm 1\}] = \frac{1}{2P} \qquad \text{and} \qquad P[\{\xi_n^i \neq 0\} \cap \{\xi_n^j \neq 0\}] = 0, \quad i \neq j.$$

Moreover, $N_n = \sum_{i=0}^{n-1} \hat{N}_i$, and $\hat{N}_i$ and Z_i, $i \in \mathbb{N}_0$, are i. i. d. sequences of stochastic variables taking values in $\{0, 1\}$ and in $\mathbb{R}^N$ and representing the number of jumps and the corresponding z-jumps (length and direction) of the Poisson measure $\tilde{N}$ in $[t_{i-1}, t_i)$. We set $N_0 = 0$ and $Z_0 = 0$ and assume that $\hat{N}_i$ have the probability distribution

$$P[\hat{N}_i = 0] = e^{-\Delta t \bar{\nu}} \qquad \text{and} \qquad P[\hat{N}_i = 1] = 1 - e^{-\Delta t \bar{\nu}}$$

for $i \in \mathbb{N}_0$, i. e. they are weak approximations of the Poisson process with intensity $\Delta t \bar{\nu}$, while the Z_i, $i \in \mathbb{N}$, are i. i. d. random variables with probability density $\frac{\nu}{\bar{\nu}}$.

Between jumps the process evolves like an approximation to a random walk, and when the process jumps there is no random walk part. This is called interlacing and also holds for the associated continuous time process. The probability to jump (once!) in a time interval is $1 - e^{-\bar{\nu}\Delta t}$, and this is only a good approximation if $\bar{\nu}\Delta t$

is small. In the current case we can easily compute the expectation and find that

$$E(\phi(X_1)) = e^{-\bar{\nu}\Delta t}\frac{1}{2P}\sum_{i=1}^{P}\Big(\phi(x+\Delta t b(x,t)+\sqrt{P\Delta t}\sigma_i(x,t)) + \phi(x+\Delta t b(x,t)-\sqrt{P\Delta t}\sigma_i(x,t))\Big) + (1-e^{-\bar{\nu}\Delta t})\int_{|z|>0}\phi(x+j(x,z,t))\frac{\nu(dz)}{\bar{\nu}}.$$

Following the derivation of the local scheme, we are again lead to the scheme (9) but now with a modified $\tilde{L}$:

$$\begin{aligned}\tilde{L}\phi(x) = {} & e^{-\bar{\nu}\Delta t}\frac{1}{2}\sum_{m=1}^{d}\Big(\frac{\phi(x+\Delta t b(x,t)+\sqrt{P\Delta t}\sigma_m(x,t))}{P\Delta t} \\ & \qquad - \frac{2\phi(x)+\phi(x+\Delta t b(x,t)-\sqrt{P\Delta t}\sigma_m(x,t))}{P\Delta t}\Big) \\ & + \frac{1-e^{-\bar{\nu}\Delta t}}{\bar{\nu}\Delta t}\int_{|z|>0}\phi(x+j(x,z,t))-\phi(x)\ \nu(dz).\end{aligned}$$

Again we see that $-\tilde{L}\circ\mathcal{I}$ is nonnegative and the scheme is monotone if $\mathcal{I}$ is monotone.

Remark 6.9. When $j\equiv 0$, the operator $\tilde{L}$ is different from the $\tilde{L}$ of Section 3, and hence gives a new scheme also for purely local equation. In fact by carefully designing the weak Euler approximation of (4) a whole family of schemes can be obtained, schemes that will have different strong and weak sides.

4.2. *The scheme in the nonlinear case under (18)*

A similar derivation using the dynamical programming principle would give the following scheme for (5) (compare with (9)):

$$\frac{U(t_m,x_\alpha)-U(t_{m-1},x_\alpha)}{\Delta t} = \sup_{\alpha\in\mathcal{A}}\Big\{\tilde{L}^\alpha[\mathcal{I}U](t_{m-1},x_\alpha)+c^\alpha(t_{m-1},x_\alpha)U(t_{m-1},x_\alpha)+f(t_{m-1},x_\alpha)\Big\}, \tag{19}$$

for $m\geq 0, \alpha\in\mathbb{Z}^N$ where $\tilde{L}^\alpha$ is defined as $\tilde{L}$ except that σ, b, and η now depend also on α.

Note that the scheme so far is explicit, and if $\mathcal{I}$ is a monotone interpolation operator, then the scheme is monotone in the sense of [4]. See Remark 6.1 for a remark on the lack of CFL condition. If linear interpolation (a monotone $O(\Delta x^2)$ interpolation) is used, then the scheme is monotone with a truncation error of the form

$$O\Big(\Delta t+\Delta t+\frac{\Delta x^2}{\Delta t}+\bar{\nu}(\Delta t+\Delta x^2)\Big).$$

Here the three first terms are as in the local case (see Remark 6.1), the fourth term comes from

$$e^{-\bar{\nu}\Delta t}, \frac{1-e^{-\bar{\nu}\Delta t}}{\bar{\nu}\Delta t} \approx 1 + O(\bar{\nu}\Delta t),$$

and the last term from the ν-integrated interpolation error.

In view of these properties, and under standard assumptions on the coefficients, we easily find that the scheme is L^∞-stable and we can then use a nonlocal version of the Barles-Souganidis result (see Section 1) to prove local uniform convergence of the scheme (19) to the (not necessarily smooth viscosity) solution of (5). To summarize, the scheme (19) has the following properties:

(1) it is *consistent* and *monotone*,

(2) it is *first order* accurate with *linear CFL condition* $\Delta t = C\Delta x$,

(3) it is L^∞*-stable* and *locally uniformly convergent* under general and natural assumptions which include degenerate problems and nonsmooth solutions,

(4) robust *a priori error estimates* can be obtained under general and natural assumptions which include degenerate problems and nonsmooth solutions.

These properties were proved in the stationary case in [10], and these proofs can be adapted to the present time dependent case without too much difficulties.

Remark 6.10.

(1) The above scheme is not fully discrete. It remains to discretize the supremum and to introduce numerical quadrature to evaluate the ν-integral. This can be done in such a way that the above properties are preserved, see [10] for a more thorough discussion of these issues.
(2) When (18) does not hold and $\int \nu(dz) = \infty$, we can reduce to the case $\int \nu(dz) < \infty$ by truncation, i. e. by replacing $\nu(dz)$ by $1_{|z|>\delta}\nu(dz)$. For δ small, the error we make in the equation and schemes will be small. Moreover, we can improve the approximation (increase the order) by adding suitably chosen local diffusion (i. e. modifying σ). In the latter case, we still have a problem which we can solve with the above scheme. We refer again to [10] for a detailed discussion on this.
(3) As in the previous section on local equations, error estimates can be obtained in the general case by a regularization argument using Krylov's method of shaking the coefficients [25]. We refer again to [10] for actual results, but see also [24; 5].

5. Appendix: Well-posedness of the Bellman equation

In this section we give some relevant well-posedness and regularity results for the Bellman equation (5) and initial condition (2). To this end, we impose the following assumptions:

(A.1) The control set $\mathcal{A}$ is a separable metric space. For any $\alpha \in \mathcal{A}$, $a^\alpha = \frac{1}{2}\sigma^\alpha\sigma^{\alpha T}$, and σ^α, b^α, c^α, f^α, η^α are continuous.

(A.2) There is a positive constant K such that for all $\alpha \in \mathcal{A}$ and $t \in [0,T]$,

$$|g|_1 + |\sigma^\alpha|_1 + |b^\alpha|_1 + |c^\alpha|_1 + |f^\alpha|_1 \leq K,$$

where $|\phi|_1 = \sup_{(t,x)\in Q_T} |\phi(x,t)| + \sup_{(x,t)\neq(y,s)} \frac{|\phi(x,t)-\phi(y,s)|}{|x-y|+|t-s|^{1/2}}$ is a space-time Lipschitz/Hölder-norm.

(A.3) ν is a positive Radon measure on $\mathbb{R}^M \setminus \{0\}$ satisfying

$$\int_{0<|z|\leq 1} |z|^2\nu(dz) + \int_{|z|\geq 1} \nu(dz) \leq K.$$

(A.4) η^α is continuous in (x,t), and for every $\alpha \in \mathcal{A}$ and $(y,s),(x,t) \in Q_T$,

$$\int_{|z|<1} |\eta^\alpha(x,t,z)|^2\nu(dz) + \int_{|z|>0} \frac{|\eta^\alpha(x,t,z) - \eta^\alpha(y,s,z)|^2}{|x-y|^2 + |t-s|}\nu(dz)$$
$$+ \int_{|z|\geq 1} \frac{|\eta^\alpha(x,t,z) - \eta^\alpha(y,s,z)|}{|x-y| + |t-s|^{1/2}}\nu(dz) \leq K.$$

Assumptions (A.1)–(A.4) are standard and pretty general. E. g. they imply that the associated SDEs have unique strong solutions for fixed nice controls. They can be relaxed in many cases to allow for e. g. growth in the coefficients and solutions, less regular c and f, and less regularity in time. Most models in finance are covered by the assumptions, including most non-local ones [12].

Under these assumptions the following results hold:

Proposition 6.1. *Assume (A.1)–(A.4).*

(a) There exists a unique bounded continuous viscosity solution u of the initial value problem (5)–(2) *satisfying* $|u|_1 < \infty$.

(b) If u_1 and u_2 are respectively viscosity sub and supersolutions of (5) *satisfying $u_1(0,\cdot) \leq u_2(0,\cdot)$, then $u_1 \leq u_2$.*

For the precise definition of viscosity solutions, we refer to e.g. [2]. The proof of Proposition 6.1 can be obtained from interpolating between the proofs of [2] and [22], see also [23].

Bibliography

[1] Barles, G. (1997). Convergence of numerical schemes for degenerate parabolic equations arising in finance theory, in *Numerical methods in finance*, Publ. Newton Inst. (Cambridge Univ. Press, Cambridge), pp. 1–21.

[2] Barles, G. and Imbert, C. (2008). Second-order elliptic integro-differential equations: viscosity solutions' theory revisited, *Ann. Inst. H. Poincaré Anal. Non Linéaire* **25**, 3, pp. 567–585, doi:10.1016/j.anihpc.2007.02.007, URL `http://dx.doi.org/10.1016/j.anihpc.2007.02.007`.

[3] Barles, G. and Jakobsen, E. R. (2002). On the convergence rate of approximation schemes for Hamilton-Jacobi-Bellman equations, *M2AN Math. Model. Numer. Anal.* **36**, 1, pp. 33–54, doi:10.1051/m2an:2002002, URL `http://dx.doi.org/10.1051/m2an:2002002`.

[4] Barles, G. and Souganidis, P. E. (1991). Convergence of approximation schemes for fully nonlinear second order equations, *Asymptotic Anal.* **4**, 3, pp. 271–283.

[5] Biswas, I. H., Jakobsen, E. R. and Karlsen, K. H. (2010). Difference-quadrature schemes for nonlinear degenerate parabolic integro-PDE, *SIAM J. Numer. Anal.* **48**, 3, pp. 1110–1135, doi:10.1137/090761501, URL `http://dx.doi.org/10.1137/090761501`.

[6] Bokanowski, O., Bruder, B., Maroso, S. and Zidani, H. (2009). Numerical approximation for a superreplication problem under gamma constraints, *SIAM J. Numer. Anal.* **47**, 3, pp. 2289–2320, doi:10.1137/080725222, URL `http://dx.doi.org/10.1137/080725222`.

[7] Bonnans, J. F., Ottenwaelter, É. and Zidani, H. (2004). A fast algorithm for the two dimensional HJB equation of stochastic control, *M2AN Math. Model. Numer. Anal.* **38**, 4, pp. 723–735, doi:10.1051/m2an:2004034, URL `http://dx.doi.org/10.1051/m2an:2004034`.

[8] Bonnans, J. F. and Zidani, H. (2003). Consistency of generalized finite difference schemes for the stochastic HJB equation, *SIAM J. Numer. Anal.* **41**, 3, pp. 1008–1021, doi:10.1137/S0036142901387336, URL `http://dx.doi.org/10.1137/S0036142901387336`.

[9] Camilli, F. and Falcone, M. (1995). An approximation scheme for the optimal control of diffusion processes, *RAIRO Modél. Math. Anal. Numér.* **29**, 1, pp. 97–122.

[10] Camilli, F. and Jakobsen, E. R. (2009). A finite element like scheme for integro-partial differential Hamilton-Jacobi-Bellman equations, *SIAM J. Numer. Anal.* **47**, 4, pp. 2407–2431, doi:10.1137/080723144, URL `http://dx.doi.org/10.1137/080723144`.

[11] Capuzzo Dolcetta, I. (1983). On a discrete approximation of the Hamilton-Jacobi equation of dynamic programming, *Appl. Math. Optim.* **10**, 4, pp. 367–377, doi:10.1007/BF01448394, URL `http://dx.doi.org/10.1007/BF01448394`.

[12] Cont, R. and Tankov, P. (2004). *Financial modelling with jump processes*, Chapman & Hall/CRC Financial Mathematics Series (Chapman & Hall/CRC, Boca Raton, FL), ISBN 1-5848-8413-4.

[13] Crandall, M. G., Ishii, H. and Lions, P.-L. (1992). User's guide to viscosity solutions of second order partial differential equations, *Bull. Amer. Math. Soc. (N.S.)* **27**, 1, pp. 1–67, doi:10.1090/S0273-0979-1992-00266-5, URL `http://dx.doi.org/10.1090/S0273-0979-1992-00266-5`.

[14] Crandall, M. G. and Lions, P.-L. (1996). Convergent difference schemes for nonlinear parabolic equations and mean curvature motion, *Numer. Math.* **75**, 1, pp. 17–41, doi:10.1007/s002110050228, URL `http://dx.doi.org/10.1007/s002110050228`.

[15] Davis, T. A. (2004). Algorithm 832: UMFPACK V4.3—an unsymmetric-pattern multifrontal method, *ACM Trans. Math. Software* **30**, 2, pp. 196–199, doi:10.1145/992200.992206, URL `http://dx.doi.org/10.1145/992200.992206`.

[16] Debrabant, K. and Jakobsen, E. R. (Accepted for publication). Semi-Lagrangian schemes for linear and fully non-linear diffusion equations, *Math. Comp.* URL `http://arxiv.org/abs/0910.1046`.

[17] Dong, H. and Krylov, N. V. (2005). On the rate of convergence of finite-difference approximations for Bellman equations with constant coefficients, *Algebra i Analiz* **17**, 2, pp. 108–132, doi:10.1090/S1061-0022-06-00905-8, URL `http://dx.doi.org/10.1090/S1061-0022-06-00905-8`.

[18] Falcone, M. (1987). A numerical approach to the infinite horizon problem of deterministic control theory, *Appl. Math. Optim.* **15**, 1, pp. 1–13, doi:10.1007/BF01442644, URL `http://dx.doi.org/10.1007/BF01442644`.

[19] Fleming, W. H. and Soner, H. M. (2006). *Controlled Markov processes and viscosity solutions, Stochastic Modelling and Applied Probability*, Vol. 25, 2nd edn. (Springer, New York), ISBN 978-0387-260457; 0-387-26045-5.

[20] Forsyth, P. A. (2011). A Hamilton-Jacobi-Bellman approach to optimal trade execution, *Appl. Numer. Math.* **61**, 2, pp. 241–265, doi:10.1016/j.apnum.2010.10.004, URL `http://dx.doi.org/10.1016/j.apnum.2010.10.004`.

[21] Jakobsen, E. R. (2010). *Encyclopedia of Quantitative Finance*, chap. Monotone Schemes (John Wiley & Sons, Ltd), pp. 1253–1263, doi:10.1002/9780470061602.eqf12010, URL `http://onlinelibrary.wiley.com/doi/10.1002/9780470061602.eqf12010/abstract`.

[22] Jakobsen, E. R. and Karlsen, K. H. (2005). Continuous dependence estimates for viscosity solutions of integro-PDEs, *J. Differential Equations* **212**, 2, pp. 278–318, doi:10.1016/j.jde.2004.06.021, URL `http://dx.doi.org/10.1016/j.jde.2004.06.021`.

[23] Jakobsen, E. R. and Karlsen, K. H. (2006). A "maximum principle for semicontinuous functions" applicable to integro-partial differential equations, *NoDEA Nonlinear Differential Equations Appl.* **13**, 2, pp. 137–165, doi:10.1007/s00030-005-0031-6, URL `http://dx.doi.org/10.1007/s00030-005-0031-6`.

[24] Jakobsen, E. R., Karlsen, K. H. and La Chioma, C. (2008). Error estimates for approximate solutions to Bellman equations associated with controlled jump-diffusions, *Numer. Math.* **110**, 2, pp. 221–255, doi:10.1007/s00211-008-0160-z, URL `http://dx.doi.org/10.1007/s00211-008-0160-z`.

[25] Krylov, N. V. (2000). On the rate of convergence of finite-difference approximations for Bellman's equations with variable coefficients, *Probab. Theory Related Fields* **117**, 1, pp. 1–16, doi:10.1007/s004400050264, URL `http://dx.doi.org/10.1007/s004400050264`.

[26] Kushner, H. J. and Dupuis, P. (2001). *Numerical methods for stochastic control problems in continuous time, Applications of Mathematics (New York)*, Vol. 24, 2nd edn. (Springer-Verlag, New York), ISBN 0-387-95139-3, stochastic Modelling and Applied Probability.

[27] Menaldi, J.-L. (1989). Some estimates for finite difference approximations, *SIAM J. Control Optim.* **27**, 3, pp. 579–607, doi:10.1137/0327031, URL `http://dx.doi.org/10.1137/0327031`.

[28] Merton, R. C. (1969). Lifetime portfolio selection under uncertainty: The continuous-time case, *The Review of Economics and Statistics* **51**, 3, pp. 247–257, doi:{10.2307/1926560}.

[29] Merton, R. C. (1971). Optimum consumption and portfolio rules in a continuous-time model, *J. Econom. Theory* **3**, 4, pp. 373–413.

[30] Motzkin, T. S. and Wasow, W. (1953). On the approximation of linear elliptic differential equations by difference equations with positive coefficients, *J. Math. Physics* **31**, pp. 253–259.

[31] Schoutens, W. (2003). *Lévy Processes in Finance: Pricing Financial Derivatives (Wiley Series in Probability and Statistics)*, 1st edn. (Wiley), ISBN 0470851562, URL `http://www.worldcat.org/isbn/0470851562`.

[32] Wang, J. and Forsyth, P. A. (2008). Maximal use of central differencing for Hamilton-Jacobi-Bellman PDEs in finance, *SIAM J. Numer. Anal.* **46**, 3, pp. 1580–1601, doi: 10.1137/060675186, URL `http://dx.doi.org/10.1137/060675186`.
[33] Witte, J. H. and Reisinger, C. (2011). A penalty method for the numerical solution of Hamilton-Jacobi-Bellman (HJB) equations in finance, *SIAM J. Numer. Anal.* **49**, 1, pp. 213–231, doi:10.1137/100797606, URL `http://dx.doi.org/10.1137/100797606`.

Chapter 7

Derivative-free weak approximation methods for stochastic differential equations in finance

Kristian Debrabant and Andreas Rößler

University of Southern Denmark, Department of Mathematics and Computer Science, Campusvej 55, 5230 Odense M, Denmark

Universität zu Lübeck, Institut für Mathematik, Ratzeburger Allee 160, 23562 Lübeck, Germany

1. Introduction and examples

In finance, quantities such as asset prices, interest rates, and their derivatives are often modeled by systems of stochastic differential equations (SDEs). Some examples are the correlated exponential Ornstein-Uhlenbeck stochastic volatility model [25; 20]

$$\begin{aligned} \mathrm{d}S(t) &= \mu S(t)\Delta t + m e^{Z(t)} S(t)\mathrm{d}W_1(t), \\ \mathrm{d}Z(t) &= -\alpha Z(t)\Delta t + k\big(\rho \mathrm{d}W_1(t) + \sqrt{1-\rho^2}\mathrm{d}W_2(t)\big), \end{aligned}$$

and the quadratic volatility model [24]

$$\begin{aligned} \mathrm{d}S(t) &= \mu S(t)\Delta t + S(t)Z(t)\mathrm{d}W_1(t), \\ \mathrm{d}Z(t) &= l(\overline{\omega} - Z(t))S(t)\Delta t + \beta Z(t)^2\big(\rho \mathrm{d}W_1(t) + \sqrt{1-\rho^2}\mathrm{d}W_2(t)\big), \end{aligned}$$

with positive parameters μ, m, α, k respectively μ, l, $\overline{\omega}$, β, and correlation $\rho \in [-1,1]$. In both models, S and Z are stochastic processes describing an asset price respectively characterizing its volatility, with initial conditions $S(t_0)$, $Z(t_0)$. Further, there exist interest rate models like e.g. the Hull-White two-factor model

$$\begin{aligned} \mathrm{d}f(r(t)) &= (\theta(t) + u(t) - af(r(t)))\,\Delta t + \sigma_1 \mathrm{d}W_1(t), \\ \mathrm{d}u(t) &= -bu(t)\Delta t + \sigma_2\big(\rho \mathrm{d}W_1(t) + \sqrt{1-\rho^2}\mathrm{d}W_2(t)\big), \end{aligned}$$

having additive noise with correlation ρ, constants a, b, σ_1 and σ_2, a functional form $f(r)$, prescribed initial value $u(0) = 0$ and initial condition $f(r(0))$, and $\theta(t)$ a function depending on the time chosen such that the model fits the initial term structure (see [12] for details).

More generally, we denote in the following by $X = (X(t))_{t \in I}$ the solution of the d-dimensional Itô SDE defined by

$$\mathrm{d}X(t) = a(t, X(t))\Delta t + \sum_{j=1}^{m} b_j(t, X(t))\mathrm{d}W_j(t), \qquad X(t_0) = x_0, \tag{1}$$

with an m-dimensional Wiener process $(W(t))_{t \geq 0}$ and $I = [t_0, T]$. We assume that the Borel-measurable coefficients $a, b_j : I \times \mathbb{R}^d \to \mathbb{R}^d$, $j = 1, \ldots, m$ are chosen such that a unique solution exists, see e. g. [9].

Closed form solutions of SDEs are usually not available, so numerical approximation is necessary. Therefore, let a discretization $I_h = \{t_0, t_1, \ldots, t_N\}$ with $t_0 < t_1 < \ldots < t_N = T$ of the considered time interval $[t_0, T]$ with step sizes $h_n = t_{n+1} - t_n$ for $n = 0, 1, \ldots, N-1$ be given. Further, define $h = \max_{0 \leq n < N} h_n$ as the maximum step size. Then we aim at constructing a numerical approximation process $Y^h = (Y^h(t))_{t \in I_h}$ which approximates in some sense $X(t)$ at some discrete times $t \in I_h$. Thereby, one is often not primarily interested in accurate approximations of individual solution paths, i .e. strong approximation, but rather in weak approximation, i. e. good approximations of the expectation of functionals of the solution, like the expectation of some payoff function in option pricing. An example for such a payoff function is

$$P = N_{var} \cdot \left(\frac{1}{T} \int_0^T V^2(t)\Delta t - K_{var} \right) \tag{2}$$

for a variance swap option with variance strike K_{var} and variance notional N_{var}, where $V(t) = me^{Z(t)}$ for the correlated Ornstein-Uhlenbeck stochastic volatility model and $V(t) = Z(t)$ in the case of the quadratic volatility model.

The quality of the approximation method is amongst others judged by its order. For the usually used definition of the weak order of convergence, let $C_P^l(\mathbb{R}^d, \mathbb{R})$ denote the space of all $g \in C^l(\mathbb{R}^d, \mathbb{R})$ fulfilling a polynomial growth condition, i. e. that there are constants $\tilde{C} > 0$ and $\kappa > 0$ such that $|\partial_x^i g(x)| \leq \tilde{C}(1 + \|x\|^\kappa)$ for any partial derivative of order $i \leq l$ and all $x \in \mathbb{R}^d$, see [13]. We say that Y^h converges weakly with order p to X as $h \to 0$ on the grid I_h if for each $f \in C_P^{2(p+1)}(\mathbb{R}^d, \mathbb{R})$ there exist a constant C_f and a finite $\delta_0 > 0$ such that

$$\max_{t \in I_h} |\mathbb{E}(f(X(t))) - \mathbb{E}(f(Y^h(t)))| \leq C_f \, h^p \tag{3}$$

holds for each $h \in \,]0, \delta_0[$.

In recent years, the development of numerical methods for the weak approximation of SDEs has become a field of increasing interest, see e. g. [13; 19] and references therein. Amongst modern approaches are stochastic multi-step methods, stochastic Taylor methods and stochastic Runge-Kutta (SRK) methods, the latter ones playing the major role, as they are easier to work with than the first ones, and do not rely on the computation of higher order derivatives of the coefficient functions of the SDE like the second ones. Second order SRK methods for the weak approximation of SDEs were proposed amongst others by [13; 14; 18; 26; 4]. However, these methods were not suitable for problems with high numbers m of Wiener processes, because for these methods the number of function evaluations per step increases quadratically in m. Recently, new classes of SRK methods were introduced by [22; 23] which overcome this problem. An alternative has later been proposed by [16]. An explicit third order weak SRK method for autonomous SDEs with scalar (i. e. one-dimensional) noise has been given in [13], a class of third order weak schemes for multi-dimensional additive noise can be found in [2].

Usually the expectation of a considered payoff function evaluated at the numerical approximation process cannot be calculated explicitly. Therefore, one often resorts to Monte Carlo simulation, i e. simulating M stochastically independent realizations $Y^{h,j} = (Y^h(t))^j_{t \in I_h}$, $j = 1, \ldots, M$, of the approximation $Y^h = (Y^h(t))_{t \in I_h}$ and applying the payoff function P to each of the realizations. Then, applying the arithmetic mean

$$\widehat{E}_M(P(Y^h)) = \frac{1}{M} \sum_{j=1}^{M} P(Y^{h,j})$$

as an estimator we obtain an approximation for the desired expectation of the payoff.

In the following, we introduce in Section 2 a class of efficient SRK methods for second order weak approximation of general multi-dimensional SDEs and present in Subsection 2.1 coefficients defining an explicit SRK method that possesses minimized leading error terms. However, sometimes stiff SDEs have to be solved where explicit methods would need very small step sizes for reasonable approximations. In this case, implicit methods are an appropriate choice and coefficients for a weak second order drift-implicit SRK method are presented in Subsection 2.2 and it is described how to solve the occurring implicit equations. In Section 3 we introduce a class of SRK methods which are suitable for the third order weak approximation of multi-dimensional SDEs with additive noise. Because some payoff functions may need approximation values not only at the prescribed discretization points but also at some intermediate time points, continuous SRK methods are presented in Section 4 that allow to get approximations at any time with negligible computational costs. This feature is of special importance if the step size control algorithm described in Section 5 is applied in order to control the approximation errors. Finally, in Section 6 we give an outlook to the topic of optimally balancing of step sizes and the number of Monte Carlo simulations as well as some remarks to the application of the multi-level Monte Carlo simulation recently proposed by Giles [10]. The aim of the presented approximation methods and the corresponding add-ons is to provide a rich toolbox for practitioners who are confronted with the problem of numerical weak approximation in mathematical finance. Here, we want to emphasize that this introduction is far away from being exhaustive and that there is a big amount of special literature that is steadily increasing due to active research in this area (see, e.g., [13] and the references therein).

2. Efficient second order stochastic Runge-Kutta methods

We consider the class of SRK methods introduced in [23] for the second order weak approximation of SDE (1). The d-dimensional approximation process Y^h with $Y_n = Y^h(t_n)$ of an s-stage SRK method is thus defined by $Y_0 = x_0$ and

$$\begin{aligned} Y_{n+1} = Y_n &+ \sum_{i=1}^{s} \alpha_i \, a(t_n + c_i^{(0)} h_n, H_i^{(0)})\, h_n \\ &+ \sum_{i=1}^{s} \sum_{k=1}^{m} \beta_i^{(1)} \, b_k(t_n + c_i^{(1)} h_n, H_i^{(k)})\, \hat{I}_{(k),n} \\ &+ \sum_{i=1}^{s} \sum_{k=1}^{m} \beta_i^{(2)} \, b_k(t_n + c_i^{(1)} h_n, H_i^{(k)})\, \frac{\hat{I}_{(k,k),n}}{\sqrt{h_n}} \\ &+ \sum_{i=1}^{s} \sum_{k=1}^{m} \beta_i^{(3)} \, b_k(t_n + c_i^{(2)} h_n, \hat{H}_i^{(k)})\, \hat{I}_{(k),n} \\ &+ \sum_{i=1}^{s} \sum_{k=1}^{m} \beta_i^{(4)} \, b_k(t_n + c_i^{(2)} h_n, \hat{H}_i^{(k)})\, \sqrt{h_n} \end{aligned} \tag{4}$$

for $n = 0, 1, \dots, N-1$ with stage values

$$\begin{aligned} H_i^{(0)} &= Y_n + \sum_{j=1}^{s} A_{ij}^{(0)} \, a(t_n + c_j^{(0)} h_n, H_j^{(0)})\, h_n \\ &\quad + \sum_{j=1}^{s} \sum_{l=1}^{m} B_{ij}^{(0)} \, b_l(t_n + c_j^{(1)} h_n, H_j^{(l)})\, \hat{I}_{(l),n} \\ H_i^{(k)} &= Y_n + \sum_{j=1}^{s} A_{ij}^{(1)} \, a(t_n + c_j^{(0)} h_n, H_j^{(0)})\, h_n \\ &\quad + \sum_{j=1}^{s} B_{ij}^{(1)} \, b_k(t_n + c_j^{(1)} h_n, H_j^{(k)})\, \sqrt{h_n} \end{aligned} \tag{5a}$$

and

$$\begin{aligned} \hat{H}_i^{(k)} &= Y_n + \sum_{j=1}^{s} A_{ij}^{(2)} \, a(t_n + c_j^{(0)} h_n, H_j^{(0)})\, h_n \\ &\quad + \sum_{j=1}^{s} \sum_{\substack{l=1 \\ l \neq k}}^{m} B_{ij}^{(2)} \, b_l(t_n + c_j^{(1)} h_n, H_j^{(l)})\, \frac{\hat{I}_{(k,l),n}}{\sqrt{h_n}} \end{aligned} \tag{5b}$$

for $i = 1, \dots, s$ and $k = 1, \dots, m$. Here, $\alpha, \beta^{(1)}, \dots, \beta^{(4)}, c^{(q)} \in \mathbb{R}^s$ and $A^{(q)}$, $B^{(q)} \in \mathbb{R}^{s \times s}$ for $q \in \{0, 1, 2\}$ are the vectors and matrices of coefficients of the SRK method where $c^{(q)} = A^{(q)}e$ for $q \in \{0,1,2\}$ with a vector $e = (1,\dots,1)^T$. They can be gathered in the following extended Butcher tableau:

$c^{(0)}$	$A^{(0)}$	$B^{(0)}$	
$c^{(1)}$	$A^{(1)}$	$B^{(1)}$	
$c^{(2)}$	$A^{(2)}$	$B^{(2)}$	
	α^T	${\beta^{(1)}}^T$	${\beta^{(2)}}^T$
		${\beta^{(3)}}^T$	${\beta^{(4)}}^T$

The SRK method needs the simulation of independent realizations of the random variables $\hat{I}_{(k),n}$, $k = 1,\dots,m$, in each step. They are three-point distributed with

$$\mathrm{P}(\hat{I}_{(k),n} = \pm\sqrt{3\,h_n}) = \frac{1}{6} \qquad \text{and} \qquad \mathrm{P}(\hat{I}_{(k),n} = 0) = \frac{2}{3}.$$

Further, independent realizations of the random variables $\hat{I}_{(k,l),n}$ for $k,l = 1,\dots,m$ that are defined by

$$\hat{I}_{(k,l),n} = \begin{cases} \frac{1}{2}(\hat{I}_{(k),n}\hat{I}_{(l),n} - \sqrt{h_n}\tilde{I}_{(k),n}) & \text{if } k < l \\ \frac{1}{2}(\hat{I}_{(k),n}\hat{I}_{(l),n} + \sqrt{h_n}\tilde{I}_{(l),n}) & \text{if } l < k \\ \frac{1}{2}(\hat{I}^2_{(k),n} - h_n) & \text{if } k = l \end{cases} \tag{6}$$

have to be simulated in each step. They are defined by two point distributed independent random variables $\tilde{I}_{(k),n}$ satisfying

$$\mathrm{P}(\tilde{I}_{(k),n} = \pm\sqrt{h_n}) = \frac{1}{2}$$

for $k = 1,\dots,m-1$. Thus, $2m-1$ realizations of discretely distributed independent random variables have to be simulated each step.

By the application of the multi–colored rooted tree analysis ([21], see also [3]) order conditions for the coefficients of the SRK method (4) – (5) can be easily determined. For a detailed analysis of convergence, we refer to Theorem 5.1 in [23] that gives order conditions for the SRK method (4) – (5) of weak order one and two.

2.1. *An efficient explicit second order SRK method*

Explicit SRK methods in the class (4) – (5) are characterized by $A_{ij}^{(q)} = B_{ij}^{(q)} = 0$ for $j \geq i$ and $q \in \{0,1\}$, which means that the equations (5a) for the stages and thus also the others are explicitly solvable. One can show that explicit order one SRK methods need at least $s = 1$ stage while order two SRK methods need at least $s = 3$ stages.

Table 1. Coefficients of the explicit weak second order SRK scheme DRI1

0									
$\frac{1}{2}$	$\frac{1}{2}$			$\frac{6-\sqrt{6}}{10}$					
1	-1	2		$\frac{3+2\sqrt{6}}{5}$	0				
0									
$\frac{342}{491}$	$\frac{342}{491}$			$3\sqrt{\frac{38}{491}}$					
$\frac{342}{491}$	$\frac{342}{491}$	0		$-3\sqrt{\frac{38}{491}}$	0				
0				0	0	0			
0	0			$-\frac{214}{513}\sqrt{\frac{1105}{991}}$	$-\frac{491}{513}\sqrt{\frac{221}{4955}}$	$-\frac{491}{513}\sqrt{\frac{221}{4955}}$			
0	0	0		$\frac{214}{513}\sqrt{\frac{1105}{991}}$	$\frac{491}{513}\sqrt{\frac{221}{4955}}$	$\frac{491}{513}\sqrt{\frac{221}{4955}}$			
	$\frac{1}{6}$	$\frac{2}{3}$	$\frac{1}{6}$	$\frac{193}{684}$	$\frac{491}{1368}$	$\frac{491}{1368}$	0	$\frac{1}{6}\sqrt{\frac{491}{38}}$	$-\frac{1}{6}\sqrt{\frac{491}{38}}$
				$-\frac{4955}{7072}$	$\frac{4955}{14144}$	$\frac{4955}{14144}$	0	$-\frac{1}{8}\sqrt{\frac{4955}{221}}$	$\frac{1}{8}\sqrt{\frac{4955}{221}}$

In [8], we gave a full classification of the coefficients of all explicit methods with minimal stage number. Based on this classification, we calculated the coefficients of an extension of the well-known RK32 scheme of Kutta [1] to an SRK method with minimized leading local error term, resulting in the method DRI1 shown in Table 1. This SRK method has weak order two, however the inherent RK32 scheme assures order three for the SRK method if it is applied to ODEs, i.e. SDE (1) with $b_j(t,x) \equiv 0$ for $j = 1, \dots, m$.

2.2. *Implicit SRK methods*

When numerically integrating an SDE, one sometimes experiences that explicit methods suffer from huge stability problems, in the sense that only excessively small step sizes lead to reasonable approximations, whereas implicit methods do not suffer from these problems. Such (systems of) equations are called stiff. Second order weak approximation methods for stiff systems have been considered in [13; 15]. In [7] we considered suitable implicit methods within the proposed class of efficient SRK methods and analyzed their stability properties, suggesting amongst others the coefficients of DDIRDI5 given in Table 2.

The drawback of implicit SRK methods is that their stage values are only given implicitly. E. g., for the methods in the class (4) - (5), if not all of the matrices $A^{(q)}$ and $B^{(q)}$, $q \in \{0, 1\}$, are lower triangular, then the stage values $H^{(k)}$, $k = 0, \dots, m$, are only given implicitly by (5a). So, in practice these implicit equations have to be solved approximately, which is usually done by iterative Newton-type methods.

Table 2. Coefficients of the implicit weak second order SRK scheme DDIRDI5

$\frac{1}{2}+\frac{1}{6}\sqrt{3}$	$\frac{1}{2}+\frac{1}{6}\sqrt{3}$	0	0	0	0	0			
$\frac{1}{2}-\frac{1}{6}\sqrt{3}$	$-\frac{1}{3}\sqrt{3}$	$\frac{1}{2}+\frac{1}{6}\sqrt{3}$	0	1	0	0			
0	0	0	0	0	0	0			
0	0	0	0	0	0	0			
1	1	0	0	1	0	0			
1	1	0	0	-1	0	0			
0	0	0	0	0	0	0			
0	0	0	0	1	0	0			
0	0	0	0	-1	0	0			
	$\frac{1}{2}$	$\frac{1}{2}$	0	$\frac{1}{2}$	$\frac{1}{4}$	$\frac{1}{4}$	0	$\frac{1}{2}$	$-\frac{1}{2}$
				$-\frac{1}{2}$	$\frac{1}{4}$	$\frac{1}{4}$	0	$\frac{1}{2}$	$-\frac{1}{2}$

These yield for (5a) the linear and thus directly solvable system

$$
\begin{aligned}
H_{i,r}^{(0)} &= Y_n + h_n \sum_{j=1}^{s} A_{ij}^{(0)} \Big(a(t_n + c_j^{(0)} h_n, H_{j,r-1}^{(0)}) + J_{j,r-1}^{(0)} (H_{j,r}^{(0)} - H_{j,r-1}^{(0)}) \Big) \\
&\quad + \sum_{l=1}^{m} \hat{I}_{(l),n} \sum_{j=1}^{s} B_{ij}^{(0)} \Big(b_l(t_n + c_j^{(1)} h_n, H_{j,r-1}^{(l)}) + J_{j,r-1}^{(l)} (H_{j,r}^{(l)} - H_{j,r-1}^{(l)}) \Big) \\
H_{i,r}^{(k)} &= Y_n + h_n \sum_{j=1}^{s} A_{ij}^{(1)} \Big(a(t_n + c_j^{(0)} h_n, H_{j,r-1}^{(0)}) + J_{j,r-1}^{(0)} (H_{j,r}^{(0)} - H_{j,r-1}^{(0)}) \Big) \\
&\quad + \sqrt{h_n} \sum_{j=1}^{s} \Big(B_{ij}^{(1)} b_k(t_n + c_j^{(1)} h_n, H_{j,r-1}^{(k)}) + J_{j,r-1}^{(k)} (H_{j,r}^{(k)} - H_{j,r-1}^{(k)}) \Big)
\end{aligned}
\tag{7}
$$

for $i = 1, \ldots, s$, $k = 1, \ldots, m$ and $r = 1, \ldots, R$ with R being the iteration number, some approximations $J_{j,r-1}^{(0)}$ and $J_{j,r-1}^{(l)}$ to the Jacobians of $a(t_n + c_j^{(0)} h_n, H_{j,r-1}^{(0)})$ and $b_l(t_n + c_j^{(1)} h_n, H_{j,r-1}^{(l)})$, and predictors (initial values for the iteration) $H_{i,0}^{(0)}$ and $H_{i,0}^{(k)}$.

For the approximations $J_{j,r-1}^{(0)}$ and $J_{j,r-1}^{(l)}$ to the Jacobian of $a(t_n + c_j^{(0)} h_n, H_{j,r-1}^{(0)})$ and $b_l(t_n + c_j^{(1)} h_n, H_{j,r-1}^{(l)})$ there exist several common choices. If we choose all $J_{j,r-1}^{(0)}$ and $J_{j,r-1}^{(l)}$ to be the exact Jacobians $a'(t_n + c_j^{(0)} h_n, H_{j,r-1}^{(0)})$ and $b_l'(t_n + c_j^{(1)} h_n, H_{j,r-1}^{(l)})$, then we obtain the classical Newton iteration method for solving (5a), which is locally convergent of order two and will be denoted in the following as full Newton iteration. If we choose instead $J_{j,r-1}^{(0)} = a'(t_n, Y_n)$ and $J_{j,r-1}^{(l)} = b_l'(t_n, Y_n)$, then we obtain the so called modified Newton iteration method, which is locally convergent of order one. Here, $J_{j,r-1}^{(0)}$ and $J_{j,r-1}^{(l)}$ are independent of the iteration number r and the stage values. Thus their computation is much cheaper and simpler than in the full Newton iteration case, and in Runge-Kutta implementations this is usually the method of choice.

The third and simplest possibility is to choose all $J^{(0)}_{j,r-1} = J^{(l)}_{j,r-1} = 0$. In this case we do not even have to solve a linear system for the stage values. This iteration method is called simple iteration method or predictor corrector method. Its disadvantage is that it leads to an explicit method, so it is not suitable for stiff systems. Setting $H_i^{(k)} = H_{i,R}^{(k)}$, $k = 0, \ldots, m$, (4), (7) and (5b) form a so-called iterated Runge-Kutta method. In [3] the order of these has been determined in dependence on the type of the SDE, the numerical approximation scheme used to solve the implicit equations, on the number of iterations and on the predictor. It turns out that for the weak approximation of Itô SDEs, when using the trivial predictor, i.e. $H_{i,0}^{(k)} = Y_n$, $k = 0, \ldots, m$, the number of modified Newton steps needed is one less than the order of the method, so for second order methods only one iteration step is necessary to preserve the overall order of convergence.

3. Third order SRK methods for SDEs with additive noise

For SDEs having additive noise, i.e., the functions $b_j(t,x) \equiv b_j$, $j = 1, \ldots, m$, in (1) are constant, like the Hull-White two-factor model, in [2] the following class of SRK methods for third order weak approximation has been proposed:

$$Y_{n+1} = Y_n + h_n \sum_{i=1}^{s} \alpha_i a(t_n + c_i h_n, H_i) + \sum_{l=1}^{m} b_l \hat{J}_{(l),n}, \tag{8a}$$

$$H_i = Y_n + h_n \sum_{j=1}^{s} A_{ij} a(t_n + c_j h_n, H_j) + \sum_{l=1}^{m} b_l (d_{1,i} \hat{J}_{(l),n} + d_{2,i} \hat{J}_{(m+l),n}), \tag{8b}$$

which defines a d-dimensional approximation process Y^h with $Y^h(t_n) = Y_n$. Here, $\hat{J}_{(k),n}$, $k = 1, \ldots, 2m$, are independent random variables which depend on h_n and whose moments all exist. Further, $\alpha = (\alpha_1, \ldots, \alpha_s)^\top$, $A = (A_{ij})_{i,j=1,\ldots,s}$, $c = (c_1, \ldots, c_s)^\top$, $d_1 = (d_{1,1}, \ldots, d_{1,s})^\top$, and $d_2 = (d_{2,1}, \ldots, d_{2,s})^\top$ are the coefficients of the SRK method, which can be arranged in an extended Butcher tableau of the form

$$\begin{array}{c|c|c|c} c & A & d_1 & d_2 \\ \hline & \alpha^T & & \end{array}.$$

In the following, we choose $c = Ae$ with $e = (1, \ldots, 1)^\top \in \mathbb{R}^s$. Applying B-series theory ([3], see also [21]) order conditions for orders one, two, and three have been determined, see Theorem 4 in [2]. To obtain an order three method, the moments of the random variables $\hat{J}_{(k),n}$ have to coincide up to the seventh moment with the ones of $N(0, h_n)$ for $k = 1, \ldots, m$ and up to the fifth moment for $k = m+1, \ldots, 2m$.

Table 3. Coefficients of the explicit weak third order SRK scheme AN3D1

0	0	0	0	0	d_1	d_2
1	1	0	0	0		
1/2	3/8	1/8	0	0		
1	−0.4526683126055039	−0.4842227708685013	1.9368910834740051	0		
	1/6	−0.005430430675258792	2/3	0.1720970973419255		

with

$$d_1 = (-0.01844540496323970, 0.8017012756521233, ...\\ 0.5092227024816198, 0.9758794209767762)^\top$$
$$d_2 = (-0.1866426386543421, -0.8575745885712401, ...\\ -0.4723392695015512, 0.3060354860326548)^\top$$

So, one can e. g. choose $\hat{J}_{(k),n}$ such that

$$\mathrm{P}(\hat{J}_{(k),n} = \pm\sqrt{6h_n}) = \frac{1}{30}, \quad \mathrm{P}(\hat{J}_{(k),n} = \pm\sqrt{h_n}) = \frac{3}{10}, \quad \mathrm{P}(\hat{J}_{(k),n} = 0) = \frac{1}{3}$$

for $k = 1, \dots, m$ and

$$\mathrm{P}(\hat{J}_{(k),n} = \pm\sqrt{3h_n}) = \frac{1}{6}, \quad \mathrm{P}(\hat{J}_{(k),n} = 0) = \frac{2}{3}$$

for $k = m+1, \dots, 2m$.

Whereas in the deterministic case we would only need three stages to construct an explicit third order method, here we need four stages to fulfill the order conditions. But with four stages, there are some remaining degrees of freedom. Determining these by requiring in addition that the method fulfills also the deterministic order four conditions and that the order four coefficients of the local error in the Euclidean norm assuming two dimensional noise are minimized, a numerical optimization yields the scheme AN3D1 presented in Table 3. AN3D1 needs $2m$ random variable and four drift evaluations per step. Compared to Platen's third order method [13], which is only applicable in the case $m = 1$, but then also for general noise, we need two random variable and three drift evaluations less.

4. Continuous weak approximation

The methods considered so far converge with some given order at the discretization points. However, in some applications higher order continuous time approximation methods guaranteeing uniform orders of convergence not only at the discretization points but also at any arbitrary time point within the approximation interval are of advantage.

Classical time discrete methods are inefficient in this case where the number of output points has to be very large because this forces the step size to be very small. Therefore, in [6; 5] we developed continuous time approximation methods guaranteeing uniform weak orders of convergence on the complete time interval, i. e., (3) can be replaced by

$$\max_{t \in I} |\mathbb{E}(f(X(t))) - \mathbb{E}(f(Y^h(t)))| \leq C_f \, h^p. \tag{9}$$

Especially, we were interested in continuous sample trajectories of the applied SRK methods. These have negligible additional computational complexity compared to the time discrete SRK but allow e. g. the use of an individual discretization for each sample trajectory which needs not necessarily to contain some common discretization points for all trajectories in order to be able to calculate the expectation at these common time points, a property which is useful e. g. when using adaptive time stepping in parallel computing. For this reason, we extended the well known convergence theorem due to [19] to the continuous case and developed continuous extensions of weak order two SRK methods both in the Itô and in the Stratonovich case.

Applying the theory developed there to the SRK method (4) – (5), we obtain continuous second order approximation formulas by replacing (4) with

$$\begin{aligned}
Y(t_n + \theta h_n) = Y_n &+ \theta \sum_{i=1}^{s} \alpha_i \, a(t_n + c_i^{(0)} h_n, H_i^{(0)}) \, h_n \\
&+ \sqrt{\theta} \sum_{i=1}^{s} \sum_{k=1}^{m} \beta_i^{(1)} \, b_k(t_n + c_i^{(1)} h_n, H_i^{(k)}) \, \hat{I}_{(k)} \\
&+ \sum_{i=1}^{s} \sum_{k=1}^{m} \beta_i^{(2)} \, b_k(t_n + c_i^{(1)} h_n, H_i^{(k)}) \, \frac{\hat{I}_{(k,k)}}{\sqrt{h_n}} \\
&+ \sum_{i=1}^{s} \sum_{k=1}^{m} \beta_i^{(3)} \, b_k(t_n + c_i^{(2)} h_n, \hat{H}_i^{(k)}) \, \hat{I}_{(k)} \\
&+ \sum_{i=1}^{s} \sum_{k=1}^{m} \beta_i^{(4)} \, b_k(t_n + c_i^{(2)} h_n, \hat{H}_i^{(k)}) \, \sqrt{h_n}
\end{aligned}$$

for $\theta \in [0, 1]$.

A continuous extension of the third order method AN3D1 is obtained when α in (8) is replaced by the vector function

$$\alpha(\theta) = \begin{pmatrix} 0.6958702845944792\theta - 0.5292036179278126\theta^2 \\ 0.0006311211304501030\theta - 0.006061551805708955\theta^2 \\ 0.608260215774515 8\theta + 0.05840645089215089\theta^2 \\ -0.3047614325400084\theta + 0.4768585298819339\theta^2 \end{pmatrix},$$

and if the left hand side of (8a) is replaced by $Y(t_n + \theta h_n)$ for $\theta \in [0, 1]$.

5. Adaptive step size selection

Often, the solution of an SDE turns out to highly vary or oscillate over some periods of time while it behaves much more stable for other periods of time. Then, the precision of the numerical approximation can be increased by spending more computational work for the periods with a highly varying solution which means to use small step sizes there and to increase step sizes for the remaining time periods. This leads to an error control algorithm with adaptive step size selection where the approximation error is estimated by a pair of embedded SRK methods of different orders of convergence as proposed in [17]. Note that in this section we focus on the second order methods introduced in Section 2, as the third order method AN3D1 does not possess an embedded second order method: If one restricts to four stages in (8), then the order two conditions determine the weights α uniquely from the remaining parameters. So, to obtain a local error estimator in this case, one can either construct methods with five stages, or use local Richardson extrapolation, which however roughly would double the computational work.

5.1. *Embedded stochastic Runge-Kutta methods*

In the following we denote by $p = (p_D, p_S)$ the order of the SRK method (4), where p_D and p_S with $p_D \geq p_S$ indicate the order of convergence if the SRK method is applied to a deterministic or a stochastic differential equation, respectively. Thus, the order p_S is guaranteed in any case.

We consider a second SRK approximation $\hat{Y}_{n+1}$ which only differs from Y_{n+1} in the coefficients α_i and $\beta_i^{(r)}$ for $r = 1, 2, 3, 4$ with $i = 1, \ldots, s$ and can thus be written in the form

$$
\begin{aligned}
\hat{Y}_{n+1} = Y_n &+ \sum_{i=1}^{s} \hat{\alpha}_i \, a(t_n + c_i^{(0)} h_n, H_i^{(0)}) \, h_n \\
&+ \sum_{i=1}^{s} \sum_{k=1}^{m} \hat{\beta}_i^{(1)} \, b_k(t_n + c_i^{(1)} h_n, H_i^{(k)}) \, \hat{I}_{(k),n} \\
&+ \sum_{i=1}^{s} \sum_{k=1}^{m} \hat{\beta}_i^{(2)} \, b_k(t_n + c_i^{(1)} h_n, H_i^{(k)}) \, \frac{\hat{I}_{(k,k),n}}{\sqrt{h_n}} \qquad (10) \\
&+ \sum_{i=1}^{s} \sum_{k=1}^{m} \hat{\beta}_i^{(3)} \, b_k(t_n + c_i^{(2)} h_n, \hat{H}_i^{(k)}) \, \hat{I}_{(k),n} \\
&+ \sum_{i=1}^{s} \sum_{k=1}^{m} \hat{\beta}_i^{(4)} \, b_k(t_n + c_i^{(2)} h_n, \hat{H}_i^{(k)}) \, \sqrt{h_n}
\end{aligned}
$$

for $n = 0, 1, \ldots, N-1$ using the same stage values given in (5). The order $\hat{p} = (\hat{p}_D, \hat{p}_S)$ of this approximation is assumed to fulfill $\hat{p}_D < p_D$ and $\hat{p}_S < p_S$. Together with the original SRK method it defines then an embedded SRK scheme of order $p(\hat{p}) = (p_D, p_S)((\hat{p}_D, \hat{p}_S))$.

For step size control, we have to calculate the two approximations Y_{n+1} and $\hat{Y}_{n+1}$ in each step, both using the same stage values given in (5a) and (5b). Therefore, the additional computational effort to calculate $\hat{Y}_{n+1}$ is negligible.

We will consider the embedded SRK schemes of order $p(\hat{p}) = (3,2)((2,1))$ obtained by extending the coefficients of the DRI1 scheme with the weights

$$\hat{\alpha} = [0\ 1\ 0]^T, \quad \hat{\beta}^{(1)} = [1\ 0\ 0]^T, \quad \hat{\beta}^{(2)} = \hat{\beta}^{(3)} = \hat{\beta}^{(4)} = [0\ 0\ 0]^T$$

for the embedded weak SRK approximation $\hat{Y}$ of order $(2,1)$. Alternatively, we can apply the implicit embedded SRK schemes of order $p(\hat{p}) = (2,2)((1,1))$ by using the coefficients of the DDIRDI5 scheme however with the weights

$$\hat{\alpha} = [1\ 0\ 0]^T, \quad \hat{\beta}^{(1)} = [1\ 0\ 0]^T, \quad \hat{\beta}^{(2)} = \hat{\beta}^{(3)} = \hat{\beta}^{(4)} = [0\ 0\ 0]^T$$

to get the implicit embedded weak SRK approximation $\hat{Y}$ of order $\hat{p} = (1,1)$.

5.2. *Step size control algorithm*

For step size control, we describe the algorithm proposed in [17] that turned out to give good results. For the problem of weak approximation, a suitable step size selection has to depend on the expectation of the solution rather than on single trajectories. The expectation gives much more regularity by smoothing the solution which is essential for the local error estimates. For each step, we calculate two approximations by the embedded SRK method based on (4) and (10) with stage values (5) based on Y_n. Then, we obtain a more precise approximation Y_{n+1} and a less precise one $\hat{Y}_{n+1}$. Suppose that we have calculated M realizations Y_n^k, $k = 1, \dots, M$, that are needed for the Monte Carlo simulation at time t_n. Then, we calculate the corresponding approximations Y_{n+1}^k and $\hat{Y}_{n+1}^k$ with the proposed step size h. After applying the payoff function $f : \mathbb{R}^d \to \mathbb{R}$ to each approximation value, we get the two estimates for the expectations

$$\widehat{E}_M(f(Y_{n+1})) \approx E(f(Y_{n+1})) \text{ and } \widehat{E}_M(f(\hat{Y}_{n+1})) \approx E(f(\hat{Y}_{n+1})).$$

Then $|\widehat{E}_M\big(f(Y_{n+1})\big) - \widehat{E}_M\big(f(\hat{Y}_{n+1})\big)|$ is used as an estimator for the local error of the less precise approximation. Now, we calculate the total tolerance *tol* as

$$tol = Atol + \max\{|\widehat{E}_M\big(f(Y_n)\big)|, |\widehat{E}_M\big(f(Y_{n+1})\big)|\} Rtol \tag{11}$$

with some prescribed positive absolute tolerance *Atol* and relative tolerance *Rtol* for the error. Thus, we can calculate an approximately optimal step size such that

$$err = \frac{|\widehat{E}_M\big(f(Y_{n+1})\big) - \widehat{E}_M\big(f(\hat{Y}_{n+1})\big)|}{tol} \approx 1 \tag{12}$$

is fulfilled. Using the local error estimate (see [17] for details), we get that

$$h_{opt} = h \left(\frac{1}{err} \right)^{\frac{1}{\hat{p}_S + 1}}$$

is the optimal step size proposal to calculate the current approximation Y_{n+1}. In case of deterministic ordinary differential equations, Hairer, Nθrsett and Wanner [11] propose a multiplication of h_{opt} by a safety factor $fac < 1$ (e.g. $fac = 0.8$) to prevent strong oscillations of the step size. Further, in order to avoid too fast growth or reduction of the step size we also use factors $facmax$ and $facmin$, so that finally the new step size is calculated as

$$h_{new} = h \cdot \min\left(facmax, \max\left(facmin, fac \cdot \left(\frac{1}{err}\right)^{\frac{1}{\hat{p}_S+1}} \right)\right). \tag{13}$$

Finally, if $err \leq 1$ is fulfilled then the current step with step size h is accepted for all M realizations of Y_{n+1} and the next step is calculated with step size h_{new}. However if $err > 1$, then the current step is rejected and has to be recalculated with the smaller step size h_{new} for all M realizations.

5.3. *Simulation of the conditional distributed random variables*

The weak second order SRK method makes use of the random variables $\hat{I}_{(k),n}$ and $\tilde{I}_{(l),n}$ for $k = 1, \ldots, m$ and $l = 1, \ldots, m-1$. If the step size control algorithm is applied, we sometimes need to reject calculated steps based on some proposed step size if the local error estimate is too large. Then, we have to recalculate the last step with a smaller step size.

Let us suppose that we already have simulated the approximation Y_n at time t_n and we calculate the next approximation at time $t_n + h$ with some proposed step size h. Let us denote the used random variables $\hat{I}_{(k),n} = \hat{I}_{(k);t_n,t_n+h}$ and $\tilde{I}_{(l),n} = \tilde{I}_{(l);t_n,t_n+h}$ in order to emphasize the corresponding time interval. If the new approximation at $t_n + h$ is rejected by the step size control algorithm, then we have to repeat the calculation of Y_{n+1} with some smaller step size $h_{new} < h$, i.e. we have to divide the integration interval $[t_n + h]$ into $[t_n + h_{new}] \cup [t_n + h_{new}, t_n + h]$. In this case, we have to simulate random variables $\hat{I}_{(k);t_n,t_n+h_{new}}$ and $\tilde{I}_{(l);t_n,t_n+h_{new}}$ for the approximation at time $t_n + h_{new}$ with a conditional distribution because the realization of the underlying random processes at time $t_n + h$ are already known.

Suppose that $\hat{i}_{(k)}$ and $\tilde{i}_{(l)}$ denote the already known realizations of the random variables $\hat{I}_{(k);t_n,t_n+h}$ and $\tilde{I}_{(l);t_n,t_n+h}$, respectively. Now, one has to simulate realizations of the independent random variables $\hat{Z}_{(k)}$ and $\tilde{Z}_{(l)}$ for $k = 1, \ldots, m$ and $l = 1, \ldots, m-1$ with distribution

$$\mathrm{P}(\hat{Z}_{(k)} = \pm\sqrt{3h_{new}}) = \frac{1}{6}, \quad \mathrm{P}(\hat{Z}_{(k)} = 0) = \frac{2}{3}$$

$$\text{and} \qquad \mathrm{P}(\tilde{Z}_{(l)} = \pm\sqrt{h_{new}}) = \frac{1}{2}.$$

Then, we have to apply random variables with a conditional distribution that are simulated as follows:

$$\hat{I}_{(k);t_n,t_n+h_{new}} = \sqrt{\frac{h-h_{new}}{h}}\hat{Z}_{(k)} + \frac{h_{new}}{h}\hat{i}_{(k)} \tag{14}$$

$$\tilde{I}_{(l);t_n,t_n+h_{new}} = \sqrt{\frac{h-h_{new}}{h}}\tilde{Z}_{(l)} + \frac{h_{new}}{h}\tilde{i}_{(l)} \tag{15}$$

and we directly obtain that

$$\hat{I}_{(k);t_n+h_{new},t_n+h} = -\sqrt{\frac{h-h_{new}}{h}}\hat{Z}_{(k)} + \frac{h-h_{new}}{h}\hat{i}_{(k)} \tag{16}$$

$$\tilde{I}_{(l);t_n+h_{new},t_n+h} = -\sqrt{\frac{h-h_{new}}{h}}\tilde{Z}_{(l)} + \frac{h-h_{new}}{h}\tilde{i}_{(l)}. \tag{17}$$

Because $\hat{I}_{(k,l),n}$ can be calculated from $\hat{I}_{(k),n}$, $\hat{I}_{(l),n}$, $\tilde{I}_{(k),n}$ and $\tilde{I}_{(l),n}$, all needed random variables for the SRK method in case of a rejected step size are given. Clearly, if the new step size h_{new} is rejected as well, we have to try another step size less than h_{new} and again we have to simulate random variables conditionally that the realizations at time $t_n + h_{new}$ are known. As a result of this, all realizations have to be stored until an approximation at some later time point is accepted by the step size control algorithm.

6. Adaptive selection of Monte Carlo sample number and time stepping

6.1. *Classical setting*

Until now, we considered $|\mathbb{E}(f(X(t))) - \mathbb{E}(f(Y^h(t)))|$ as a measure of the error. However, in practice it is often not possible to evaluate the expectation $\mathbb{E}(f(Y^h(t)))$ exactly, and one often resorts then to Monte-Carlo methods, calculating M independent realizations $Y^{h,j}$, $j = 1, \ldots, M$, and using the estimator

$$\mathbb{E}(f(Y^h(t))) \approx \widehat{E}_M(f(Y^h(t))) = \frac{1}{M}\sum_{j=1}^{M} f(Y^{h,j}(t))$$

as an approximation for the desired expectation.

The resulting error $|\mathbb{E}(f(X(t))) - \widehat{E}_M(f(Y^h(t)))|$ is, however, itself a stochastic variable. It is often assessed in the mean-square sense, thus looking at

$$e_{ms}(t) = \mathbb{E}\Big(\big(\mathbb{E}(f(X(t))) - \widehat{E}_M(f(Y^h(t))\big)^2 \Big),$$

which can be rewritten as the sum of the squared systematic error of the weak approximation method and the variance of the Monte-Carlo estimator, i.e.

$$e_{ms}(t) = \Big(\mathbb{E}\big(f(X(t)) - f(Y^h(t)) \big) \Big)^2 + \underbrace{\operatorname{Var} \widehat{E}_M(f(Y^h(t))}_{=\frac{1}{M}\operatorname{Var} f(Y^h(t))}.$$

Thus, to obtain an estimator with a mean-square error smaller than some prescribed ε^2, one can choose M, and, in the case of fixed time stepping, h, otherwise $Atol_i$ and $Rtol_i$, such that suitable estimators for

$$\left|\mathbb{E}\Big(f(X(t)) - f(Y^h(t))\Big)\right| \qquad \text{and} \qquad \frac{1}{M} \operatorname{Var} f(Y^h(t))$$

are not larger than $\frac{\varepsilon}{\sqrt{2}}$. Whereas the systematic error can be estimated by Richardson extrapolation techniques, the variance is usually estimated as sample variance. To simplify the presentation, in the following we restrict to the case of constant step sizes and controlling the error only at the end point T. Then, similar to the multi-level Monte Carlo algorithm in [10], the following algorithm can be applied to determine M and h, given some rough initial guesses:

Step 1 Choose initial N and M.

Step 2 Define $h = \frac{T-t_0}{N}$.

Step 3 Calculate the unbiased sample variance s^2 of M samples of $f(Y^h(T))$.

Step 4 Choose new M according to $M = \lceil \frac{\sqrt{2}s^2}{\varepsilon} \rceil$. Evaluate extra samples as needed for new M, and calculate $\widehat{E}_M(f(Y^h(T)))$.

Step 5 Calculate $\widehat{E}_M\Big(f(Y^{\frac{h}{2}}(T))\Big)$ and the extrapolated value

$$F = \frac{2^p \widehat{E}_M\Big(f(Y^{\frac{h}{2}}(T))\Big) - \widehat{E}_M\Big(f(Y^h(T))\Big)}{2^p - 1},$$

where p denotes the weak approximation order of the method.

Step 6 Calculate the error indicator

$$\chi = \frac{\sqrt{2}}{\varepsilon} |F - \widehat{E}_M\Big(f(Y^h(T))\Big)|.$$

If $\chi < 1$, F can be used as approximation with the desired accuracy, so we are done.

Otherwise, calculate the new number of time steps as

$$N_{new} = \lceil N \sqrt[p]{\chi} \cdot \lambda \rceil,$$

where the safety factor λ can be chosen e. g. equal to 1.25. Let $N = N_{new}$ and restart with Step 2.

6.2. *Short outlook to multi-level Monte Carlo simulation*

The presented SRK methods are weak order two schemes in the sense of (3). However, for the approximation of the expectation we applied the arithmetic mean as an estimator given a particular number of independent sample trajectories. In order to reduce the variance of such an estimator and to optimally balance the computational costs for the number of samples and the computation of each approximation, Giles proposed the multi-level Monte Carlo simulation (see [10] for details). Here, we want to mention that although multi-level Monte Carlo simulation originally relies on pathwise approximation properties, this technique can be applied to weak approximation schemes like the presented SRK methods as well. In this case, the random variables $\hat{I}_{(k),n}$ and $\tilde{I}_{(l),n}$ that are used by the SRK method have to be chosen from infinitely divisible distribution, e. g. the normal distribution, to ensure that simulations on a coarse grid can be build from the accumulated realizations of the random variables that are used on a fine grid. Of special interest is the combination of a weak order one scheme on all levels except of the finest level, where a higher order weak approximation should be applied. In this case, the computational effort can be significantly reduced compared to multi-level Monte Carlo simulation based on the same approximation scheme on each level. Further details will be available in a forthcoming report. Multi-level Monte Carlo simulation is in the focus of active current research.

Bibliography

[1] Butcher, J. C. (2003). *Numerical methods for ordinary differential equations.* (Chichester: Wiley. xiv, 425 p.).

[2] Debrabant, K. (2010). Runge-Kutta methods for third order weak approximation of SDEs with multidimensional additive noise, *BIT* **50**, 3, pp. 541–558, doi:10.1007/s10543-010-0276-2.

[3] Debrabant, K. and Kvaerno, A. (2008/09). B-series analysis of stochastic Runge-Kutta methods that use an iterative scheme to compute their internal stage values, *SIAM J. Numer. Anal.* **47**, 1, pp. 181–203, doi:10.1137/070704307.

[4] Debrabant, K. and Rößler, A. (2008a). Classification of stochastic Runge-Kutta methods for the weak approximation of stochastic differential equations, *Math. Comput. Simulation* **77**, 4, pp. 408–420, doi:10.1016/j.matcom.2007.04.016.

[5] Debrabant, K. and Rößler, A. (2008b). Continuous Runge-Kutta methods for Stratonovich stochastic differential equations, in *Monte Carlo and quasi-Monte Carlo methods 2006* (Springer, Berlin), ISBN 978-3-540-74495-5, pp. 237–250, doi:10.1007/978-3-540-74496-2_13.

[6] Debrabant, K. and Rößler, A. (2008c). Continuous weak approximation for stochastic differential equations, *J. Comput. Appl. Math.* **214**, 1, pp. 259–273, doi:10.1016/j.cam.2007.02.040.

[7] Debrabant, K. and Rößler, A. (2009a). Diagonally drift-implicit Runge-Kutta methods of weak order one and two for Itô SDEs and stability analysis, *Appl. Numer. Math.* **59**, 3-4, pp. 595–607, doi:10.1016/j.apnum.2008.03.011.

[8] Debrabant, K. and Rößler, A. (2009b). Families of efficient second order Runge-Kutta methods for the weak approximation of Itô stochastic differential equations, *Appl. Numer. Math.* **59**, 3-4, pp. 582–594, doi:10.1016/j.apnum.2008.03.012.

[9] Gard, T. C. (1988). *Introduction to stochastic differential equations, Monographs and Textbooks in Pure and Applied Mathematics*, Vol. 114 (Marcel Dekker Inc., New York), ISBN 0-8247-7776-X.

[10] Giles, M. B. (2008). Multilevel Monte Carlo path simulation, *Oper. Res.* **56**, 3, pp. 607–617, doi:10.1287/opre.1070.0496, URL `http://dx.doi.org/10.1287/opre.1070.0496`.

[11] Hairer, E., Norsett, S. P. and Wanner, G. (2010). *Solving ordinary differential equations. I: Nonstiff problems. 2nd revised ed., 3rd corrected printing.* (Springer Series in

Computational Mathematics 8. Berlin: Springer. xv, 528 p.).

[12] Hull, John C. (2009). Options, futures, and other derivatives, *Prentice-Hall International Editions*. Upper Saddle River, NJ: Prentice Hall.

[13] Kloeden, P. E. and Platen, E. (1999). *Numerical solution of stochastic differential equations*, *Applications of Mathematics*, Vol. 21, 2nd edn. (Springer-Verlag, Berlin).

[14] Komori, Y. (2007). Weak second-order stochastic Runge–Kutta methods for non-commutative stochastic differential equations, *J. Comput. Appl. Math.* **206**, 1, pp. 158–173, doi:10.1016/j.cam.2006.06.006.

[15] Komori, Y. (2008). Weak first- or second-order implicit Runge-Kutta methods for stochastic differential equations with a scalar Wiener process, *J. Comput. Appl. Math.* **217**, 1, pp. 166–179, doi:10.1016/j.cam.2007.06.024.

[16] Komori, Y. and Burrage, K. (2011). Supplement: Efficient weak second order stochastic Runge-Kutta methods for non-commutative Stratonovich stochastic differential equations [supplement to mr2333843], *J. Comput. Appl. Math.* **235**, 17, pp. 5326–5329, doi:10.1016/j.cam.2011.04.021, URL `http://dx.doi.org/10.1016/j.cam.2011.04.021`.

[17] Küpper, D., Lehn, J. and Rößler, A. (2007). A step size control algorithm for the weak approximation of stochastic differential equations, *Numer. Algorithms* **44**, 4, pp. 335–346, doi:10.1007/s11075-007-9108-0, URL `http://dx.doi.org/10.1007/s11075-007-9108-0`.

[18] Mackevičius, V. and Navikas, J. (2001). Second order weak Runge–Kutta type methods of Itô equations, *Math. Comput. Simulation* **57**, 1-2, pp. 29–34, doi:10.1016/S0378-4754(00)00284-6.

[19] Milstein, G. N. (1995). *Numerical integration of stochastic differential equations*, *Mathematics and its Applications*, Vol. 313 (Kluwer Academic Publishers Group, Dordrecht), translated and revised from the 1988 Russian original.

[20] Perelló, J., Sircar, R. and Masoliver, J. (2008). Option pricing under stochastic volatility: the exponential Ornstein–Uhlenbeck model, *Journal of Statistical Mechanics: Theory and Experiment* **2008**, 06, p. P06010, URL `http://stacks.iop.org/1742-5468/2008/i=06/a=P06010`.

[21] Rößler, A. (2006). Rooted tree analysis for order conditions of stochastic Runge–Kutta methods for the weak approximation of stochastic differential equations, *Stoch. Anal. Appl.* **24**, 1, pp. 97–134, doi:10.1080/07362990500397699.

[22] Rößler, A. (2007). Second order Runge–Kutta methods for Stratonovich stochastic differential equations, *BIT* **47**, 3, pp. 657–680, doi:10.1007/s10543-007-0130-3.

[23] Rößler, A. (2009). Second order Runge–Kutta methods for Itô stochastic differential equations, *SIAM J. Numer. Anal.* **47**, 3, pp. 1713–1738 (electronic), doi:10.1137/060673308.

[24] Schmitz Abe, K. and Giles, M. (2008). Pricing exotic options using strong convergence properties, in *Progress in industrial mathematics at ECMI 2006*, *Math. Ind.*, Vol. 12 (Springer, Berlin), pp. 614–629, doi:10.1007/978-3-540-71992-2_103, URL `http://dx.doi.org/10.1007/978-3-540-71992-2_103`.

[25] Scott, L. O. (1987). Option pricing when the variance changes randomly: Theory, estimation, and an application, *The Journal of Financial and Quantitative Analysis* **22**, 4, pp. 419–438, URL `http://www.jstor.org/stable/2330793`.

[26] Tocino, Á. and Vigo-Aguiar, J. (2002). Weak second order conditions for stochastic Runge–Kutta methods, *SIAM J. Sci. Comput.* **24**, 2, pp. 507–523 (electronic), doi:10.1137/S1064827501387814.

Chapter 8

Wavelet solution of degenerate Kolmogoroff forward equations for exotic contracts in finance

Oleg Reichmann and Christoph Schwab*

Seminar für Angewandte Mathematik, ETH Zürich. Rämistrasse 101, 8092 Zürich, Switzerland.

Abstract We review deterministic numerical methods for option pricing which are based on discretizations of the (Kolmogoroff forward) pricing equations. Specifically, wavelet Galerkin discretizations in (log) state space G of the stochastic process are based on the Dirichlet forms corresponding to the stochastic process S, and exponentially convergent numerical time stepping of discontinuous Galerkin type is used, based on the analyticity of the semigroup generated by the process.

We illustrate the general concepts in particular for option pricing under market models beyond the Black-Scholes-Merton setup, where standard path simulation methods are known to require special attention, due to the degeneracies of the process S, typically near the origin in the state space.

Here, the infinitesimal generators' domains are known to coincide with certain Sobolev spaces with singular weights and, possibly, nonconstant order of differentiation, leading to nonstandard partial integrodifferential equations. Specific models considered here are the CEV model, the Ait-Sahalia model as well as the Heston stochastic volatility model. In each of these cases, we exhibit the corresponding evolution triple consisting of a (weighted) L^2 space, identified with its own dual, and the domain of the Dirichlet form of the stochastic process, and its dual. While the chapter of Kloeden and Neuenkirch focuses on the simulation of the paths of the process, the methods presented here allow the *deterministic* approximation of the conditional expectations. The arising (Kolmogoroff forward) pricing equations are degenerate parabolic. We show well-posedness of their variational formulations for a number of contracts (Barrier-, American-, Compound- and Swing contracts) on certain weighted Sobolev spaces. Their discretization is carried out using wavelets in space and an hp-discontinuous Galerkin (dG) timestepping scheme for European contracts. The computation of Greeks for these models is also addressed.

*CS was supported by the European Research Council (ERC) under the project ERC AdG 247277.

1. Financial Modelling

A large class (but not all) models in quantitative finance assume that the dynamics of market ("spot") prices $S(t)$ of the risky assets (such as, for example, stocks) are governed by an Ito stochastic ordinary differential equation (SODE for short). Specifically, the dynamics of the risky asset given by stochastic differential equation

$$dS(t) = b(S,t)dt + \sigma(S,t)dW(t) \,, \quad 0 \leq t \leq T < \infty \,. \tag{1}$$

Here, $W(t)$ is a standard Brownian motion, and (1) is to be understood as an Ito SODE, and the coefficients $b(s,t)$ and $\sigma(s,t)$ are assumed to be such that (1) admits unique, strong solutions $\{S(t)\}_{t\geq 0}$ which are adapted to the natural filtration of the driving process $W = \{W(t)\}_{t\geq 0}$.

The archetype of market models like (1) is the classical Black-Scholes (BS) model, where the parameters b and σ are linear with respect to $S(t)$, i.e.

$$dS(t) = bS(t)dt + \sigma S(t)dW(t) \,. \tag{2}$$

In (1), the coefficient functions $b(s,t)$ and $\sigma(s,t)$ in (1) are assumed to depend on various parameters which can be used to calibrate the model (1) to market data. Typically, this is done by calibrating *predicted, arbitrage-free option prices* computed from (1) for particular instances of the parameter values to observed prices of derivative contracts in the markets.

Evidently, for the efficiency of this methodology, it is necessary to have efficient numerical solvers for (1) for a given set of parameters, in order to *compute the arbitrage free prices and their sensitivities* from (1).

Under the assumptions made above, the price process $S = \{S(t)\}_{t\geq 0}$ is a Markov process, and we speak of (1) as *parametric Makovian market model*; we refer to [22] for details. We emphasize that not all models which are currently in use are parametric, Markovian diffusions governed by Ito SODEs like (1): it has been argued that (1) is too specific, and that the class of "driving processes", i.e. W in (1), should be widened to include also processes with jumps, such as Lévy processes: indeed, no arbitrage pricing theory is available for even more general, non-Markovian models (see, e.g., [13] and the references there), and in the Markovian case, the inclusion of driving processes with jumps in (1) is widely practised for about a decade by now (see, e.g. [7]) and the numerical methodologies which are discussed below are available also in this setting (see [35] and the references there).

Arguments against the inclusion of driving processes with jumps have been that the completeness of market models is then lost, and, from a numerical simulation perspective, that path simulation will become difficult in cases where sampling from the exact increments of the driving process is not possible anymore.

In the present note, we consider *computational option pricing methods for diffusion driven market models* such as (1) where, however, the coefficient functions $b(s,t)$ and $\sigma(s,t)$ exhibit various *degeneracies* which render the numerical solution and path simulation for the SODE (1) nonstandard. Rather than elaborating on path simulation, we consider here the evaluation of the option prices by the *numerical solution of the (deterministic) Kolmogoroff equations.* These will take the generic form

$$\partial_t u = \mathcal{A}u \ , \quad 0 < t \leq T \ , \qquad\qquad u|_{t=0} = u_0 \ . \tag{3}$$

Here, $\mathcal{A}$ denotes the *infinitesimal generator* of the semigroup associated to the price process $\{S(t)\}_{t\geq 0}$, with domain $D(\mathcal{A}) \subset \mathcal{H}$, and G denotes the range of admissible values which can be taken by the price process S. For degenerate diffusions, the pivot space $\mathcal{H}$ depends on the market model and could be a (weighted) L^2 space over G.

Our numerical approach will *not* be based on path simulation, but rather on the *variational approximation* of *weak solutions of* (3). These are given by

$$\forall v \in \mathcal{V}: \quad (v, \partial_t u) + a(u, v) = 0 \ , \quad 0 < t \leq T \ , \qquad u|_{t=0} = u_0 \ . \tag{4}$$

Here, the *bilinear form* $a(\cdot, \cdot)$ denotes the so-called *Dirichlet form* of the process S, and $\mathcal{V} = D(\mathcal{A}^{1/2}) \subset \mathcal{H}$

Since the parabolic forward equation (3), resp. (4) is deterministic and parabolic, its solution operator has strong smoothing properties: for example, it forms in all cases considered here, an *analytic semigroup* which implies *time-analyticity* of the dependence of the solutions $u(t)$ on t. It is, as we shall show, therefore possible to design *high-order convergent discretization schemes* for (4), both with respect to time-integration and with respect to the discretization in the spot price s. The methods which we present here are *mesh based*, i.e they require meshing the state domain G of the price process S. While this causes technical difficulties in the case when S in (1) is vector-valued (as, e.g., in basket contracts) due to the curse of dimensionality, it is a substantial advantage in that *prices of compound and of multi-period contracts can be computed without any modification of the basic numerical methodology.*

The price to be paid for this is, of course, the *solution of large linear systems of equations in each timestep.* As we will show, *these systems can be preconditioned perfectly by means of multiresolution (i.e. wavelet) Galerkin discretizations in G*, also in cases when the generator $\mathcal{A}$ is a degenerate, second order differential operator; this case is, in financial models, typical, and standard preconditioning methods such as multilevel or multigrid preconditioning are not directly applicable. The use of multiresolution bases for Galerkin discretizations has, moreover, also the advantage to render the matrices arising from processes with jumps numerically sparse; this aspect is not developed here, and we refer to [35] and the references there for more on this.

Here, we consider only contracts of *finite maturity*, i.e. $T < \infty$. Pricing of perpetual contracts will also allow the use of our techniques, but for brevity of exposition, we will not elaborate here. The outline of this chapter is as follows: a general setup of Markovian market models and of financial derivative contracts are introduced in Section 2. Well-posedness, i.e. existence of a unique *weak* solution, of variational formulations of the forward equations is subsequently discussed in Section 3. The space and time discretization of the pricing problems is presented and analyzed in Section 4. In particular, we propose a wavelet discretization in space and an *hp*-dG timestepping scheme. Finally, numerical examples and an outlook are given in Section 5.

2. Pricing of Derivative Contracts

In the pricing of derivative contracts enter two classes of ingredients: a) the market model and b) the conditions from the contracts' term sheets.

2.1. *Models*

In the following we present some examples of SDEs with degenerate coefficients, where by degenerate we mean that the coefficients *do not* satisfy a *global Lipschitz condition*, i.e.,

$$|b(x) - b(y)| + |\sigma(x) - \sigma(y)| \leq C\,|x - y|\,, \tag{5}$$

for some functions b and σ, $x, y \in \mathbb{R}$ and C independent of x and y.

Example 1 (CEV model). *Let S be given as the solution of*

$$\mathrm{d}S(t) = bS(t)\mathrm{d}t + \sigma S(t)^{\rho}\mathrm{d}W(t), \quad S(0) = s,$$

$b \in \mathbb{R}$, $\sigma, s \in \mathbb{R}_+$ *and* $\rho \in (0,1)$. *Note that in the limiting case $\rho \uparrow 1$ we obtain the classical Black–Scholes model and for $\rho = 0.5$ a square-root process, see [10; 14]. The generator $\mathcal{A}^{CEV}$ for the CEV model is given as*

$$\mathcal{A}^{CEV}(s; \partial_s) = bs\partial_s + \frac{1}{2}\sigma^2 s^{2\rho}\partial_{ss}\,.$$

Example 2 (Heston model). *Let $S = (S(t))_{t\geq 0}$ be given as $S(t) = \exp(X(t))$. Then (the log price process) $X = (X(t))_{t\geq 0}$ is the solution of*

$$\begin{aligned}\mathrm{d}X(t) &= (b - \frac{1}{2}Y(t))\mathrm{d}t + \sqrt{Y(t)}S(t)\mathrm{d}W_1(t), \quad X(0) = x,\\ \mathrm{d}Y(t) &= \alpha(m - Y(t))\mathrm{d}t + \rho\beta\sqrt{Y(t)}W_1(t) + \sqrt{1-\rho^2}\sqrt{Y(t)}W_2(t),\\ &\qquad\qquad\qquad\qquad\qquad\qquad Y(0) = Y_0\,.\end{aligned}$$

Here, $\alpha, b, m \in \mathbb{R}$, $Y_0, \beta, s \in \mathbb{R}_+$ and $\rho \in [0,1]$ and W_1, W_2 denote two independent Brownian motions, see [19]. The generator $\mathcal{A}$ for the Heston model is given as

$$\begin{aligned}\mathcal{A}^H(x, y; \partial_x, \partial_y) := &\frac{1}{2}y\partial_{xx} + \beta\rho y\partial_{xy} + \frac{1}{2}\beta^2 y\partial_{yy}\\ &+ \left(b - \frac{1}{2}y\right)\partial_x + \alpha(m - y)\partial_y\,.\end{aligned}$$

Note that we set the market price of volatility risk to zero for simplicity. A non-vanishing market price of volatility risk would lead to a different drift component for the volatility process, see also [2; 19].

Example 3 (Ait-Sahalia model). *Let S be given as the solution of*

$$\begin{aligned}\mathrm{d}S(t) &= (\alpha_{-1}S(t)^{-1} - \alpha_0 + \alpha_1 S(t) - \alpha_2 S(t)^{\nu})\mathrm{d}t \\ &\quad + \sigma S(t)^{\rho}\mathrm{d}W(t), \quad S(0) = s,\end{aligned}$$

for $\alpha_i > 0$, $i = -1, \ldots, 2$, $\sigma, \nu, \rho, s > 0$, see [1; 40] for details. The generator $\mathcal{A}^{AS}$ for the Ait-Sahalia model is given as

$$\mathcal{A}^{AS}(s; \partial_s) = (\alpha_{-1}s^{-1} - \alpha_0 + \alpha_1 s - \alpha_2 s^{\nu})\partial_s + \frac{1}{2}\sigma^2 s^{2\rho}\partial_{ss} \,,$$

in the following we set $\alpha_0 = \alpha_2 = 0$ for simplicity.

2.2. *Contracts*

In this section we derive pricing expressions, i.e., PDEs, PDIs or sequences of PD(I)s for the valuation of plain vanilla and exotic contracts under general market models as described in the previous section. Therefore let $S = (S(t))_{t\geq 0}$ be an admissible market model and $\mathcal{A}$ the corresponding generator as in Example 1-3.

2.2.1. *European*

The value $V(t, s)$ of a European type contract with (reasonable) payoff g is given as

$$V(t, s) = \mathbb{E}[e^{-r(T-t)}g(S(T))|S(t) = s], \tag{6}$$

where the expectation is computed under some (possibly non-unique) martingale measure $\mathbb{Q}$ and $r \geq 0$ denotes the risk free interest rate. Under some regularity assumptions on the option price $V(t, s)$ we formally obtain the following result.

Theorem 1. *Let $V \in C^{1,2}((0, T) \times G) \cap C^0([0, T] \times G)$ be given as Eq. (6), then V satisfies the following equation*

$$\partial_t V + \mathcal{A}V - rV = 0 \quad \text{in } (0, T) \times G, \quad V(T, s) = g(s) \text{ in } G, \tag{7}$$

where $\mathcal{A}$ is the generator of S in the CEV model resp. in the Ait-Sahalia model, or the generator of $Z = (X, Y)$ in the Heston model, and where G denotes the state space of the process S, respectively Z.

Barrier options differ from vanillas in the sense that the option contract is triggered if the price of the underlying hits some barrier $B > 0$. Recall that a *stopping time* τ for a given filtration $\mathcal{F}$ is a random variable taking values in $(0, \infty)$ and satisfying

$$\{\tau \leq t\} \in \mathcal{F}_t, \quad \forall t \geq 0.$$

Let $\tau_{(B,\infty)}$ denote the first hitting time of the complement set $(-\infty, B]$ by the stochastic process S. Then the value of a down-and-out option is then given by

$$V(t,s) = \mathbb{E}[e^{-r(T-t)}g(S(T))\mathbb{1}_{\tau_{(B,\infty)}>T}|S(t)=s]. \tag{8}$$

Under some regularity assumptions on the option price $V(t,s)$ we obtain the following result.

Theorem 2. *Let $V \in C^{1,2}((0,T)\times G_B)\cap C^0([0,T]\times G_B)$ be given as Eq. (8), then V satisfies the following equation*

$$\partial_t V + \mathcal{A}V - rV = 0 \quad in\ (0,T)\times G_B, \quad V(T,s) = g(s)\ in\ G_B, \tag{9}$$

and boundary conditions

$$V(t,s) = 0, \ \ in\ J\times G_B^c.$$

with $\mathcal{A}$ as in Theorem 1 and $G_B = (B,\infty)$ for the CEV model and for the Ait-Sahalia model and $G = (\log(B),\infty)\times\mathbb{R}_+$ for the Heston model. Analogous formulations can be obtained for up-and-out as well as knock in options. Note that the value of the corresponding plain vanilla contract is given as the sum of the knock-in and knock-out option. We refer to [16] for a rigorous derivation of the pricing equation for Barrier options under certain additive market models.

2.2.2. *Compound*

Compound options are options on options. Let $V_1(t,s)$ be the option price of a European option with payoff $g_1(s)$ and maturity $T_1 > 0$. The value of a compound option V with payoff $g(s)$ and maturity $0 < T < T_1$ is given by

$$V(t,s) = \mathbb{E}\big[e^{r(t-T)}g(V_1(T,S_T)) \mid S(t)=s\big].$$

Applying Theorem 1 we can obtain the value V_1 of the underlying option by solving the partial differential equation

$$\partial_t V_1 + \mathcal{A}V_1 - rV_1 = 0, \quad \text{in } (0,T_1)\times G, \quad V_1(T_1,s) = g_1(s), \quad \text{in } G,$$

In a second step we get the price of the compound option by solving

$$\partial_t V + \mathcal{A}V - rV = 0, \quad \text{in } (0,T)\times G, \quad V(T,s) = g(V_1(T,s)), \quad \text{in } G.$$

2.2.3. *American*

Denoting by $\mathcal{T}_{t,T}$ the set of all stopping times for S with values in the interval (t,T), the value of an American option is given by

$$V(t,s) := \sup_{\tau\in\mathcal{T}_{t,T}} \mathbb{E}\big[e^{-r(\tau-t)}g(S_\tau) \mid S_t = s\big]\ . \tag{10}$$

As for the European vanilla style contracts, there is a close connection between the probabilistic representation (10) of the price and a deterministic, PDE based representation of the price. Discretization of the arising variational inequalities leads to a system of linear complementarity problems (LCPs) which can be solved employing the projected SOR (PSOR) algorithm, see [11] for details. We follow a different approach in the following and employ a semismooth Newton method, see [23; 24] for details. The application of this method to multidimensional diffusion models is discussed in [18] and the case of certain bivariate Lévy models is analyzed in [36]. The derivation of a system of inequalities satisfied by $V(t, s)$ can be justified rigorously for different market models. We refer to [26] for the Black-Scholes model and to [29] for Lévy type market models. The corresponding formulations for the Ait-Sahalia model and the Heston model are given in Section 3.3.

2.2.4. *Swing*

Although swing options appear in various forms in applications, most of them are mathematically optimal multiple stopping problems. For example, in energy markets the delivery of a commodity is limited by capacity constraints usually resulting in a pre-specified refracting time for contracts with several exercise rights. It can be agreed that the refraction period δ which is greater than the minimal delivery time is constant. This separation of two exercise times not only represents an important contract constraint, but also prevents the case of single optimal stopping time problems where all rights are exercised at once, see [3]. Let us denote by $\mathcal{T}_{t,T}$ the set of all stopping times for S with values in (t, T) and by $\mathcal{T}_{t,\infty}$ the set of all stopping times with values greater or equal than t. For stopping time problems with $p \in \mathbb{N}$ exercise rights, constant refracting period δ and maturity T the following sets are defined.

Definition 8.1. The set of admissible stopping time vectors with length $p \in \mathbb{N}$ and refracting time $\delta > 0$ is defined by

$$\begin{aligned}\mathcal{T}_t^{(p)} := \{\tau^{(p)} = (\tau_1, \dots, \tau_p) \mid \tau_i \in \mathcal{T}_{t,\infty} \text{ with } \tau_1 \leq T \\ \text{a.s. and } \tau_{i+1} - \tau_i \geq \delta \text{ for } i = 1, \dots, p-1\}.\end{aligned}$$

Consider a Lipschitz continuous time-dependent payoff function $g : \mathbb{R}_+ \times \mathbb{R}_+ \to \mathbb{R}_+$ where we assume that $g(t, \cdot) = 0$ for $t > T$. The finite horizon multiple stopping time problem with maturity T and $p \in \mathbb{N}$ exercise rights is defined as

$$V^{(p)}(t, s) := \sup_{\tau^{(p)} \in \mathcal{T}_t^{(p)}} \mathbb{E}\left[\sum_{i=1}^{p} e^{-r(\tau_i - t)} g(\tau_i, S_{\tau_i}) \mid S_t = s\right]. \tag{11}$$

It is shown in [5] for the Black-Scholes model that the multiple stopping time problem can be reduced to a cascade of single stopping time problems. In particular we have

$$V^{(p)}(t, s) = \sup_{\tau \in \mathcal{T}_{t,T}} \mathbb{E}\left[e^{-r(\tau - t)} g^{(p)}(\tau, S_\tau) \mid S_t = s\right], \tag{12}$$

with

$$g^{(p)}(t,s) := \begin{cases} g(t,s) + e^{-r\delta}\mathbb{E}\left[V^{(p-1)}(t+\delta, S_{t+\delta}) \mid S_t = s\right] & \text{if } t \leq T-\delta, \\ g(t,s) & \text{if } t \in (T-\delta, T], \end{cases}$$

$$V^{(0)}(t,s) := 0.$$

We remark that positivity of the refraction time δ is essential in the analysis of [5]; it is, however, possible to have period-dependent refraction times $\delta_i > 0$. In the following we consider the cascade of single stopping time problems for the market models in Example 1-3. We remark that the connection between Eq. (11) and Eq. (12) is formal for these market models to our knowledge. For the description of a Finite Element based pricing algorithm for swing options we refer to [42]. We point out that pricing multiperiod contracts such as swing options by path simulation is costly: To start path simulation in period $i+1$, prices at the end of period i must be known, approximately, *in all points of the state space or, at least, on a sufficiently fine triangulation in the state space of the process.* This requires numerical pricing in period i by path simulation for a large number of spot prices, see for example [27]. For the presently considered mesh based pricing methods, valuation of contracts on meshes in (log) price space in period i is naturally performed, so that the computational methodology and the computational complexity versus numerical accuracy is not affected by the transition from single- to multiperiod contracts.

3. Well-posedness of PDEs and PDIs

3.1. *General results*

We consider the standard parabolic set up in the following. Let the separable Hilbert spaces $\mathcal{V} \subset \mathcal{H}$ be given with continuous and dense embedding. We identify $\mathcal{H}$ with its dual and obtain the Gelfand triple

$$\mathcal{V} \subset \mathcal{H} \equiv \mathcal{H}^* \subset \mathcal{V}^*.$$

The following abstract well-posedness result holds.

Theorem 3. *Assume that the bilinear form $a(\cdot,\cdot) : \mathcal{V} \times \mathcal{V} \to \mathbb{R}$ satisfies continuity and a Garding inequality, i.e. the following properties. There exist constants $C_1, C_2 > 0$ and $C_3 \geq 0$ such that for all $u, v \in \mathcal{V}$ there holds*

$$|a(u,v)| \leq C_1 \|u\|_{\mathcal{V}} \|v\|_{\mathcal{V}}, \tag{13}$$

$$a(u,u) \geq C_2 \|u\|_{\mathcal{V}}^2 - C_3 \|u\|_{\mathcal{H}}^2. \tag{14}$$

Then the following abstract parabolic problem is uniquely solvable.
Find $u \in L^2((0,T);\mathcal{V}) \cap H^1((0,T);\mathcal{V}^*)$ such that

$$(\partial_t u, v)_{\mathcal{V}^*,\mathcal{V}} + a(u,v) = (f,v)_{\mathcal{V}^*,\mathcal{V}}, \forall v \in \mathcal{V}, \ a.e. \ in \ (0,T), \tag{15}$$

$$u(0) = g, \tag{16}$$

with $g \in \mathcal{H}$, $f \in L^2((0,T);\mathcal{V}^*)$ and $T > 0$.

Proof. See Theorem 4.1 in [30]. □

For the study of optimal stopping problems which arise e.g. in the contact of American options or swing options we require variational formulations of parabolic variational inequalities. To this end, let $\emptyset \neq \mathcal{K} \subset \mathcal{V}$ be a closed, non-empty and convex subset of $\mathcal{V}$ with indicator functions

$$\mathcal{I}_{\mathcal{K}}(v) = \begin{cases} 0, & \text{if } v \in \mathcal{K}, \\ +\infty, & \text{else.} \end{cases}$$

We denote by $\overline{\mathcal{K}}^{\|\cdot\|_{\mathcal{H}}}$ the closure of $\mathcal{D}(\mathcal{I}_{\mathcal{K}})$ in $\mathcal{H}$ and consider the following variational problem: given $f \in L^2((0,T);\mathcal{V}^*)$, $u_0 \in \overline{\mathcal{K}}^{\|\cdot\|_{\mathcal{H}}}$, find $u \in L^2((0,T);\mathcal{V}) \cap H^1((0,T);\mathcal{V}^*)$ such that $u(t,\cdot) \in \mathcal{D}(\mathcal{I}_{\mathcal{K}})$ a.e. in $(0,T)$ and

$$\begin{aligned}(\partial_t u, u-v)_{\mathcal{V}^*,\mathcal{V}} + a(u,u-v) - (f,u-v)_{\mathcal{V}^*,\mathcal{V}} + \mathcal{I}_{\mathcal{K}}(u) - \mathcal{I}_{\mathcal{K}}(v) \geq 0,\\ \forall v \in \mathcal{D}(\mathcal{I}_{\mathcal{K}}) \text{ a.e. in } (0,T).\end{aligned} \tag{17}$$

The existence and uniqueness results can be obtained for the formulation Eq. (17) under certain regularity assumptions on f and on the initial condition u_0 for bilinear forms $a(\cdot,\cdot)$ satisfying Eq. (13)-Eq. (14) from Chapter 6, Theorem 2.1 in [17], see also the monograph [28].

In the following we consider bilinear forms arising in the context of option pricing under the CEV market model and the Ait-Sahalia model and prove Garding inequalities and continuity of the corresponding bilinear forms on weighted Sobolev spaces. As described above, this is the main step to the proof of well-posedness of the corresponding pricing equation. We consider the bilinear form $a^{\mathrm{CEV}}(\cdot,\cdot) : W_\rho(G) \times W_\rho(G) \to \mathbb{R}$ given as

$$\begin{aligned} a^{\mathrm{CEV}}(\phi,\psi) := \frac{1}{2}\sigma^2 \int_0^R s^{2\rho}\partial_s\varphi\partial_s\phi \mathrm{d}s + \rho\sigma^2 \int_0^R s^{2\rho-1}\partial_s\varphi\phi \mathrm{d}s \\ -r\int_0^R s\partial_s\varphi\phi \mathrm{d}s + r\int_0^R \varphi\phi \mathrm{d}s\,, \end{aligned}$$

for $\rho \in [0,1/2)$, $G = (0,R)$, $R > 0$ and $W_\rho(G)$ is $\overline{C_0^\infty(G)}^{W_\rho(G)}$, where $\|\phi\|_{W_\rho(G)}$ is given as

$$\|\phi\|_\rho^2 = \|\phi\|_{W_\rho(G)}^2 = \int_0^R s^{2\rho}\left|\partial_s\phi\right|^2 + |\phi|^2\, ds.$$

Then we have the following result.

Lemma 1. *For $\rho \in (0,1/2)$ the bilinear form $a^{\mathrm{CEV}}(\cdot,\cdot)$ is continuous and satisfies a Garding inequality on $W_\rho(G)$.*

Proof. Let $\varphi \in C_0^\infty(G)$. By Hardy's inequality, for $\varepsilon \neq 1$, $\varepsilon > 0$, and any $R > 0$

$$\Big(\int_0^R s^{\varepsilon-2}|\varphi|^2 \mathrm{d}s\Big)^{\frac{1}{2}} \leq \frac{2}{|\varepsilon - 1|}\Big(\int_0^R s^{\varepsilon}|\partial_s\varphi|^2 \mathrm{d}s\Big)^{\frac{1}{2}} \tag{18}$$

we find with $\varepsilon = 2\rho \neq 1$, that

$$\Big|\int_0^R s^{2\rho-1}\partial_s\varphi\phi \mathrm{d}s\Big| \leq \Big(\int_0^R s^{2\rho}(\partial_s\varphi)^2 \mathrm{d}s\Big)^{\frac{1}{2}} \Big(\int_0^R s^{2\rho-2}\phi^2 \mathrm{d}s\Big)^{\frac{1}{2}}$$

$$\leq \|\varphi\|_\rho \frac{2}{|2\rho - 1|}\|\phi\|_\rho$$

and, by the Cauchy-Schwarz inequality,

$$\Big|\int_0^R s\partial_s\varphi\phi \mathrm{d}s\Big| \leq \|\varphi\|_\rho\Big(\int_0^R s^{2-2\rho}\phi^2 \mathrm{d}s\Big)^{\frac{1}{2}} \leq \|\varphi\|_\rho \, R^{1-\rho} \, \|\phi\|_{L^2(G)} \,.$$

Thus, for $\varphi, \phi \in C_0^\infty(G)$, $\rho \neq \frac{1}{2}$, there holds

$$|a_\rho^{\mathrm{CEV}}(\varphi, \phi)| \leq C(\rho, \sigma, r)\|\varphi\|_\rho\|\phi\|_\rho \,.$$

Hence, we may extend the bilinear form $a_\rho^{\mathrm{CEV}}(\cdot,\cdot)$ from $C_0^\infty(G)$ to $W_\rho(G)$ by continuity for $\rho \in [0,1]\backslash\{\frac{1}{2}\}$. Furthermore, we have

$$\begin{aligned} a_\rho^{\mathrm{CEV}}(\varphi, \varphi) = \frac{1}{2}\sigma^2\|s^\rho\partial_s\varphi\|_{L^2(G)}^2 + \frac{1}{2}\rho\sigma^2 \int_0^R s^{2\rho-1}\partial_s(\varphi^2)\mathrm{d}s \\ - \frac{1}{2}r\int_0^R s\partial_s(\varphi^2)\mathrm{d}s + r\int_0^R \varphi^2\mathrm{d}s \,. \end{aligned}$$

Integrating by parts, we get, for $0 \leq \rho \leq \frac{1}{2}$,

$$\int_0^R s^{2\rho-1}\partial_s(\varphi^2)\mathrm{d}s = -(2\rho - 1)\int_0^R s^{2\rho-2}\varphi^2\mathrm{d}s \geq 0 \,.$$

Analogously, $\frac{1}{2}\int_0^R s\partial_s(\varphi^2)\mathrm{d}s = -\frac{1}{2}\int_0^R \varphi^2\mathrm{d}s$, hence we get for $0 \leq \rho \leq \frac{1}{2}$

$$a_\rho^{\mathrm{CEV}}(\varphi, \varphi) \geq \frac{1}{2}\sigma^2\|s^\rho\partial_s\varphi\|_{L^2(G)}^2 + \frac{3}{2}r\|\varphi\|_{L^2(G)}^2 \geq \frac{1}{2}\min\{\sigma^2, 3r\}\|\varphi\|_\rho^2 \,.$$

□

For $\rho \in [1/2, 1)$ we consider a different formulation. We introduce the spaces $W_{\rho,\mu}(G)$ as closure of $C_0^\infty(G)$ with respect to the norm

$$\|\varphi\|_{\rho,\mu}^2 := \int_0^R \big(s^{2\rho+2\mu}|\partial_s\varphi|^2 + s^{2\mu}|\varphi|^2\big)\mathrm{d}s \,. \tag{19}$$

We consider the bilinear form $a_\mu^{\mathrm{CEV}}(\cdot,\cdot) : W_{\rho,\mu}(G) \times W_{\rho,\mu}(G) \to \mathbb{R}$ is defined by

$$\begin{aligned} a_\mu^{\mathrm{CEV}}(\varphi, \phi) := \frac{1}{2}\sigma^2 \int_0^R s^{2\rho+2\mu}\partial_s\varphi\partial_s\phi\mathrm{d}s + \sigma^2(\rho+\mu)\int_0^R s^{2\rho+2\mu-1}\partial_s\varphi\phi\mathrm{d}s \\ -r\int_0^R s^{1+2\mu}\partial_s\varphi\phi\mathrm{d}s + r\int_0^R s^{2\mu}\varphi\phi\mathrm{d}s \,. \end{aligned} \tag{20}$$

Proposition 8.1. *Assume* $0 \le \rho \le 1$ *and select*

$$\begin{cases} \mu = 0 & \text{if } 0 \le \rho < \frac{1}{2},\ \rho = 1\,, \\ -\frac{1}{2} < \mu < \frac{1}{2} - \rho & \text{if } \frac{1}{2} \le \rho < 1\,. \end{cases} \tag{21}$$

Then there exists $C_1, C_2 > 0$ *such that* $\forall \varphi, \phi \in W_{\rho,\mu}(G)$ *there holds*

$$|a^{\mathrm{CEV}}_{\rho,\mu}(\varphi,\phi)| \le C_1 \|\varphi\|_{\rho,\mu}\|\phi\|_{\rho,\mu}\,, \tag{22}$$

$$a^{\mathrm{CEV}}_{\rho,\mu}(\varphi,\varphi) \ge C_2 \|\varphi\|^2_{\rho,\mu}\,. \tag{23}$$

Proof. The continuity (22) of $a^{\mathrm{CEV}}_{\rho,\mu}$ in $W_{\rho,\mu}(G) \times W_{\rho,\mu}(G)$ follows from the Cauchy-Schwarz inequality and by Hardy's inequality (18) with $\varepsilon = 2(\rho+\mu) \neq 1$

$$\begin{aligned} |a^{\mathrm{CEV}}_{\rho,\mu}(\varphi,\phi)| &\le \frac{1}{2}\sigma^2 \|\varphi\|_{\rho,\mu}\|\phi\|_{\rho,\mu} + \sigma^2(\rho+\mu)\|\varphi\|_{\rho,\mu}\left(\int_0^R s^{2\rho+2\mu-2}\phi^2 \mathrm{d}s\right)^{1/2} \\ &\quad + r\|\varphi\|_{\rho,\mu}\left(\int_0^R s^{2+2\mu-2\rho}\phi^2 \mathrm{d}s\right)^{1/2} \\ &\le \left(\frac{\sigma^2}{2} + \frac{2\sigma^2(\rho+\mu)}{|2\rho+2\mu-1|} + rR^{1-\rho}\right)\|\varphi\|_{\rho,\mu}\|\phi\|_{\rho,\mu}. \end{aligned}$$

Let $\varphi \in C_0^\infty(G)$. We calculate

$$\begin{aligned} a^{\mathrm{CEV}}_{\rho,\mu}(\varphi,\varphi) &= \frac{1}{2}\sigma^2\|s^{\rho+\mu}\partial_s\varphi\|^2_{L^2(G)} + \frac{1}{2}\sigma^2(\rho+\mu)\int_0^R s^{2\rho+2\mu-1}\partial_s(\varphi^2)\mathrm{d}s \\ &\quad - \frac{1}{2}r\int_0^R s^{1+2\mu}\partial_s(\varphi^2)\mathrm{d}s + r\|s^\mu\varphi\|^2_{L^2(G)} \\ &= \frac{1}{2}\sigma^2\|s^{\rho+\mu}\varphi_s\|^2_{L^2(G)} - \frac{1}{2}\sigma^2(\rho+\mu)(2\rho+2\mu-1)\times \\ &\quad \int_0^R s^{2\rho+2\mu-2}\varphi^2\mathrm{d}s + \frac{1}{2}r(1+2\mu)\int_0^R s^{2\mu}\varphi^2\mathrm{d}s + r\|s^\mu\varphi\|^2_{L^2(G)}\,. \end{aligned}$$

Given $1/2 \le \rho < 1$, we now choose μ such that $-1/2 \le \mu < 1/2 - \rho$. Then, $2\rho + 2\mu - 1 < 0$, $1 + 2\mu \ge 0$, $\rho + \mu \ge 0$ and we get

$$a^{\mathrm{CEV}}_{\rho,\mu}(\varphi,\varphi) \ge \frac{1}{2}\sigma^2\|s^{\rho+\mu}\partial_s\varphi\|^2_{L^2(G)} + r\|s^\mu\varphi\|^2_{L^2(G)} \ge \frac{1}{2}\min\{\sigma^2, 2r\}\|\varphi\|^2_{\rho,\mu}$$

By density of $C_0^\infty(0,R)$ in $W_{\rho,\mu}(G)$, we have shown (23). □

In the following we obtain a variational formulation for the Ait-Sahalia short rate model, let us consider the space $\widetilde{W}_{\rho,\mu}(G)$ given as $\overline{C_0^\infty(G)}^{\widetilde{W}_{\rho,\mu}(G)}$ with the norm $\|\cdot\|_{\widetilde{W}_{\rho,\mu}(G)}$

$$\|u\|^2_{\widetilde{W}_{\rho,\mu}(G)} = \left\|s^{2(\rho+\mu)}\,|\partial_s u|^2\right\|^2_{L^2(G)} + \left\|s^{-2\rho-2+2\mu}u^2\right\|^2_{L^2(G)} \tag{24}$$

and the pivot space $\mathcal{H}_{\rho,\mu}(G)$ with norm $\|\cdot\|_{\mathcal{H}_{\rho,\mu}(G)}$ given as

$$\|u\|^2_{\mathcal{H}_{\rho,\mu}(G)} = \left\|s^{-\rho-1+\mu}u\right\|^2_{L^2(G)}\,.$$

We consider in the following the bilinear form $a_{\mu}^{\mathrm{AS}} : \widetilde{W}_{\mu,\rho}(G) \times \widetilde{W}_{\mu,\rho}(G) \to \mathbb{R}$ given by

$$\begin{aligned} a_{\rho,\mu}^{\mathrm{AS}}(\varphi,\phi) := \frac{1}{2}\sigma^2 \int_0^R s^{2\rho+2\mu}\partial_s\varphi\partial_s\phi \mathrm{d}s + \sigma^2(\rho+\mu)\int_0^R s^{2\rho+2\mu-1}\partial_s\varphi\phi \mathrm{d}s \\ -\alpha_{-1}\int_0^R s^{-1+2\mu}\partial_s\varphi\phi \mathrm{d}s - \alpha_1 \int_0^R s^{1+2\mu}\partial_s\varphi\phi \mathrm{d}s \\ +\int_0^R s^{2\mu+1}\varphi\phi \mathrm{d}s\,. \end{aligned} \tag{25}$$

Proposition 8.2. *Assume* $0 \le \rho \le 1$ *and select*

$$\mu \ge \rho + 1. \tag{26}$$

Assume also $r > 0$. *Then there exists* $C_1, C_2 > 0$ *and* $C_3 \ge 0$ *such that* $\forall \varphi, \phi \in W_{\rho,\mu}(G)$ *there holds*

$$|a_{\rho,\mu}^{\mathrm{AS}}(\varphi,\phi)| \le C_1 \|\varphi\|_{\rho,\mu}\|\phi\|_{\rho,\mu}\,, \tag{27}$$

$$a_{\rho,\mu}^{\mathrm{AS}}(\varphi,\varphi) \ge C_2\|\varphi\|_{\rho,\mu}^2 - C_3\,\|\varphi\|_{\mathcal{H}_{\rho,\mu}(G)}^2\,, \tag{28}$$

with $\|u\|^2_{\mathcal{H}_{\rho,\mu}(G)} = \left\|s^{-\rho-1+\mu}u\right\|^2_{L^2(G)}$

Proof. The continuity (27) of $a_{\rho,\mu}^{\mathrm{AS}}$ in $W_{\rho,\mu}(G) \times W_{\rho,\mu}(G)$ follows from the Cauchy-Schwarz inequality and by Hardy's inequality (18) with $\varepsilon = 2(\rho+\mu) \ne 1$

$$\begin{aligned} |a_{\rho,\mu}^{\mathrm{AS}}(\varphi,\phi)| &\le \frac{1}{2}\sigma^2\|\varphi\|_{\rho,\mu}\|\phi\|_{\rho,\mu} + \sigma^2(\rho+\mu)\|\varphi\|_{\rho,\mu}\left(\int_0^R s^{2\rho+2\mu-2}\phi^2\mathrm{d}s\right)^{1/2} \\ &\quad + \alpha_{-1}\|\varphi\|_{\rho,\mu}\left(\int_0^R s^{-2+2\mu-2\rho}\phi^2\mathrm{d}s\right)^{1/2} \\ &\quad + \alpha_1\|\varphi\|_{\rho,\mu}\left(\int_0^R s^{2+2\mu-2\rho}\phi^2\mathrm{d}s\right)^{1/2} + R^{2+\rho}\|\varphi\|_{\rho,\mu}\|\phi\|_{\rho,\mu} \\ &\le \left(\frac{\sigma^2}{2} + \frac{2\sigma^2(\rho+\mu)}{|2\rho+2\mu-1|} + \alpha_{-1} + \alpha_1 R^2 + R^{2+\rho}\right)\|\varphi\|_{\rho,\mu}\|\phi\|_{\rho,\mu}. \end{aligned}$$

Let $\varphi \in C_0^\infty(G)$. We calculate

$$\begin{aligned} a_{\rho,\mu}^{\mathrm{AS}}(\varphi,\varphi) &= \frac{1}{2}\sigma^2\|s^{\rho+\mu}\partial_s\varphi\|^2_{L^2(G)} + \sigma^2(\rho+\mu)\int_0^R s^{2\rho+2\mu-1}\partial_s(\varphi^2)\mathrm{d}s \\ &\quad + \alpha_{-1}\int_0^R s^{-1+2\mu}\partial_s(\varphi^2)\mathrm{d}s + \alpha_1\int_0^R s^{1+2\mu}\partial_s(\varphi^2)\mathrm{d}s \\ &\quad + \|s^{\mu+1}\varphi\|^2_{L^2(G)} \end{aligned}$$

$$\geq \frac{1}{2}\sigma^2 \|s^{\rho+\mu}\varphi_s\|^2 - \sigma^2(\rho+\mu)(2\rho+2\mu-1)\int_0^R s^{2\rho+2\mu-2}\varphi^2 \mathrm{d}s$$
$$- \alpha_{-1}(-1+2\mu)\int_0^R s^{-2+2\mu}\varphi^2 \mathrm{d}s - \alpha_1(1+2\mu)\int_0^R s^{2\mu}\varphi^2 \mathrm{d}s$$
$$+ \|s^{\mu+1}\varphi\|^2_{L^2(G)}$$
$$\geq \frac{1}{2}\sigma^2 \|s^{\rho+\mu}\varphi_s\|^2 - R^{4\rho}\sigma^2(\rho+\mu)\,\|\phi\|^2_{\mathcal{H}_{\rho,\mu}(G)}$$
$$- \alpha_{-1}(-1+2\mu)R^{2\rho}\,\|\phi\|^2_{\mathcal{H}_{\rho,\mu}(G)} - \alpha_1(1+2\mu)R^{2\rho+2}\,\|\phi\|^2_{\mathcal{H}_{\rho,\mu}(G)}$$
$$\geq C_2\,\|\phi\|^2_{\rho,\mu} - C_3\,\|\phi\|^2_{\mathcal{H}_{\rho,\mu}(G)}\,.$$

By density of $C_0^\infty(G)$ in $\widetilde{W}_{\rho,\mu}(G)$, we have shown (23). □

3.2. *Contracts of European type*

Well-posedness results for European contracts in different market models described above are given in this section.

3.2.1. *Univariate models*

We consider the problem Eq. (7) on a bounded domain $G = (0, R)$ and impose homogeneous Dirichlet conditions, either due to a (knock-out) barrier condition or due to localization. The localization can be justified rigorously for a wide range of models, see e.g. [20; 31]. In financial modelling terms, localization corresponds to the approximation of a plain vanilla option by the corresponding double barrier contract. For justification and mathematical analysis of localization (and for the meaning of barrier contracts) and its errors in Lévy type market models, we refer to [9; 8; 34].

The arising variational formulation for a European option reads: find $u \in L^2((0,T);\mathcal{V}) \cap H^1((0,T);\mathcal{V}^*)$ such that

$$(\partial_t u, v)_{\mathcal{V}^*,\mathcal{V}} + a(u,v) = (f,v)_{\mathcal{V}^*,\mathcal{V}}, \forall v \in \mathcal{V}, \text{ a.e. in } (0,T), \tag{29}$$
$$u(0) = g, \tag{30}$$

where $\mathcal{V} = D(\mathcal{A}^{1/2})$ denotes the domain of the corresponding bilinear form. The dual paring $(\cdot,\cdot)_{\mathcal{V}^*,\mathcal{V}}$ is understood as the continuous extension of the $\mathcal{H}$ scalar product and the bilinear form $a(\cdot,\cdot)$ can be any of the forms in the set $\{a^{AS}_{\rho,\mu}, a^H, a^{CEV}\}$ (or, more generally, any form for which continuity and a Garding inequality on a closed subspace of $D(\mathcal{A}^{1/2})$ can be verified). We have the following result which is a direct consequence of Theorem 3 and of Proposition 8.2 or of Proposition 8.1.

Theorem 4. *The formulation Eq. (29)-Eq. (30) admits a unique solution.*

3.2.2. *Stochastic volatility models*

Note that the option price is a function of $z = (x, y)$ and not just x, since we have to condition on $(X(t), Y(t)) = Z(t) = z$ and not just on $X(t) = x$. The reason for this is that the process Z is Markovian, but the process X is not. We find that the function $v(t, z) := V(T - t, z)$ is solution of the PDE

$$\begin{aligned} \partial_t v - \mathcal{A}^{\mathrm{H}} v + rv = 0 \ \text{ in } J \times \mathbb{R} \times \mathbb{R}_{\geq 0} \\ v(0, z) = v_0(z) := g(e^x) \ \text{ in } \mathbb{R} \times \mathbb{R}_{\geq 0}, \end{aligned} \tag{31}$$

where the infinitesimal generator $\mathcal{A}^{\mathrm{H}}$ appearing in the pricing equation (31) is given in log-price by

$$\begin{aligned} (\mathcal{A}^{\mathrm{H}} f)(z) := \frac{1}{2} y \partial_{xx} f(x, y) + \beta\rho y \partial_{xy} f(x, y) + \frac{1}{2}\beta^2 y \partial_{yy} f(x, y) \\ + \left(r - \frac{1}{2} y\right) \partial_x f(x, y) + \alpha(m - y)\partial_y f(x, y). \end{aligned} \tag{32}$$

To cast the pricing equation (31) corresponding to the Heston model in a variational formulation and to establish its well-posedness, we change variables

$$\widetilde{v}(t, x, \widetilde{y}) := v(t, x, 1/4\widetilde{y}^2). \tag{33}$$

The pricing equation for $\widetilde{v}$ becomes,

$$\begin{aligned} \partial_t \widetilde{v} - \widetilde{\mathcal{A}}^{\mathrm{H}} \widetilde{v} + r\widetilde{v} = 0 \ \text{ in } J \times \mathbb{R} \times \mathbb{R}_{\geq 0} \\ \widetilde{v}_0 = g(e^x) \ \text{ in } \mathbb{R} \times \mathbb{R}_{\geq 0}, \end{aligned} \tag{34}$$

with

$$\begin{aligned} (\widetilde{\mathcal{A}}^{\mathrm{H}} f)(x, \widetilde{y}) := \frac{1}{8}\widetilde{y}^2 \partial_{xx} f(x, \widetilde{y}) + \frac{1}{2}\rho\beta\widetilde{y}\partial_{x\widetilde{y}} f(x, \widetilde{y}) + \frac{1}{2}\beta^2 \partial_{\widetilde{y}\widetilde{y}} f(x, \widetilde{y}) \\ + \left(r - \frac{1}{8}\widetilde{y}^2\right)\partial_x f(x, \widetilde{y}) + \frac{1}{2}\left(- \alpha\widetilde{y} + \frac{4\alpha m - \beta^2}{\widetilde{y}}\right)\partial_{\widetilde{y}} f(x, \widetilde{y}). \end{aligned} \tag{35}$$

We consider the weak formulation of the transformed pricing equation (34). It is convenient to multiply the value of the option v in (34) with an exponentially decaying factor, i.e., we consider

$$w := ve^{-\eta}, \tag{36}$$

where $\eta \in C^2(\mathbb{R}^{n_v+1})$ is assumed to be at most polynomially growing at infinity. For notational simplicity, we drop "$\widetilde{\ }$" in $\widetilde{v}$ and $\widetilde{\mathcal{A}}$. Consider the change of variables (36) with $\eta = \eta(x, y) := \frac{1}{2}\kappa y^2$, $\kappa > 0$. It follows that the pricing equation for $w := (v - v_0)e^{-\eta}$ in the Heston model becomes

$$\begin{aligned} \partial_t w - \mathcal{A}^{\mathrm{H}}_\kappa w + rw = f^{\mathrm{H}}_\kappa \ \text{ in } J \times \mathbb{R} \times \mathbb{R}_{\geq 0} \\ w(0, x, y) = 0 \ \text{ in } \mathbb{R} \times \mathbb{R}_{\geq 0}, \end{aligned} \tag{37}$$

where $f_\kappa^{\mathrm{H}} := e^{-\kappa/2y^2}(\mathcal{A}^{\mathrm{H}}v_0 - rv_0)$ and

$$\begin{aligned}(\mathcal{A}_\kappa^{\mathrm{H}} f)(z) := &\frac{1}{8}y^2\partial_{xx}f(x,y) + \frac{1}{2}\rho\beta y\partial_{xy}f(x,y) + \frac{1}{2}\beta^2\partial_{yy}f(x,y) \\ &+ \left(r - \frac{1}{8}y^2 + \frac{1}{2}\beta\kappa\rho y^2\right)\partial_x f(x,y) \\ &+ \frac{1}{2}\left((2\beta^2\kappa - \alpha)y + \frac{4\alpha m - \beta^2}{y}\right)\partial_y f(x,y) \\ &+ \left(\frac{1}{2}y^2\kappa(\beta^2\kappa - \alpha) + 2\alpha\kappa m\right)f(x,y).\end{aligned} \tag{38}$$

Let $G := \mathbb{R} \times \mathbb{R}_{\geq 0}$ and denote by $(\cdot,\cdot)$ the $L^2(G)$-inner product, i.e., $(\varphi,\phi) = \int_G \varphi\phi \mathrm{d}x\mathrm{d}y$. We associate to $-\mathcal{A}_\kappa^{\mathrm{H}} + r$ the bilinear form $a_\kappa^H(\cdot,\cdot)$ via

$$a_\kappa^H(\varphi,\phi) := \big((-\mathcal{A}_\kappa^{\mathrm{H}} + r)\varphi, \phi\big), \quad \varphi,\phi \in C_0^\infty(G).$$

Integration by parts yields

$$\begin{aligned}a_\kappa^{\mathrm{H}}(\varphi,\phi) = &\frac{1}{8}\big(y\partial_x\varphi, y\partial_x\phi\big) + \frac{1}{2}\beta^2\big(\partial_y\varphi,\partial_y\phi\big) + \frac{1}{2}\rho\beta\big(y\partial_x\varphi,\partial_y\phi\big) \\ &+ \frac{1}{2}[\rho\beta - 2r]\big(\partial_x\varphi,\phi\big) + \frac{1}{8}[1 - 4\beta\kappa\rho]\big(y\partial_x\varphi, y\phi\big) \\ &- \frac{1}{2}[2\beta^2\kappa - \alpha]\big(y\partial_y\varphi,\phi\big) - \frac{1}{2}[4\alpha m - \beta^2]\big(y^{-1}\partial_y\varphi,\phi\big) \\ &- \frac{1}{2}\kappa[\beta^2\kappa - \alpha]\big(y\varphi, y\phi\big) - [2\kappa\alpha m - r]\big(\varphi,\phi\big) \\ =: &\sum_{k=1}^{9} b_k(\varphi,\phi).\end{aligned} \tag{39}$$

Define the weighted Sobolev space

$$\mathcal{V} := \overline{C_0^\infty(G)}^{\|\cdot\|_{\mathcal{V}}}, \tag{40}$$

where the closure is taken with respect to the norm $\|\cdot\|_{\mathcal{V}}$ given by

$$\|v\|_{\mathcal{V}}^2 := \|y\partial_x v\|_{L^2(G)}^2 + \|\partial_y v\|_{L^2(G)}^2 + \|\sqrt{1+y^2}v\|_{L^2(G)}^2. \tag{41}$$

Theorem 5. *Assume that* $0 < \kappa < \alpha/\beta^2$ *and that*

$$1 - 2|4\alpha m/\beta^2 - 1| > \rho^2.$$

Then, there exist constants $C_i > 0$, $i = 1,2,3$, *such that for all* $\varphi,\phi \in \mathcal{V}$ *there holds*

$$\begin{aligned}|a_\kappa^{\mathrm{H}}(\varphi,\phi)| &\leq C_1\|\varphi\|_{\mathcal{V}}\|\phi\|_{\mathcal{V}} \\ a_\kappa^{\mathrm{H}}(\varphi,\varphi) &\geq C_2\|\varphi\|_{\mathcal{V}}^2 - C_3\|\varphi\|_{L^2(G)}^2.\end{aligned}$$

Proof. For the proof of this statement we refer to [20] . □

By the abstract well-posedness result Theorem 3 in the triple of spaces $\mathcal{V} \subset L^2(G) = \mathcal{H} \equiv \mathcal{H}^* \subset V^*$ we conclude that the weak formulation to the (transformed) Heston model (37),

$$\begin{aligned}&\text{Find } w \in L^2(J;\mathcal{V}) \cap H^1(J;L^2(G)) \text{ such that}\\ &(\partial_t w, v) + a_\kappa^{\mathrm{H}}(w,v) = \langle f_\kappa^{\mathrm{H}}, v\rangle_{\mathcal{V}^*,\mathcal{V}},\ \forall v \in \mathcal{V}, \quad \text{a.e. in } J,\\ &w(0) = 0.\end{aligned} \tag{42}$$

admits a unique solution for every $f_\kappa^{\mathrm{H}} \in \mathcal{V}^*$. For localization estimates we refer to [20] . For details on the discretization as well as the well-posedness and preconditioning we refer to [21]

3.3. *Contracts of American type*

3.3.1. *Univariate models*

The value of an American option is given as

$$V(t,s) := \sup_{\tau \in \mathcal{T}_{t,T}} \mathbb{E}\left[e^{-r(T-t)} g(S(T)) | S(t) = s\right],$$

where $\mathcal{T}_{t,T}$ denotes the set of all stopping times for S. A similar result to the Black-Scholes case is not available for the Ait-Sahalia model due to the degeneracy of the coefficients in the SDE. We therefore make the following assumption.

Assumption 8.1. Let $v(t,z)$ be a sufficiently smooth solution of the following system of inequalities

$$\begin{aligned}\partial_t v - \mathcal{A}^{\mathrm{AS}} v + rv \geq 0 &\quad \text{in } J \times \mathbb{R}_+,\\ v(t,s) \geq g(s) &\text{ in } J \times \mathbb{R}_+,\\ (\partial_t v - \mathcal{A}^{\mathrm{AS}} v + rv)(g - v) = 0 &\quad \text{in } J \times \mathbb{R}_+,\\ v(0,s) = g(s) &\text{ in } \mathbb{R},\end{aligned} \tag{43}$$

where $\mathcal{A}^{\mathrm{AS}}$ is the infinitesimal generator of the process S. Then, $V(T-t,s) = v(t,s)$.

We consider the problem Eq. (43) on the domain G in excess-to-payoff coordinates imposing homogeneous Dirichlet conditions. The arising cone of positive solutions is given by

$$\mathcal{K}_{+,R} = \{v \in \widetilde{W}_{\rho,\mu}(G) | v \geq 0 \text{ a.e. in } G\}.$$

The formulation reads

$$\begin{aligned}&\text{Find } u_R \in L^2(J;\mathcal{V}) \cap H^1(J;\mathcal{V}^*) \text{ such that } u_R(t,\cdot) \in \mathcal{K}_{+,R} \text{ and}\\ &(\partial_t u_R, v - u_R) + a^{\mathrm{AS}}(u_R, v - u_R) \geq -a^{\mathrm{AS}}(g, v - u_R) \quad \forall v \in \mathcal{K}_{+,R},\\ &u_R(0) = 0.\end{aligned} \tag{44}$$

The well-posedness of this formulation follows under certain regularity assumptions on the payoff g from [17] .

3.3.2. *Stochastic volatility models*

American options in stochastic volatility models can be obtained similar to the Ait-Sahalia model. The value of an American option is given as

$$V(t,z) := \sup_{\tau \in \mathcal{T}_{t,T}} \mathbb{E}\left[e^{-r(T-t)} g(e^{X_T}) | Z_t = z\right],$$

where $\mathcal{T}_{t,T}$ denotes the set of all stopping times for Z. A similar result to the Black-Scholes case is not available, to our knowledge, for general stochastic volatility models due to the possible degeneracy of the coefficients in the SDE. We therefore make the following assumption.

Assumption 8.2. Let $v(t,z)$ be a sufficiently smooth solution of the following system of inequalities

$$\begin{aligned} \partial_t v - \mathcal{A}^{\mathrm{H}} v + rv &\geq 0 && \text{in } J \times \mathbb{R} \times \mathbb{R}_+, \\ v(t,z) &\geq g(e^x) && \text{in } J \times \mathbb{R} \times \mathbb{R}_+, \\ (\partial_t v - \mathcal{A}^{\mathrm{H}} v + rv)(g - v) &= 0 && \text{in } J \times \mathbb{R} \times \mathbb{R}_+, \\ v(0,z) &= g(e^x) && \text{in } \mathbb{R} \times \mathbb{R}_+, \end{aligned} \tag{45}$$

where $\mathcal{A}^{\mathrm{H}}$ is the infinitesimal generator of the process Z. Then, $V(T-t,z) = v(t,z)$.

We perform the same transformations as described in Section 3.2.2 and obtain the following system of inequalities for $w := (v - v_0)e^{-\eta}$ in the Heston model

$$\begin{aligned} \partial_t w - \mathcal{A}_\kappa^{\mathrm{H}} w + rw &\geq f_\kappa^{\mathrm{H}} && \text{in } J \times \mathbb{R} \times \mathbb{R}_+, \\ w(t,x,y) &\geq 0 && \text{in } J \times \mathbb{R} \times \mathbb{R}_+, \\ (\partial_t w - \mathcal{A}_\kappa^{\mathrm{H}} w + rw)w &= 0 && \text{in } J \times \mathbb{R} \times \mathbb{R}_+, \\ w(0,x,y) &= 0 && \text{in } \mathbb{R} \times \mathbb{R}_+. \end{aligned} \tag{46}$$

The set of admissible solutions in the Heston model for the variational form of (46) is the convex set $\mathcal{K}_0$ given as

$$\mathcal{K}_0 := \{v \in \mathcal{V} | v \geq 0 \text{ a.e. } z \in \mathbb{R} \times \mathbb{R}_+\},$$

where V is given in (40). The variational formulation of (46) reads:

$$\begin{aligned} &\text{Find } w \in L^2(J;\mathcal{V}) \cap H^1(J;\mathcal{V}^*) \text{ such that } w(t,\cdot) \in \mathcal{K}_0 \text{ and} \\ &(\partial_t w, v - w) + a_\kappa^{\mathrm{H}}(w, v - w) \geq \langle f_\kappa^{\mathrm{H}}, v - w \rangle_{\mathcal{V}^*,\mathcal{V}}, \ \forall v \in \mathcal{K}_0, \quad \text{a.e. in } J, \\ &w(0) = 0. \end{aligned} \tag{47}$$

Since the bilinear form $a_\kappa^{\mathrm{H}}(\cdot,\cdot)$ is continuous and satisfies a Garding inequality in $\mathcal{V}$ by Theorem 5, problem (47) admits a unique solution for every payoff $g \in L^\infty(\mathbb{R})$ by [17] . We localize the problem to a bounded domain $G_R = (-R_1, R_1) \times (-R_2, R_2)$, $R_1, R_2 > 0$, as in Section 3.2.2 and obtain the following problem for

$$\mathcal{K}_{+,R} := \left\{v \in \widetilde{\mathcal{V}} | v \geq 0 \text{ a.e. } z \in G_R\right\}.$$

$$
\begin{aligned}
&\text{Find } w \in L^2(J;\widetilde{V}) \cap H^1(J;L^2(G_R)) \text{ such that } w(t,\cdot) \in \mathcal{K}_{+,R} \text{ and} \\
&(\partial_t w, v-w) + a^{\mathrm{H}}(w, v-w) \geq \langle f_\kappa^{\mathrm{H}}, v-w\rangle_{\widetilde{\mathcal{V}}^*,\widetilde{\mathcal{V}}}, \ \forall v \in \mathcal{K}_{+,R}, \quad \text{a.e. in } J, \\
&w(0) = 0,
\end{aligned}
\tag{48}
$$

where $\widetilde{\mathcal{V}} := \overline{C_0^\infty(G_R)}^{\|\cdot\|_{\mathcal{V}}}$. Well-posedness results for the variational formulations associated to the values of swing options and compound options in the models considered above can be obtained similarly.

3.4. *Greeks*

A key task in financial engineering is the fast and accurate calculation of sensitivities of market models with respect to model parameters. This becomes necessary for example in model calibration, risk analysis and in the pricing and hedging of certain derivative contracts. Classical examples are variations of option prices with respect to the spot price or with respect to time-to-maturity, the so-called "Greeks" of the model. For classical, diffusion type models and plain vanilla type contracts, the Greeks can be obtained analytically. For more complicated contracts, closed form solutions are generally not available for pricing and calibration. Thus, prices and model sensitivities have to be approximated numerically. We distinguish two classes of sensitivities.

i) The sensitivity of the solution to a variation of an input parameter. Typical examples are the Greeks Vega, Rho and Vomma. Other sensitivities which are not so commonly used in the financial community are the sensitivity of the price with respect to the correlation of the stock and the volatility.
ii) The sensitivity of the solution to a variation of arguments t, x. Typical examples are the Greeks Theta, Delta and Gamma.

We refer to [22] for details on the computation of Greeks in both cases. The computation of sensitivities as in i) requires an additional solve of the pricing equation with a non-trivial right hand side, while Greeks of the type ii) can be obtained via post processing of the solution of the pricing problem. In both cases, we obtain the same convergence results as for the option price, see [22] .

4. Discretization

Since we discretize the parabolic equations in $(0,T) \times G$ in the spatial variable with spline wavelet bases for $\mathcal{V}$, we briefly recapitulate basic definitions and results on wavelets from e.g. [6] and the references there. For specific spline wavelet constructions on a bounded interval, we refer to e.g. [33], [15] and [41]. We discretize the domain $G = (a,b)$ by equidistant mesh points

$$a = x_0 < x_1 < x_2 < \cdots < x_{N_L+1} = b,$$

where we assume the number N_L satisfies $N_L = 2^{L+1} - 1$ with $L \in \mathbb{N}_0$ and use the notation $V_L = V_{N_L}$. Then, we have the nested spaces with $2, 4, \ldots, 2^{L+1}$ subintervals

$$V_0 \subset V_1 \subset \cdots \subset V_L,$$

and $\dim V_\ell = 2^{\ell+1} - 1 =: N_\ell$. Here, we write $b_{\ell,j}$ to indicate the refinement level. Wavelets constitute a so-called *hierarchical* or *multiscale* basis. We start with $\{\psi_{0,1}\}$ for the space V_0. Then, we add basis functions $\{\psi_{1,1}, \psi_{1,2}\}$ such that $\operatorname{span}\{\psi_{0,1}, \psi_{1,1}, \psi_{1,2}\} = V_1$. Similarly, we add again basis functions $\{\psi_{2,1}, \psi_{2,2}, \psi_{2,3}, \psi_{2,4}\}$ such that $\operatorname{span}\{\psi_{0,1}, \psi_{1,1}, \psi_{1,2}, \psi_{2,1}, \psi_{2,1}, \psi_{2,3}, \psi_{2,4}\} = V_2$ and so on. Therefore, we introduce for $\ell \in \mathbb{N}_0$ the complement spaces $W_\ell = \operatorname{span}\{\psi_{\ell,k} : k \in \nabla_\ell\}$ where $\nabla_\ell := \{1, \ldots, 2^\ell\}$ such that $V_\ell = V_{\ell-1} \oplus W_\ell$, $\ell \geq 1$ and $V_0 = W_0$. This decomposition is illustrated in Figure 1.

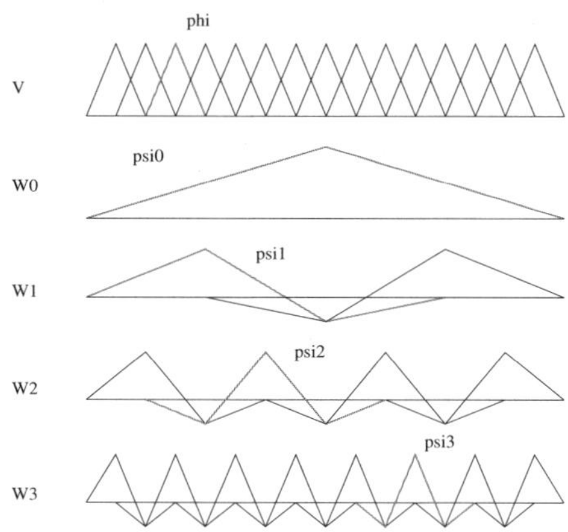

Figure 1. Single-scale space V_L and its decomposition into multiscale wavelet spaces W_ℓ for $L = 3$

We assume that the wavelets $\psi_{\ell,k}$ have compact support $|\operatorname{supp} \psi_{\ell,k}| \leq C2^{-\ell}$, are normalized in $L^2(G)$, i.e., $\|\psi_{\ell,k}\|_{L^2} = 1$, and $\Phi_\ell := \{b_{\ell,j}(x) : 1 \leq j \leq N_\ell\}$ has approximation order p. In addition, we associated with Φ_ℓ a dual basis, $\widetilde{\Phi}_\ell = \{\widetilde{b}_{\ell,j} : 1 \leq j \leq N_\ell\}$, i.e., one has $\langle b_{\ell,j}, \widetilde{b}_{\ell,j'}\rangle = \delta_{j,j'}$, $1 \leq j, j' \leq N_\ell$. The approximation order of $\widetilde{\Phi}_\ell$ is denoted by $\widetilde{p}$ and we assume $p \leq \widetilde{p}$.

Example 4 (Piecewise linear wavelets). *We define the wavelet functions $\psi_{\ell,k}$ as the following piecewise linear functions. Let $h_\ell = 2^{-\ell-1}(b-a)$ and $c_\ell := \sqrt{3}/2 \cdot (2h_\ell)^{-1/2}$. For $\ell = 0$ we have $N_0 = 1$ and $\psi_{0,1}$ is the function with value $2c_0$ at $x = a + h_0$. For $\ell \geq 1$ the wavelet $\psi_{\ell,1}$ has the values $\psi_{\ell,1}(a + h_\ell) = 2c_\ell$, $\psi_{\ell,1}(a + 2h_\ell) = -c_\ell$ and zero at all other nodes. The wavelet $\psi_{\ell,2^\ell}$ has the values $\psi_{\ell,2^\ell}(b - h_\ell) = 2c_\ell$, $\psi_{\ell,2^\ell}(b - 2h_\ell) = -c_\ell$ and zero at all other nodes. The wavelet $\psi_{\ell,k}$ with $1 < k < 2^\ell$ has the values $\psi_{\ell,k}(a + (2k-2)h_\ell) = -c_\ell$, $\psi_{\ell,k}(a + (2k-1)h_\ell) = 2c_\ell$, $\psi_{\ell,k}(a + 2kh_\ell) = -c_\ell$ and zero at all other nodes. For $\ell = 0, \ldots, 3$ these wavelets are plotted in Figure 1. The constants c_ℓ are chosen such that the wavelets $\psi_{\ell,k}$ are* *normalized in* $L^2(G)$. *Note that these biorthogonal wavelets* W_ℓ *are not orthogonal on* $V_{\ell-1}$. *But the inner wavelets* $\psi_{\ell,k}$ *with* $1 < k < 2^\ell$ *have two vanishing moments, i.e.,* $\int \psi_{\ell,k}(x)x^n \mathrm{d}x = 0$ *for* $n = 0, 1$. *The approximation order of* V_ℓ *is* $p = 2$.

Since $V_L = \operatorname{span}\{\psi_{\ell,k} : 0 \le \ell \le L,\, k \in \nabla_\ell\}$ we have a unique decomposition

$$u = \sum_{\ell=0}^{L} u_\ell = \sum_{\ell=0}^{L} \sum_{k\in\nabla_\ell} u_{\ell,k}\psi_{\ell,k},$$

for any $u \in V_L$ with $u_\ell \in W_\ell$. Furthermore, any $u \in \widetilde{H}^s(G)$, $0 \le s \le p$ admits a representation as an infinite wavelet series,

$$u = \sum_{\ell=0}^{\infty} u_\ell = \sum_{\ell=0}^{\infty} \sum_{k\in\nabla_\ell} u_{\ell,k}\psi_{\ell,k}, \tag{49}$$

which converges in $\widetilde{H}^s$. The coefficients $u_{\ell,k}$ are the so-called wavelet coefficients of the function u.

4.1. *Wavelet transformation*

For $u \in V_L$ we want to show how to obtain the multi-scale wavelet coefficients $d_{\ell,k} := u_{\ell,k}$ in $u = \sum_{\ell=0}^{L}\sum_{k\in\nabla_\ell} d_{\ell,k}\psi_{\ell,k}$, from the single-scale coefficients $c_{\ell,j}$ in $u = \sum_{j=1}^{N_L} c_{\ell,j} b_{\ell,j}$.

For the wavelets $\psi_{\ell,k}$ as in Example 4 and hat functions $b_{\ell,j}$ we have

$$\begin{aligned}
\psi_{\ell,k} &= -0.5b_{\ell,2k-2} + b_{\ell,2k-1} - 0.5b_{\ell,2k}, \quad 1 < k < 2^\ell,\\
\psi_{\ell,1} &= b_{\ell,1} - 0.5b_{\ell,2},\\
\psi_{\ell,2^\ell} &= -0.5b_{\ell,2^{\ell+1}-2} + b_{\ell,2^{\ell+1}-1},\\
b_{\ell-1,j} &= 0.5b_{\ell,2j-1} + b_{\ell,2j} + 0.5b_{\ell,2j+1},
\end{aligned}$$

where we set the normalization factors $c_\ell = 0.5$ for simplicity. For $u = \sum_{j=1}^{N_L} c_{L,j} b_{L,j}$ we have

$$\sum_{j=1}^{N_{\ell+1}} c_{\ell+1,j} b_{\ell+1,j} = \sum_{j=1}^{N_\ell} c_{\ell,j} b_{\ell,j} + \sum_{k\in\nabla_{\ell+1}} d_{\ell+1,k}\psi_{\ell+1,k},$$

and therefore

$$\begin{aligned}
c_{\ell+1,2j+1} &= 0.5c_{\ell,j} + 0.5c_{\ell,j+1} + d_{\ell+1,j+1}, \quad j = 2,\dots,N_\ell - 1\\
c_{\ell+1,2j} &= c_{\ell,j} - 0.5d_{\ell+1,j} - 0.5d_{\ell+1,j+1}, \quad j = 2,\dots,N_\ell\\
c_{\ell+1,1} &= 0.5c_{\ell,1} + d_{\ell+1,1},\\
c_{\ell+1,N_{\ell+1}} &= 0.5c_{\ell,N_\ell} + d_{\ell+1,2^{\ell+1}}.
\end{aligned} \tag{50}$$

This can be written in matrix form

$$\begin{pmatrix} c_{\ell+1,1} \\ c_{\ell+1,2} \\ \vdots \\ \vdots \\ c_{\ell+1,N_{\ell+1}-1} \\ c_{\ell+1,N_{\ell+1}} \end{pmatrix} = \begin{pmatrix} 1 & 0.5 & & & & \\ -0.5 & \ddots & -0.5 & & & \\ & 0.5 & \ddots & 0.5 & & \\ & & \ddots & \ddots & \ddots & \\ & & & -0.5 & \ddots & -0.5 \\ & & & & 0.5 & 1 \end{pmatrix} \begin{pmatrix} d_{\ell+1,1} \\ c_{\ell,1} \\ \vdots \\ \vdots \\ c_{\ell,N_\ell} \\ d_{\ell+1,2^{\ell+1}} \end{pmatrix}$$

Now, starting with $u = \sum_{j=1}^{N_L} c_{L,j} b_{L,j}$ we can compute using the decomposition algorithm (50), the coefficients $c_{L-1,j}$ and $d_{L,k}$. Iteratively, we decompose $c_{\ell+1,j}$ into $c_{\ell,j}$ and $d_{\ell+1,j}$ until we have the series representation $u = \sum_{\ell=0}^{L} \sum_{k \in \nabla_\ell} d_{\ell,k} \psi_{\ell,k}$. Similarly, we can obtain the single-scale coefficients $c_{L,j}$ from the multi-scale wavelet coefficients $d_{\ell,k}$.

4.2. *Norm equivalences*

For preconditioning of the large systems which are solved at each time step, we require *wavelet norm equivalences.* These are analogous to the classical Parseval relation in Fourier analysis which allow to express Sobolev norms of a periodic function u in terms of (weighted) sums of its Fourier coefficients. Wavelets allow for analogous statements: the Parseval equation is replaced by appropriate inequalities and the function u need not to be periodic. For $u \in L^2(G)$ there holds

$$\|u\|_{L^2(G)}^2 \sim \sum_{\ell=0}^{\infty} \|u_\ell\|_{L^2(G)}^2 \sim \sum_{\ell=0}^{\infty} \sum_{k \in \nabla_\ell} |u_{\ell,k}|^2 .$$

The mapping $u \mapsto u_0 + \cdots + u_\ell$ defines a continuous projector $\mathcal{P}_\ell : L^2(G) \to V_\ell$. For general Sobolev spaces $\widetilde{H}^s(G)$, we have the direct (or Jackson type) estimate,

$$\|u - \mathcal{P}_\ell u\|_{L^2(G)} \leq C 2^{-ls} \|u\|_{\widetilde{H}^s(G)}, \quad 0 \leq s \leq p.$$

For $u \in V_\ell$ we also have the inverse (or Bernstein-type) estimates,

$$\|u\|_{\widetilde{H}^s(G)} \leq C 2^{ls} \|u\|_{L^2(G)}, \quad s < p - 1/2.$$

Using the inverse estimate and the series representation (49) we have

$$\|u\|_{\widetilde{H}^s(G)}^2 \sim \sum_{\ell=0}^{\infty} \|u_\ell\|_{\widetilde{H}^s(G)}^2 \leq C \sum_{\ell=0}^{\infty} \sum_{k \in \nabla_\ell} 2^{2ls} |u_{\ell,k}|^2 , \quad 0 \leq s < p - 1/2.$$

In the other direction we have for $u \in \widetilde{H}^p(G)$

$$\begin{aligned} \|u_\ell\|_{L^2(G)} &= \|\mathcal{P}_\ell u - \mathcal{P}_{\ell-1} u\|_{L^2(G)} \leq \|\mathcal{P}_\ell u - u\|_{L^2(G)} + \|\mathcal{P}_{\ell-1} u - u\|_{L^2(G)} \\ &\leq C(2^{-\ell p} + 2^p 2^{-\ell p})) \|u\|_{\widetilde{H}^p(G)}, \end{aligned}$$

and therefore

$$\sum_{\ell=0}^{L}\sum_{k\in\nabla_\ell} 2^{2lp}\left|u_{\ell,k}\right|^2 \le C'L\left\|u\right\|^2_{\widetilde{H}^p(G)}.$$

Unfortunately, we do not quite obtain the required bound. This estimate is sharp. For $0\le s<p$, one can modify this argument and obtain

$$\left\|u\right\|^2_{\widetilde{H}^s(G)} \sim \sum_{\ell=0}^{\infty}\left\|u_\ell\right\|^2_{\widetilde{H}^s(G)} \sim \sum_{\ell=0}^{\infty}\sum_{k\in\nabla_\ell} 2^{2ls}\left|u_{\ell,k}\right|^2, \quad 0\le s<p-1/2. \tag{51}$$

4.3. *Weighted spaces*

We can obtain similar results to Eq. (51) for weighted spaces as given in Eq. (19) and Eq. (24). In the case of weighted spaces with degenerate weights we need special wavelet constructions, see [12]. For the proof of norm equivalences we refer to [4]. In addition to the requirements given above, the boundary wavelets and the dual wavelets need to satisfy certain assumptions, see [4] . The norm equivalences can be used to define diagonal preconditioners for the arising mass and stiffness matrices, therefore allowing for an efficient solution of the linear systems. For sparse tensor product discretizations in stochastic volatility models we refer to [21].

4.4. *Space discretization*

Let $\mathcal{T}_0=\{x_0=-R<x_1=0<x_2=R\}$ be a coarse partition of G. Furthermore, define the mesh $\mathcal{T}_\ell$, for $\ell\in\mathbb{N}$, recursively by bisection of each interval in $\mathcal{T}^{\ell-1}$. We denote our computational mesh obtained in this way as $\mathcal{T}_L$, for some $L\in\mathbb{N}_0$, with mesh size $h=R2^{-L}$. The finite element space V_ℓ used for the spatial discretization is the space of all continuous piecewise polynomials of approximation order p on the triangulation $\mathcal{T}_\ell$ which vanish on the boundary ∂G.

The semi-discrete problem corresponding to Eq. (29)-Eq. (30) reads:

$$\begin{aligned}&\text{Find } u_L\in H^1(J;V_L) \text{ such that}\\ &(\partial_t u_L, v_L)+a(u_L,v_L)=0,\ \forall v_L\in V_L,\ \text{ a.e. in } J,\quad u_L(0)=u_{L,0},\end{aligned} \tag{52}$$

where $u_{L,0}=\mathcal{P}_{N_L}$ is the L^2 projection of u_0 onto V_L. We have the following *a priori* result on the spatial semi-discretization which is shown in [32].

Theorem 6. *We consider the problem Eq. (29)-Eq. (30). Then, for $t>0$, there holds the error estimate*

$$\|u(t)-u_L(t)\|_{L^2(G)}\le C\min\{1,h^p t^{-\frac{p}{2}}\}.$$

Here, $C>0$ is a constant independent of h and t, and u, u_L are the solutions of Eq. (29)-Eq. (30) and (52) *respectively.*

4.5. *Discontinuous Galerkin time discretization*

We refer to [37; 38; 39] for details on discontinuous Galerkin timestepping schemes. For $0 < T < \infty$ and $M \in \mathbb{N}$, let $\mathcal{M} = \{J_m\}_{m=1}^M$ be a partition of $J = (0, T)$ into M subintervals $J_m = (t_{m-1}, t_m)$, $m = 1, \ldots, M$, with

$$0 = t_0 < t_1 < t_2 < \ldots < t_M = T.$$

Moreover, denote by $k_m = t_m - t_{m-1}$ the length of J_m. For $u \in H^1(\mathcal{M}, V_L) = \{v \in L^2(J, V_L) : v|_{J_m} \in H^1(J_m, V_L),\ m = 1, \ldots, M\}$, define the one-sided limits

$$\begin{aligned} u_+^m &= \lim_{s\to 0^+} u(t_m + s), \qquad m = 0, \ldots, M-1, \\ u_-^m &= \lim_{s\to 0^+} u(t_m - s), \qquad m = 1, \ldots, M, \end{aligned}$$

and the jumps

$$[\![u]\!]_m = u_+^m - u_-^m, \qquad m = 1, \ldots, M-1.$$

To each time step J_m we associate an approximation order $r_m \geq 0$. The orders are collected in the degree vector $\mathbf{r} = (r_1, \ldots, r_M)$. We introduce the following space of functions which are discontinuous in time

$$S^{\mathbf{r}}(\mathcal{M}, V_L) = \{u \in L^2(J, V_L) \ : \ u|_{J_m} \in S^{r_m}(J_m, V_L),\ m = 1, \ldots, M\},$$

where $S^{r_m}(J_m)$ denotes the space of polynomials of degree at most r_m on J_m.

Consider the problem (52). For test function $w \in C^1(J, V_L)$ with $w(T) = 0$ we integrate the variational formulation with respect to the time variable t and use integration by parts to obtain

$$\begin{aligned} &\int_J \Big((\partial_t u_L, w) + a(u_L, w) \Big)\, \mathrm{d}t = 0, \\ \Rightarrow \quad &\int_J \Big(-(u_L, w') + a(u_L, w) \Big)\, \mathrm{d}t = (u_{L,0}, w(0)). \end{aligned}$$

Replacing u_L by a function $U \in S^{\mathbf{r}}(\mathcal{M}, V_L)$ and integrating by parts in each J_m, we obtain with $w^m = w(t_m)$

$$\begin{aligned} -\int_J (u_L, w')\, \mathrm{d}t &= -\sum_{m=1}^{M} \Big((U, w)|_{t_{m-1}}^{t_m} - \int_{J_m} (U', w) \mathrm{d}t \Big) \\ &= \int_J (U', w)\, \mathrm{d}t + \sum_{m=1}^{M-1} ([\![U]\!]_m, w^m) + (U_+^0, w^0). \end{aligned}$$

Therefore, we obtain the fully discrete scheme: Find $U \in S^{\mathbf{r}}(\mathcal{M}, V_L)$ such that for all $W \in S^{\mathbf{r}}(\mathcal{M}, V_L)$

$$\int_J \Big((U', W) + a(U, W) \Big)\, \mathrm{d}t + \sum_{m=1}^{M-1} ([\![U]\!]_m, W_+^m) + (U_+^0, W_+^0) = (u_{L,0}, W_+^0). \quad (53)$$

The solution operator of the parabolic problem is assumed to generate a holomorphic semi-group; holomorphy of the semigroup must be verified on a case-by-case basis, for example by verifying for the Dirichlet form $a(\cdot,\cdot)$ the so-called *Sector Condition*, see Definition 4.7.12 in [25]. Therefore, the solution $u(t)$ is analytic with respect to t for all $t > 0$. However, due to the non-smoothness of the initial data, the solution may be singular at $t = 0$. By the use of so-called geometric time discretization, the low regularity of the solution at $t = 0$ can be resolved.

Definition 8.2. We call a partition $\mathcal{M}_{M,\gamma} = \{J_m\}_{m=1}^M$ of the time interval $J = (0,T)$, $0 < T < \infty$, *geometric* with M time steps $J_m = (t_{m-1}, t_m)$, $m = 1,\ldots,M$, and grading factor $\gamma \in (0,1)$, if

$$t_0 = 0, \qquad t_m = T\gamma^{M-m}, \quad 1 \le m \le M.$$

A polynomial degree vector $\mathbf{r} = (r_1,\ldots,r_M)$ is called *linear* with slope $\mu > 0$ on $\mathcal{M}_{M,\gamma}$, if $r_1 = 0$ and $r_m = \lfloor \mu m \rfloor$, $m = 2,\ldots,M$, where $\lfloor \mu m \rfloor = \max\{q \in \mathbb{N}_0 \,:\, q \le \mu m\}$.

We have the following *a priori* error estimate on the *hp*-dG scheme [32] .

Theorem 7. *Let $u_0 \in \widetilde{H}^s(G)$, $0 < s \le 1$. Then, there exist $\mu_0, m_0 > 0$ such that for all geometric partitions $\mathcal{M}_{M,\gamma}$ with $M \ge m_0|\log h|$, and all polynomial degree vectors $\mathbf{r}$ on $\mathcal{M}_{M,\gamma}$ with slope $\mu > \mu_0$, the fully discrete solution U obtained by* (53) *satisfies*

$$\|u(T) - U(T)\|_{L^2(G)} \le Ch^p, \tag{54}$$

where $C > 0$ is a constant independent of mesh width h, and u is the solution of the parabolic problem Eq. (29)-Eq. (30).

We now study the linear systems resulting from the *hp*-dG method (53). We show that they may be solved iteratively, without causing a loss in the rates of convergence in the error estimate (54), by the use of an incomplete GMRES method. Furthermore, we prove that the overall complexity is linear (up to logarithmic terms).

4.5.1. *Derivation of the linear systems*

The *hp*-dG time stepping scheme (53) corresponds to a linear system of size $(r_m + 1)N_L$ to be solved in each time step $m = 1,\ldots,M$. Let $\{\widehat{\varphi}_j \,:\, j = 0,\ldots,r_m\}$ be a basis of the polynomial space $S^{r_m}(-1,1)$. We also refer to $\widehat{\varphi}_j$ as the reference time shape functions. On the time interval $J_m = (t_{m-1}, t_m)$ the time shape functions $\varphi_{m,j}$ are then defined as $\varphi_{m,j} = \widehat{\varphi}_j \circ F_m^{-1}$, where F_m is the mapping from the reference interval $(-1,1)$ to J_m given by

$$F_m(\widehat{t}) = \frac{1}{2}(t_{m-1} + t_m) + \frac{1}{2}k_m\widehat{t}.$$

Since the semidiscrete approximation $U|_{J_m}$ and the test function $W|_{J_m}$ in (53) are both in $S^{r_m}(J_m, V_L)$, they can be written in terms of the basis $\{\varphi_{m,j} : j = 0, \ldots, r_m\}$,

$$U|_{J_m}(x,t) = \sum_{j=0}^{r_m} U_{m,j}(x)\varphi_{m,j}(t), \quad W|_{J_m}(x,t) = \sum_{j=0}^{r_m} W_{m,j}(x)\varphi_{m,j}(t).$$

We choose normalized Legendre polynomials as reference time shape functions, i.e.,

$$\widehat{\varphi_j}(\widehat{t}) = \sqrt{j+1/2} \cdot L_j(\widehat{t}), \quad j \in \mathbb{N}_0, \tag{55}$$

where L_j are the usual Legendre polynomials of degree j on $(-1,1)$.

Example 5. The first four reference time shape functions of the form (55) are

$$\widehat{\varphi}_0(\widehat{t}) = \sqrt{1/2},\ \widehat{\varphi}_1(\widehat{t}) = \sqrt{3/2} \cdot \widehat{t},\ \widehat{\varphi}_2(\widehat{t}) = \sqrt{5/2} \cdot (3\widehat{t}^2 - 1)/2,$$

$$\widehat{\varphi}_3(\widehat{t}) = \sqrt{7/2} \cdot (5\widehat{t}^3 - 3\widehat{t})/2\ .$$

The variational formulation (53) then reads: For $m = 1, \ldots, M$, find $(U_{m,j})_{j=0}^{r_m} \in V_L^{r_m+1}$ such that there holds for all $(W_{m,i})_{i=0}^{r_m} \in V_L^{r_m+1}$,

$$\sum_{i,j=0}^{r_m} \mathbb{C}_{ij}^m (U_{m,j}, W_{m,i}) + \frac{k_m}{2} \sum_{i,j=0}^{r_m} \mathbf{I}_{ij}^m\, a(U_{m,j}, W_{m,i}) = \sum_{i=0}^{r_m} f_{m,i},$$

where $f_{m,i} = \widehat{\varphi}_i^+(-1)(U_{m-1}(t_{m-1}), W_{m,i})$, with $U_0(t_0) = u_{L,0} \in V_L$, and for $i, j = 1, \ldots, r_m$,

$$\mathbb{C}_{ij}^m = \int_{-1}^{1} \widehat{\varphi}_j' \widehat{\varphi}_j \mathrm{d}\widehat{t} + \widehat{\varphi}_j^+(-1)\widehat{\varphi}_i^+(-1), \quad \mathbf{I}_{ij}^m = \int_{-1}^{1} \widehat{\varphi}_j \widehat{\varphi}_i \mathrm{d}\widehat{t} = \delta_{ij}.$$

The matrices $\mathbb{C}^m$ and $\mathbf{I}^m$, $m = 1, \ldots, M$, are independent of the time step and can be calculated in a preprocessing step. Their size, however, depends on the corresponding approximation order r_m.

Denoting by $\mathbf{M}$ and $\mathbf{A}$ the mass and (wavelet compressed) stiffness matrix with respect to $(\cdot,\cdot)$ and $a(\cdot,\cdot)$, (53) takes the matrix form

$$\begin{aligned} &\text{Find } \underline{u}^m \in \mathbb{R}^{(r_m+1)N_L} \text{ such that for } m = 1, \ldots, M \\ &(\mathbb{C}^m \otimes \mathbf{M} + \frac{k}{2}\mathbf{I}^m \otimes \mathbf{A})\underline{u}^m = (\underline{\varphi}^m \otimes \mathbf{M})\underline{u}^{m-1}, \\ &\underline{u}^0 = \underline{u}_0\ . \end{aligned} \tag{56}$$

where $\underline{u}^m$ denotes the coefficient vector of $U|_{J_m}$ and $\underline{\varphi}^m := (\widehat{\varphi}_1^+(-1), \ldots, \widehat{\varphi}_{r_m+1}^+(-1))^\top \in \mathbb{R}^{r_m+1}$. Furthermore, $\underline{u}_0 \in \mathbb{R}^{N_L}$ is the coefficient vector of $u_{L,0}$ with respect to the wavelet basis of V_L.

For notational simplicity, we consider for the rest of this section a generic time step, omit the index m and write $\mathbb{C}$ and $\mathbf{I}$ for the matrices $\mathbb{C}^m$ and $\mathbf{I}^m$, respectively, $\underline{u}, \underline{\varphi}$ for the vectors appearing in (56), and r for the approximation order. Furthermore, we denote the right hand side by $\underline{f} = (\underline{\varphi}^m \otimes \mathbf{M})\underline{u}^{m-1}$.

Theorem 8. *Let the assumptions of Theorem 7 hold. Then, choosing the number and order of time steps such that $M = r = \mathcal{O}(|\log h|)$ and in each time step $n_G = \mathcal{O}(|\log h|)^5$ GMRES iterations, implies that*

$$\left\| u(T) - U^{dG}(T) \right\|_{L^2(G)} \leq Ch^p, \tag{57}$$

where U^{dG} denotes the (perturbed) hp-dG approximation of the exact solution u toEq. (29)-Eq. (30) obtained by the incomplete GMRES(m_0) method.

5. Numerical examples

In this section several numerical examples are given to illustrate the theoretical results of the previous sections.

Example 6 (CEV). *We consider the CEV model as in Example 1 with variational formulation as in Proposition 8.1. The discretization was carried out using linear finite elements in space and the θ-scheme in time. The convergence rates for different choices of ρ can be observed in Figure 2. We use the following parameters: $T = 1$, $r = 0.05$, $\sigma = 0.3$ and $K = 1$.*

Example 7 (Heston). *We consider the Heston model as introduced in Example 2, the weak formulation is given in Eq. (42). We discretize using linear finite elements in space and the θ-scheme in time. The price of a European put option is computed as well as the sensitivity of the option price with respect to the correlation coefficient ρ. We observe a convergence rate of approximately $\mathcal{O}(N^{-1})$ for the sensitivity as well as for the option price, see Figure 3. The following parameters were chosen: $(\alpha, m, \beta, \rho, \kappa, r, T, K) = (2.5, 0.025, 0.5, -0.4, 0.8, 0, 0.5, 100)$.*

Example 8 (Swing). *We price a swing option in the CEV model with $K = 1$, $T = 1$, $\sigma = 0.3$, $\rho = 0.5$, $r = 0.05$, $\delta = 0.1$ and $p = 5$. The price of the swing options and the corresponding exercise boundary are depicted in Figure 5.*

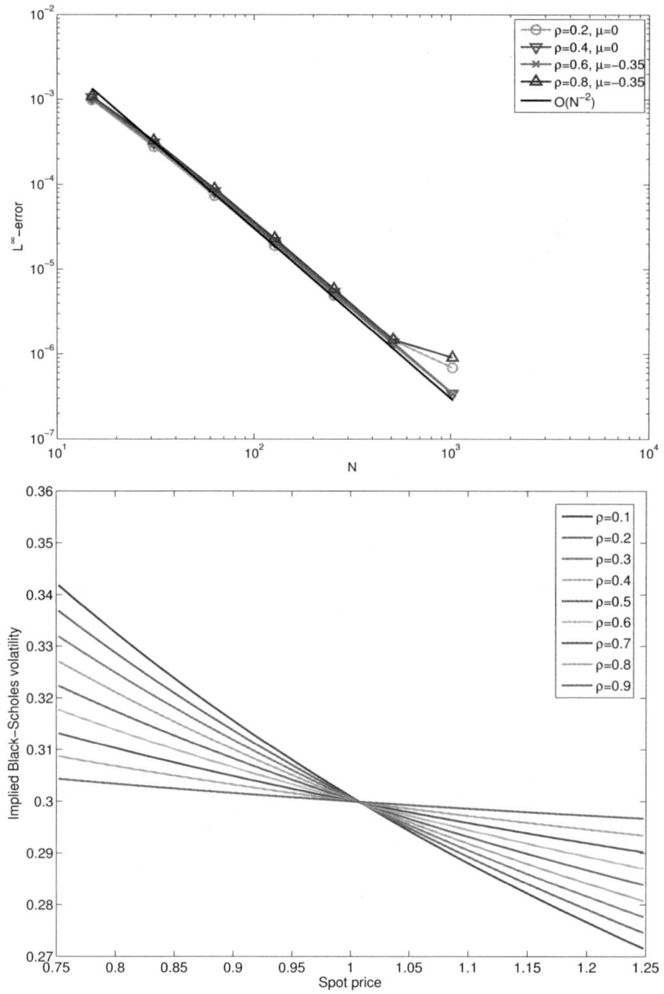

Figure 2. Convergence rates in the L^∞-norm on the domain of interest $(K/2, 3/2K)$ for a European put option in the CEV model (left). Implied Black-Scholes volatilities in the CEV model (right).

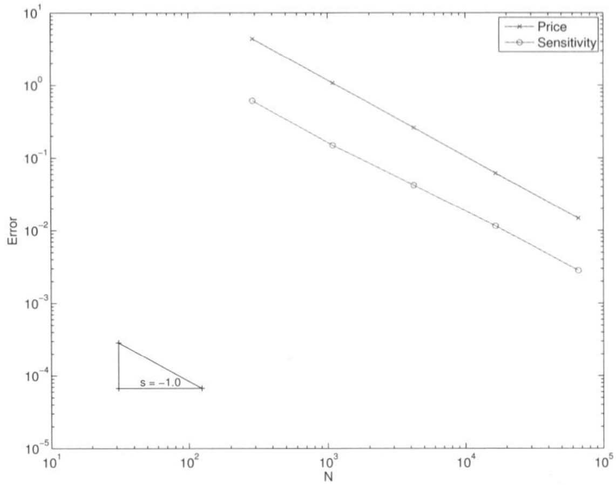

Figure 3. Convergence rates in the L^∞-norm on the domain of interest $(K/2, 3/2K) \times (0.04, 1)$ for a European call option in the Heston model.

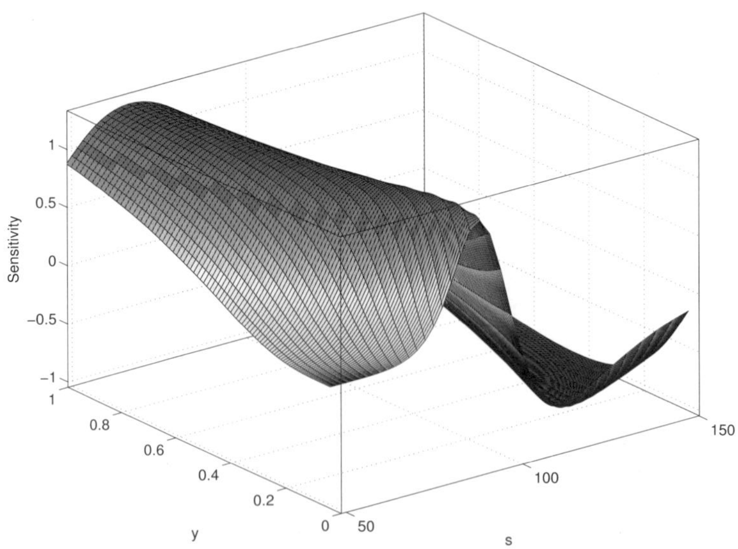

Figure 4. Computed sensitivity of a European call in the Heston model with respect to the parameter ρ.

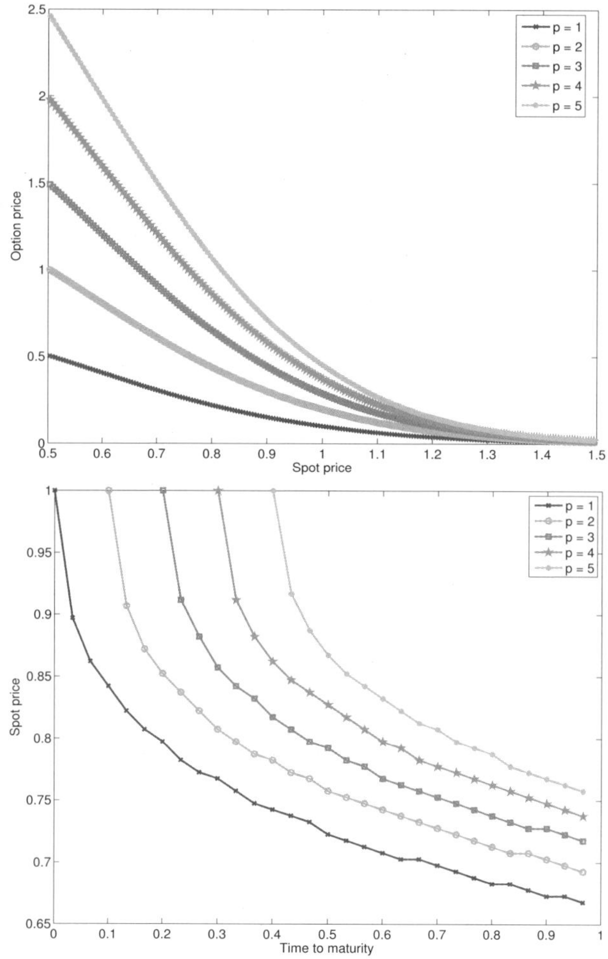

Figure 5. Computed Swing option price (left) and exercise boundary (right) in the CEV.

Bibliography

[1] Ait-Sahalia, A.: Testing continuous-time models of the spot interest rate. *Rev. Financ. Stud.*, 9(2):385–426, 1196.

[2] Ball, C.A. and Roma, A.: Stochastic volatility option pricing. *J. Fin. Quant. Anal.*, 29(4):589–607, 1994.

[3] Benth, F. E., Lempa, J. and Nilssen, T. K.: On the optimal exercise of swing options in electricity markets. *Journal of Energy Markets*, 4(4):3–28, 2012.

[4] Beuchler, S., Schneider, R. and Schwab, C.: Multiresolution weighted norm equivalences and applications. *Numerische Mathematik*, 98:67–97, 2004.

[5] Carmona, R. and Touzi, N.: Optimal multiple stopping and valuation of swing options. *Math. Fin.*, 18:239–268, 2008.

[6] Cohen, A.: *Numerical Analysis of Wavelet Methods*, volume 32. North Holland-Elsevier, Amsterdam, 2003.

[7] Cont, R. and Tankov, P.: *Financial Modelling with Jump Processes.* Financial Mathematics Series. Chapman & Hall/CRC, Boca Raton, FL, 2004.

[8] Cont, R. and Voltchkova, E:. A finite difference scheme for option pricing in jump diffusion and exponential Lévy models. *SIAM J. Numer. Anal.*, 43(4):1596–1626, 2005.

[9] Cont, R. and Voltchkova, E.: Integro-differential equations for option prices in exponential Lévy models. *Finance and Stochastics*, 9(3):299–325, 2005.

[10] Cox, J.C.: Notes on option pricing I: constant elasticity of diffusions. Technical report, Stanford University, Stanford, CA., 1975.

[11] Cryer, C.W.: The solution of a quadratic programming problem using systematic overrelaxation. *SIAM J. Control*, 9:385–392, 1971.

[12] Dahmen, W., Kunoth, A. and Urban, K.: Biorthogonal spline wavelets on the interval - stability and moment conditions. *Applied and Computational Harmonic Analysis*, 6:132–196, 1999.

[13] Delbaen, F. and Schachermayer, W.: *The Mathematics of Arbitrage.* Springer Finance. Springer-Verlag, Heidelberg, 2nd edition, 2008.

[14] Delbaen, F. and Shirakawa, H.: A note on option pricing for the constant elasticity of variance model. *Asia-pacific Financial Markets*, 9:85–99, 2002.

[15] Dijkema, T.J.: *Adaptive tensor product wavelet methods for solving PDEs.* PhD thesis, University of Utrecht, 2009.

[16] Glau, K.: *Feynman-Kac-Darstellungen zur Optionspreisbewertung in Lévy-Modellen.* PhD thesis, University of Freiburg, 2010. `http://www.freidok.uni-freiburg.de/volltexte/7724/pdf/Glau_Dissertation.pdf`.

[17] Glowinski, R., Lions, J.-L. and Trémolières, R.: *Numerical Analysis of Variational Inequalities*, volume 8 of *Studies in Mathematics and its Applications.* North-Holland Publishing Co., Amsterdam, 1981.

[18] Hager, C., Hüeber, S. and Wohlmuth, B.: Numerical techniques for the valuation of basket options and their greeks. *Journal of Computational Finance*, 13(4):3–33, 2010.

[19] Heston, S.L.: A closed-form solution for options with stochastic volatility, with applications to bond and currency options. *Rev. Fin.*, 6:327–343, 1993.

[20] Hilber, N.: *Stabilized Wavelet Method for Pricing in High Dimensional Stochastic Volatility Models.* PhD thesis, SAM, ETH Dissertation No. 18176, 2009. `http://e-collection.ethbib.ethz.ch/view/eth:41687`.

[21] Hilber, N., Matache, A.M. and Schwab, C.: Sparse wavelet methods for option pricing under stochastic volatility. *J. Comp. Fin.*, 8(4):1–42, 2005.

[22] Hilber, N., Schwab, C. and Winter, C.: Variational sensitivity analysis of parametric Markovian market models. In L. Stettner, editor, *Advances in Mathematics of Finance*, volume 83, pages 85–106. Banach Center Publ., 2008.

[23] Ito, K. and Kunisch, K.: Semi-smooth Newton Methods for variational inequalities of the first kind. *Mathematical Modelling and Numerical Analysis*, 37:41–62, 2003.

[24] Ito, K. and Kunisch, K.: Parabolic variational inequalities: The Lagrange multiplier approach. *J. Mat. Pures Appl.*, 85:415–449, 2005.

[25] Jacob, N.: *Pseudo-Differential Operators and Markov Processes Vol. 1: Fourier Analysis and Semigroups.* Imperial College Press, London, 2001.

[26] Jaillet, P., Lamberton, D. and Lapeyre, B.: Variational inequalities and the pricing of American options. *Acta Appl. Math.*, 21(3):263–289, 1990.

[27] Kiesel, R. and Gernhard, J.: Valuation of commodity-based swing options. *The Journal of Energy Markets*, 3(3):91–112, 2010.

[28] Kinderlehrer, D. and Stampacchia, G.: *An Introduction to Variational Inequalities*

and their Applications. Academic Press INC. LTD., London, 1980.

[29] Lamberton, D. and Mikou, M.: The critical price for the American put in an exponential Lévy model. *Fin. Stoch.*, 12(4):561–581, 2008.

[30] Lions, J.-L. and Magenes, E.: *Problèmes aux limites non homogènes et applications*, volume 1 of *Travaux et Recherches Mathématiques*. Dunod, Paris, 1968.

[31] Matache, A.-M., Petersdorff, T. and Schwab, C.: Fast deterministic pricing of options on Lévy driven assets. *M2AN Math. Model. Numer. Anal.*, 38(1):37–71, 2004.

[32] Matache, A.-M., Schwab, C. and Wihler, T.P.: Linear complexity solution of parabolic integro-differential equations. *Numer. Math.*, 104(1):69–102, 2006.

[33] Primbs, M.: *Stabile biorthogonale Spline-Waveletbasen auf dem Intervall.* PhD thesis, University of Duisburg-Essen, 2006.

[34] Reich, N., Schwab, C. and Winter, C.: On Kolmogorov equations for anisotropic multivariate Lévy processes. *Finance and Stochastics*, 14(4):527–567, 2010.

[35] Reichmann, O. and Schwab, C.: Numerical analysis of additive, Lévy and Feller processes with applications to option pricing. *Lévy Matters I, Lecture Notes in Mathematics*, 2001:137–196, 2010.

[36] Reichmann, O., Schwab, C. and Tops, R.: Numerical pricing of American options for general bivariate Lévy models (working title). Technical report, SAM, ETH, 2012. in preparation.

[37] Schötzau, D.: *hp-DGFEM for parabolic Evolution Problems.* PhD thesis, ETH Zürich, 1999.

[38] Schötzau, D. and Schwab, C.: An *hp* a priori error analysis of the DG time-stepping method for initial value problems. *Calcolo*, 37(4):207–232, 2000.

[39] Schötzau, D. and Schwab, C.: *hp*-discontinuous Galerkin time-stepping for parabolic problems. *C. R. Acad. Sci. Paris Sér. I Math.*, 333(12), 2001.

[40] Szpruch, L., Mao, X., Higham, D. and Pan, J.: Numerical simulation of a strongly nonlinear Ait-Sahalia-type interest rate model. *BIT Numer. Math.*, 51:405–425, 2011.

[41] Urban, K.: *Wavelet Methods for Elliptic Partial Differential Equations.* Oxford University Press, Oxford, 2009.

[42] Wilhelm, M. and Winter, C.: Finite element valuation of swing options. *J. Comp. Fin.*, 11(3):107–132, 2008.

Chapter 9

Randomized multilevel quasi-Monte Carlo path simulation

Thomas Gerstner and Marco Noll

Institut für Mathematik, Goethe-Universität, Robert-Mayer-Strasse 10, D-60325 Frankfurt am Main, Germany

Abstract This paper presents theoretical and numerical results with regard to convergence rates, for a combination of the multilevel Monte Carlo method with quasi-Monte Carlo path simulation. A complexity theorem, applicable for all path simulation methods that use the multilevel approach, shows that for many payoff functions the convergence rate is improved significantly by using quasi-random numbers instead of random numbers. The theoretical results are confirmed by numerical examples with several types of options.

1. Introduction

For many traded financial options it is not possible to attain an exact price and therefore numerical methods are needed. In this paper, the expected value of an option depends on the stock price given by the stochastic differential equation (SDE),

$$dS(t) = a(S(t), t)dt + b(S(t), t)dW(t). \tag{1}$$

The function a denotes the drift and b the volatility of the underlying. The simplest discretization methods for the SDE (1) are the Euler scheme

$$\hat{S}(t_{n+1}) = \hat{S}(t_n) + a(\hat{S}(t_n), t_n)h + b(\hat{S}(t_n), t_n)\Delta W_n, \tag{2}$$

and the Milstein scheme

$$\begin{aligned}\hat{S}(t_{n+1}) = \hat{S}(t_n) &+ a(\hat{S}(t_n), t_n)h + b(\hat{S}(t_n), t_n)\Delta W_n \\ &+ \frac{1}{2}\frac{\partial b(\hat{S}(t_n), t_n)}{\partial S} b(\hat{S}(t_n), t_n)(\Delta W_n - h)^2,\end{aligned} \tag{3}$$

where $h = T/N$ is the timestep for $N \in \mathbb{N}$ steps on the grid $t_n = nh$, $n = 0, 1, ..., N$ and $\Delta W_n = W(t_n) - W(t_{n-1})$ denotes the increments of a standard Brownian motion which is discretized by a Brownian bridge [7].

Let $f : \mathbb{R} \longrightarrow \mathbb{R}$ be a function, so that $f(S(T))$ is the discounted payoff of an option. For European or digital options the value $f(S(T))$ only depends on the stock price at the final time T, but in the case of Asian or barrier options the entire path $S(t), 0 < t < T$, needs to be considered.

The standard Monte Carlo method requires a computational complexity of $C = O(\varepsilon^3)$, i.e a convergence rate $\varepsilon = O(C^{1/3})$, to achieve a mean square error of $MSE = O(\varepsilon^2)$. For the quasi-Monte Carlo method (QMC) it is well established that, in the best case, a convergence rate of $\varepsilon = O(C^{1/2})$ [7] can be achieved.

After introducing the multilevel Monte Carlo method, we examine in section 3 if the randomized multilevel quasi-Monte Carlo method (RMLQMC) improves the convergence rate. The next section gives an extension of the complexity theorem in [3], which can be used for every simulation method that use the multilevel approach. Finally, we compare the RMLQMC method with the QMC method. The contribution of this paper are theoretical results for the RMLQMC method, the extended complexity theorem in section 4 and numerical results for several types of options.

2. Multilevel Monte Carlo method

The multilevel Monte Carlo method, as proposed by Giles [3], improves the convergence rate by simulating the stock prices for different timesteps. The payoff of the option on the finest level L is rewritten as a telescope sum

$$E[\hat{P}_L] = E[\hat{P}_0] + \sum_{l=1}^{L} E[\hat{P}_l - \hat{P}_{l-1}],$$

where $P = f(S(T))$ is the option value and $\hat{P}_l$ is its approximation with timestep $h_l = M^{-l}T$, where $M \in \mathbb{N}$. The expected values of $\hat{P}_l - \hat{P}_{l-1}$ are then simulated for $l = 1, ..., L$ independently.

It is important that $\hat{P}_l$ and $\hat{P}_{l-1}$ are calculated from the same Brownian path. We first generate a path with timestep h_l and then add M Brownian increments of the fine path to calculate $\hat{P}_{l-1}$.

From the estimator

$$\hat{Y}_l = N_l^{-1} \sum_{i=1}^{N_l} (\hat{P}_l^{(i)} - \hat{P}_{l-1}^{(i)}), \quad l = 1, ..., L,$$

for $E[\hat{P}_l - \hat{P}_{l-1}]$ and the estimator

$$\hat{Y}_0 = N_0^{-1} \sum_{i=1}^{N_0} \hat{P}_0^{(i)}$$

for $E[\hat{P}_0]$ it follows that $\hat{Y} = \sum_{l=0}^{L} \hat{Y}_l$ approximates $E[\hat{P}_L]$. The variance of $\hat{Y}_l$ is given by $Var[\hat{Y}_l] = N_l^{-1} V_l$, where $V_l = Var[(\hat{P}_l - \hat{P}_{l-1})]$ for $l = 1, ..., L$ and

$V_0 = Var[\hat{P}_0]$. This results in

$$Var[\hat{Y}] = Var\Big[\sum_{l=0}^{L} \hat{Y}_l\Big] = \sum_{l=0}^{L} Var[\hat{Y}_l] = \sum_{l=0}^{L} N_l^{-1} V_l.$$

The variance is minimised if $N_l \sim \sqrt{V_l h_l}$. The computational complexity depends on the number of simulations for every timestep and is therefore given by $C = \sum_{l=1}^{L} N_l h_l^{-1}$. With the requirement of a Lipschitz bounded payoff function and $a(S,t)$ and $b(S,t)$ satisfying certain conditions [11] $V_l = O(h_l)$ is obtained from strong convergence. From $N_l = O(\varepsilon^{-2}(L+1)h_l)$ and $L = \frac{\log \varepsilon^{-1}}{\log M} + O(1)$ then follows the complexity estimate $C = O\big(\varepsilon^{-2}(\log \varepsilon)^2\big)$. The weak convergence of the Euler scheme finally leads to $E[P_L - P] = O(\varepsilon)$ which implies $MSE = O\big(\varepsilon^2\big)$.

The complexity theorem from Giles [3] generalizes the previous results. It can also be applied to non-Lipschitz payoff functions and the Euler method can be replaced by any other discretization method [3].

3. Randomized multilevel quasi-Monte Carlo method

This section examines to what extent the multilevel approach can also improve the convergence rate of the quasi-Monte Carlo method. Numerical examples in [5] have already shown that this approach works quite well. In the following, we try to justify these results theoretically.

First, we apply the randomization approach of Tuffin [17] to estimate the variance of the quasi-Monte Carlo method. In the following

$$\hat{Y}_l = \frac{1}{I} \sum_{i=1}^{I} Z_i^{(l)}$$

denotes the estimator which is calculated by I independent copies of $Z^{(l)}$

$$Z^{(l)} = \frac{1}{N_l} \sum_{j=1}^{N_l} f_l(\{x_l^j + \xi^{(j,l)}\}),$$

where $(\xi^{(j,l)})_{j\in\mathbb{N}}$ denotes a low-discrepancy sequence, x_l^j is an in $[0,1]^d$ uniformly distributed random variable, $\{x_l^j + \xi^{(j,l)}\}$ denotes the fractional part of the random variable, so that $\{x_l^j + \xi^{(j,l)}\} \in [0,1]^d$ and f_l is the payoff function of the multilevel method, i.e. $f_l = P_l - P_{l-1}$ for $l = 1, ..., L$.

The key point here is that the sequence $(\xi^{(j,l)})_{j\in\mathbb{N}}$ is the same for every $i = 1, ..., I$ for each estimator $\hat{Y}_l$. Only the uniformly distributed random variable x_l^j is different for each i. Then, the multilevel estimator is given by

$$\hat{Y} = \sum_{l=0}^{L} \hat{Y}_l.$$

In the following V_{HK} denotes the Hardy-Krause variation, $D^*_{N_l}$ the star discrepancy, $Q_{x_l} = (\{x_l + \xi^{(n,l)}\})_{n\in\mathbb{N}}$ the set of randomized points of x_l and $\mathbb{P} = (\xi^{(n,l)})_{n\in\mathbb{N}}$ the set of low-discrepancy points. Now we are able to calculate the variance

$$\begin{aligned} Var[\hat{Y}_l] &= \frac{1}{I} Var\Big[\frac{1}{N_l}\sum_{j=1}^{N_l} f_l(\{x_l^j + \xi^{(j,l)}\})\Big] \\ &= \frac{1}{I} \int\limits_{[0,1]^d} \Big(\frac{1}{N_l}\sum_{j=1}^{N_l} f_l(\{x_l^j + \xi^{(j,l)}\}) - E[f_l]\Big)^2 dx_l \\ &\leq \frac{1}{I} \int\limits_{[0,1]^d} \Big(V_{HK}(f_l) D^*_{N_l}(Q_{x_l})\Big)^2 dx_l \\ &\leq \Big(4^d V_{HK}(f_l) D^*_{N_l}(\mathbb{P})\Big)^2. \end{aligned}$$

The third step uses the Koksma-Hlawka Inequality [7], and for the last step, $D^*_{N_l}(Q_{x_l}) \leq 4^d D^*_{N_l}(\mathbb{P})$ is needed [12]. For a low-discrepancy sequence the star discrepancy is given by the estimate $D^*_{N_l}(\mathbb{P}) = O(N_l^{-1}(\log N_l)^d)$. Because of $(\log N)^d \geq 4^d \ \forall \ N \geq 55$ we can replace 4^d in $D^*_{N_l}(\mathbb{P})$ by doubling the dimension. For many constructions it is sufficient to consider a star discrepancy of $D^*_{N_l}(\mathbb{P}) = O\big(N_l^{-1+\varepsilon}\big) \ \forall \ \varepsilon > 0$ [12]. Applying this consideration leads to a variance of

$$Var[\hat{Y}_l] = O\Big(\big(V_{HK}(f_l)\big)^2 N_l^{-2+\varepsilon}\Big).$$

The next step is to examine $V_{HK}(f_l)$. The standard assumption is that $V_{HK}(f_l)$ is bounded. However in the multilevel approach we have the situation that the payoff function is the difference of two calculated option values to different timesteps. The difference of the timesteps becomes smaller with higher levels, therefore the difference of P_l and P_{l-1} decreases. That leads to the assumption that $V_{HK}(f_l)$ decreases if the timestep decreases. It is also expected that the strong convergence of the discretization method influences the variation. Thus it seems to be a decent assumption that the variation acts like the variance of $P_l - P_{l-1}$ of the multilevel Monte Carlo method which leads to the assumption

$$Var[\hat{Y}_l] = O\Big(h_l^{2\beta} N_l^{-2+\varepsilon}\Big), \tag{4}$$

where β denotes the strong convergence rate of the discretization method. In higher dimensions it is difficult to estimate the variation, therefore we examine if this assumption holds in the numerical results.

Next we analyze the computational complexity under these assumptions. To do that, we use results from Giles in [3] for the MLMC method and modify them for the RMLQMC method. As discretization method the Euler scheme is used. As before, we want to achieve an accurancy of $MSE = (E[Y - \hat{Y}])^2 + Var[\hat{Y}] = O(\varepsilon^2)$.

From weak convergence follows $E[\hat{P}_L - P] = O(h_L)$. Setting the maximum level $L = \frac{\log \varepsilon^{-1}}{\log M} + O(1)$ and letting $\varepsilon \to 0$ leads to $h_L = O(\varepsilon)$ for the finest timestep. Because $\hat{Y}$ is an unbiased estimatior of P_L we have $E[\hat{Y} - Y] = O(\varepsilon)$. To simplify the variance we neglect $\varepsilon > 0$, so that

$$Var[\hat{Y}_l] = O\left(h_l^{2\beta} N_l^{-2}\right). \tag{5}$$

To minimize the variance for a fixed computaional complexity we treat N_l as a continuous variable. Then we have an extremal value problem with $f(N) = \sum_{l=0}^{L} N_l^{-2} h_l^{2\beta}$ and $g(N) = \sum_{l=0}^{L} N_l h_l^{-1} - K$. Thereby, K is chosen such that $g(N) = 0$ with $N = (N_0, ..., N_L)$. N is a minimal if there are Lagrange multipliers $\lambda_0, ..., \lambda_L \in \mathbb{R}$ that satisfy

$$\nabla f(N) = \lambda_0 \nabla g_0(N) + ... + \lambda_L \nabla g_L(N).$$

This leads to

$$-2h_l^{2\beta} N_l^{-3} = \lambda_l h_l^{-1}, \qquad \text{for } \; l = 0, ..., L,$$

from which follows $N_l = O\big(h_l^{\frac{1+2\beta}{3}}\big)$. For the Euler discretization we have $\beta = 1/2$ and in order to achieve $Var[\hat{Y}] = O(\varepsilon^2)$ we choose

$$N_l = O(\varepsilon^{-7/6}(L+1)^{1/2} h_l^{2/3}).$$

With (5) we have

$$\begin{aligned} Var[\hat{Y}] = & \sum_{l=0}^{L} c_1 N_l^{-2} h_l = \sum_{l=0}^{L} c_1 h_l (\varepsilon^{-7/6}(L+1)^{1/2} h_l^{2/3})^{-2} \\ = & \sum_{l=0}^{L} c_1 h_l (\varepsilon^{7/3}(L+1)^{-1} h_l^{-4/3}) \le c_1 \varepsilon^{7/3} \varepsilon^{-1/3} \le c_1 \varepsilon^2, \end{aligned}$$

with a constant c_1. That leads to $MSE = O(\varepsilon^2)$ and the computational complexity is thus

$$\begin{aligned} C &= c_1 \sum_{l=0}^{L} N_l h_l^{-1} &&= c_1 \sum_{l=0}^{L} (\varepsilon^{-7/6}(L+1)^{1/2} h_l^{2/3}) h_l^{-1} \\ &= c_1 \sum_{l=0}^{L} \varepsilon^{-7/6}(L+1)^{1/2} h_l^{-1/3} &&\le c_1 \varepsilon^{-7/6}(L+1)^{3/2} h_L^{-1/3} \\ &\le c_1 \varepsilon^{-9/6}(L+1)^{3/2} &&= O\big(\varepsilon^{-3/2}(\log \varepsilon^{-1})^{3/2}\big). \end{aligned}$$

4. Extended complexity theorem

The following theorem gives general results for the RMLQMC method. It is an extension of the complexity theorem in [3] which can be used for every simulation method that use the multilevel approach. This is achieved by extending the complexity theorem by an parameter δ that depends on the simulation method. For the MLMC method we usually have $\delta = 1$. For the RMLQMC method we can at best achieve $\delta = 2 - \varepsilon$.

Theorem 1. *Let P denote a functional of the solution of stochastic differential equation (1) for a given Brownian path $W(t)$, and let $\hat{P}_l$ denote the corresponding approximation using a numerical discretization with timestep $h_l = M^{-l}T$.*
If there exist independent estimators $\hat{Y}_l$ based on N_l simulations and positive constants $\alpha \geq 1/2, \beta, \delta \geq 1, c_1, c_2, c_3$ such that

i) $E[\hat{P}_l - P] \leq c_1 h_l^\alpha$

ii) $E[\hat{Y}_l] = \begin{cases} E[\hat{P}_l], & l = 0 \\ E[\hat{P}_l - \hat{P}_{l-1}], & l > 0 \end{cases}$

iii) $Var[\hat{Y}_l] \leq c_2 N_l^{-\delta} h_l^\beta$

iv) C_l, the computational complexity of $\hat{Y}_l$, is bounded by $C_l \leq c_3 N_l h_l^{-1}$,

then there exists a positive constant $c_4 > 0$ such that for any $\varepsilon < e^{-1}$ there are values L and N_l for which the multilevel estimator

$$\hat{Y} = \sum_{l=0}^{L} \hat{Y}_l$$

has a mean square error with bound

$$MSE = E[(\hat{Y} - E[P])^2] < \varepsilon^2$$

with a computational complexity C with bound

$$C \leq \begin{cases} c_4 \max\left(\varepsilon^{-\frac{2}{\delta}} (\log \varepsilon^{-1})^{1+\frac{1}{\delta}}, \varepsilon^{-\frac{1}{\alpha}}\right), & \beta \geq \delta, \\ c_4 \max\left(\varepsilon^{-\frac{2}{\delta} - \frac{1}{\alpha} + \frac{\beta}{\alpha\delta}} (\log \varepsilon^{-1})^{1+\frac{1}{\delta}}, \varepsilon^{-\frac{1}{\alpha}}\right), & \beta < \delta. \end{cases}$$

Proof. The proof is similar to the one of the complexity theorem in [3]. Let $\lceil x \rceil$ be the integer n satisfying $x \leq n < x + 1$. First, we choose L to satisfy

$$\frac{1}{\sqrt{2}} M^{-\alpha} \varepsilon \; < c_1 h_L^\alpha \; \leq \frac{1}{\sqrt{2}} \varepsilon, \tag{6}$$

which leads to

$$L = \left\lceil \frac{\log(\sqrt{2}c_1 T^{\alpha}\varepsilon^{-1})}{\alpha \log M} \right\rceil. \tag{7}$$

Because of i), ii) and (6), the bias is bounded by

$$(E[\hat{Y}] - E[P])^2 \leq \frac{1}{2}\varepsilon^2. \tag{8}$$

This result is needed later to estimate the MSE. For $L \to \infty$, we have

$$\sum_{l=0}^{L} h_l^{-1} < h_L^{-1} \frac{M}{M-1}.$$

Because of (6) we also have $h_L^{-1} < M\left(\frac{\varepsilon}{\sqrt{2}c_1}\right)^{-\frac{1}{\alpha}}$, so that with $\varepsilon^{-\frac{1}{\alpha}} \leq \varepsilon^{-2}$ for $\alpha \geq \frac{1}{2}, \varepsilon < e^{-1}$, we have

$$\sum_{l=0}^{L} h_l^{-1} < \frac{M^2}{M-1}\left(\frac{\varepsilon}{\sqrt{2}c_1}\right)^{-\frac{1}{\alpha}}. \tag{9}$$

It's also easy to see that

$$\left(h_L^{-1}\right)^{1-\frac{1}{\delta}} = h_L^{-1+\frac{1}{\delta}} \leq M^{1-\frac{1}{\delta}}\left(\frac{\varepsilon}{\sqrt{2}c_1}\right)^{\frac{-1+1/\delta}{\alpha}}, \tag{10}$$

and

$$\left(h_L^{-1}\right)^{1-\frac{\beta}{\delta}} = h_L^{-1+\frac{\beta}{\delta}} \leq M^{1-\frac{\beta}{\delta}}\left(\frac{\varepsilon}{\sqrt{2}c_1}\right)^{-\frac{1}{\alpha}\left(1-\frac{\beta}{\delta}\right)}. \tag{11}$$

These resuls are used later to calculate an upper bound for the computational complexity. We first choose

$$N_l = \left\lceil \left(2\varepsilon^{-2}(L+1)c_2 h_l^{\beta}\right)^{\frac{1}{\delta}} \right\rceil \tag{12}$$

to achieve $MSE < \varepsilon^2$. This follows from (8) combined with

$$Var[\hat{Y}] = \sum_{l=0}^{L} Var[\hat{Y}_l] \overset{iii)}{\leq} \sum_{l=0}^{L} c_2 N_l^{-\delta} h_l^{\beta} \leq \frac{\varepsilon^2}{2}.$$

To calculate the computational complexity we need to find an upper bound for the maximum level L. To do that we use $1 < \log \varepsilon^{-1}$ for $\varepsilon < e^{-1}$,

$$\begin{aligned}
L+1 &\overset{(7)}{\leq} \frac{\log(\sqrt{2}c_1 T^\alpha \varepsilon^{-1})}{\alpha \log M} + 2 \\
&\leq \frac{\log \varepsilon^{-1}}{\alpha \log M} + \frac{\log(\sqrt{2}c_1 T^\alpha)}{\alpha \log M} + 2 \\
&\leq \log \varepsilon^{-1} \Big(\frac{1}{\alpha \log M} + \frac{\log(\sqrt{2}c_1 T^\alpha)}{\alpha \log M \log \varepsilon^{-1}} + \frac{2}{\log \varepsilon^{-1}} \Big) \\
&\overset{1<\log \varepsilon^{-1}}{\leq} \log \varepsilon^{-1} \underbrace{\Big(\frac{1}{\alpha \log M} + \max\Big(0, \frac{\log(\sqrt{2}c_1 T^\alpha)}{\alpha \log M}\Big) + 2 \Big)}_{=:c_4} \\
&= c_4 \log \varepsilon^{-1}.
\end{aligned}$$

With $\delta \geq 1$ it follows for $c_5 = (c_4)^{1+\frac{1}{\delta}}$ that

$$(L+1)^{1+\frac{1}{\delta}} \leq c_5 \big(\log \varepsilon^{-1} \big)^{1+\frac{1}{\delta}}. \tag{13}$$

For further calculations we need to consider thus different cases for β.

case 1: $\beta \geq \delta$

We need

$$h_l^{-1+\frac{\beta}{\delta}} \leq (M^{-l}T)^{-1+\frac{\beta}{\delta}} \leq T^{-1+\frac{\beta}{\delta}} \leq \max\{T^{-1}, T^\beta\} < \infty, \tag{14}$$

for $\beta \geq \delta \geq 1$, $l = 0, ..., L$ and $T < \infty$. Now we can estimate the computational complexity by

$$\begin{aligned}
C &\overset{iv),\,(12)}{\leq} \sum_{l=0}^{L} c_3 h_l^{-1} \Big(\big(2\varepsilon^{-2}(L+1)c_2 h_l^\beta\big)^{\frac{1}{\delta}} + 1 \Big) \\
&\overset{(9),\,(14)}{\leq} c_7 \varepsilon^{-\frac{2}{\delta}} (L+1)^{1+\frac{1}{\delta}} + \frac{M^2}{M-1} (\sqrt{2}c_1)^{\frac{1}{\alpha}} \varepsilon^{-\frac{1}{\alpha}} \\
&\overset{(13)}{\leq} c_7 \varepsilon^{-\frac{2}{\delta}} (c_5 \log(\varepsilon^{-1})^{1+\frac{1}{\delta}}) + c_8 \varepsilon^{-\frac{1}{\alpha}} \\
&\leq c_9 \max\Big(\varepsilon^{-\frac{2}{\delta}} (\log \varepsilon^{-1})^{1+\frac{1}{\delta}}, \varepsilon^{-\frac{1}{\alpha}} \Big),
\end{aligned}$$

with constants c_6, c_7, c_8, c_9.

case 2: $\delta > \beta$

In this case we get an upper bound similarly

$$
\begin{aligned}
C &\overset{(12)}{\leq} \sum_{l=0}^{L} c_3 h_l^{-1}\Big(\big(2\varepsilon^{-2}(L+1)c_2 h_l^{\beta}\big)^{\frac{1}{\delta}}+1\Big)\\
&\overset{h_l^{-1+\frac{\beta}{\delta}} \leq h_L^{-1+\frac{\beta}{\delta}}}{\leq} c_6\varepsilon^{-\frac{2}{\delta}}(L+1)^{1+\frac{1}{\delta}} h_L^{-1+\frac{\beta}{\delta}} + \frac{M^2}{M-1}(\sqrt{2}c_1)^{\frac{1}{\alpha}}\varepsilon^{-\frac{1}{\alpha}}\\
&\overset{(11)}{\leq} c_7\varepsilon^{-\frac{2}{\delta}}(\log(\varepsilon^{-1}))^{1+\frac{1}{\delta}}\varepsilon^{-\frac{1}{\alpha}\left(1-\frac{\beta}{\delta}\right)} + c_8\varepsilon^{-\frac{1}{\alpha}}\\
&\leq c_9 \max\left(\varepsilon^{-\frac{2}{\delta}-\frac{1}{\alpha}+\frac{\beta}{\alpha\delta}}(\log\varepsilon^{-1})^{1+\frac{1}{\delta}}, \varepsilon^{-\frac{1}{\alpha}}\right),
\end{aligned}
$$

with new constants c_6, c_7, c_8, c_9. □

For the multilevel Monte Carlo method ($\delta = 1$) it is possible to improve the upper bound in the theorem. This was already shown in [3]. In order to avoid too many cases this case is not considered seperately here.

If we have a Lipschitz bounded payoff function and use the Euler discretization then $\alpha = 1$ and $\beta = 1$. With $\delta = 2$ for the RMLQMC method we obtain the same computational complexity as calculated in section 3. Replacing the Euler scheme by the Milstein scheme gives us $\beta = 2$, from which a computational complexity of $C = O\big(\varepsilon^{-1}(\log\varepsilon^{-1})^{3/2}\big)$ can be obtained. It's remarkable that almost a convergence rate of 1 is achieved if the payoff function is smooth enough.

To apply this theorem we first need to determine the parameters α, β, δ. The parameter α follows from the weak convergence of the discretization method. The parameter β is numerically easy to determine and depends on the payoff function. For smooth payoff functions it is expected that $\beta/2$ is equivalent to the strong convergence rate. The parameter δ needs to be determined numerically. Later, we examine how δ depends on the payoff function. Also has to be analyzed if δ depends on the dimension for the RMLQMC method.

In the following numerical examples we hence first determine the necessary parameters for the extended complexity theorem, then we compare the theoretical convergence rate with the numerical calculated convergence rate.

5. Numerical results

Now we compare the QMC method with the RMLQMC method. It has to be emphasised that the QMC method is not randomized, because this decreases the convergence rate. For the RMLQMC method we use the algorithm suggested in [5], but we replace the rank-1 lattice rule by a Sobol sequence [16]. This algorithm aims to achieve a RMSE of $O(\varepsilon)$.

Both methods perform significantly better then pure MC or MLMC, which are, therefore not considered in the following.

For the numerical results, the option value and the computational costs are simulated 100 times. These data are used to determine the RMSE, the variance and the bias. We also assume that the stock price corresponds to a simple geometric Brownian motion,

$$dS(t) = rSdt + \sigma SdW(t),$$

with rate of interest $r = 0.05$ and volatility $\sigma = 0.2$. For all examples we further use $S(0) = 1$, $K = 1$ and $T = 1$.

5.1. *European option*

First, we consider a European call option with discounted payoff function

$$P = \exp(-rT)\big(S(T) - K\big)^+.$$

For calculating α we choose a fixed number of samples (here $N_l = 5000$) and plot $E[\hat{P}_l - P]$ versus the timestep $h_l = 2^{-l}$ for $l = 2, ..., 8$. We consider the Euler and the Milstein discretization, but no difference can be expected, since α only depends on the weak convergence rate. Figure 1 shows an α of 1 for both methods.

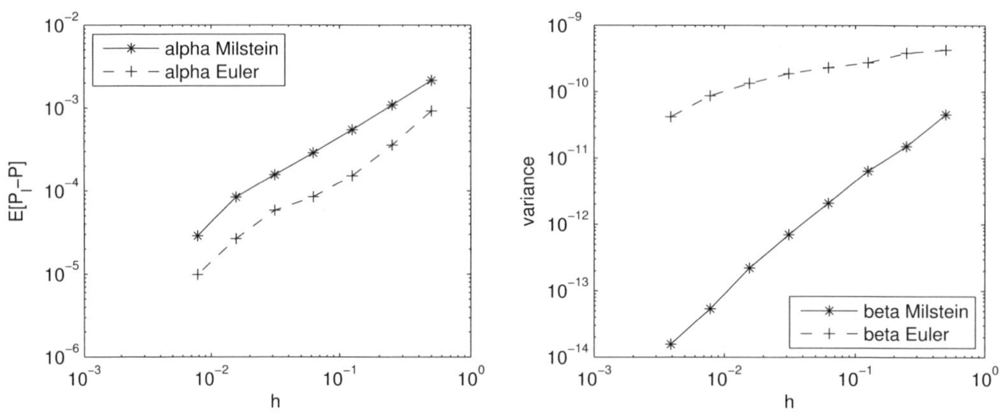

Figure 1. The parameters α (left) and β (right) for varying mesh widths h for a European option

The parameter β is calculated for a fixed number of samples by comparing the variance of $\hat{Y}_l$ for different timesteps h_l. The right plot shows a β of 1.65 for the Milstein scheme and $\beta = 0.48$ for the Euler scheme.

Determing δ is more difficult, because it represents the convergence rate of the simulation method. For the RMLQMC method δ is expected to depend on the dimension which increases as the timestep decreases. Hence we determine δ for different timesteps. In Figure 2 we examine δ for $l = 3, 5, 7$. Level 7 already presents the worst-case scenario, because no higher levels were needed. For level 3 the Euler and Milstein scheme have a δ of 1.6. It gets worse for level 5 with $\delta = 1.5$ and for level 7 δ reduces to 1.3. As expected we see that δ depends on the dimension.

Using the determined parameters in the complexity theorem gives us a convergence rate between $\varepsilon = O(C^{-0.65})$ and $\varepsilon = O(C^{-0.8})$ for the Milstein discretization and between $\varepsilon = O(C^{-0.45})$ and $\varepsilon = O(C^{-0.5})$ for the Euler discretization.

The next step is to compare these results with the numerical calculated convergence rates. The top left plot in Figure 3 shows the behavior of the RMSE for the QMC and RMLQMC method. The QMC method achieves a RMSE convergence rate of 0.5, i.e. $RMSE = O(C^{-0.5})$. Combining the RMLQMC method with the Milstein discretization the RMSE convergence rate is strikingly increased to 0.75. Using the Euler discretization the convergence rate couldn't be improved because the variance wasn't reduced. This shows the bottom plot in Figure 3. The convergence rate of the variance is only 1. The Milstein discretization improves this rate to 1.5.

The top right plot gives the behavior of the squared bias versus the costs. The QMC method has a convergence rate of the squared bias of 0.75, whereas the RMLQMC method improves this rate to 1.5. In this example, the RMSE is almost equally determined by the variance and the bias. The numerically calculated computational costs are in the range of the computational costs determined by the complexity theorem.

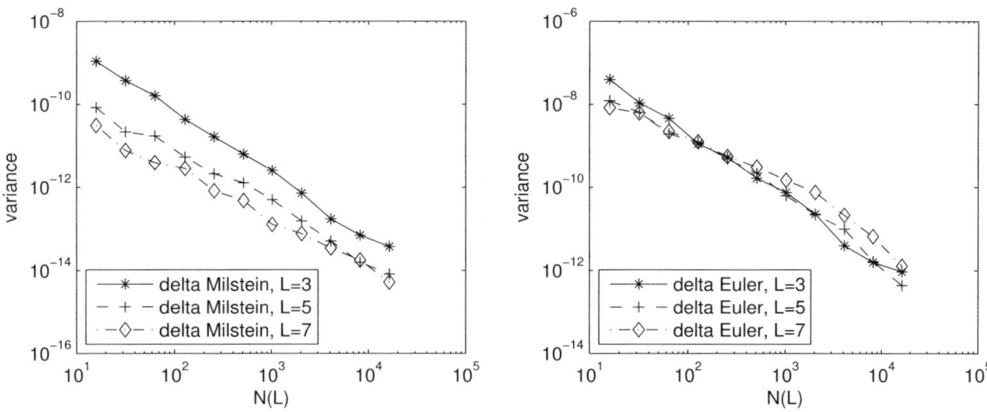

Figure 2. The parameter δ for a varying number of samples $N(L)$ for a European option

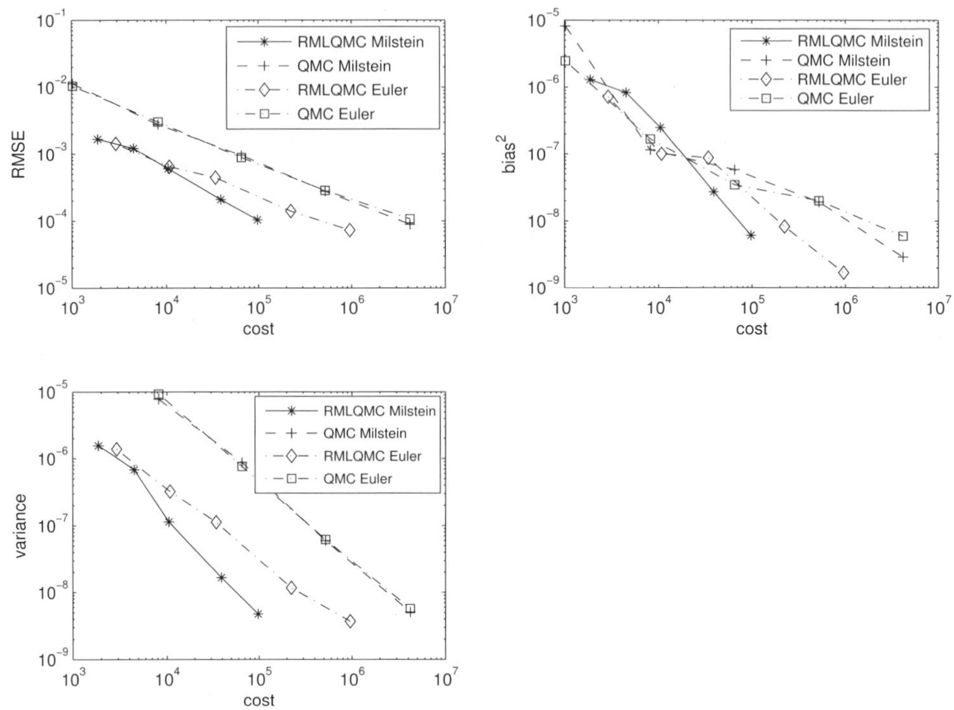

Figure 3. Convergence rates for a European option: RMSE (top left), squared bias (top right) and variance (bottom left)

5.2. *Asian option*

Next, we consider an Asian option with payoff function

$$P = \exp(-rT)\max(0, \overline{S} - K),$$

where

$$\overline{S} = \int_0^T S(t)dt$$

is approximated by $\overline{S_l} = \sum_{n=1}^{N_l} \frac{1}{2}\big(\hat{S_n} + \hat{S}_{n-1}\big)h_l$ for each level l.

We first determine α, β and δ. The left plot in Figure 4 shows an α of at least 1. The right plot shows $\beta = 1.1$ for the Euler scheme and $\beta = 1.6$ for the Milstein scheme.

Figure 5 has the results for δ. For level 3 $\delta = 1.5$, for level 5 and 7 $\delta = 1.2$. These results are similar to those of the European option, but approximately 0.1 worse.

That indicates that δ is not independent of the payoff function. A reason for this is that the variance of $\hat{Y}_l$ depends on the payoff function. With these values we get a convergence rate between $\varepsilon = O(C^{-0.6})$ and $\varepsilon = O(C^{-0.75})$ from the extended complexity theorem.

The top left plot in Figure 6 shows a convergence rate of $\varepsilon = O(C^{-0.6})$ for all considered methods. Therefore the RMLQMC method couldn't improve the convergence rate. One reason might be that the approximation of $\overline{S}$ by $\overline{S_l}$ is too slow. In particular the convergence rate of the variance couldn't be improved. For all methods we have a convergence rate of the variance of 1.3. Thus the computational complexity is only reduced by a constant. For this option we also see that the convergence rate of the squared bias has no impact on the RMSE convergence.

Obviously the Milstein discretization does not work better than the Euler discretization. The reason for this is again that the approximation of $\overline{S}$ by $\overline{S_l}$ is not good enough. But we also see in Figure 4 that the Euler method already works very well. For the first 5 levels both discretization methods achieve the same results.

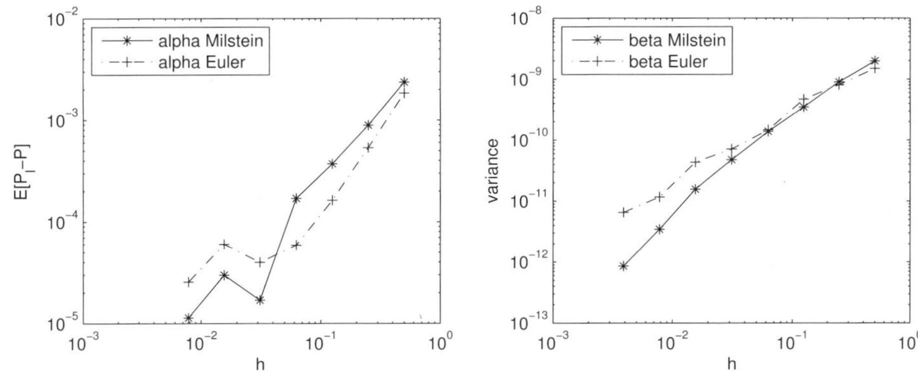

Figure 4. The parameters α (left) and β (right) for varying mesh widths h for an Asian option

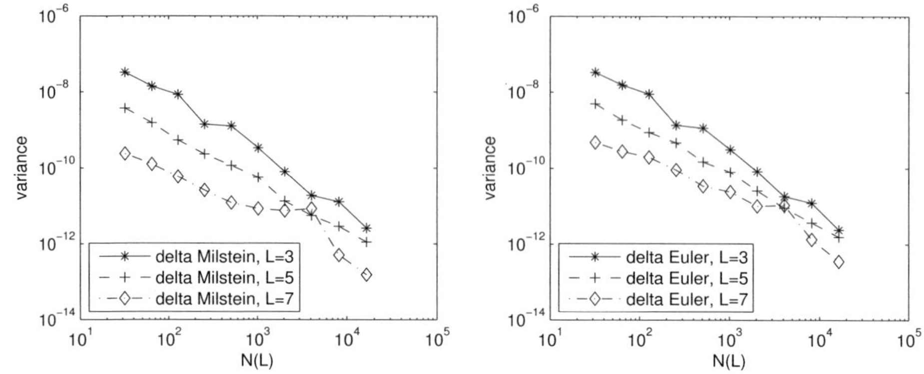

Figure 5. The parameter δ for a varying number of samples $N(L)$ for an Asian option

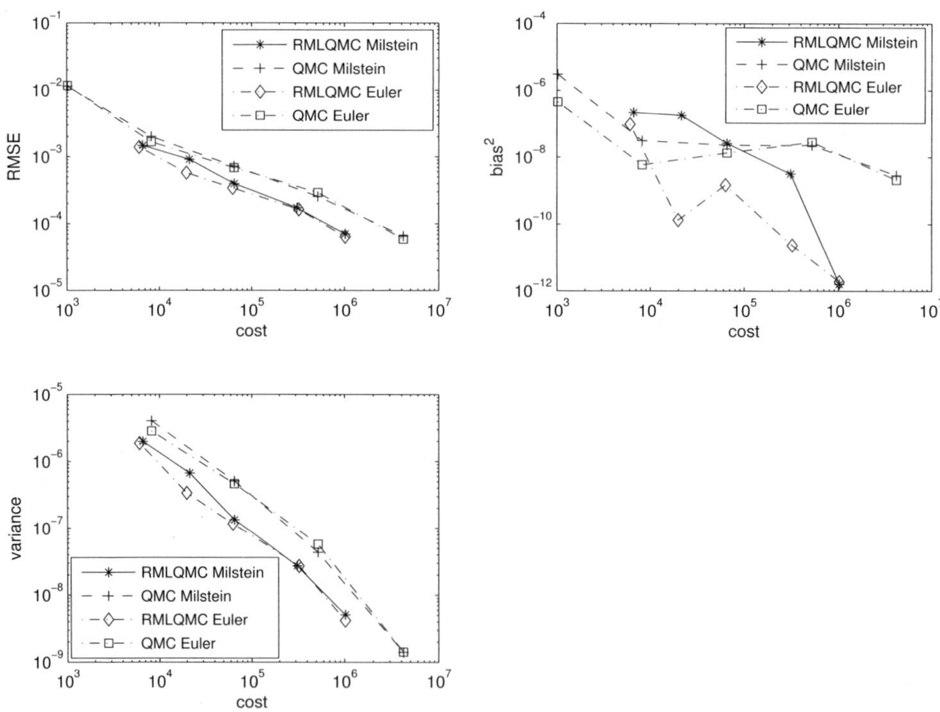

Figure 6. Convergence rates for an Asian option: RMSE (top left), squared bias (top right) and variance (bottom left)

5.3. *Barrier option*

In the following a barrier option is considered with payoff function

$$P = \exp(-rT)\big(S(T) - K\big)^{+}\mathbf{1}_{\tau>T},$$

where $\tau = \inf_{t>0}\{S(t) < B\}$ denotes the barrier crossing time. The barrier B is chosen to be 0.85.

To improve the convergence rate of the variance, a standard approach is used [4; 7] which redefines the payoff on the fine Level $\hat{P}_l$ to

$$\exp(-rT)\Big((\hat{S}^f_{n_T} - K)^{+} \prod_{n=0}^{n_T-1} \hat{p}^f_n\Big),$$

where $n_T = T/h$ describes the number of timesteps used in this approximation. The probability that the path did not cross the barrier during the nth timestep is denoted by $\hat{p}^f_n$. This probability can be expressed by

$$\hat{p}^f_n = 1 - \exp\Big(\frac{-2(\hat{S}^f_n - B)^{+}(\hat{S}^f_{n+1} - B)^{+}}{(\sigma\hat{S}^f_n)^2 h^f}\Big),$$

where $\hat{S}^f_n$ denotes the approximated asset price for the fine-path at the nth timestep. For the payoff on the coarse level we first construct an interpolated midpoint

$$\hat{S}_{m+\frac{1}{2}} = \frac{1}{2}(\hat{S}_{m+1} + \hat{S}_m) + \sigma\hat{S}_m\Big(W_{m+\frac{1}{2}} - \frac{1}{2}(W_{m+1} + W_m)\Big).$$

The key point here is that $W_{m+\frac{1}{2}}$ was already calculated for the fine-path approximation. The probability that the path did not cross the barrier during the mth timestep is [4]

$$\begin{aligned}\hat{p}^c_m = &\Big\{1 - \exp\Big(\frac{-2(\hat{S}^c_m - B)^{+}(\hat{S}^c_{m+1/2} - B)^{+}}{(\sigma\hat{S}^c_m)^2 h^c}\Big)\Big\} \\ &\times \Big\{1 - \exp\Big(\frac{-2(\hat{S}^c_{m+1/2} - B)^{+}(\hat{S}^c_{m+1} - B)^{+}}{(\sigma\hat{S}^c_m)^2 h^c}\Big)\Big\}.\end{aligned}$$

The payoff of the coarse level is given by

$$\exp(-rT)\Big((\hat{S}^c_{m_T} - K)^{+} \prod_{m=0}^{m_T-1} \hat{p}^c_m\Big).$$

Important is that with this construction the expected payoff is not changed, so that $E[\hat{P}^f_l] = E[\hat{P}^c_l]$.

The left plot in Figure 7 shows $\alpha = 1$ for both discretization methods. The right plot gives an average β of 0.75 for the Milstein discretization. For the Euler discretization only a β of 0.2 could be achieved.

Figure 8 shows the results for δ. For $L = 3$ δ is 1.3 and for $L = 7$ only a δ of 1 is attained. This indicates that the RMLQMC method is not better than the MLMC method if the fine timesteps are important. With these values we attain a computational complexity between $\varepsilon = O(C^{-0.45})$ and $\varepsilon = O(C^{-0.75})$ for the Milstein scheme and between $\varepsilon = O(C^{-0.35})$ and $\varepsilon = O(C^{-0.55})$ for the Euler scheme.

The top left plot in Figure 9 shows a RMSE convergence rate of 0.35 for the Euler discretization, the Milstein discretization improves this to 0.55. But the QMC method also achieves this convergence rate. It is clear that the Euler discretization does not work well for the RMLQMC method because of the low value for β that leads to a lower convergence rate of the variance.

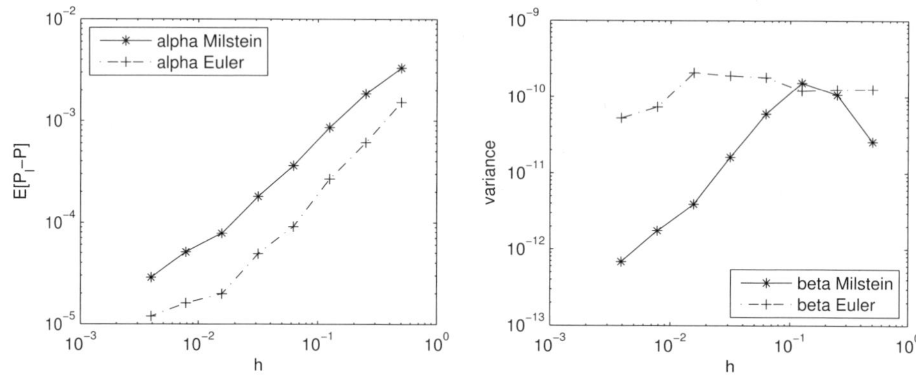

Figure 7. The parameters α (left) and β (right) for varying mesh widths h for a barrier option

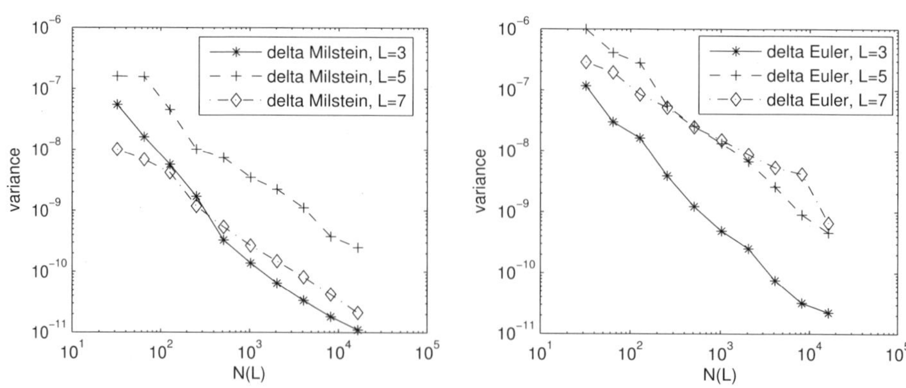

Figure 8. The parameter δ for a varying number of samples $N(L)$ for a barrier option

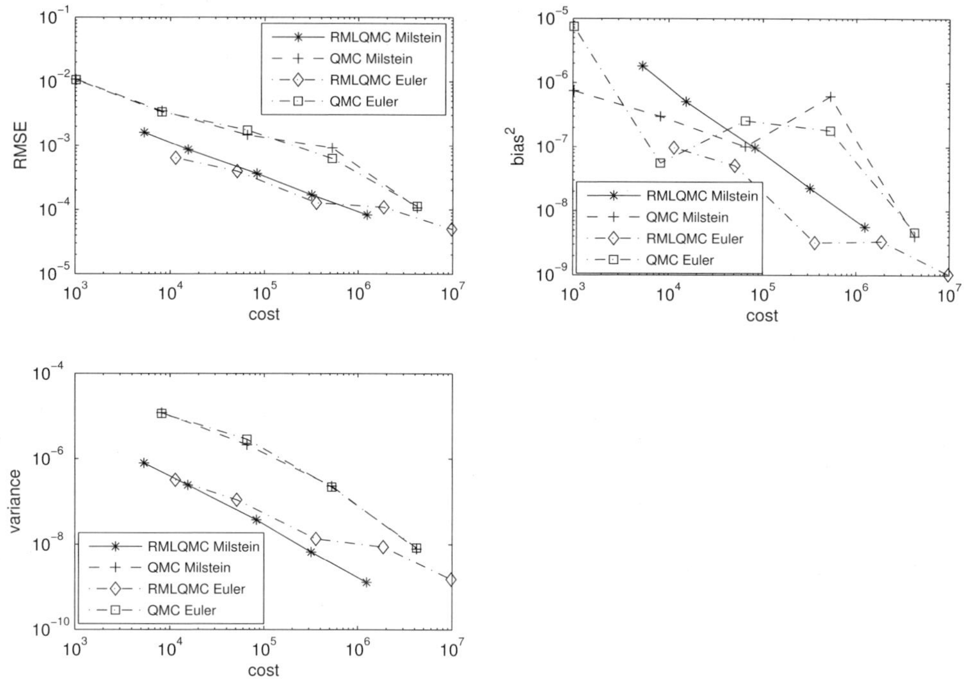

Figure 9. Convergence rates for a barrier option: RMSE (top left), squared bias (top right) and variance (bottom left)

5.4. *Digital option*

The considered digital option is defined by

$$P = \exp(-rT)\mathbf{1}_{S(T)>K}.$$

Now we use the technique of conditional expectation [4; 7] to improve the convergence rate of the variance. This can be achieved by terminating the path calculations at time $n_T - 1$. The number of timesteps is described by $n_T = T/h$. For the stock price approximated on the fine level at this time $\hat{S}^f_{n_T-1}$, we calculate the probability of the event $\{\hat{S}^f_{n_T} > K\}$ which is given by

$$\hat{p}^f = \Phi\Big(\frac{\hat{S}^f_{n_T-1} + r\hat{S}^f_{n_T-1}h^f - K}{\sigma\hat{S}^f_{n_T-1}\sqrt{h^f}}\Big),$$

where Φ is the cumulative Normal distribution. The payoff on the fine level is

$$P^f_l = \exp(-rT)\hat{p}^f.$$

For the coarse path simulation we use that the Brownian increments ΔW on the fine path are already calculated. Here we have $n_T/2$ timesteps and the probability of the event $\{\hat{S}^c_{n_T/2} > K\}$ is

$$\hat{p}^c = \Phi\Big(\frac{\hat{S}^c_{n_T/2-1} + r\hat{S}^c_{n_T/2-1}h^c + \sigma\hat{S}^c_{n_T/2-1}\Delta W_{n_T-1} - K}{\sigma\hat{S}^c_{n_T/2-1}\sqrt{h^c}}\Big),$$

where ΔW_{n_T-1} describes the Brownian increment of the fine path-calculation at time $n_T - 1$. The payoff is $P^c_l = \exp(-rT)\hat{p}^c$ [4]. We again have $E[\hat{P}^f_l] = E[\hat{P}^c_l]$.

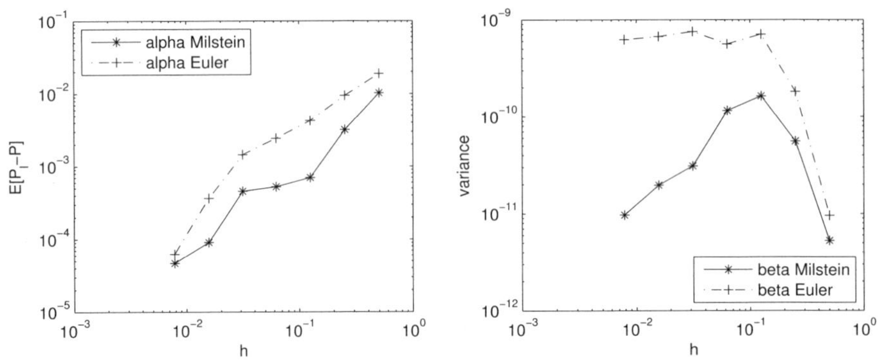

Figure 10. The parameters α (left) and β (right) for varying mesh widths h for a digital option

In the left plot in Figure 10 we see that α is 1. The right plot shows β. In this example it is difficult to determine a value for β. The Euler discretization even seems to have a negative β. Hence we neglect the first two points and assume a β of 0. A similar problem appears when simulating with the Milstein scheme. Neglecting the first two points gives a β of 0.8, but this value seems too good, so that we assume $\beta = 0.5$.

In Figure 11 we attain for level 3 an value for δ of 1.5. For level 7 we still have $\delta = 1.2$. There is no significant difference between Euler and Milstein discretization. The extended complexity theorem gives a computational complexity between $\varepsilon = O(C^{-0.375})$ and $\varepsilon = O(C^{-0.43})$ for the Euler discretization and between $\varepsilon = O(C^{-0.45})$ and $\varepsilon = O(C^{-0.7})$ for the Milstein discretization. But in this example those values are not reliable because we couldn't attain a clear value for β. The numerically obtained convergence rates are presented in Figure 12. The QMC method achieves a RMSE convergence rate of 0.35. The RMLQMC method improves the convergence rate to 0.4 using the Euler scheme, and the Milstein scheme shows a convergence rate of 0.5. Obviously the convergence rate of the squared bias has no impact on the RMSE convergence for the QMC method. For the RMLQMC method we have a similar result, the RMSE convergence is clearly dominated by the variance.

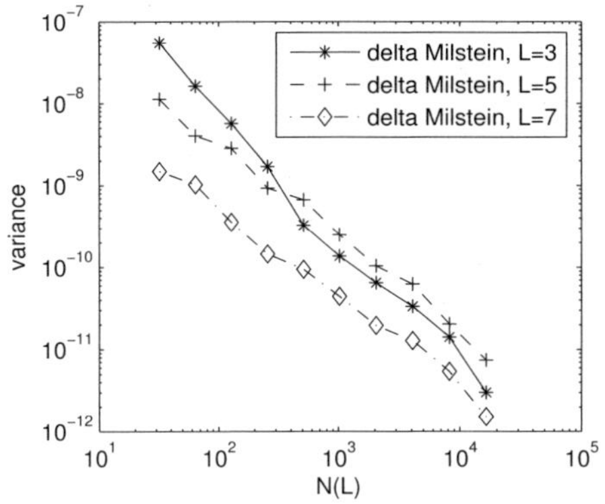

Figure 11. The parameter δ for a varying number of samples $N(L)$ for a digital option

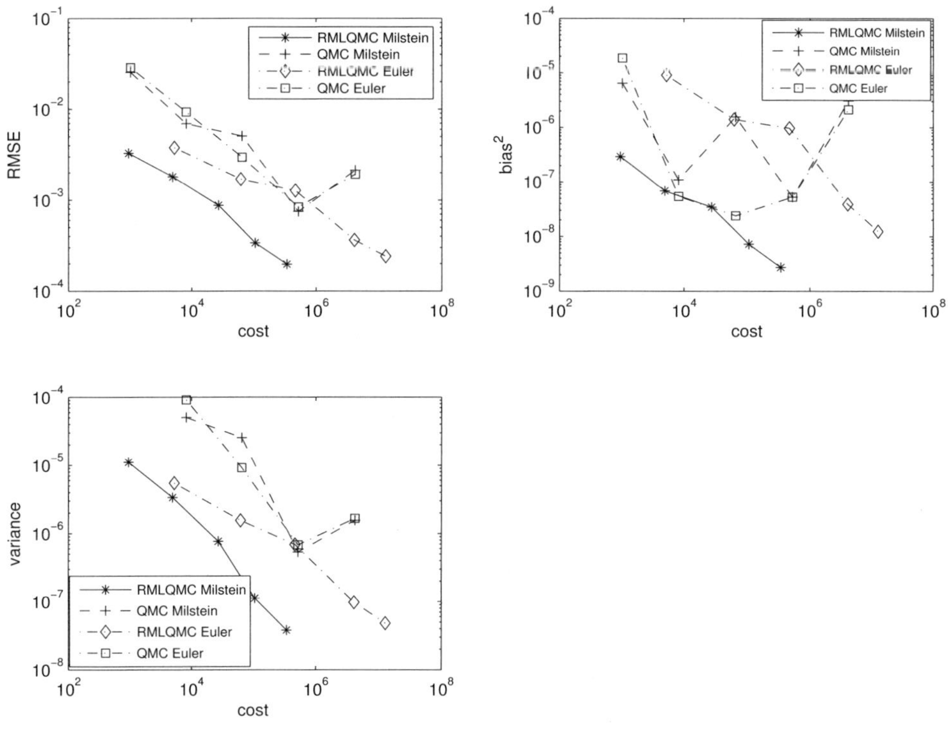

Figure 12. Convergence rates for a digital option: RMSE (top left), squared bias (top right) and variance (bottom left)

6. Conclusion

This chapter gives theoretical and numerical results for the RMLQMC method. First, we established a theorem which can be used for all simulation methods that use the multilevel approach. In this context we analyzed numerical examples. These showed partially clear improvements. The best result was achieved for a European option, where the RMLQMC method with Milstein discretization improved the convergence rate to 0.75, whereas the QMC method only had a convergence rate of 0.5. In the case of an Asian option the RMLQMC, independent of the discretization method, achieved the same convergence rate as the standard QMC method. No improvement was achieved because the used approximation formula for the payoff which is given by an integral was too slow, therefore there was no advantage using the Milstein scheme instead of the Euler scheme. However, it is this combination that achieves better results than the standard QMC method. For a barrier and a digital option the results were similar. The RMLQMC method worked very well for the digital option, whereas for the barrier option the convergence rate of the QMC method could not be improved.

The weak point of the extended complexity theorem is the calculation of the parameter δ. For simulation methods which depend on the dimension, δ also depends on the dimension. Hence it is difficult to determine δ which seems to play a crucial role. However the extended complexity theorem shows the possibilities of the multilevel approach if a better simulation method is used. That is the key point of future work.

Bibliography

[1] Bally, V., Talay, D.: The law of the Euler scheme for stochastic differential equations, I: convergence rate of the distribution function, *Probability Theory and Related Fields*, 104(1):43-60, 1995.

[2] Cambanis, S., Hu, Y.: Exact Convergence Rate of the Euler-Maruyama Scheme, with Application to Sampling Design, Stochastics, *An International Journal of Probability and Stochastic Processes*, Volume 59, Issue 3 and 4, pp. 211-240, 1996.

[3] Giles, M. B.: Multilevel Monte Carlo path Simulation, *operations Research*, Vol. 56, No. 3, pp. 607-617, 2008.

[4] Giles, M. B.: Improved multilevel Monte Carlo convergence using the Milstein scheme, *Monte Carlo and Quasi-Monte Carlo Methods 2006*, Springer, 2007

[5] Giles, M. B., Waterhouse, B.J.: Multilevel quasi-Monte Carlo path simulation, *Radon Series on Computational and Applied Mathematics*, Springer, 2009.

[6] Giles, M. B., Higham, D. J., Mao X.: Analysing Multi-level Monte Carlo for Options with Non-globally Lipschitz Payoff, *Finance and Stochastics*, 2009.

[7] Glassermann, P.: *Monte Carlo Methods in Financial Engineering*, Springer, New York, 2004.

[8] Günther, M., Jüngel, A.: *Finanzderivate mit MATLAB: Mathematische Modellierung und numerische Simulation*, Vieweg, 1.Auflage, Wiesbaden, 2003.

[9] Haslett J., Joshi, C., Geveli, R.: On Efficient Estimation of the Variance of the Randomised Quasi Monte Carlo Estimate, Trinity College Dublin, 2011.

[10] Hull, J. C.: *Optionen, Futures und andere Derivate,* Pearson Studium, 7.te aktualisierte Auflage, 2009.

[11] Kloeden, P. E., Platen, E.: *Numerical Solution of Stochastic Differential Equations,* Springer, 1992.

[12] Niederreiter, H.: *Random Number Generation and Quasi-Monte Carlo Methods,* SIAM, 1992.

[13] Niederreiter, H., Keller, A., Heinrich, S.: *Monte Carlo and Quasi-Monte Carlo Methods,* Springer-Verlag, 2006.

[14] Novak, E., Ritter, K.: *High dimensional integration of smooth functions over cubes,* Springer-Verlag, 1996.

[15] Owen, B.: Quasi-Monte Carlo for Integrands with Point Singularities at Unknown Locations, *Technical Report No. 2004-26*, 2004.

[16] Sobol, I. M.: Distribution of points in a cube and approximate evaluation of integrals, Akademija Nauk SSSR. Zurnal VycislitelÂ´noi Mathematiki i Matematiceskoi Fiziki 7 (1967) pp. 784-802.

[17] Tuffin, B.: Randomization of Quasi-Monte Carlo Methods for Error Estimation: Survey and Normal Approximation, *Monte Carlo Methods and Applications*, Vol.10, Num.3-4, pages 617-628, 2004.

[18] Tuffin, B.: Variance Reduction Applied to Product-Form Multi-Class Queuing Networks, *Journal ACM Transactions on Modeling and Computer Simulation*, 1997.

[19] Tuffin, B., L'Ecuyer, P., Lecot, C.: A Randomized Quasi-Monte Carlo Method for Markov Chains, *Operations Research*, 2007.

[20] Tuffin, B., Okten, G., Burago, V.: A central limit theorem and improved error bounds for a hybrid-Monte Carlo sequence with applications in computational finance, *Journal of Complexity*, Vol. 22, pages 435-458, 2006.

PART 3
Applications

Chapter 10

Drift-Free Simulation methods for pricing cross-market derivatives with LIBOR Market Model

J.L. Fernández, M.R. Nogueiras, M. Pou and C. Vázquez

Dep. de Matemáticas, Universidad Autónoma de Madrid, Francisco Tomás y Valiente, 7, 28049, Madrid, Spain

Dep. de Matemáticas, Universidade da Coruña, Área de Valoración y Precios, Banesto, c/ Messena, 80, 28033 Madrid, Spain

Dep. de Matemáticas, Universidade da Coruña, Campus Elviña s/n, 15071, A Coruña, Spain

Dep. de Matemáticas, Universidade da Coruña, Campus Elviña s/n, 15071, A Coruña, Spain

Abstract In this work we present an efficient procedure to simulate the dynamics of the LMM mainly avoiding the use of the drift dependent paths in Monte Carlo simulation. We extend this simulation procedure to the case of the LMM for two economies. In the case of two currencies we distinguish the domestic and foreign forward LIBOR rates plus the forward foreign exchange rate. In the model proposed here, we introduce the current value of the discount factors and the volatilities for both economies directly from the market. We also adjust the correlations by extending Rebonato's formula to the cross currency case. Notice that the existence of Quanto products mainly motivates the need of cross-currency models, therefore we present some numerical results concerning to a quanto pricing example.

1. Introduction

The LIBOR Market Model (LMM) has become very popular since its introduction in the late nineties [6; 16]. This is mainly due to the fact that LMM prices caps by means of the standard Black's formula have employed in the markets. Also, the model avoids arbitrage among bonds and keeps all rates positive. As the LMM assumes lognormal dynamics for each forward LIBOR rate, it is also referred as the Lognormal Forward-LIBOR Model (LFM) in [6], for example. Thus, the model assumes deterministic diffusion coefficients for the logarithmic forward rates (i.e. lognormal volatility), although each forward rate becomes a martingale under an appropriate probability measure which is not the same for all rates [13]. As indicated for example in [6], several common measures can be chosen (terminal, spot, etc), each one of them associated to a particular choice of the numeraire. The dynamics of each forward rate under this common measure can be obtained by the change of numeraire technique. Under this common numeraire almost all the forward LIBOR dynamics include the presence of a drift term depending on several forward rates. This dependence of the drift on the forward rates prevents us from having an exact solution of the system of stochastic differential equations governing the dynamics of the set of LIBOR rates.

LMM models the whole Zero Coupon curve, so it is specially well suited to price derivatives on a single interest rate curve by using Monte Carlo simulation. Clearly, this approach requires the discretization of continuous models. The discretized approximation should maintain the desired properties of the continuous model: guaranteeing no arbitrage among different bonds, preserving positivity of forward rates and pricing caplets according to Black's formula.

The most immediate Euler scheme in Monte Carlo simulation discretizes the stochastic differential equations verified by LIBOR rates under the common measure [11]. This approach requires the discretization of the drift terms so a discretization error is introduced. Particularly, in order to guarantee that the discretized model remains arbitrage free, all discretized deflated bonds must remain martingales. Discretization of the forward LIBOR stochastic equations require some not easy drift free adjustment [1]. In this chapter we follow an alternative approach that mainly consists of first simulating certain martingales and next computing the LIBOR rates in terms of them (see [12], for example). This methodology is referred to as the drift free simulation (DFS) technique.

In the present work we describe an efficient method, proposed in [9], to simulate the dynamics of forward LIBOR rates in the LMM for a single interest rate curve when using as the numeraire a bond maturing at any LIBOR tenor date (forward measure) or the bank account (spot measure). This method starts from some ideas developed in [12] that replace the simulation of the drifts associated to LIBOR rates by the simulation of appropriate martingales. In this setting, we first describe three drift free simulation (DFS) techniques. The first one is implicit and requires a sequential implementation of the simulation of the involved martingales and does not guarantee the positivity of the computed martingales at discrete level. The second technique is based on a log-Euler simulation technique leading to a non recursive simulation algorithm which allows to parallelize in a certain sense the simulation of all the involved martingales. Although the positivity of martingales is guaranteed in these case, the forward LIBOR rates can be negative. In both of these previous methods, the martingales are defined as the discounted values of certain bonds in terms of a chosen numeraire. Next, as a third technique, we recall the method proposed by Glasserman and Zhao in [12], in which the differences between particular deflated bonds are used as martingales. These martingales are obtained via a particular change of coordinates applied to the deflated bonds, which preserves the martingale property of those bonds at discrete level, although negative deflated bonds may appear thus leading to arbitrage among different bonds, as indicated in [4]. In [4], alternatives to the Glasserman-Zhao method to avoid this limitation are presented, although all of them exhibit some further particular disadvantages (negative forward rates, an additional discretization bias, no arbitrage free,...).

An efficient method proposed in [9] simulates forward LIBOR rates according to the LMM model avoiding negative deflated bonds and negative forward rates. The method is based upon a new parameterization of the martingales introduced by Glasserman and Zhao. The numerical results illustrate the better performance of the proposed method when pricing caplets. Also the proposed adjustment for ensuring the martingale property at discrete level provides slightly better results with the methods proposed here. Although different alternative probability measures have been analyzed, here we just present two examples of numeraires: discount bond maturing at an intermediate tenor date and the bank account. At that point, we also consider a combination of these two: up to maturity of the bond we use the bond and after that the bank account starting at that date.

Moreover, we develop a model for pricing cross-market derivatives. A first example comes from the two currencies framework where we have domestic rates, foreign rates and the FX rate. A second example arises in the inflation derivatives setting in which we distinguish nominal rates, real rates and the inflation index. Thus, in the general cross-market setting, our aim is to state models for the Zero Coupon Curve evolution of a market A, for the Zero Coupon Curve evolution of a market B and for the evolution of the exchange rate between both markets. Both Zero Coupon Curves will be represented by their respective forward rates. In particular, we model rates of each economy by a LMM, so we can construct a market model that assumes lognormality for forward LIBOR rates. When addressing the LMM methodology for cross-currency derivatives, the assumption that all forward foreign exchange (FX) rates are lognormal results in inconsistent with the lognormal hypothesis about domestic and foreign rates for all maturities [16]. In [21] consistent combinations of lognormality assumptions are analyzed, amongst them we will choose to assume lognormality just for one maturity in forward FX rates. This choice has also been considered in [16], where the simulation procedure proposed in [12] for one currency is extended to the cross-currency case. The proposed extension takes advantage of the drift-free procedures to simulate the dynamics of the forward rates in each economy. Here we mainly present the two currency case, so we model the evolution of domestic Zero Coupon Curve, foreign Zero Coupon Curve, and a forward FX rate. Thus, we apply the drift-free simulation methods to the two currencies case and then use it to obtain the value of Quanto Caplets. Finally, we compare the values obtained with the different DFS techniques with the value obtained with the market formula.

This chapter is organized as follows. Section 2 describes the proposed drift-free simulation methods for LMM. In Section 3 we present the dynamics for forward rates in the domestic and the foreign markets and the FX rate under a general probability measure and a particular one. In Section 4 we extend the methodology proposed in Section 2 to the two currencies case. Section 5 is devoted to lognormal approximation of swap rates. Section 6 deals with the model calibration procedure. Section 7 introduces particular Quanto Caplets derivatives and Section 8 presents the obtained numerical results for Quanto Caplets pricing.

2. Drift-Free simulations methods for LIBOR Market Model (LMM)

In this section we recall the Drift-Free Simulation (DFS) methodologies for LMM that were presented in [12] and a new parameterization technique described and discussed in [9]. For this purpose, we first introduce the usual notation related to LMM.

We consider a tenor structure $\mathcal{T}{:=}\{T_0, T_1, ..., T_{N+1}\}$, with $T_0 = 0$, $T_j < T_k, 0 \leq j < k \leq N+1$, and the corresponding accruals $\delta_j = T_{j+1} - T_j$, $0 \leq j \leq N$. For $j = 0, ..., N+1$, let $B_j(t)$ denote the price at time t of a zero-coupon bond that matures at the tenor date $T_j \geq t$. For $j = 0, ..., N$, we denote by $L_j(t)$ the value at time $t \leq T_j$ of a forward LIBOR rate for the accrual period $[T_j, T_{j+1}]$. These forward LIBOR rates can be obtained in terms of quotients of zero coupon bond prices, as follows:

$$L_j(t) = \frac{B_j(t) - B_{j+1}(t)}{\delta_j B_{j+1}(t)} \,. \tag{1}$$

The bank account, β, is the asset whose value at times $t \in (T_{i-1}, T_i]$ is given by

$$\beta(t) = B_i(t) \prod_{l=0}^{i-1} \left(1 + \delta_l L_l(T_l)\right) . \tag{2}$$

We shall denote by $W(t)$ a (correlated) N-dimensional Wiener process in a certain probability space $(\Omega, \mathcal{F}, \mathbf{P})$ with a correlation matrix $\Sigma = (\Sigma_{i,j})$. The natural filtration generated by $W(t)$ is denoted by $\mathcal{F}_t$. Every stochastic process we consider is defined on this probability space. In the sequel several particular probability measures that correspond to particular choices of numeraire are considered.

LMM assumes lognormal forward LIBOR rates under $\mathbf{P}$. Hence, for $j = 1, ..., N$, we assume the following dynamics for forward LIBOR rates:

$$dL_j(t) = \mu_j(t)L_j(t)dt + \sigma_j(t)L_j(t)dW_j(t), \tag{3}$$

where $\mu_j(t)$ denotes the drift term and $\sigma_j(t)$ denotes the volatility term. The instantaneous correlation coefficient $< dL_j(t), dL_k(t) >$ is then $\Sigma_{j,k}dt$.

As previously indicated, we first present DFS procedures that allow to simulate these forward LIBOR rates avoiding the calculation of the drifts appearing in (3). Once we have chosen a numeraire Num, the drift free simulation procedures are based on the simulation of the following deflated bonds

$$M_j(t) \doteq \frac{B_j(t)}{\text{Num}(t)}, \qquad t \leq T_j, \qquad j = 1, \ldots, N+1, \tag{4}$$

which are martingales under an appropriate equivalent probability measure $\mathbf{Q}$, which, of course, depends on the choice of Num.

The DFS techniques are based on the following (recurrence) relation

$$M_j(t) = \big(1 + \delta_j L_j(t)\big) M_{j+1}(t)\,, \tag{5}$$

connecting consecutive deflated bonds and a LIBOR rate.

Remark 10.1. As we have already pointed out we shall only consider two types of numeraries: discount bonds of any (tenor) maturity and the bank account. It is a simple fact, but quite relevant in the sequel, that for any of those numeraries and for each interval $[T_j, T_{j+1})$ there exists an index k such that $dM_k(t) = 0$ along that interval. If the numeraire is the bond B_n, $n \leq N+1$, then this index k is always $k = n$, while for the bank account as numeraire the index k is $j+1$ within the interval $[T_j, T_{j+1})$.

2.1. *Implicit Drift-Free Simulation (IDFS)*

Taking into account that $M_j(t)$ is a martingale under $\mathbf{Q}$ and the recurrence relation (5), we obtain the following (equivalent) expressions relating the dynamics of consecutive M_j:

$$dM_j(t) \;=\; \delta_j\sigma_j(t)L_j(t)M_{j+1}(t)\,dW_j(t) \;+\; \big(1+\delta_j L_j(t)\big)\,dM_{j+1}(t), \tag{6}$$

$$dM_{j+1}(t) \;=\; \frac{-\delta_j\sigma_j(t)L_j(t)}{\big(1+\delta_j L_j(t)\big)^2} M_j(t)\,dW_j(t) \;+\; \frac{1}{1+\delta_j L_j(t)}\,dM_j(t). \tag{7}$$

These expressions are relevant for our purposes, because as pointed out in Remark 10.1, in each time interval $(t, t + dt]$, $dM_k(t) = 0$ for some k, and then we may compute recursively the $dM_j(t)$, for $j = k-1, k-2, ...$, using (6), and for $j = k+1, k+2, ...$, by (7), just in terms of the brownian jumps and the previous levels of the LIBOR rates and the deflated bonds.

The simulation algorithm goes as follows. For simplicity, we assume the same constant accrual $\delta_j = \Delta T$ for $j = 1, \ldots, N$. For the IDFS procedure we consider a constant simulation time step δt, which is a fraction* of the constant accrual ΔT. We consider $t_i = i\delta t$ as a generic discretization time.

The steps of the IDFS algorithm are the following:

1. Initialize the values $\{B_j(0)\}_{j=0}^{N+1}$, $\{L_j(0)\}_{j=0}^{N}$ and $\{M_j(0)\}_{j=0}^{N+1}$.

2. Simulation of the correlated Brownian motions $\{W_j(t_i)\}_{j=1}^{N}$ with $t_i \leq T_j$.

3. For each $i = 0, 1, 2, \cdots, (N+1)(\delta t/\Delta T) - 1$, the computation of the martingale structure and LIBOR rates at time t_{i+1} follows two steps.

3.1. First, calculate all the martingale structure at time t_{i+1}, starting from $dM_k(t_i) = 0$, and $M_k(t_{i+1}) = M_k(t_i) + dM_k(t_i)$. Thus, for $j = k-1, k-2, \ldots$, with $T_j \geq t_{i+1}$, using the Euler discretization, we set

$$M_j(t_{i+1}) = M_j(t_i) + dM_j(t_i), \tag{8}$$

where $dM_j(t_i)$ is given by (6). Next, for $j = k, k+1, \ldots, N-1, N$ we set

$$M_{j+1}(t_{i+1}) = M_{j+1}(t_i) + dM_{j+1}(t_i), \tag{9}$$

where $dM_{j+1}(t_i)$ is obtained with (7).

3.2. Secondly, calculate the forward LIBOR rates at t_{i+1} by using the following equivalent expression to (1):

$$L_j(t_{i+1}) = \frac{M_j(t_{i+1}) - M_{j+1}(t_{i+1})}{\Delta T M_{j+1}(t_{i+1})}, \quad \text{with } j \text{ such that } T_j \geq t_{i+1}.$$

Next, go back to Step 3.1. for the following value of i.

Once we have run the algorithm, we can recover the numeraire and discount bonds values only at tenor dates by means of

$$\text{Num}(T_j) = \frac{1}{M_j(T_j)}, \quad j = 1, ..., N+1, \tag{10}$$

and

$$B_j(T_i) = M_j(T_i)\text{Num}(T_i), \quad i \leq j. \tag{11}$$

*In this way we guarantee that tenor dates are simulation dates, thus avoiding the use of additional complex interpolation procedures.

The constraint of recovering only for tenor dates is not a problem, because for pricing we just need the values of martingales, forward rates, numeraire and discount bounds at tenor dates. Actually, the choice of δt small is only for convergence purposes.

The above IDFS procedure combined the Euler discretization results with a recursive method: to pass from time t_i to t_{i+1} we choose an appropriate order in j (first decreasing from $k-1$ to 0 and next increasing from $k+1$ to $N+1$) to calculate the M_j at time t_{i+1}. Note that Euler discretization does not guarantee the required positivity of the martingales M_j. In order to avoid these two disadvantages (recursiveness and non positiveness) we introduce another simulation procedure.

2.2. *Explicit Drift-Free Simulation (EDFS)*

In this section we introduce and describe an alternative log-Euler discretization at Step 3 of the IDFS algorithm leading to a non recursive simulation procedure which we call Explicit Drift-Free Simulation (EDFS) algorithm and describe next.

First, by using (6), we have

$$\frac{dM_j(t)}{M_j(t)} = \sigma_j(t)\left(1 - \frac{M_{j+1}(t)}{M_j(t)}\right) dW_j(t) + \frac{dM_{j+1}(t)}{M_{j+1}(t)}. \tag{12}$$

Again, as indicated in Remark 10.1, since $dM_k(t) = 0$ for some k, we have that

$$\frac{dM_j(t)}{M_j(t)} = \varepsilon_{k,j} \sum_{l\in\theta_{k,j}} \left(\frac{dM_l(t)}{M_l(t)} - \frac{dM_{l+1}(t)}{M_{l+1}(t)}\right), \tag{13}$$

where

$$\varepsilon_{k,j} = \begin{cases} -1, \text{ if } j > k, \\ 0, \quad \text{if } j = k, \\ +1, \text{ if } j < k, \end{cases} \qquad \theta_{k,j} = \begin{cases} \{k, k+1, ..., j-1\}, \text{ if } j > k, \\ \emptyset, \qquad\qquad\qquad \text{if } j = k, \\ \{j, j+1, ..., k-1\}, \text{ if } j < k. \end{cases} \tag{14}$$

Next, by combining (12) and (13), we get

$$\frac{dM_j(t)}{M_j(t)} = \varepsilon_{k,j} \sum_{l\in\theta_{k,j}} \sigma_l(t)\left(1 - \frac{M_{l+1}(t)}{M_l(t)}\right) dW_l(t), \tag{15}$$

or equivalently,

$$\frac{dM_j(t)}{M_j(t)} = \sigma_j^M(t) dW_j^M(t), \tag{16}$$

where

$$\sigma_j^M(t) = \varepsilon_{k,j}\Big[\sum_{l_1,l_2\in\theta_{k,j}}\Big(1-\frac{M_{l_1+1}(t)}{M_{l_1}(t)}\Big)\Big(1-\frac{M_{l_2+1}(t)}{M_{l_2}(t)}\Big)\sigma_{l_1}(t)\sigma_{l_2}(t)\Sigma_{l_1,l_2}\Big]^{\frac{1}{2}}, \quad (17)$$

and $W_j^M(t)$ is an appropriate new Brownian motion.

Therefore, given (15) and (17), and applying the log-Euler discretization, we have the following algorithm for the martingale structure:

$$M_j(t_{i+1}) = M_j(t_i)\exp\left(\frac{dM_j(t_i)}{M_j(t_i)} - \frac{1}{2}\sigma_j^M(t_i)^2\delta t\right), \quad j = 1,\ldots,N+1, \quad (18)$$

as an alternative to Step 3.1 in the IDFS algorithm.

The rest of the steps of the algorithm are analogous to the previously described IDFS procedure.

In this EDFS method, once we know all the martingales at time t_i we can compute independently the martingales at time t_{i+1} this is why it is termed *explicit*.

In the EDFS method, the positivity of the deflated bonds M_j is guaranteed by construction, although this is not the case for the forward LIBOR rates, thus raising the possibility of discount bonds values greater than one. In order to overcome this disadvantage we present another simulation procedure proposed by Glasserman and Zhao in [12].

2.3. *Glasserman-Zhao Drift-Free Simulation (GZDFS)*

In order to guarantee the positivity of the forward LIBOR rates, Glasserman and Zhao in [12] have introduced the following set of martingales:

$$X_j(t) \doteq M_j(t) - M_{j+1}(t) = \Delta T L_j(t) M_{j+1}(t), \quad t \le T_j \quad j = 1,\ldots,N. \quad (19)$$

Taking into account that X_j is a martingale under $\mathbf{Q}$, we deduce the following dynamics:

$$\frac{dX_j(t)}{X_j(t)} = \frac{dM_{j+1}(t)}{M_{j+1}(t)} + \sigma_j(t)dW_j(t). \quad (20)$$

Next, by combining (15) and (20), we get

$$\frac{dX_j(t)}{X_j(t)} = \varepsilon_{k,j+1}\sum_{l\in\theta_{k,j+1}}\sigma_l(t)\left(1-\frac{M_{l+1}(t)}{M_l(t)}\right)dW_l(t) + \sigma_j(t)dW_j(t), \quad (21)$$

or equivalently,

$$\frac{dX_j(t)}{X_j(t)} = \sigma_j^X(t)dW_j^X(t), \quad (22)$$

where

$$\sigma_j^X(t) = \left[\sum_{l_1,l_2\in\theta_{k,j+1}} \left(1 - \frac{M_{l_1+1}(t)}{M_{l_1}(t)}\right)\left(1 - \frac{M_{l_2+1}(t)}{M_{l_2}(t)}\right)\sigma_{l_1}(t)\sigma_{l_2}(t)\Sigma_{l_1,l_2} \right.$$

$$\left. +\sigma_j(t)^2 + 2\sigma_j(t)\varepsilon_{k,j+1} \sum_{l\in\theta_{k,j+1}} \left(1 - \frac{M_{l+1}(t)}{M_l(t)}\right)\sigma_l(t)\Sigma_{l,j} \right]^{\frac{1}{2}}, \tag{23}$$

and $W_j^X(t)$ is a new Brownian motion.

Therefore, given (21) and (23), and by applying the log-Euler discretization, we have the following algorithm for simulating the X_j's:

$$X_j(t_{i+1}) = X_j(t_i)\exp\left(\frac{dX_j(t)}{X_j(t)} - \frac{1}{2}\sigma_j^X(t_i)^2\delta t\right), \quad j = 1, \dots, N. \tag{24}$$

Taking into account the previous dynamics, Step 3 of GZDFS algorithm can be written as follows:

3. For each $i = 0, 1, 2, \cdots, (N+1)(\delta t/\Delta T)-1$, the computation of the martingale structure and LIBOR rates at time t_{i+1} follows three steps.

3.1. First, calculate the values of $X_j(t_{i+1})$, with $T_j \geq t_{i+1}$, by (24).

3.2. Secondly, calculate all the martingale structure at time t_{i+1}, starting from $dM_k(t_i) = 0$, and $M_k(t_{i+1}) = M_k(t_i) + dM_k(t_i)$. Then, for $j = k-1, k-2, \dots,$ with $T_j \geq t_{i+1}$,

$$M_j(t_{i+1}) = M_{j+1}(t_{i+1}) + X_j(t_{i+1}). \tag{25}$$

Next, for $j = k+1, k+2, \dots, N, N+1$,

$$M_j(t_{i+1}) = M_{j-1}(t_{i+1}) - X_{j-1}(t_{i+1}). \tag{26}$$

3.3. Thirdly, calculate the forward LIBOR rates at t_{i+1} by using the following equivalent expression to (1):

$$L_j(t_{i+1}) = \frac{X_j(t_{i+1})}{\Delta T M_{j+1}(t_{i+1})}, \quad \text{with } j \text{ such that } T_j \geq t_{i+1}.$$

Next, go back to Step 3.1. for the following value of i.

Note that the positivity is guaranteed for X_j for $j = 1, \dots, N$ and for M_j for $j = 1, \dots, k$, but not for $j = k+1, \dots, N+1$.

As indicated in [4], the main disadvantage of Glasserman-Zhao method is that for a wide set of terminal measures and for the spot measure the deflated bonds in the discrete model could become negative, thus leading to internal arbitrage of the model. Moreover, these deflated bonds are used to obtain the forward LIBOR rates. This undesirable property can be avoided, for example, by using a terminal measure. Nevertheless, other measures as the bank account offer advantages from the accuracy of Monte Carlo simulation point of view. In [4] several alternative strategies are proposed to overcome these difficulties and the different pros and cons of each one are highlighted.

In order to overcome the pointed out drawback of the Glasserman-Zhao method and the difficulties indicated in the alternatives proposed in [4], in the next subsection we describe a new parameterized DFS method that was proposed in [9]. In the final section this method will be compared with the previous ones when pricing a two currency derivative product.

2.4. *A new Parameterized Drift-Free Simulation (PDFS) algorithm*

In this subsection we present a parameterized algorithm that guarantees the positivity of the deflated bonds and of the forward LIBOR rates under any forward or spot probability measure (see [9]).

This procedure is based on adding the dynamic of the martingale M_{N+1} to the system of stochastic differential equations proposed by Glasserman and Zhao in [12], dX_j, $j = 1, ..., N$. In this way, if we ensure that $X_1, ..., X_N, M_{N+1}$ are positive, then we can compute each martingale M_j as a sum of positive terms, $M_j = M_{j+1} + X_j$, for $j = N, N-1, ..., 1$.

We have the following new system of $N+1$ dynamics

$$\begin{cases} \dfrac{dX_j(t)}{X_j(t)} = \varepsilon_{k,j+1} \displaystyle\sum_{l \in \theta_{k,j+1}} \sigma_l(t) \left(1 - \frac{M_{l+1}(t)}{M_l(t)}\right) dW_l(t) + \sigma_j(t) dW_j(t), \quad j = 1, ..., N, \\ \dfrac{dM_{N+1}(t)}{M_{N+1}(t)} = -\displaystyle\sum_{l=k}^{N} \sigma_l(t) \left(1 - \frac{M_{l+1}(t)}{M_l(t)}\right) dW_l(t), \end{cases} \tag{27}$$

to which we impose the following additional constraints

i) $X_j > 0$, for $j = 1, ..., N$, and $M_{N+1} > 0$. This condition guarantees positive martingales and forward LIBOR rates.
ii) $X_k + X_{k+1} + ... + X_N + M_{N+1} = M_k$, where M_k is a value that we know in each time, taking into account that $dM_k = 0$. This condition is necessary to ensure the compatibility of the system.

We propose to ensure these conditions using the following parameterization of

X_j, $j = 1, ..., N$, and M_{N+1}:

$$X_j(t) = \exp\big(u_j(t)\big), \quad j = 1, ..., k-1,$$

$$X_j(t) = \frac{\exp\big(u_j(t)\big) M_k(t)}{\sum_{m=k}^{N+1} \exp\big(u_m(t)\big)}, \quad j = k, ..., N, \tag{28}$$

$$M_{N+1}(t) = \frac{\exp\big(u_{N+1}(t)\big) M_k(t)}{\sum_{m=k}^{N+1} \exp\big(u_m(t)\big)},$$

where $u_k(t) = 0$.

Equivalently, we have

$$u_j(t) = \begin{cases} \ln\big(X_j(t)\big), & \text{if } j = 1, ..., k-1, \\ 0, & \text{if } j = k, \\ \ln\big(X_j(t)\big) - \ln\big(X_k(t)\big), & \text{if } j = k+1, ..., N, \\ \ln\big(M_{N+1}(t)\big) - \ln\big(X_k(t)\big), & \text{if } j = N+1. \end{cases} \tag{29}$$

Thus, we can simulate the values of u_j, $j = 1, ..., N+1$, by an Euler discretization of the dynamics

$$du_j(t) = \begin{cases} \dfrac{dX_j(t)}{X_j(t)} - \dfrac{1}{2}\sigma_j^X(t)^2 \delta t, & \text{if } \ j = 1, ..., k-1, \\ 0, & \text{if } \ j = k, \\ \left(\dfrac{dX_j(t)}{X_j(t)} - \dfrac{dX_k(t)}{X_k(t)}\right) - \dfrac{1}{2}\left(\sigma_j^X(t)^2 - \sigma_k^X(t)^2\right)\delta t, & \text{if } \ j = k+1, ..., N, \\ \left(\dfrac{dM_{N+1}(t)}{M_{N+1}(t)} - \dfrac{dX_k(t)}{X_k(t)}\right) - \dfrac{1}{2}\left(\sigma_{N+1}^M(t)^2 - \sigma_k^X(t)^2\right)\delta t, & \text{if } \ j = N+1, \end{cases} \tag{30}$$

where $\dfrac{dM_{N+1}}{M_{N+1}}$, σ_{N+1}^M, $\dfrac{dX_j}{X_j}$ y σ_j^X are given by (15), (17), (21) and (23), respectively. Then, we compute $X_1, ..., X_N, M_{N+1}$ by (28), and finally we obtain the rest of the martingales M_j, for $j = N, N-1, ..., 1$, as follows

$$M_j(t) = M_{j+1}(t) + X_j(t). \tag{31}$$

The other steps of the algorithm are analogous to the previously described GZDFS procedure.

Note that now, since the forward LIBOR rates are calculated as quotients of positive terms, they are positive for any forward or spot probability measure.

2.5. *Martingale adjustment*

In order to preserve in the discrete implementation the martingale property of each continuous model, we also propose an appropriate martingale adjustment at simulation level. We distinguish the IDFS, EDFS, GDFS cases and the PDFS one.

2.5.1. *Martingale adjustment for the IDFS, the EDFS and the GDFS algorithms*

In the previous discretization procedures the martingale property for M_j and X_j can be lost in practice. So, following [12], we introduce an adjustment at the simulation level. The martingale adjustment consists of multiplying each simulated martingale by the ratio of its value at time zero and its mean. That is, by denoting N_S the number of simulations, the values obtained in Step 3 are replaced by the following values

$$M_j(t_{i+1})[p] = M_j(t_{i+1})[p] \, \frac{M_j(0)[p]}{\dfrac{1}{N_S}\displaystyle\sum_{p=1}^{N_S} M_j(t_{i+1})[p]} \tag{32}$$

in the EDFS and IDFS cases, and by

$$X_j(t_{i+1})[p] = X_j(t_{i+1})[p] \, \frac{X_j(0)[p]}{\dfrac{1}{N_S}\displaystyle\sum_{p=1}^{N_S} X_j(t_{i+1})[p]} \tag{33}$$

in the GZDFS case.

The previous adjustments guarantee the martingale property for M_j and X_j at discrete level that appeared in the previous simulation algorithms.

2.5.2. *Martingale adjustment for the PDFS algorithm*

As before, the martingale property for X_j and M_{N+1} can be lost with the discretization, so that we need an adjustment procedure for u_j to ensure that X_j and M_{N+1} are martingales. Since the dynamics of u_j are coupled for $j = k+1, ..., N+1$, in this case the adjustment is not analogous to that one of the other previous DFS methods. As we explain below, the adjustment for these terms will be an approximation.

- **Adjustment for u_j, $j < k$:**

By denoting N_S the number of simulations, if $j < k$ the value obtained for each u_j is replaced by the following value

$$u_j(t_{i+1})[p] = u_j(t_{i+1})[p] + \alpha_j(t_{i+1}), \tag{34}$$

where

$$\alpha_j(t_{i+1}) = \ln\left(\frac{X_j(0)}{\frac{1}{N_S}\sum_{p=1}^{N_S}\exp\left(u_j(t_{i+1})[p]\right)}\right). \tag{35}$$

Note that, as in the previous DFS methods, this is an exact adjustment for each j, but this is not the case for the adjustment when $j > k$ in next paragraph.

- **Adjustment for u_j, $j > k$:**

In this case, since the dynamics of u_j are coupled for $j = k+1, ..., N+1$, we have to calculate all adjustments $\alpha_{k+1}(t_{i+1}), ..., \alpha_{N+1}(t_{i+1})$ at the same time. So, for each t_{i+1}, we look for the vector $\overrightarrow{\alpha}(t_{i+1}) = \left(\alpha_{k+1}(t_{i+1}), ..., \alpha_{N+1}(t_{i+1})\right)$ such that

$$F_{i+1}\left(\overrightarrow{\alpha}(t_{i+1})\right) = \left(X_{k+1}(0), ..., X_N(0), M_{N+1}(0)\right)^t$$

where $F_{i+1} : \mathbb{R}^{N+1-k} \longrightarrow \mathbb{R}^{N+1-k}$ is defined by

$$F_{i+1}\left(\overrightarrow{\alpha}(t_{i+1})\right) = \begin{pmatrix} \frac{1}{N_S}\sum_{p=1}^{N_S}\frac{\exp\left(u_{k+1}(t_{i+1})[p] + \alpha_{k+1}(t_{i+1})\right)M_k(t_{i+1})[p]}{\sum_{m=k}^{N+1}\exp\left(u_m(t_{i+1})[p] + \alpha_m(t_{i+1})\right)} \\ \vdots \\ \frac{1}{N_S}\sum_{p=1}^{N_S}\frac{\exp\left(u_{N+1}(t_{i+1})[p] + \alpha_{N+1}(t_{i+1})\right)M_k(t_{i+1})[p]}{\sum_{m=k}^{N+1}\exp\left(u_m(t_{i+1})[p] + \alpha_m(t_{i+1})\right)} \end{pmatrix} \tag{36}$$

Since we know that $\overrightarrow{\alpha}(t_{i+1})$ will be close to the zero vector of $\mathbb{R}^{N+1-k}$, $\overrightarrow{\Theta} = (0, 0, ..., 0)$, by using just one iteration of Newton method and starting from $\overrightarrow{\Theta} = (0, 0, ..., 0)$, we can calculate the approximated adjustment by

$$\begin{aligned} \overrightarrow{\alpha}(t_{i+1}) &\approx JF_{i+1}(\overrightarrow{\Theta})^{-1} \\ &\times\left(\left(X_{k+1}(0), ..., X_N(0), M_{N+1}(0)\right)^t - F_{i+1}(\overrightarrow{\Theta})\right), \end{aligned} \tag{37}$$

where $JF_{i+1}(\overrightarrow{\Theta})$ is the Jacobian matrix of F_{i+1} in $\overrightarrow{\Theta}$, that is given by

$$(JF_{i+1})_{j_1,j_2} = \begin{cases} \dfrac{1}{N_S}\displaystyle\sum_{p=1}^{N_S}\left(\dfrac{M_k(t_{i+1})[p]^2 \exp\left(u_{j_1}(t_{i+1})[p]\right) \displaystyle\sum_{l=k,l\neq j_1}^{N+1} \exp\left(u_l(t_{i+1})[p]\right)}{\left(\displaystyle\sum_{m=k}^{N+1} \exp\left(u_m(t_{i+1})[p]\right)\right)^2}\right), & \text{if } j_1 = j_2, \\ -\dfrac{1}{N_S}\displaystyle\sum_{p=1}^{N_S}\left(\dfrac{M_k(t_{i+1})[p]^2 \exp\left(u_{j_1}(t_{i+1})[p]\right) \exp\left(u_{j_2}(t_{i+1})[p]\right)}{\left(\displaystyle\sum_{m=k}^{N+1} \exp\left(u_m(t_{i+1})[p]\right)\right)^2}\right), & \text{if } j_1 \neq j_2. \end{cases} \tag{38}$$

3. Two currencies LMM formulation

In this section we formulate the LMM for the two currencies setting. For this purpose, we use the one currency case notation introduced in the previous section for the domestic currency, and we extend the notation to the foreign currency accordingly.

Thus, for $j = 0, ..., N+1$, let $B_j^*(t)$ denote the price at time t of a discount bond in foreign currency that matures at the tenor date $T_j > t$. For $j = 0, ..., N$, we denote by $L_j^*(t)$ the value at time t of the foreign forward LIBOR rate for the period $[T_j, T_{j+1}]$. Analogously to the domestic case, these foreign rates can be obtained using quotients of their respective discount bond prices as follows:

$$L_j^*(t) = \frac{B_j^*(t) - B_{j+1}^*(t)}{\delta_j B_{j+1}^*(t)}. \tag{39}$$

Moreover, we denote by $FX(t)$ the FX rate at time t between domestic and foreign economies. For $j = 0, 1, ..., N+1$, let $FXfwd_j(t)$ denote the value at time t of a forward FX rate that matures at time $T_j > t$. So, we have:

$$FXfwd_j(t) = \frac{FX(t)B_j^*(t)}{B_j(t)}. \tag{40}$$

Remark 10.2. In what follows, when convenient we use the notation $A^{(*)}$ to refer to both A and A^* respectively, where A represents a term associated to the domestic economy and A^* is the respective term associated to the foreign one.

Note that we have N domestic forward LIBOR rates, N foreign forward LIBOR rates and $N+1$ forward exchange rates, i.e., we can choose between $3N+1$ rates to formulate our cross-currency model. If we represent all these rates in a graph G as indicated in Figure 1, where

$$\frac{1}{1+\delta_i L_i} FXfwd_{i+1} = FXfwd_i \frac{1}{1+\delta_i L_i^*},$$

by using graph theory we can demonstrate that we have to choose for our model the rates corresponding to the $2N+1$ edges of any of the possible spanning trees of G, see [10; 2]. In this way, we get a model without redundancies (because we do not have cycles), so that the other remaining N rates can be calculated (because all nodes are connected).

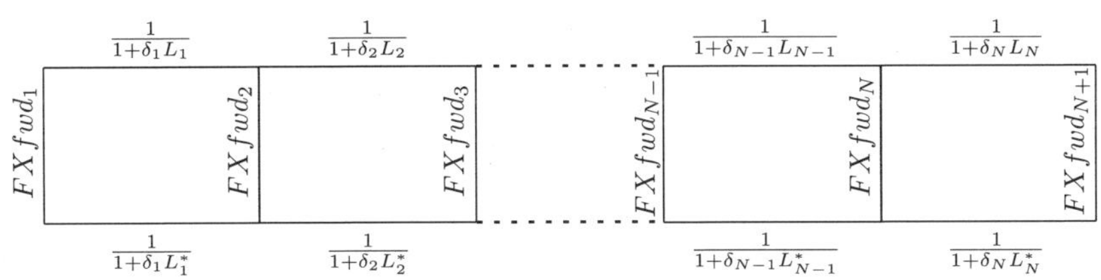

Figure 1. Graph associated to the rates appearing in LMM for two currencies

As we are in the context of LMM, we choose the domestic and foreign forward LIBOR rates and one of the forward exchange rate. Namely, we select the forward exchange rate that is martingale because we are interested in the simulation of these rates avoiding the use of the drift terms. Thus, we choose $FXfwd_k$ where k is n if the numeraire is the bond B_n and k is $j+1$ along the interval $[T_j, T_{j+1})$ if the numeraire is the bank account.

In order to formulate the model we denote by $Z(t) = (W_1(t), ..., W_N(t), W_1^*(t), ..., W_N^*(t), W_k^{FX}(t))$ a (correlated) $(2N+1)$-dimensional Wiener process in a certain probability space $(\Omega, \mathcal{F}, \mathbf{P})$ with a correlation matrix

$$C = \begin{pmatrix} \Sigma & \rho & \gamma \\ \rho & \Sigma^* & \gamma^* \\ \gamma & \gamma^* & 1 \end{pmatrix} \quad \in \mathcal{M}^{(2N+1)\times(2N+1)},$$

so $\Sigma^{(*)}$, ρ, $\gamma^{(*)}$ denote the correlations among domestic (foreign) forward LIBOR rates, among domestic and foreign forward LIBOR rates and among domestic (foreign) forward LIBOR rates and the forward exchange rate. The natural filtration spanned by $Z(t)$ is denoted by $\mathcal{F}_t$. Every stochastic process we consider is defined on this probability space. In the sequel several particular probability measures that correspond to particular choices of numeraire are considered.

LMM assumes lognormal forward LIBOR rates in both currencies under any probability measure $\mathbf{P}$. Hence, for $j = 1, ..., N$, we assume the following dynamics for forward LIBOR rates in both economies:

$$dL_j^{(*)}(t) = \mu_j^{(*)}(t)L_j^{(*)}(t)dt + \sigma_j^{(*)}(t)L_j^{(*)}(t)dW_j^{(*)}(t), \tag{41}$$

where $\mu_j^{(*)}(t)$ denotes the drift term, $\sigma_j^{(*)}(t)$ denotes the volatility term and $W_j^{(*)}(t)$ denotes a Brownian motion under $\mathbf{P}$, for the domestic (foreign) forward LIBOR rate. The instantaneous correlation coefficient $< dL_j^{(*)}(t), dL_k^{(*)}(t) >$ is then $\Sigma_{j,k}^{(*)}dt$. (The previous remark is already applied in this notation.)

Moreover, we can assume lognormality for the chosen forward FX rate, $FXfwd_k$. That is,

$$dFXfwd_k(t) = \mu_k^{FX}(t)FXfwd_k(t)dt + \sigma_k^{FX}(t)FXfwd_k(t)dW_k^{FX}(t), \tag{42}$$

where $\mu_k^{FX}(t)$ and $\sigma_k^{FX}(t)$ denote the drift and volatility of the forward FX rate while $W_k^{FX}(t)$ denotes a Brownian motion under $\mathbf{P}$.

As in the one currency case, the drift term depends on the chosen probability measure. Thus, if we consider Q^n then $L_{n-1}^{(*)}(t)$ and $FXfwd_n(t)$ are martingales, and if we consider Q^β the the terms that are martingales depend on the time interval, that is, $L_j^{(*)}(t)$ and $FXfwd_{j+1}(t)$ are martingales if $t \in (T_j, T_{j+1}]$. As we say, we are interested in the simulation of these rates avoiding the use of the drift terms. For this purpose, in the next section we adapt the DFS methodology to the two currencies case.

4. DFS for two currencies LMM

In this section, we extend the DFS procedures introduced in Section 2 to the cross-currency setting.

As in the one currency case, once we have chosen the numeraire Num, the DFS procedures are based on the simulation of the following deflated bonds

$$M_j(t) \doteq \frac{B_j(t)}{\mathrm{Num}(t)}, \quad M_j^*(t) \doteq \frac{B_j^*(t)TC(t)}{\mathrm{Num}(t)}, \qquad t \le T_j, \qquad j = 1, \ldots, N+1, \tag{43}$$

which are martingales under an appropriate equivalent probability measure $\mathbf{Q}$, which, of course, depends on the choice of Num.

The DFS techniques are based on the following (recurrence) relations

$$M_j^{(*)}(t) = \left(1 + \delta_j L_j^{(*)}(t)\right) M_{j+1}^{(*)}(t)\,, \tag{44}$$

connecting consecutive deflated bonds and a LIBOR rate in both economies.

Then, the DFS procedure (with or without adjustment) for the domestic part is carried out as indicated in Section 2. The construction of the foreign structure is totally analogous with one main difference: while the calculus of the martingale structure in the domestic currency starts from the known martingale dynamics $dM_k(t) = 0$ (remember Remark 10.1), the calculus in the foreign currency starts from the martingale dynamics $dM_k^*(t) = dFXfwd_k(t)^*$. Therefore, a previous step to the calculation of the foreign martingale is the computation of $FXfwd_k(t)$ from its drift-free dynamics, which is given by

$$dFXfwd_k(t) = \sigma_k^{FX}(t)FXfwd_k(t)dW_k^{FX}(t). \tag{45}$$

As in the case of martingale M_j, we can apply the analogous adjustment of this dynamics to guarantee the martingale property of $FXfwd_k(t)$ at discrete level.

Next, once the FX rate has been simulated, we apply the previous described methodology, although taking into account that now $dM_k^*(t) = dFXfwd_k(t)$ and replacing (13) and (15) by

$$\begin{aligned}\frac{dM_j^*(t)}{M_j^*(t)} &= \frac{dFXfwd_k(t)}{FXfwd_k(t)} + \varepsilon_{k,j} \sum_{l\in\Theta_{k,j}} \left(\frac{dM_l^*(t)}{M_l^*(t)} - \frac{dM_{l+1}^*(t)}{M_{l+1}^*(t)}\right)\\ &= \frac{dFXfwd_k(t)}{FXfwd_k(t)} + \varepsilon_{k,j} \sum_{l\in\Theta_{k,j}} \sigma_l^* \left(1 - \frac{M_{l+1}^*(t)}{M_l^*(t)}\right) dW_l^*(t).\end{aligned} \tag{46}$$

Moreover, while in the domestic case we calculate the numeraire at the tenor dates with (10), in the foreign case we compute the FX rate at time T_j with expression

$$FX(T_j) = M_j^*(T_j)\mathrm{Num}(T_j), \quad \forall j = 1, ..., N+1. \tag{47}$$

Next, we compute the foreign Zero Coupon curve at tenor dates with

$$B_j^*(T_i) = \frac{M_j^*(T_i)\mathrm{Num}(T_i)}{FX(T_i)}, \quad i \le j. \tag{48}$$

5. Swap rates and lognormal approximation model

In view of their utility in model calibration and derivatives pricing, we introduce the swap rates in both economies. For this purpose, first let us consider a swap contract as an agreement between two parties whereby at each time T_l, with $l = j+1, \ldots, k$, the following exchange occurs:

$$K\delta_{l-1} \quad \leftrightarrow \quad L_{l-1}^{(*)}(T_{l-1})\delta_{l-1},$$

*Note that we just need to assume lognormality for one forward exchange rate, which is consistent with assuming lognormality for all forward LIBOR rates (see [16], for example), as we said before.

where $\delta_{l-1} = T_l - T_{l-1}$. We denote by $S_{j,k}^{(*)}(t)$ the value at time t of a domestic (foreign) forward swap rate for the time interval $[T_j, T_k]$, that is, the rate K that makes the present value of the swap contract equal to zero. These forward swap rates can be obtained in terms of discounted bonds as

$$S_{j,k}^{(*)}(t) = \frac{B_j^{(*)}(t) - B_k^{(*)}(t)}{A_{j+1,k}^{(*)}(t)}, \tag{49}$$

where

$$A_{j+1,k}^{(*)}(t) = \sum_{l=j+1}^{k} \delta_l B_l^{(*)}(t).$$

Due to inconsistency with the arbitrage-free hypothesis, when using an LMM we cannot assume also lognormality for the swap rates. Nevertheless, note that in [7] it is pointed out the existence of empirical works showing that forward swap rates obtained by LMM are nearly lognormal under the appropriate measure (see [3], for example). However, a lognormal approximation of swap rates is required so that we extend the one proposed in [20].

For this purpose, we assume that $S_{j,j+s}(t)$ and $S_{j,j+s}^{*}(t)$ can be lognormally approximated under any probability measure Q, so that

$$\frac{dS_{j,j+s}^{(*)}(t)}{S_{j,j+s}^{(*)}(t)} \approx \mu_{j,j+s}^{S^{(*)}}(t)dt + \sigma_{j,j+s}^{S^{(*)}}(t)dW_{j,j+s}^{S^{(*)}}(t). \tag{50}$$

Hence, we have the approximation

$$\left(\frac{dS_{j,j+s}^{(*)}(t)}{S_{j,j+s}^{(*)}(t)}\right)\left(\frac{dS_{j,j+s}^{(*)}(t)}{S_{j,j+s}^{(*)}(t)}\right) \approx \sigma_{j,j+s}^{S^{(*)}}(t)^2 dt. \tag{51}$$

Moreover, in [20] the approximation

$$S_{j,j+s}^{(*)}(t) \approx \sum_{l=j+1}^{j+s} \omega_{j+1,j+s^{(*)}}^{l}(0) L_{l-1}^{(*)}(t), \tag{52}$$

is proposed, where

$$\omega_{j+1,j+s^{(*)}}^{l}(t) = \frac{\delta_l B_l^{(*)}(t)}{A_{j+1,j+s}^{(*)}(t)}. \tag{53}$$

Thus, from (52) and by freezing the forward LIBOR and swap rates to their value at time 0, we have

$$\left(\frac{dS_{j,j+s}^{(*)}(t)}{S_{j,j+s}^{(*)}(t)}\right)^2 \approx$$

$$\frac{\sum_{r,p=j+1}^{j+s} \omega^r_{j+1,j+s^{(*)}}(0)\omega^p_{j+1,j+s^{(*)}}(0)L^{(*)}_{r-1}(0)L^{(*)}_{p-1}(0)\Sigma^{(*)}_{r-1,p-1}\sigma^{(*)}_{r-1}(t)\sigma^{(*)}_{p-1}(t)}{S^{(*)}_{j,j+s}(0)^2}dt. \tag{54}$$

Then, from approximations (51) and (54), for all $t \leqslant T_j$ we obtain

$$\int_t^{T_j} \sigma^{S^{(*)}}_{j,j+s}(z)^2 dz$$

$$\approx \sum_{r,p=j+1}^{j+s} \frac{\omega^r_{j+1,j+s^{(*)}}(0)\omega^p_{j+1,j+s^{(*)}}(0)L^{(*)}_{r-1}(0)L^{(*)}_{p-1}(0)\Sigma^{(*)}_{r-1,p-1}}{S^{(*)}_{j,j+s}(0)^2}$$

$$\times \int_t^{T_j} \sigma^{(*)}_{r-1}(z)\sigma^{(*)}_{p-1}(z)dz, \tag{55}$$

where $\Sigma^{(*)}_{r-1,p-1} = \langle W^{(*)}_{r-1}, W^{(*)}_{p-1}\rangle$. Approximation (55) is known as Rebonato's formula, which relates the volatilities of forward and swap rates with the correlations between forward rates.

For practical purposes, we are interested in the expression for the dynamics of swap rates under $Q^{A_{j+1,j+s}}$ (probability measure with $A_{j+1,j+s}(t)$ as numeraire). Note that under this measure we have that

- $S_{j,j+s}(t)$ is a martingale, so that
$$\frac{dS_{j,j+s}(t)}{S_{j,j+s}(t)} \approx \sigma^S_{j,j+s}(t)dW^S_{j,j+s}(t), \tag{56}$$
where $\sigma^S_{j,j+s}(t)$ is given by (55).
- $S^*_{j,j+s}(t)$ is not a martingale, so that
$$\frac{dS^*_{j,j+s}(t)}{S^*_{j,j+s}(t)} \approx \mu^{S^*}_{j,j+s}(t)dt + \sigma^{S^*}_{j,j+s}(t)dW^{S^*}_{j,j+s}(t), \tag{57}$$
where $\sigma^{S^*}_{j,j+s}(t)$ is given by (55).

In order to approximate the drift in this second case, we first use (52) to obtain

$$E_{Q^{A_{j+1,j+s}}}[S^*_{j,j+s}(T_j)|\mathcal{F}_t] \approx \sum_{l=j+1}^{j+s} \omega^l_{j+1,j+s^*}(0)E_{Q^{A_{j+1,j+s}}}[L^*_{l-1}(T_j)|\mathcal{F}_t]. \tag{58}$$

On the other hand, from approximation (57) we get

$$E_{Q^{A_{j+1,j+s}}}[S^*_{j,j+s}(T_j)|\mathcal{F}_t] \approx S^*_{j,j+s}(t)e^{\int_t^{T_j}\mu^{S^*}_{j,j+s}(z)dz}. \tag{59}$$

Therefore, by identifying approximations (58) and (59), the computations detailed in the Annexe lead to the following approximation of the drift term:

$$\int_t^{T_j} \mu^{S^*}_{j,j+s}(z)dz$$
$$\approx \ln\left(\frac{\sum_{l=j+1}^{j+s}\left[\omega^l_{j+1,j+s^*}(0)L^*_{l-1}(t)\sum_{k=j+1}^{j+s}\omega^k_{j+1,j+s}(t)e^{\int_t^{T_j}X_{l,k}(z)dz}\right]}{S^*_{j,j+s}(t)}\right), \tag{60}$$

with

$$X_{l,k}(z) = -\gamma^*_{l-1,l}\sigma^*_{l-1}(z)\sigma^{FX}_l(z) + \sigma^*_{l-1}(z)\varepsilon_{l,k}\sum_{m\in\theta_{l,k}}\frac{\delta_m\sigma_m(0)L_m(0)}{1+\delta_m L_m(0)}\rho_{m,l-1}, \tag{61}$$

where $\gamma^*_{l-1,l}$ represents the correlation between $L^*_{l-1}(t)$ and $FXfwd_l(t)$ and $\rho_{m,l-1}$ represents the correlation between $L_m(t)$ and $L^*_{l-1}(t)$.

6. Model calibration

We are in the context of market models so we have to introduce into our model as many parameters as market information allows. As indicated in Setp 1 of the algorithm, we first introduce the following market data:

$$\{B_j(0)\}_{j=0}^{N+1}, \quad \{B^*_j(0)\}_{j=0}^{N+1}, \quad FX(0).$$

Other free parameters (volatilities and correlations) are considered in the next paragraphs so that the model parameters are adjusted to the market.

6.1. *Adjustment of volatilities*

The structure of volatilities is adjusted to market by imposing that:

$$\int_0^{T_i}\sigma_i(t)^2dt = T_i(\sigma_i^{mar})^2, \quad \int_0^{T_i}\sigma^*_i(t)^2dt = T_i(\sigma_i^{*mar})^2,$$

$$\int_0^{T_j}\sigma_j^{FX}(t)^2dt = T_j(\sigma_j^{FX,mar})^2,$$

where $\sigma_i^{(*)mar}$ are quoted volatilities of caplets of each economy and $\sigma_j^{FX,mar}$ are quoted volatilities of call options on forward exchange rates.

In practice, for simplicity, we choose constant volatilities, that is:

$$\sigma_i(t) = \sigma_i^{mar}, \quad \sigma^*_i(t) = \sigma_i^{*,mar}, \quad \sigma_i^{FX}(t) = \sigma_i^{FX,mar}, \quad \forall t.$$

6.2. *Adjustment of correlations*

When working under the probability measure Q^{n+1}, we have to adjust the following correlation matrix:

$$C = \begin{pmatrix} \Sigma_{11} & \dots & \Sigma_{1N} & \rho_{11} & \dots & \rho_{1N} & \gamma_{1,n+1} \\ \vdots & & \vdots & \vdots & & \vdots & \vdots \\ \Sigma_{N1} & \dots & \Sigma_{NN} & \rho_{N1} & \dots & \rho_{NN} & \gamma_{N,n+1} \\ \rho 11 & \dots & \rho_{N1} & \Sigma^*_{11} & \dots & \Sigma^*_{1N} & \gamma^*_{1,n+1} \\ \vdots & & \vdots & \vdots & & \vdots & \vdots \\ \rho 1N & \dots & \rho_{NN} & \Sigma^*_{N1} & \dots & \Sigma^*_{NN} & \gamma^*_{N,n+1} \\ \gamma_{1,n+1} & \cdots & \gamma_{N,n+1} & \gamma^*_{1,n+1} & \cdots & \gamma^*_{N,n+1} & \Theta_{n+1,n+1} \end{pmatrix}$$

where:

- $\Sigma^{(*)}_{ij}$ is the correlation between $L^{(*)}_i$ and $L^{(*)}_j$,
- ρ_{ij} is the correlation between L_i and L^*_j,
- $\gamma^{(*)}_{i,n+1}$ is the correlation between $L^{(*)}_i$ and $FXfwd_{n+1}$,
- $\Theta_{n+1,n+1}$ is the correlation between $FXfwd_{n+1}$ and $FXfwd_{n+1}$, so that $\Theta_{n+1,n+1} = 1$.

First, we adjust the correlations within each economy, $(\Sigma_{ij})^N_{i,j=1}$ and $(\Sigma^*_{ij})^N_{i,j=1}$, by using swaps volatilities from the market in each economy and taking into account (55).

As the other correlations cannot be calculated analogously, we obtain them from historical data and assume they are constant, that is:

$$\rho_{ij} = \rho, \quad \gamma_{i,n+1} = \gamma, \quad \gamma^*_{i,n+1} = \gamma^*, \quad \forall i, j = 1, ..., N.$$

Remark 10.3. If matrix C built as above is not positive semidefinite then we have to approximate it by another one that is positive semidefinite (and therefore a correlation matrix). We can do this by parameterizing correlations with some method, as the ones proposed in [19], and then trying to minimize the error between the matrix C and its approximated correlation matrix.

7. Cross-currency derivatives: Quanto Caplets

Nowadays there exists a lot of financial derivatives on rates of different currency markets (see [6], for example). They can be classified into

- Standard cross-currency products (Cross-currency swap, Cross-crurrency swaption,...): These contracts are agreements between two parts so that each part pays terms of one currency in this currency.
- Quanto products (Quanto fra, Quanto caplet/floorlet, Quanto swap, Quanto Swaption,...): They consists of products that pay terms of one

currency economy in another currency. As will be explained later, in the pricing of these derivatives there is one term that does not affect the pricing of the corresponding derivatives on the one currency (fra, caplet/floorlet, swap, swaption,...). This term is known as quanto adjustment.

In view of the previous description, Quanto products clearly depend on the domestic and foreign Zero coupon curves, as well as on the FX rate. Thus, they justify the need of two currencies models. Here we present a Quanto Caplet pricing methodology.

Let $QCaplet_j(t)$ denote the price at time t of a Quanto Caplet that matures at time T_j, so that it pays at T_{j+1}.

We distinguish two kinds of Quanto Caplets:

i) The Quanto Caplet that pays at T_{j+1} the amount $N(L_j^*(T_j) - K^*)_+\delta_j$ in the domestic currency, where N is the invested notional that is paid back at the payment date. The payoff of this product is:

$$QCaplet_j(T_j) = N\delta_j(L_j^*(T_j) - K^*)_+B_{j+1}(T_j). \qquad (62)$$

Hence, for all $t < T_j$, the Quanto Caplet price is given by

$$QCaplet_j(t) = B_{j+1}(t)N\delta_j E_{Q^{j+1}}[(L_j^*(T_j) - K^*)_+|\mathcal{F}_t]. \qquad (63)$$

Moreover, we know that under Q^{j+1} measure we have

$$\frac{dL_j^*(t)}{L_j^*(t)} = -\gamma_{j,j+1}^*\sigma_j^*(t)\sigma_{j+1}^{FX}(t)dt + \sigma_j^*(t)dW_j^*(t), \qquad (64)$$

where $\gamma_{j,j+1}^*$ denotes the correlation between L_j^* and $FXfwd_{j+1}$. So, by a proposition in [6], p. 919, we obtain the Black formula

$$QCaplet_j(t) = B_{j+1}(t)N\delta_j\left(L_j^*(t)e^{\Omega_j^*}\phi(d_+) - K^*\phi(d_-)\right), \qquad (65)$$

with

$$\Omega_j^* = \int_t^{T_j} -\gamma_{j,j+1}^*\sigma_j^*(z)\sigma_{j+1}^{FX}(z)dz,$$

$$d_\pm = \frac{\ln\left(\frac{L_j^*(t)}{K^*}\right) + \Omega_j^* \pm \frac{1}{2}\int_t^{T_j}\sigma_j^*(z)^2dz}{\left(\int_t^{T_j}\sigma_j^*(z)^2dz\right)^{\frac{1}{2}}}.$$

ii) The Quanto Caplet that pays at T_{j+1} the amount $N(L_j(T_j) - K)_+\delta_j$ in the foreign currency, where N denotes the invested notional. Therefore, the payoff of this derivative is:

$$QCaplet_j(T_j) = N\delta_j(L_j(T_j) - K)_+B_{j+1}^*(T_j)FX(T_j). \qquad (66)$$

Hence, for all $t < T_j$, the Quanto Caplet price is given by

$$QCaplet_j(t) = B_{j+1}(t)N\delta_j E_{Q^{j+1}}[(L_j(T_j) - K)_+ FXfwd_{j+1}(T_j)|\mathcal{F}_t]. \quad (67)$$

Moreover, we know that under Q^{j+1} measure we have

$$\frac{dL_j(t)FXfwd_{j+1}(t)}{L_j(t)FXfwd_{j+1}(t)}$$

$$= \gamma_{j,j+1}\sigma_j(t)\sigma_{j+1}^{FX}(t)dt + \sigma_j dW_j(t) + \sigma_{j+1}^{FX}dW_{j+1}^{FX}(t), \quad (68)$$

where $\gamma_{j,j+1}$ is the correlation between L_j and $FXfwd_{j+1}$. So, using the same proposition of [6] that in previous case, we can deduce the Black formula

$$QCaplet_j(t) = B_{j+1}(t)N\delta_j FXfwd_{j+1}(t)\left(L(t)e^{\Omega_j}\phi(d_+) - K\phi(d_-)\right), \quad (69)$$

with

$$\Omega_j = \int_t^{T_j} \gamma_{j,j+1}\sigma_j(z)\sigma_{j+1}^{FX}(z)dz,$$

$$d_\pm = \frac{\ln\left(\frac{L_j(t)}{K}\right) + \Omega_j \pm \frac{1}{2}\int_t^{T_j}\sigma_j(z)^2dz}{\left(\int_t^{T_j}\sigma_j(z)^2dz\right)^{\frac{1}{2}}}.$$

Furthermore, we can obtain the value of Quanto Caplets at time t by using simulation under Q (probability measure with Num as numeraire) and taking into account that

$$QCaplet_j(t) = Num(t)E_Q\left[\frac{QCaplet_j(T_j)}{Num(T_j)}|\mathcal{F}_t\right]. \quad (70)$$

In next section we propose an efficient procedure to simulate the joint dynamics that mainly avoids the use of the drift dependent paths.

8. Numerical results

In this section, we present the same data used for the different DFS methods described in previous sections and the obtained results for martingales and caplet pricing in each currency (dollar and euro). We also present the results obtained for the pricing of one Quanto Caplet. Thus, in Tables 1 to 10 we consider the dollar as the domestic currency and the euro as the foreign one.

Table 1 shows the data for the Monte Carlo simulation procedure. More precisely, we consider the case $N = 10$. The accrual period is constant and equal to one year. The time step and the number of simulations are also indicated. The domestic and foreign market data are given in Tables 2 and 3, respectively. They include the Zero Coupon Curves, the volatilities of the involved forward LIBOR and swap rates

Table 1. Parameters of the simulation procedure.

N	ΔT	δt	Number of simulations (N_S)
10	1	0.25	400.000

Table 2. Domestic market data.

i	$B_i(0)$	σ_i	$\sigma^S_{i,i+1}$	$\sigma^S_{i,i+2}$	$\sigma^S_{i,i+3}$	$\sigma^S_{i,i+4}$	$\sigma^S_{i,i+5}$	$\sigma^S_{i,i+6}$	$\sigma^S_{i,i+7}$	$\sigma^S_{i,i+8}$	$\sigma^S_{i,i+9}$	$\sigma^S_{i,i+10}$
0	1,00000000											
1	0,990761	0,7666	0,7666	0,5925	0,5015	0,4501	0,4240	0,3938	0,3764	0,3684	0,3565	0,3404
2	0,980002	0,5095	0,5095	0,4445	0,4063	0,3842	0,3632	0,3534	0,3386	0,3308	0,3221	
3	0,954435	0,3950	0,3950	0,3650	0,3509	0,3319	0,3232	0,3165	0,3031	0,2998		
4	0,922703	0,3378	0,3378	0,3272	0,3102	0,3032	0,2945	0,2876	0,2783			
5	0,887878	0,3132	0,3132	0,2966	0,2893	0,2820	0,2736	0,2692				
6	0,851701	0,2893	0,2893	0,2781	0,2695	0,2622	0,2546					
7	0,815186	0,2655	0,2655	0,2596	0,2497	0,2425						
8	0,779710	0,2485	0,2485	0,2413	0,2336							
9	0,744830	0,2315	0,2315	0,2230								
10	0,710892	0,2145	0,2145									
11	0,677054											

Table 3. Foreign market data.

i	$B^*_i(0)$	σ^*_i	$\sigma^{S^*}_{i,i+1}$	$\sigma^{S^*}_{i,i+2}$	$\sigma^{S^*}_{i,i+3}$	$\sigma^{S^*}_{i,i+4}$	$\sigma^{S^*}_{i,i+5}$	$\sigma^{S^*}_{i,i+6}$	$\sigma^{S^*}_{i,i+7}$	$\sigma^{S^*}_{i,i+8}$	$\sigma^{S^*}_{i,i+9}$	$\sigma^{S^*}_{i,i+10}$
0	1,000000											
1	0,987689	0,5152	0,5152	0,3844	0,3402	0,3088	0,2847	0,2689	0,2568	0,2495	0,2434	0,238100
2	0,966393	0,3566	0,3566	0,2896	0,2702	0,2549	0,2428	0,2370	0,2325	0,2289	0,2256	
3	0,938774	0,2837	0,2837	0,2395	0,2281	0,2190	0,2121	0,2098	0,2081	0,2065		
4	0,907973	0,2367	0,2367	0,2086	0,2019	0,1971	0,1931	0,1917	0,1903			
5	0,874979	0,2015	0,2015	0,1854	0,1826	0,1802	0,1780	0,1767				
6	0,840833	0,1851	0,1851	0,1732	0,1711	0,1690	0,1673					
7	0,806538	0,1687	0,1687	0,1610	0,1596	0,1579						
8	0,772759	0,1602	0,1602	0,1538	0,1528							
9	0,739546	0,1518	0,1518	0,1467								
10	0,707055	0,1433	0,1433									
11	0,674961											

in both economies. Table 4 shows data associated to FX rate. As the correlation matrix is completed with constant input data, these data are shown in Table 14.

We consider as numeraire:

$$\mathrm{Num}(t) = \begin{cases} B_5(t), & \text{if } t \leq T_5, \\ \beta_5(t), & \text{if } t > T_5, \end{cases}$$

Table 4. Forward foreign exchange (FX) rate data.

$FX(0)$	σ_1^{FX}	σ_2^{FX}	σ_3^{FX}	σ_4^{FX}	σ_5^{FX}	σ_6^{FX}	σ_7^{FX}	σ_8^{FX}	σ_9^{FX}	σ_{10}^{FX}	σ_{11}^{FX}
1.511	0,1365	0,1311	0,1256	0,1202	0,1147	0,1106	0,1064	0,1023	0,0981	0,0940	0,0899

Table 5. Correlation data.

ρ	γ	γ^*
0.7	0.2	0.3

where $\beta_5(t)$ corresponds to a the bank account starting at time T_5, that is

$$\beta_5(t) = B_i(t) \prod_{l=5}^{i-1} \left(1 + \Delta T L_l(T_l)\right), \quad t \in (T_{i-1}, T_i], \quad i = 6, \ldots, N+1. \tag{71}$$

In this case the numeraire is B_5 until T_5 and the spot measure starting after T_5 for $t > T_5$. Although this choice neither corresponds to a forward measure nor to a pure spot one, the proposed methodology can be applied.

Taking into account this information, in each currency we compare:

- the value of the martingales at time $t = 0$ jointly with the expected values obtained with different simulation methods (see Tables 6 and 7 and Figures 2 and 3),
- the Black value of the caplets (see, for example [6]) and the one obtained with the simulations at time $t = 0$ (see Tables 8 and 9 and Figures 4 and 5).

Finally, Table 10 shows the market price and the simulated prices in Case 1 and Case 2 for both the Quanto Caplet and the Quanto Floorlet that mature at time T_5 (so that they pay at time T_6), without and with adjustment when using the previously described DFS methods.

For all DFS methods, Tables 6 and 7 illustrate how the proposed adjustment guarantees the martingale property for M_j and M_j^* at discrete level. On the other hand, Figures 2 and 3 show that the numerical results are better in the PDFS method for the domestic currency (even when compared with the recently introduced GZDFS method) and in the IDFS one for the foreign currency. In any case, the results for the foreign martingale with the IDFS method are very close to those ones with the PDFS method. As illustrated by Figures 4 and 5, the PDFS method exhibits the best behavior for caplet pricing in both currencies. Finally, the PDFS method provides the best results in the quanto pricing for the Caplet 2 and Floorlet 1, and it results to be the second best method in the Caplet 1 case. So, taking

Table 6. Expected value of the domestic martingales, $E(M_j(T_j))$, obtained from the different Drift-Free Simulation methods without and with adjustment (WA), compared with $M_j(0)$.

j	$M_j(0)$	IDFS	IDFS WA	EDFS	EDFS WA	GZDFS	GZDFS WA	PDFS	PDFS WA
1	1,115875	1,115886	1,115875	1,115888	1,115875	1,115910	1,115875	1,115910	1,115875
2	1,103757	1,103853	1,103757	1,103855	1,103757	1,103869	1,103757	1,103869	1,103757
3	1,074962	1,074947	1,074962	1,074949	1,074962	1,074963	1,074962	1,074963	1,074962
4	1,039223	1,039260	1,039223	1,039260	1,039223	1,039269	1,039223	1,039269	1,039223
5	1,000000	1,000000	1,000000	1,000000	1,000000	1,000000	1,000000	1,000000	1,000000
6	0,959254	0,959235	0,959254	0,959235	0,959254	0,959227	0,959254	0,959253	0,959254
7	0,918129	0,918048	0,918129	0,918050	0,918129	0,918036	0,918129	0,918083	0,918129
8	0,878172	0,878017	0,878172	0,878020	0,878172	0,878005	0,878172	0,878066	0,878171
9	0,838888	0,838771	0,838888	0,838775	0,838888	0,838752	0,838888	0,838820	0,838888
10	0,800663	0,800441	0,800663	0,800447	0,800663	0,800419	0,800663	0,800492	0,800664
11	0,762553	0,762317	0,762553	0,762323	0,762553	0,762293	0,762553	0,762367	0,762556

Table 7. Expected value of the foreign martingales, $E(M_j^*(T_j))$, obtained from the different Drift-Free Simulation methods without and with adjustment (WA), compared with $M_j^*(0)$.

j	$M_j^*(0)$	IDFS	IDFS WA	EDFS	EDFS WA	GZDFS	GZDFS WA	PDFS	PDFS WA
1	1,680858	1,680860	1,680858	1,680885	1,680858	1,680895	1,680858	1,680895	1,680858
2	1,644618	1,645066	1,644618	1,645107	1,644618	1,645110	1,644618	1,645110	1,644618
3	1,597615	1,597917	1,597615	1,597981	1,597615	1,597975	1,597615	1,597975	1,597615
4	1,545198	1,545197	1,545198	1,545230	1,545198	1,545226	1,545198	1,545226	1,545198
5	1,489049	1,488996	1,489049	1,489033	1,489049	1,489033	1,489049	1,489033	1,489049
6	1,430939	1,431101	1,430939	1,431127	1,430939	1,431123	1,430939	1,431129	1,430939
7	1,372574	1,372859	1,372574	1,372900	1,372574	1,372900	1,372574	1,372910	1,372574
8	1,315089	1,315772	1,315089	1,315815	1,315089	1,315814	1,315089	1,315828	1,315089
9	1,258567	1,259181	1,258567	1,259225	1,258567	1,259226	1,258567	1,259243	1,258567
10	1,203273	1,204043	1,203273	1,204094	1,203273	1,204095	1,203273	1,204114	1,203273
11	1,148656	1,149099	1,148656	1,149138	1,148656	1,149137	1,148656	1,149158	1,148656

into account its results and its positiveness preserving property, in general PDFS method should be chosen for the cross-currency simulation.

ANNEXE: Approximation of drift terms

In this Annexe we develop the calculations leading to the approximation of the drift term, $\mu^{S^*}_{j,j+s}$, provided by expression (60). For this purpose, we first identify approximations (58) and (59), so that

$$S^*_{j,j+s}(t)e^{\int_t^{T_j}\mu^{S^*}_{j,j+s}(z)dz} = \sum_{l=j+1}^{j+s}\omega^l_{j+1,j+s^*}(0)E_{Q^{A_{j+1,j+s}}}[L^*_{l-1}(T_j)|\mathcal{F}_t]. \tag{72}$$

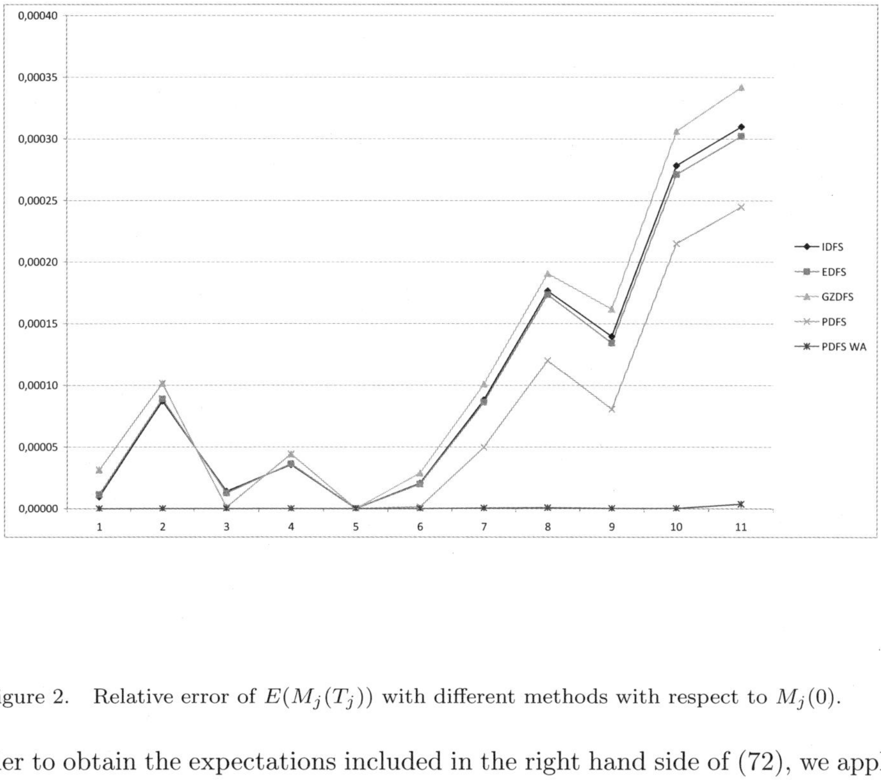

Figure 2. Relative error of $E(M_j(T_j))$ with different methods with respect to $M_j(0)$.

In order to obtain the expectations included in the right hand side of (72), we apply the following change of numeraire formula:

$$\frac{A_{j+1,j+s}(t)}{A_{j+1,j+s}(T_j)}\, dQ^{A_{j+1,j+s}} = \frac{B_l(t)}{B_l(T_j)}\, dQ^l \tag{73}$$

for $l = j+1, \ldots, j+s$. Therefore, we obtain

$$E_{Q^{A_{j+1,j+s}}}[L^*_{l-1}(T_j)|\mathcal{F}_t] = \frac{B_l(t)}{A_{j+1,j+s}(t)} E_{Q^l}\left[L^*_{l-1}(T_j)\frac{A_{j+1,j+s}(T_j)}{B_l(T_j)}|\mathcal{F}_t\right] \tag{74}$$

or, equivalently

$$E_{Q^{A_{j+1,j+s}}}[L^*_{l-1}(T_j)|\mathcal{F}_t] = \frac{B_l(t)}{A_{j+1,j+s}(t)} \sum_{k=j+1}^{j+s} \delta_k E_{Q^l}\left[L^*_{l-1}(T_j)\frac{B_k(T_j)}{B_l(T_j)}|\mathcal{F}_t\right]. \tag{75}$$

Next, we consider the following Proposition, which provides the expectations appearing in the right hand side of (75).

Proposition 10.1. *For $l = j+1, \ldots, j+s$, we have*

$$E_{Q^l}\left[L^*_{l-1}(T_j)\frac{B_k(T_j)}{B_l(T_j)}|\mathcal{F}_t\right] \approx L^*_{l-1}(t)\frac{B_k(t)}{B_l(t)} \exp\left(\int_t^{T_j} X_{kl}(z)dz\right), \tag{76}$$

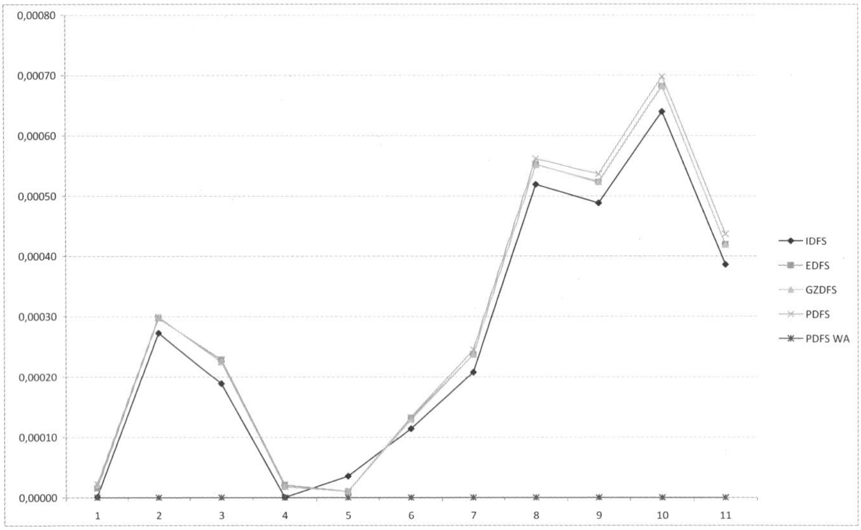

Figure 3. Relative error of $E(M_j^*(T_j))$ with different methods with respect to $M_j^*(0)$.

Table 8. Value at the zero time of domestic caplets ATM that mature at T_j. The Black values and the ones obtained from the different Drift-Free Simulation methods without and with adjustment are presented.

j	BLACK	IDFS	IDFS WA	EDFS	EDFS WA	GZDFS	GZDFS WA	PDFS	PDFS WA
1	0,003212	0,003423	0,003429	0,003402	0,003408	0,003220	0,003219	0,003220	0,003219
2	0,007193	0,007492	0,007464	0,007449	0,007420	0,007242	0,007202	0,007242	0,007202
3	0,008495	0,008665	0,008673	0,008642	0,008649	0,008500	0,008506	0,008500	0,008506
4	0,009209	0,009361	0,009344	0,009351	0,009334	0,009240	0,009215	0,009240	0,009215
5	0,009905	0,010022	0,010014	0,010031	0,010023	0,009930	0,009914	0,009897	0,009897
6	0,010112	0,010234	0,010209	0,010254	0,010230	0,010167	0,010128	0,010138	0,010112
7	0,009739	0,009840	0,009806	0,009864	0,009832	0,009792	0,009746	0,009768	0,009733
8	0,009581	0,009611	0,009630	0,009631	0,009650	0,009571	0,009586	0,009557	0,009577
9	0,009217	0,009320	0,009272	0,009341	0,009294	0,009293	0,009226	0,009278	0,009217
10	0,008984	0,009024	0,009015	0,009040	0,009032	0,008999	0,008984	0,008989	0,008976

where $X_{kl}(z)$ is given by (61).

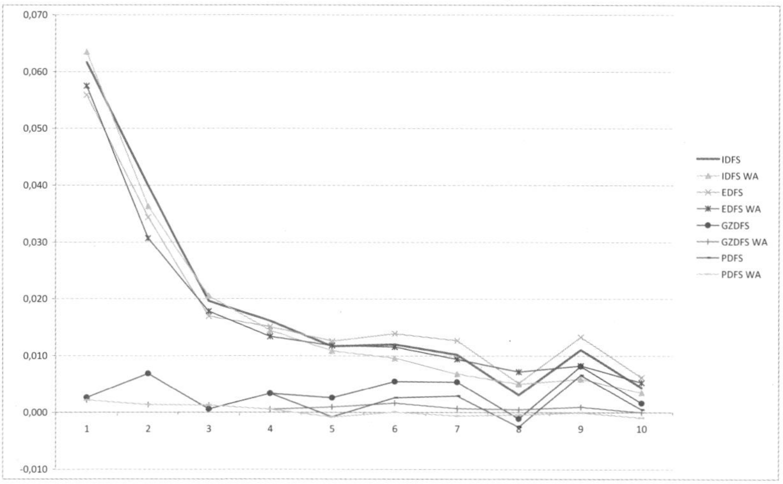

Figure 4. Relative error with respect to Black's formula for caplets in the domestic currency with different methods.

Table 9. Value at the zero time of foreign caplets ATM that mature at T_j. The Black values and the ones obtained from the different Drift-Free Simulation methods without and with adjustment are presented.

j	BLACK	IDFS	IDFS WA	EDFS	EDFS WA	GZDFS	GZDFS WA	PDFS	PDFS WA
1	0,006541	0,006804	0,006802	0,006727	0,006724	0,006558	0,006552	0,006558	0,006552
2	0,008308	0,008512	0,008515	0,008439	0,008442	0,008327	0,008327	0,008327	0,008327
3	0,009033	0,009159	0,009171	0,009105	0,009114	0,009023	0,009037	0,009023	0,009037
4	0,009328	0,009410	0,009425	0,009362	0,009377	0,009299	0,009319	0,009299	0,009319
5	0,009197	0,009268	0,009282	0,009236	0,009248	0,009195	0,009205	0,009184	0,009201
6	0,009294	0,009371	0,009368	0,009345	0,009341	0,009306	0,009303	0,009298	0,009293
7	0,009013	0,009046	0,009063	0,009027	0,009043	0,008997	0,009015	0,008990	0,009009
8	0,008997	0,009067	0,009061	0,009051	0,009043	0,009025	0,009014	0,009017	0,009011
9	0,008841	0,008883	0,008884	0,008868	0,008869	0,008845	0,008846	0,008838	0,008839
10	0,008692	0,008733	0,008729	0,008725	0,008719	0,008706	0,008699	0,008700	0,008692

Proof: First, we calculate the dynamics of the process $L^*_{l-1}(t)\frac{B_k(t)}{B_l(t)}$ under Q^l. For this purpose, we take into account that

$$\frac{dL^*_{l-1}(t)}{L^*_{l-1}(t)} = -\gamma^*_{l-1,l}\sigma_l^{TC}(t)\sigma^*_{l-1}(t)dt + \sigma^*_{l-1}(t)dW^*_{l-1}(t), \tag{77}$$

under Q^l.

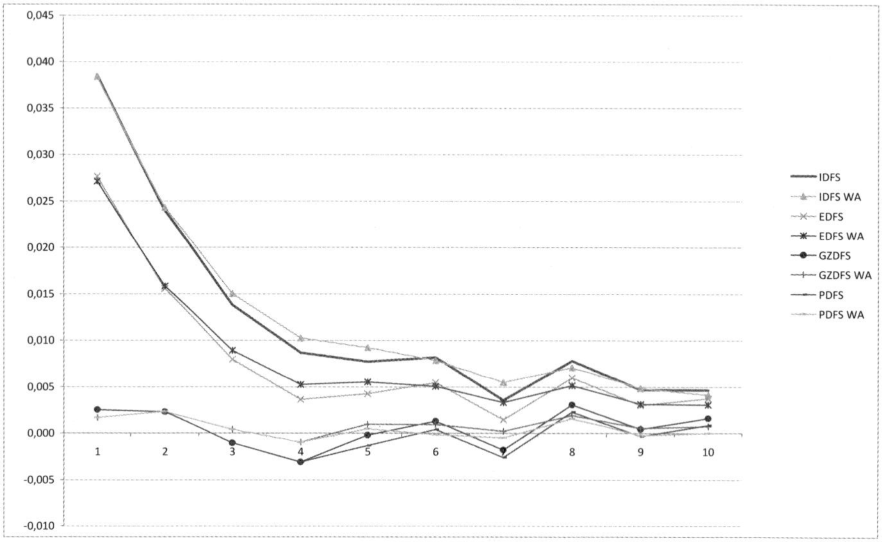

Figure 5. Relative error with respect to Black's formula for caplets in the foreign currency with different methods.

Table 10. Value at time $t = 0$ of Cases 1 and 2 of Quanto Caplets/Floorlets that mature at T_5. The market values and the ones obtained from Drift-Free Simulation methods are presented.

CASE	MARKET	IDFS	IDFS WA	EDFS	EDFS WA	GDFS	GDFS WA	PDFS	PDFS WA
Caplet 1	0,005559	0,005611	0,005620	0,005573	0,005580	0,005542	0,005549	0,005537	0,005548
Caplet 2	0,015901	0,016884	0,016870	0,016897	0,016884	0,016793	0,016765	0,016713	0,016712
Floorlet 1	0,006611	0,006724	0,006715	0,006675	0,006668	0,006641	0,006636	0,006638	0,006630
Floorlet 2	0,014157	0,014069	0,014084	0,014083	0,014097	0,013938	0,013951	0,013924	0,013925

On the other hand, for $l = j+1, \ldots, j+s$, we introduce the notation

$$N_l^k(t) = \frac{B_k(t)}{B_l(t)}. \tag{78}$$

Next, we can proceed as to obtain (15), although with $l = n+1$ and $k = j$, and we get

$$\frac{dN_l^k(t)}{N_l^k(t)} = \varepsilon_{l,k} \sum_{m \in \theta_{l,k}} \sigma_m \left(1 - \frac{N_l^{m+1}(t)}{N_l^m(t)}\right) dW_m(t) \tag{79}$$

Moreover, by taking into account that

$$\sigma_m(t)\left(1-\frac{N_l^{m+1}(t)}{N_l^m(t)}\right)=\frac{\sigma_m(t)\delta_m L_m(t)}{1+\delta_m L_m(t)}, \tag{80}$$

then we have

$$\frac{dNl^k(t)}{N_l^k(t)}=\varepsilon_{l,k}\sum_{m\in\Theta_{l,k}}\frac{\sigma_m(t)\delta_m L_m(t)}{1+\delta_m L_m(t)}dW_m(t) \tag{81}$$

under Q^l.

Notice that the volatility of the martingale N_l^k results to be stochastic. We avoid this complication by freezing that term to its value at time $t=0$:

$$\frac{dNl^k(t)}{N_l^k(t)}\approx\varepsilon_{l,k}\sum_{m\in\Theta_{l,k}}\frac{\sigma_m(0)\delta_m L_m(0)}{1+\delta_m L_m(0)}dW_m(t). \tag{82}$$

Then, by Ito's Lemma

$$\frac{d\left(L_{l-1}^*(t)\frac{B_k(t)}{B_l(t)}\right)}{L_{l-1}^*(t)\frac{B_k(t)}{B_l(t)}}$$

$$\approx\left(-\gamma_{l-1,l}^*\sigma_l^{TC}(t)\sigma_{l-1}^*(t)+\sigma_{l-1}^*(t)\varepsilon_{l,k}\sum_{m\in\Theta_{l,k}}\frac{\sigma_m(0)\delta_m L_m(0)}{1+\delta_m L_m(0)}\rho_{m,l-1}\right)dt$$

$$+\sigma_{l-1}^*dW_{l-1}^*(t)+\varepsilon_{l,k}\sum_{m\in\Theta_{l,k}}\frac{\sigma_m(0)\delta_m L_m(0)}{1+\delta_m L_m(0)}dW_m(t). \tag{83}$$

Therefore, we get (76) and the proof is concluded.

□

Next, by using the previous Proposition in (75), we have

$$E_{Q^{A_{j+1,j+s}}}[L_{l-1}^*(T_j)|\mathcal{F}_t]$$

$$\approx\frac{B_l(t)}{A_{j+1,j+s}(t)}\sum_{k=j+1}^{j+s}\delta_k L_{l-1}^*(t)\frac{B_k(t)}{B_l(t)}\exp\left(\int_t^{T_i}X_{kl}(z)dz\right), \tag{84}$$

and by introducing (84) in (72), we get

$$\int_t^{T_j}\mu_{j,j+s}^{S^*}(z)dz\approx$$

$$\ln\left(\frac{\sum_{l=j+1}^{j+s}\omega_{j+1,j+s^*}^l(0)\frac{B_l(t)}{A_{j+1,j+s}(t)}\sum_{k=j+1}^{j+s}\delta_k L_{l-1}^*(t)\frac{B_k(t)}{B_l(t)}\exp\left(\int_t^{T_j}X_{kl}(z)dz\right)}{S_{j,j+s}^*(t)}\right), \tag{85}$$

or, equivalently

$$\int_t^{T_j} \mu^{S^*}_{j,j+s}(z)dz \approx$$
$$\ln\left(\frac{\sum_{l=j+1}^{j+s}\left[\omega^l_{j+1,j+s^*}(0)L^*_{l-1}(t)\sum_{k=j+1}^{j+s}\omega^k_{j+1,j+s}(t)\exp\left(\int_t^{T_j}X_{kl}(z)dz\right)\right]}{S^*_{j,j+s}(t)}\right), \tag{86}$$

where $X_{kl}(z)$ is given by (61).

Acknowledgements

This work has been partially founded by Spanish MCIIN (Project MTM2010–21135–C02-01) and by Xunta de Galicia (Project INCITE09105339PR and Ayuda CN2011/004, cofounded with FEDER founds).

Bibliography

[1] Aubert, G., Fruchard, E. and Sang Fung, T. (2004). BGM numéraire alignment at will, *Risk International*, pp. 40-43.

[2] Biggs, N. (1997). International Finance, in L. W. Beineke and R. J. Wilson (eds.), *Graph Connections. Relationships between Graph Theory and other Areas of Mathematics* (Claredon Press, Oxford), pp. 261–279.

[3] Brace, A., Dun, T. and Barton, G. (2001). Towards a Central Interest Rate Model, in E. Jouini, J. Cvitanic and M. Musiela (eds.), *Handbooks in Mathematical Finance: Topics in Option Pricing, Interest Rates and Risk Management* (Cambridge University Press), pp.278-313.

[4] Beveridge, C., Denson, N. and Joshi, M. (2009). Comparing discretizations of the LIBOR market model in the spot measures, *Australian Actuarial Journal* **15**, pp. 231-253.

[5] Brace, A., Gatarek, D. and Musiela, M. (1997). The Market Model of Interest Rate Dynamics, *Mathematical Finance* **7**, 2, pp. 127-155 .

[6] Brigo, D. and Mercurio, F. (2007). Interest Rate Models: Theory and Practice (Springer Finance).

[7] Brigo, D. and Liinev, J. (2003). On the distributional distance between the LIBOR and the Swap market models.

[8] Duffy, D. V. (2008). Financial Instrument Pricing Using C++ (The Wiley Finance Series).

[9] Fernández, J. L., Nogueiras, M. R., Pou, M. and Vázquez, C. (2012). A new parameterization for the drift-free simulation in the LIBOR Market Model. (Preprint submitted for publication).

[10] Gallucio, S., Ly, J. M., Huang, Z. and Scaillet, O. (2007). Theory and calibration of swap market models, *Mathematical Finance* **17**, 1, pp. 111-141.

[11] Glasserman, P. (2004). Monte Carlo methods in financial engineering (Springer, New York).

[12] Glasserman, P. and Zhao, X. (2000). Arbitrage-free discretization of lognormal forward LIBOR and swap rate models, *Finance and Stochatics* **4**, pp. 35-68.

[13] Jamshidian, F. (1997). LIBOR and swap market models and measures, *Finance Stochastics* **1**, pp. 293-330.

[14] Kloeden, P. E. and Platen, E. (1999) Numerical Solution of Stochastic Differential Equations (Springer).

[15] Mercurio, F. (2005). Pricing Inflation-Indexed Derivatives, *Quantitative Finance*, **5**, 3, pp. 289-302.

[16] Mikkelsen, P. (2001). Cross-currency LIBOR market models, *CAF-Working Paper Series* **85** (Denmark).

[17] Miltersen, K., Sandmann, K. and Sondermann, D. (1997). Closed Form Solutions for Term Structure Derivatives with Lognormal Interest Rate, *Journal of Finance* **52**, 1, pp. 409-430.

[18] Press, W. H., Teukolsky, S. A., Vetterling, W. T. and Flannery, B. P. (1992). Numerical Recipes in C. The Art of Scientific Computing (Cambridge University Press).

[19] Rapisarda, F., Brigo, D. and Mercurio, F. (2007). Parameterizing correlations: a geometric interpretation, *IMA Journal of Management Matematics* **18**, pp. 55-73.

[20] Rebonato, R. (1998) Interest Rate Option Models (Wiley, Chichester).

[21] Scholgl, E. (1999). A multicurrency extension of the lognormal interest rate models, Working paper, *School of Mathematical Sciences* (University of Technology, Sydney, Australia).

Chapter 11

Application of simplest random walk algorithms for pricing barrier options

M. Krivko and M.V. Tretyakov

Department of Mathematics, University of Leicester, Leicester LE1 7RH, UK

School of Mathematical Sciences, University of Nottingham, Nottingham NG7 2RD, UK

Abstract The effectiveness of the first-order algorithm from [Milstein, Tretyakov. Theory Prob. Appl. 47 (2002), 53-68] is demonstrated in application to barrier option pricing. This algorithm uses the weak Euler approximation far from barriers and a special construction motivated by linear interpolation of the price near barriers. It is easy to implement and is universal: it can be applied to various structures of the contracts including derivatives on multi-asset correlated underlyings and can deal with various type of barriers. In contrast to the Brownian bridge techniques currently commonly used for pricing barrier options, the algorithm tested here does not require knowledge of trigger probabilities nor their estimates. We illustrate this algorithm via pricing a barrier caplet, barrier trigger swap and barrier swaption.
AMS 2000 subject classification. 65C30, 60H30, 91B28, 91B70.
Keywords. Barrier options, exotic derivatives, weak approximation of stochastic differential equations in bounded domains, Monte Carlo technique, the Dirichlet problem for parabolic partial differential equations, interest rate derivatives.

1. Introduction

Barrier option contracts are among the most traded and oldest exotic derivatives. They accommodate an investors' view about the future market behavior more closely and they are generally cheaper than the corresponding plain vanilla options. Typically, a barrier option is activated (knocked in) or deactivated (knocked out) depending on whether a vector of underlying assets or their functional has crossed a specified barrier level, which itself can be a functional of the underlying assets. Due to its attractive features, barrier optionality has been introduced in a wide range of derivatives products. In the context of credit risk, the event of default of some reference entity can be modelled as a lower barrier on the equity of the entity. This is the key idea in structural models for pricing popular credit instruments such as credit default swaps (CDSs) and credit default obligations (CDOs). More recently, barrier optionality has been also used to model contingent convertible (CoCo) bonds which were introduced to provide financial institutions with sufficient capital in times of distress or systemic risk.

Closed form solutions for barrier option prices can be obtained only in some particular settings. For instance, they are available in the case of a single underlying asset and a constant barrier within the standard Black-Scholes setup (see, e.g. [10; 19; 4; 18; 21]). In products involving a large number of dependent assets numerical approximation for pricing and hedging barrier options is usually inevitable and this can be a challenging problem.

In this paper we assume that underlying assets are modelled via multidimensional stochastic differential equations (SDEs) and we consider European-type barrier options. The arbitrage price $u(t,x)$ of such an option solves the Dirichlet problem for a linear parabolic partial differential equation. Finding this price numerically requires efficient weak approximations of diffusions in a bounded domain.

"Ordinary" numerical methods for SDEs in $\mathbf{R}^d$ on a finite time interval are based on a time discretization [9; 11; 16]. They ensure smallness of time increments at each step, but might not ensure smallness of space increments. In [13; 12; 15] (see also [16]) a number of weak-sense approximations for SDEs in a bounded domain were proposed, in which space increments are controlled at each step so that the constructed approximation belongs to the bounded domain. Approximations of [12] (see also [16]) are based on adaptive control of a time step of numerical integration of the SDEs. A step is chosen such that (of course, aside of reaching a required accuracy) the next state of a Markov chain approximating in the weak sense the SDEs' solution remains in the bounded domain with probability one. This leads to a decrease of the time step when the chain is close to the boundary of the domain. The chain is stopped in a narrow zone near the boundary so that values of the solution $u(t,x)$ (i.e., the option price at time t and underlyings' price x) in this zone can be approximated accurately by the known values of the function φ on the boundary (i.e., the value of the option at the barrier). Another type of approximation was proposed in [15] (see also [16]). In the algorithm of weak order one from [15] the step of numerical integration of the SDEs is constant for points belonging to a certain time layer $t = t_k$. Far from the boundary, a Markov chain approximating the SDEs' solution is constructed using the weak Euler scheme (i.e., using discrete random variables for approximating the Wiener increments). When a point is close to the boundary, we make an intermediate (auxiliary) step of the random walk, which preserves the point in the time layer $t = t_k$. On this auxiliary step we "flip a coin" to decide whether to terminate the chain on the boundary or jump back in the domain and continue the random walk. The construction of this step is based on the idea of linear interpolation for the solution $u(t,x)$. The algorithm is efficient and very easy to implement. Its simpler version of order 1/2 in the weak sense is also presented in [15; 16]. In Section 2 we recall these two algorithms from [15; 16].

As far as we know, despite simplicity of these weak schemes, they have not been used in financial applications. In this article we try to fill this gap and illustrate their applicability to pricing barrier options.

Currently, the popular numerical approach for pricing barrier options exploits the Brownian bridge technique [1; 2; 7; 20; 3] (also see [8] for a review and the references therein). It is based on simulation of a one-dimensional Brownian bridge extremum between time steps and computing analytically the associated probability of exiting the spatial domain for each time interval of the partition. It was proved in [7] that this approach realized along with the Euler scheme (which uses Gaussian random variables for simulating Wiener increments) results in an approximation of weak order one. The Brownian bridge technique relies on analytical formulas for trigger probabilities and can run into difficulties in the case of multiple barriers and/or correlated structure of the underlyings when there are no closed formulas for the distribution of extremum. Though some extensions to these more general and not uncommon problems have been considered, e.g. in [8; 20].

In contrast the simplest random walk algorithm of [15] displays a high degree of flexibility and can be applied to various structures of the contracts including derivatives on multi-asset correlated underlyings and can deal with various types of barriers, e.g. single, double and time dependent barriers. In comparison with the Brownian bridge techniques the method of [15] does not require the knowledge of the trigger probabilities or their estimates.

In Sections 4-6 we present three examples on how to apply the algorithm from [15] for valuation of barrier options. These contracts cover the most common types of barrier options. In the first example (Section 4), we consider an algorithm for barrier derivatives where the payoff depends on a single underlying. As an illustration, we deal with pricing a barrier caplet and provide ready-for-implementation procedure which can be easily applied to similar other contracts. The second example (Section 5) is devoted to multi-asset options with barriers imposed on all or some of the correlated underlying assets. We illustrate this case by pricing a trigger swap. The last example (Section 6) is barrier contracts written on an asset that can be expressed through some other multi-asset underlying. As a specific case, we consider valuation of a barrier swaption under the LIBOR market model (LMM). Though all three examples are for fixed-income markets, the considered algorithms are also directly applicable to barrier options in other markets.

2. Simplest random walks for stopped diffusions

Let $(\Omega, \mathcal{F}, \mathbb{P})$ be a complete probability space, $\mathcal{F}_t$, $0 \leq t \leq T$, be a filtration satisfying the usual hypotheses, $(w_t, \mathcal{F}_t)$ be an r-dimensional standard Wiener process. Let G be a bounded domain in $\mathbf{R}^d$ and $Q = [t_0, T) \times G$ be a cylinder in $\mathbf{R}^{d+1}$, $\Gamma = \bar{Q} \backslash Q$ be the part of the cylinder's boundary consisting of the upper base and lateral surface. Price of barrier options with underlying modelled by a diffusion process can usually be expressed as

$$u(t,x) = E\left[\varphi(\tau, X_{t,x}(\tau))Y_{t,x,1}(\tau) + Z_{t,x,1,0}(\tau)\right], \tag{1}$$

where $X_{t,x}(s)$, $Y_{t,x,y}(s)$, $Z_{t,x,y,z}(s)$, $s \geq t$, is the solution of the Cauchy problem for the system of SDEs:

$$dX = (b(s,X) - \sigma(s,X)\mu(s,X))\,ds + \sigma(s,X)\,dw(s), \quad X(t) = x, \tag{2}$$

$$dY = c(s,X)Y\,ds + \mu^{\mathsf{T}}(s,X)Y\,dw(s), \quad Y(t) = y, \tag{3}$$

$$dZ = g(s,X)Y\,ds + F^{\mathsf{T}}(s,X)Y\,dw(s), \quad Z(t) = z, \tag{4}$$

$(t,x) \in Q$, and $\tau = \tau_{t,x}$ is the first exit time of the trajectory $(s, X_{t,x}(s))$ to the boundary Γ. In (2)-(4), $b(s,x)$ is a d-dimensional column-vector, the $\sigma(s,x)$ is a $d \times r$ matrix, $\mu(s,x)$ and $F(s,x)$ are r-dimensional vectors, and $Y(s)$, $Z(s)$, $c(s,X)$ and $g(s,X)$ are scalars. We assume that all the coefficients in (2)-(4), the function $\varphi(t,x)$ defined on Γ and the boundary ∂G of the space domain G satisfy some regularity conditions.

We note that the value of the expectation $u(t,x)$ in (1) does not depend on a choice of functions $\mu(s,x)$ and $F(s,x)$. This flexibility can be used for reducing variance of the random variable $\varphi(\tau, X_{t,x}(\tau))Y_{t,x,1}(\tau) + Z_{t,x,1,0}(\tau)$ with the aim of reducing the statistical error in computing $u(t,x)$ via the Monte Carlo technique [11; 16]. For instance, if μ and F are such that

$$\sum_{i=1}^{d} \sigma^{ij} \frac{\partial u}{\partial x^i} + u\mu^j + F^j = 0, \; j = 1, \ldots, r, \tag{5}$$

then $Var[\varphi(\tau, X_{t,x}(\tau))Y_{t,x,1}(\tau) + Z_{t,x,1,0}(\tau)] = 0$ and $\varphi(\tau, X_{t,x}(\tau))Y_{t,x,1}(\tau) + Z_{t,x,1,0}(\tau) \equiv u(t,x)$ [14; 16]. As we see from (5), optimal μ and F require knowledge of the the solution $u(t,x)$ (i.e., the option price) to the considered problem and its derivatives (i.e., deltas) which is impractical. However, instead of the exact $u(t,x)$ in (5), one can use its approximation (e.g. price and deltas for a related option for which the closed-form solution is known) to find some suboptimal μ and F which can lead to variance reduction [6; 17].

To simulate (1)-(4), we need an approximation of the trajectory $(s, X(s))$ which satisfies some restrictions related to its nonexit from the domain $\bar{Q}$. Let us recall two algorithms for (1)-(4) from [15; 16].

We apply the weak explicit Euler approximation with the simplest simulation of noise to the system (2)-(4):

$$X_{t,x}(t+h) \approx X = x + h\,(b(t,x) - \sigma(t,x)\,\mu(t,x)) + h^{1/2}\sigma(t,x)\,\xi\,, \tag{6}$$

$$Y_{t,x,y}(t+h) \approx Y = y + hc(t,x)\,y + h^{1/2}\mu^{\mathsf{T}}(t,x)\,y\,\xi\,, \tag{7}$$

$$Z_{t,x,y,z}(t+h) \approx Z = z + hg(t,x)\,y + h^{1/2}F^{\mathsf{T}}(t,x)\,y\,\xi\,, \tag{8}$$

where $h > 0$ is a time-discretization step (a sufficiently small number), $\xi = (\xi^1, \ldots, \xi^r)^{\mathsf{T}}$, ξ^i, $i = 1, \ldots, r$, are mutually independent random variables taking the values ± 1 with probability $1/2$. Clearly, the random vector X takes 2^r different values.

Introduce the set of points close to the boundary (a boundary zone) $S_{t,h} \subset \bar{G}$ on the layer t : we say that $x \in S_{t,h}$ if at least one of the 2^r values of the vector X is outside $\bar{G}$. It is not difficult to see that due to compactness of $\bar{Q}$ there is a constant $\lambda > 0$ such that if the distance from $x \in G$ to the boundary ∂G is equal to or greater than $\lambda\sqrt{h}$ then x is outside the boundary zone and, therefore, for such x all the realizations of the random variable X belong to $\bar{G}$.

Since we should impose restrictions on an approximation of the system (2) so that it does exit from the domain $\bar{G}$, the formulas (6)-(8) can be used only for the points $x \in \bar{G}\backslash S_{t,h}$ on the layer t, and a special construction is required for points from the boundary zone. Let $x \in S_{t,h}$. Denote by $x^\pi \in \partial G$ the projection of the point x on the boundary of the domain G (the projection is unique assuming that h is sufficiently small and ∂G is smooth) and by $n(x^\pi)$ the unit vector of internal normal to ∂G at x^π. Introduce the random vector $X^\pi_{x,h}$ taking two values x^π and $x + h^{1/2}\lambda n(x^\pi)$ with probabilities $p = p_{x,h}$ and $q = q_{x,h} = 1 - p_{x,h}$, respectively, where

$$p_{x,h} = \frac{h^{1/2}\lambda}{|x + h^{1/2}\lambda n(x^\pi) - x^\pi|} \,. \tag{9}$$

This construction is motivated by the following observation [15]. If $v(x)$ is a twice continuously differentiable function with the domain of definition $\bar{G}$, then an approximation of $v(x)$ by the expectation $Ev(X^\pi_{x,h})$ corresponds to linear interpolation and

$$v(x) = Ev(X^\pi_{x,h}) + O(h) = pv(x^\pi) + qv(x + h^{1/2}\lambda n(x^\pi)) + O(h) \,. \tag{10}$$

We emphasize that the second value $x + h^{1/2}\lambda n(x^\pi)$ does not belong to the boundary zone. We also note that p is always greater than $1/2$ (since the distance from x to ∂G is less than $h^{1/2}\lambda$) and that if $x \in \partial G$ then $p = 1$ (since in this case $x^\pi = x$).

Let a point $(t_0, x_0) \in Q$. We would like to find the value $u(t_0, x_0)$. Introduce a discretization of the interval $[t_0, T]$, for definiteness the equidistant one:

$$t_0 < t_1 < \cdots < t_M = T, \quad h := (T - t_0)/M.$$

To approximate the solution of the system (2), we construct a Markov chain (t_k, X_k) which stops when it reaches the boundary Γ at a random step $\varkappa \leq M$. The resulting algorithm can be formulated as Algorithm 2.1 given below.

It is proved in [15] (see also [16]) that under appropriate regularity assumptions on the coefficients of (2)-(4), the boundary condition $\varphi(t, x)$ in (1) and on the boundary ∂G Algorithm 2.1 converges with weak order one.

The next algorithm is obtained by a simplification of Algorithm 2.1: as soon as X_k gets into the boundary domain $S_{t_k,h}$, the random walk terminates, i.e., $\varkappa = k$,

Algorithm 2.1 Algorithm of weak order one for (1)-(4).

STEP 0. $X'_0 = x_0,\ Y_0 = 1,\ Z_0 = 0,\ k = 0.$

STEP 1. If $X'_k \notin S_{t_k,h}$, then $X_k = X'_k$ and go to STEP 3.
If $X'_k \in S_{t_k,h}$, then either $X_k = X'^{\pi}_k$ with probability $p_{X'_k,h}$ or $X_k = X'_k + h^{1/2}\lambda n(X'^{\pi}_k)$ with probability $q_{X'_k,h}$.

STEP 2. If $X_k = X'^{\pi}_k$, then STOP and $\varkappa = k$, $X_\varkappa = X'^{\pi}_k,\ Y_\varkappa = Y_k,\ Z_\varkappa = Z_k$.

STEP 3. Simulate ξ_k and find $X'_{k+1},\ Y_{k+1},\ Z_{k+1}$ according to (6)-(8) for $t = t_k,\ x = X_k,\ y = Y_k,\ z = Z_k$, $\xi = \xi_k$.

STEP 4. If $k+1 = M$, STOP and $\varkappa = M$, $X_\varkappa = X'_M$, $Y_\varkappa = Y_M$, $Z_\varkappa = Z_M$, otherwise $k = k+1$ and return to STEP 1.

Algorithm 2.2 Algorithm of weak order 1/2 for (1)-(4).

STEP 0. $X_0 = x_0,\ Y_0 = 1,\ Z_0 = 0,\ k = 0.$

STEP 1. If $X_k \notin S_{t_k,h}$, then go to STEP 2.
If $X_k \in S_{t_k,h}$, then STOP and $\varkappa = k$, $\bar{X}_\varkappa = X^{\pi}_k$, $Y_\varkappa = Y_k,\ Z_\varkappa = Z_k$.

STEP 2. Simulate ξ_k and find $X_{k+1},\ Y_{k+1},\ Z_{k+1}$ according to (6)-(8) for $t = t_k,\ x = X_k,\ y = Y_k$, $z = Z_k,\ \xi = \xi_k$.

STEP 3. If $k+1 = M$, STOP and $\varkappa = M$, $\bar{X}_\varkappa = X_M$, $Y_\varkappa = Y_M$, $Z_\varkappa = Z_M$, otherwise $k = k+1$ and return to STEP 1.

and $\bar{X}_\varkappa = X^{\pi}_k$, $Y_\varkappa = Y_k$, $Z_\varkappa = Z_k$ is taken as the final state of the Markov chain. The resulting algorithm takes the form of Algorithm 2.2.

It is proved in [15; 16] that under appropriate regularity assumptions on the coefficients of (2)-(4), the boundary condition $\varphi(t,x)$ in (1) and on the boundary ∂G Algorithm 2.2 converges with weak order 1/2. We note that in one-dimension (i.e., in the case of a single underlying) Algorithm 2.2 is analogous to pricing barrier options by binary trees (see, e.g. [5]).

3. LIBOR Market Model

We will now assume that there exists an arbitrage-free market with continuous and frictionless trading taking place inside a finite time horizon $[t_0, t^*]$.

Among the most important benchmark interest rates is the London Interbank

Offered Rate (LIBOR). It is based on simple (or simply compounded) interest. The forward LIBOR rate $L(t,T,T+\delta)$ is the rate set at time t for the interval $[T,T+\delta]$, $t \le T$. If we enter into a contract at time t to borrow one unit at time T and repay it with interest at time $T+\delta$, the interest due will be $\delta L(t,T,T+\delta)$.

A simple replication argument (see, e.g., [4]) relates LIBOR rates and bond prices via the following identity

$$L(t,T,T+\delta) = \frac{1}{\delta}\left(\frac{P(t,T)}{P(t,T+\delta)} - 1\right), \tag{11}$$

where $P(t,T)$ is the price at time $t \le T$ of a default-free zero coupon bond.

For simplicity, we fix an equidistant finite set of maturities or tenor dates

$$T_0 < \cdots < T_N = T^*, \quad T_i = i\delta, \quad i = 0,\ldots,N, \tag{12}$$

where

$$\delta = (T^* - T_0)/N,$$

denotes the fixed length of the interval between tenor dates.

Let us introduce a simplified notation for the time t forward LIBOR rate for the accrual period $[T_i, T_{i+1}]$ and the payment at T_{i+1}:

$$\begin{aligned} L^i(t) \ &: = L(t,T_i,T_{i+1}), \\ & t_0 \le t \le t^* \wedge T_i, \quad t_0 < T_i \le T^*, \quad i = 0,\ldots,N-1. \end{aligned}$$

In the case of LIBOR Market Model (LMM) the arbitrage-free dynamics of $L^i(t)$ under the forward measure $\mathrm{Q}^{T_{k+1}}$ associated with the numeraire $P(t,T_{k+1})$ can be written as the following system of SDEs (see, e.g. [4; 18; 21]):

$$\frac{dL^i(t)}{L^i(t)} = \begin{cases} \sigma_i(t) \sum\limits_{j=k+1}^{i} \frac{\delta L^j(t)}{1+\delta L^j(t)} \rho_{i,j}\sigma_j(t)dt + \sigma_i(t)dW_i^{T_{k+1}}(t), \ i > k, \ t \le T_k, \\ \sigma_i(t)dW_i^{T_{k+1}}(t), \ i = k, \ t \le T_i, \\ -\sigma_i(t) \sum\limits_{j=i+1}^{k} \frac{\delta L^j(t)}{1+\delta L^j(t)} \rho_{i,j}\sigma_j(t)dt + \sigma_i(t)dW_i^{T_{k+1}}(t), \ i < k, \ t \le T_i, \end{cases} \tag{13}$$

where $W^{T_{k+1}} = (W_0^{T_{k+1}},\ldots,W_{N-1}^{T_{k+1}})^\top$ is an N-dimensional correlated Wiener process defined on a filtered probability space $\left(\Omega, \mathcal{F}, \{\mathcal{F}_t\}_{t_0 \le t \le t^*}, \mathrm{Q}^{T_{k+1}}\right)$; the instantaneous correlation structure is defined as

$$E\left[W_i^{T_{k+1}}(t) W_j^{T_{k+1}}(t)\right] = \rho_{i,j}, \quad i,j = 0,\ldots,N-1; \tag{14}$$

and $\sigma_i(t)$, $i = 0,\ldots,N-1$, are instantaneous volatilities which we assume here to be deterministic bounded functions.

Let ρ be the instantaneous correlation matrix with elements $\rho_{i,j}$. To simulate the correlated Wiener processes, we will use the pseudo-root of the correlation matrix ρ defined via the equation

$$\rho = UU^\top, \tag{15}$$

where U is an upper triangular matrix with components $U_{i,j}$, $i, j = 0, \ldots, N-1$. Using U and introducing N-dimensional standard Wiener process, one can re-write (13) in the form of (2).

In what follows we will assume that the current time t_0 is set to 0. For convenience, we also assume a unit notional value of all the contracts we introduce below. In our numerical experiments in the next sections we take the correlation function of the form:

$$\rho_{i,j} = \exp(-\beta\, |T_i - T_j|), \; i, j = 0, \ldots, N-1. \tag{16}$$

4. Barrier cap/floor

In this section we consider the Monte Carlo evaluation of barrier options written on a single underlying. We use a knock-out caplet for illustration, though our treatment is rather general and can be used to value different barrier option, for instance European and Parisian barrier options and options with different barriers including single, double and time-dependent barriers, both for fixed-income and equity markets.

An Interest Rate Cap is a security that allows its holder to benefit from low floating rates and be protected from high ones. Similarly, an Interest Rate Floor is an instrument designed to protect from low floating interest rates yet allow the holder to benefit from high rates. Formally, a cap price is obtained by summing up the prices of the underlying caplets, which are call options on successive LIBOR rates. Also, a floor is a strip of floorlets, which are put options on successive LIBOR rates.

A knock-out caplet pays the same payoff as a regular caplet as long as a prescribed barrier rate H is not reached from below by the corresponding LIBOR rate before the option expires. More specifically, the price at time $t \leq T_0$ of a knock-out caplet set at time T_{i-1} with payment date at T_i, $i \geq 1$, with strike K and unit cap nominal value is given by

$$V_{caplet}(t) = \delta P(t, T_{i+1}) E^{Q^{T_{i+1}}} \left[\left(L^i(T_i) - K\right)_+ \chi(\theta > T_i) \Big| \mathcal{F}_t \right], \tag{17}$$

where θ is the first exit time of $L^i(s)$, $s \geq t$, from the interval $G = (0, H)$. Let τ be the first exit time of the space-time diffusion $(s, L^i(s))$, $s \geq t$, from the domain $Q = [t, T_i) \times (0, H)$. Obviously, $\tau = \theta \wedge T_i$.

The dynamics of $L^i(s)$ under $Q^{T_{i+1}}$ is (see (13)):

$$\frac{dL^i(s)}{L^i(s)} = \sigma_i(s) dW_i^{T_{i+1}}(s), \; s \leq T_i. \tag{18}$$

Note that the correlation structure of (13) does not influence the price of the knock-out caplet since it does not depend on the joint dynamics of forward rates.

One can observe that the dynamics (18) coincides with the model of a stock price process under the risk-neutral measure in the case of zero interest rate. This means that by dropping the factor $\delta P(t, T_{i+1})$ in (17), the valuation of European barrier options on equity with zero interest rate and of barrier caplets under the LMM coincide.

In the considered case the price of the barrier caplet has the well-known closed-form solution:

$$\begin{aligned} V_{caplet}(t) &= V_{caplet}(t, L^i(t)) \\ &= \delta P(t, T_{i+1}) \left\{ L^i(t) \left[\Phi(\delta_+(L^i(t)/K, v_i)) - \Phi(\delta_+(L^i(t)/H, v_i)) \right] \right. \\ &\quad -K \left[\Phi(\delta_-(L^i(t)/K, v_i)) - \Phi(\delta_-(L^i(t)/H, v_i)) \right] \\ &\quad -H \left[\Phi(\delta_+(H^2/(KL^i(t)), v_i)) - \Phi(\delta_+(H/L^i(t), v_i)) \right] \\ &\quad \left. +KL^i(t) \left[\Phi(\delta_-(H^2/(KL^i(t)), v_i)) - \Phi(\delta_-(H/L^i(t), v_i)) \right] /H \right\}, \end{aligned} \tag{19}$$

where $\Phi(\cdot)$ denotes the standard normal cumulative distribution function,

$$\delta_+(x, v) = (\ln x + v^2/2)/v, \quad \delta_-(x, v) = (\ln x - v^2/2)/v, \tag{20}$$

and

$$v_i^2 = \int_t^{T_i} (\sigma_i(s))^2 \, ds.$$

This analytical result will be used in our numerical experiments to assess the performance of proposed algorithms. We note that the algorithm presented in this example can be easily extended to a more general model of underlying when the closed-form solution might be not available. In particular, there is no difficulty in including a drift term in the underlying dynamics (see also Sections 5 and 6). In the experiments we simulate

$$\tilde{V}_{caplet}(t) = V_{caplet}(t)/\delta P(t, T_{i+1}), \tag{21}$$

i.e. we drop $\delta P(t, T_{i+1})$ from (17), which does not imply any loss of generality since the caplet price can easily be recovered by multiplying $\tilde{V}_{caplet}(t)$ by the factor $\delta P(t, T_{i+1})$ observable at time t.

4.1. *Algorithm*

To preserve positivity of the LIBOR rate, we simulate the log dynamics corresponding to (18) rather than the LIBOR rate $L^i(t)$ itself. To illustrate the variance reduction technique discussed in Section 2, we complement (18) with the equation (cf. (4)):

$$dZ = F(s, L^i) dW_i^{T_{i+1}}(s), \quad Z(0) = 0, \tag{22}$$

with (see (5) and (21))

$$F(s, L^i) = -\sigma_i(s) \frac{\partial}{\partial L^i} \tilde{V}_{caplet}(s, L^i(s)) \, . \tag{23}$$

We choose a time step $h > 0$ so that $M = T_i/h$ is an integer. We set $\ln L_0^i = L^i(0)$ and $Z_0 = 0$. The weak Euler scheme (6), (8) applied to (18) in the log form and (22) takes the form:

$$\ln L_{k+1}^i = \ln L_k^i - \frac{1}{2}\left(\sigma_i(t_k)\right)^2 h + \sigma_i(t_k)\sqrt{h}\xi_{k+1}, \tag{24}$$

$$Z_{k+1} = Z_k + F(s, L_k^i)\sqrt{h}\xi_{k+1} \ , \tag{25}$$

where ξ_k are independent random variables distributed by the law $P(\xi = \pm 1) = 1/2$.

The boundary zone $S_{t,h}$ required for Algorithms 2.1 and 2.2 is chosen here as

$$S_{t_k,h} = \{L_k^i : \ \ln L_k^i \geq \ln H + \frac{1}{2}\sigma_i^2(t_k)h - \sigma_i(t_k)\sqrt{h}\}, \tag{26}$$

i.e., the condition for $\ln L_{k+1}^i$ to be inside the domain G is

$$\ln L_k^i < \ln H + \frac{1}{2}\sigma_i^2(t_k)h - \sigma_i(t_k)\sqrt{h} \ , \tag{27}$$

and the corresponding λ_k for Algorithm 2.1 is so that

$$\lambda_k\sqrt{h} = -\frac{1}{2}\sigma_i^2(t_k)h + \sigma_i(t_k)\sqrt{h} \ . \tag{28}$$

We note that instead of (26) and (28) we could take the wider boundary zone $S_{t_k,h} = \{L_k^i : \ \ln L_k^i \geq \ln H - \sigma_i(t_k)\sqrt{h}\}$ and correspondingly $\lambda_k = \sigma_i(t_k)$. A wider boundary zone usually leads to a bigger numerical integration error. Here we cannot take a boundary zone narrower than $S_{t_k,h}$ in (26) because it would not ensure that the chain $\ln L_k^i$ belongs to $\bar{G}$.

To realize Algorithm 2.1, we follow the random walk generated by (24) and at each time t_k, we check whether at the next step L_{k+1}^i cannot cross the barrier H, i.e., we check whether the condition (27) holds. If it does, we perform (24)-(25) to find $\ln L_{k+1}^i$, Z_{k+1}. Otherwise, L_k^i has reached the boundary zone $S_{t_k,h}$, where we make the auxiliary step: we either stop the chain at $\ln H$ with probability p :

$$p = \frac{\lambda_k\sqrt{h}}{\ln H - \ln L_k^i + \lambda_k\sqrt{h}}$$

or we kick the current position of the random walk $\ln L_k^i$ back into the domain to the position $\ln L_k^i - \lambda_k\sqrt{h}$ with probability $1-p$ and then carry out (24)-(25) to find $\ln L_{k+1}^i$, Z_{k+1}. If $k+1 = M$, we stop, otherwise we continue with the algorithm. The outcome of simulating each trajectory is a point $(t_\varkappa, \ln L_\varkappa^i, Z_\varkappa)$.

To realize Algorithm 2.2, we also follow the random walk generated by (24), and at each time t_k, we check whether the condition (27) holds. If it does not, L_k^i has reached the boundary zone $S_{t_k,h}$ and we stop the chain at $\ln H$. If it does, we perform (24)-(25) to find $\ln L_{k+1}^i$, Z_{k+1}. If $k+1 = M$, we stop, otherwise we continue with the algorithm. The outcome of simulating each trajectory is again a point $(t_\varkappa, \ln L_\varkappa^i, Z_\varkappa)$.

In the experiments we evaluate the expectation

$$\tilde{V}_{caplet}(0) = E^{Q^{T_{i+1}}}\left[\left(L^i(T_i) - K\right)_+ \chi(\theta > T_i) + Z(\tau)\right] \tag{29}$$
$$\approx E^{Q^{T_{i+1}}}\left[\left(\exp(\ln L^i_\varkappa) - K\right)_+ \chi(\varkappa = M) + Z_\varkappa\right].$$

The approximate equality in (29) is related to the bias due to the numerical approximation. The expectation on the right-hand side is realized via the Monte Carlo technique.

4.2. *Numerical results*

Here we present some results of numerical tests of Algorithms 2.1 and 2.2 for pricing the barrier caplet (29). We use the following parameters in the experiments: $i = 9$, $\delta = 1$, $K = 1\%$, $H = 28\%$, $L^9(0) = 13\%$. The volatility $\sigma_i(t)$ is assumed to be constant at 25% . The exact caplet price $\tilde{V}_{caplet}(0)$ with these parameters evaluated by (19) without the factor $\delta P(0, T_{i+1})$ is 6.57%. In the experiments we did 10^6 Monte Carlo runs. The results are presented in Fig. 1. We see that Algorithm 2.1 is much more accurate than Algorithm 2.2. We also observe "oscillating" convergence which is typical for binary tree methods [5].

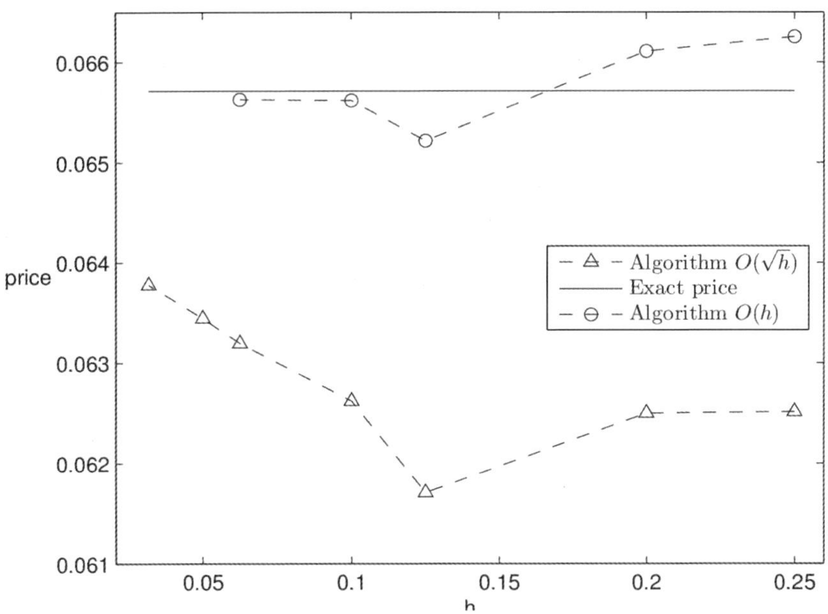

Figure 1. *Barrier caplet price.* Comparison of the results of numerical experiments for the Algorithm 2.1 (Algorithm $O(h)$) and Algorithm 2.2 (Algorithm $O(\sqrt{h})$) and the exact caplet price (solid line) evaluated for $i = 9$, $\delta = 1$, $K = 1\%$, $H = 28\%$, $L^9(0) = 13\%$, $\sigma_i(t) = 25\%$.

Let us also remark on the effect of variance reduction in these experiments. For instance, in Algorithm 2.1 for $h = 0.02$ we got the Monte Carlo error (which we define as half of the size of the confidence interval for the corresponding estimator with probability 0.95) equal to 1.11×10^{-4} in the case of $F = 0$ and 1.55×10^{-5} in the case of the optimal F from (23) (i.e., 100 time speed-up in reaching the same level of the Monte Carlo error). The use of the optimal F does not result in zero Monte Carlo error due to the error of numerical integration.

5. Trigger swap

This example is devoted to evaluation of multi-asset barrier options with barriers on all or some of the correlated underlying assets. We consider a trigger swap as a specific case, though the considered approach can be used to value other multi-asset barrier options, for instance basket options, CDOs and nth-to-default CDSs, and it can also be applied for options with single, double and time-dependent barriers.

A trigger swap is a swap on a floating reference rate that takes effect or terminates when some index rate hits a specified trigger level. Trigger swaps have a number of variations [4; 21]. Here we consider a knock-in version of a payer trigger swap with a fixed rate K whose barrier is continuously monitored. The index and reference rate both coincide with a LIBOR rate. For given trigger levels $H^0, \dots, H^{N-1}$ associated with the LIBOR rates $L^0(t), \dots, L^{N-1}(t)$, the structure of the swap under consideration is expressed as follows. Once one of the the continuously monitored LIBOR rate $L^0(t), \dots, L^{N-1}(t)$ for the first time hits the corresponding trigger level $H^0, \dots, H^{N-1}$ from below, the contract holder enters into the payer swap starting at next tenor date for the remaining time to the last tenor T_{N-1}. More specifically, let θ be the first exit time of $L^0(s), \dots, L^{N-1}(s)$, $s \geq 0$, from the domain $G = \left(0, H^0\right) \times \cdots \times \left(0, H^{N-1}\right)$, τ be the first exit time of the space-time diffusion $(s, L^0(s), \dots, L^{N-1}(s))$ from the domain $Q = [0, T_{N-1}) \times G$ (clearly $\tau = \theta \wedge T_{N-1}$), and $T_{\varrho(\tau)}$ be the closest tenor date T_i to τ from the right, i.e., $\varrho(t)$ is defined as

$$\varrho(t) = \min\left\{i,\ i = 0, 1, \dots, N-1 : t \leq T_i\right\}.$$

If $\theta \leq T_{N-1}$, then at a tenor date $T_{\varrho(\tau)}$ the contract holder enters into the contract according to which the holder pays fixed payments of δK and receives floating payments of $\delta L^{i-1}(T_{i-1})$ at the coupon dates T_i, $i = \varrho(\tau) + 1, \dots, N$; otherwise the contract expires worthless.

The value of this trigger swap at time $t = 0$ under the forward measure Q^{T_N} is given by

$$V_{trswap}(0) = P(0, T_N) E^{\mathrm{Q}^{T_N}} \left[\frac{1}{P(T_{\varrho(\tau)}, T_N)} \right. \tag{30}$$

$$\left. \times \left(1 - P(T_{\varrho(\tau)}, T_N) - K\delta \sum_{i=\varrho(\tau)+1}^{N} P(T_{\varrho(\tau)}, T_i) \right) \chi(\theta \leq T_{N-1}) \right],$$

or in terms of the LIBOR rates

$$V_{trswap}(0) = P(0,T_N)E^{Q^{T_N}}\left[\left(\prod_{j=\varrho(\tau)}^{N-1}(1+\delta L^j(T_{\varrho(\tau)})) \right.\right. \tag{31}$$

$$\left.\left. -K\delta \sum_{i=\varrho(\tau)+1}^{N-1}\prod_{j=i}^{N-1}(1+\delta L^j(T_{\varrho(\tau)})) - 1\right)\chi(\theta \leq T_{N-1})\right].$$

In order to price this contract by the Monte Carlo technique, we need to generate paths for the vector $L^{\varrho(t)}(t), \ldots, L^{N-1}(t)$. This means that the size of the vector of LIBOR rates which we need to simulate decreases over time. The dynamics of $L^i(t)$ under Q^{T_N} are described by the SDEs (cf. (13)):

$$\frac{dL^i(t)}{L^i(t)} = -\sigma_i(t)\sum_{j=i+1}^{N-1}\frac{\delta L^j(t)}{1+\delta L^j(t)}\rho_{i,j}\sigma_j(t)dt + \sigma_i(t)dW_i^{T_N}(t),$$

$$i = \varrho(t), \ldots, N-1. \tag{32}$$

5.1. *Algorithm*

Here we only apply Algorithm 2.1. For simplicity, we consider such a set of tenor dates T_i and time steps h that T_i/h are integers.

As before, we simulate dynamics of the LIBOR rates $L^{\varrho(t)}(t), \ldots, L^{N-1}(t)$ in log space according to the weak Euler scheme (cf. (6)):

$$\ln L_{k+1}^i = \ln L_k^i - \sigma_i(t_k)h\sum_{j=i+1}^{N-1}\frac{\delta L_k^j}{1+\delta L_k^j}\rho_{i,j}\sigma_j(t_k) \tag{33}$$

$$-\frac{1}{2}(\sigma_i(t_k))^2 h + \sigma_i(t_k)\sqrt{h}\sum_{j=i}^{N-1}U_{i,j}\xi_{j,k+1},$$

$$i = \varrho_{k+1}, \ldots, N-1,$$

where $\xi_{j,k}$ are mutually independent random variables distributed by the law $P(\xi = \pm 1) = 1/2$ and $\varrho_k =: \varrho(t_k)$.

The algorithm for the considered trigger swap proceeds as follows. Let $L_k = (L_k^{\varrho_k}, \ldots, L_k^{N-1})^\top$. Denote by $\varkappa$ the first exit time of (t_k, L_k) from Q. Let $M = T_{N-1}/h$. Suppose by a time step k none of the rates $L_k^{\varrho_k}, \ldots, L_k^{N-1}$ have crossed their barriers $H^{\varrho_k}, \ldots, H^{N-1}$, i.e., $\chi(\varkappa \leq k) = 0$. Then we evaluate whether at the next time step $k+1$ the event $\varkappa = k+1$ might be realized. One can see that the rate L_{k+1}^i, $i = \varrho_{k+1}, \ldots, N-1$, computed via (33) will be below the barrier H^i, i.e. inside the domain G, if the following is true

$$\ln L_k^i < \ln H^i - \lambda_k\sqrt{h}\ , \tag{34}$$

where

$$\lambda_k = \sigma_{Max}\sqrt{N - \varrho_{k+1}}$$

and $\sigma_{Max} = \max_{j,k} \sigma_j(t_k)$.

If (27) is satisfied for all the rates $L_k^{\varrho_{k+1}}, \ldots, L_k^{N-1}$ then we move to the step $k+1$, evaluate $\ln L_{k+1}^i$, $i = \varrho_{k+1}, \ldots, N-1$, according to (33) and continue with the algorithm unless $k+1 = M$ (in this case the trigger swap expires worthless).

The case when the condition (27) does not hold for a single i implies that the point L_k is in the boundary zone $S_{t_k,h}$ and is near the barrier H^i. Then we either assign $\varkappa = k$, $\ln L_\varkappa^i = \ln H^i$, $\ln L_\varkappa^j = \ln L_k^j$ for $j \neq i$, $T_{\varrho_\varkappa} = T_{\varrho_k}$ with probability

$$p^i = \frac{\lambda_k \sqrt{h}}{\ln H^i - \ln L_k^i + \lambda_k \sqrt{h}} \tag{35}$$

and carry on with simulating $\ln L_{k+1}^i$, $i = \varrho_{k+1}, \ldots, N-1$, according to (33) starting from $\ln L_\varkappa$ until time $T_{\varrho_\varkappa} = \min\{T_i : t_\varkappa \leq T_i,\ i = 0, 1, \ldots, N-1\}$ (the barriers are "removed" in simulating the remaining part of this trajectory); or we jump outside the boundary zone $S_{t_k,h}$ by changing the ith component of $\ln L_k$ from $\ln L_k^i$ to $\ln L_k^i - \lambda_k \sqrt{h}$ with probability $1 - p^i$, perform the usual step according to (33) and continue with the algorithm unless $k+1 = M$ (in this case the trigger swap expires worthless).

We note that in comparison with the original formulation of Algorithm 2.1 here we do not stop the chain L_k at its first exit time from the space domain G. Instead, when the barrier is hit, we find the trigger tenor date $T_{\varrho_\varkappa}$, and if $t_\varkappa < T_{N-1}$, we continue the simulation according to (33) until $T_{\varrho_\varkappa}$ to get the required LIBOR rates $L^j(T_{\varrho(\tau)})$ in (31).

Now we discuss the case when the condition (27) does not hold for more than one i (i.e., the random walk has reached a corner of the domain G). In this case the algorithm proceeds as follows. Let us denote by $\ell = \{l_1, \ldots, l_n\}$ the set of tenor dates corresponding to the LIBOR rates for which (27) is violated. First, we select the rate from the set $\left\{\ln L_k^{l_1}, \ldots, \ln L_k^{l_n}\right\}$, which is the closest to its boundary, i.e., l_j such that $\ln H^{l_j} - \ln L_k^{l_j}$ is minimum over $j = 1, \ldots, n$. Then, we repeat the procedure which is given above for a single i with the following difference. If $\ln L_k^i$ jumps from the boundary to $\ln L_k^i - \lambda_k \sqrt{h}$, we find the second closest rate from the set $\left\{\ln L_k^{l_1}, \ldots, \ln L_k^{l_n}\right\}$ and as before repeat for this point the routine we have presented for a single i. We follow this procedure in the outlined fashion until either the set ℓ is empty or for some l_j we reach the boundary and assign $\ln L_k^{l_j} = \ln H^{l_j}$.

The outcome of simulating each trajectory is the payer swap starting tenor date $T_{\varrho_\varkappa}$, the stopping time $\varkappa$ and the point $\ln L_\eta$ with $\eta = T_{\varrho_\varkappa}/h$, which are used for evaluating the trigger swap:

$$V_{trswap}(0) \approx P(0,T_N)E^{\mathrm{Q}^{T_N}}\left[\left(\prod_{j=\varrho_\varkappa}^{N-1}(1+\delta L_\eta^j) - K\delta\sum_{i=\varrho_\varkappa+1}^{N-1}\prod_{j=i}^{N-1}(1+\delta L_\eta^j) - 1\right)\chi(\varkappa < M)\right]$$

with the expectation simulated by the Monte Carlo technique. We present the pseudocode for simulating a single trajectory based on the algorithm we described above.

Algorithm 5.1 Pseudocode for simulating a single trajectory in pricing the barrier trigger swap.

```
SET M to T_{N-1}/h, k to 1, κ to M
WHILE k < M
    IF κ > k
        FOR j = 0 to N − 1,
            IF (27) for j is false
                calculate p^j by (35)
                form array p of p^j
            ENDIF
        ENDFOR
        sort p in descending order
        FOR n = 1 to length(p)
            generate u ~ Unif[0.1];
                IF u < p(n)
                    SET κ to k
                    SET ln L^i_k to ln H^i
                    SET M to T_{ϱκ}/h
                    BREAK
                ELSE
                    SET ln L^i_k to ln L^i_k − λ_k √h
                ENDIF
        ENDFOR
    ENDIF
    Evaluate ln L_{k+1} by (33)
    Increase k by 1
ENDWHILE
```

5.2. *Numerical results*

Let us present results of numerical experiments we performed for pricing a trigger swap using Algorithm 6.1. The parameters chosen for the experiments are $T_0 = 5$, $\delta = 1$, $N = 11$ (i.e., $T^* = T_N = 16$), $K = 0.01$, $H = 0.13$, $\beta = 0.2$. The initial LIBOR rate curve is assumed to be flat at 0.04 and the volatility $\sigma_i(t)$ is set to be constant at 0.2. In the simulations we run 10^6 Monte Carlo paths.

Since the closed-form formula for the trigger swap (31) is not available, we found the reference trigger swap price by evaluating the price using Algorithm 5.1 with $h = 0.01$ and the number of Monte Carlo runs 10^6. This reference price is 5.46×10^{-2}

with the Monte Carlo error 5.50×10^{-4}, which gives half of the size of the confidence interval for the corresponding estimator with probability 0.95.

The results of the experiments with Algorithm 5.1 are presented in Table 1. In the table the values before "±" are estimates of the bias computed as the difference between the reference price and its sampled approximation, while the values after "±" give half of the size of the confidence interval for the corresponding estimator with probability 0.95. The "mean exit time" is the average time for trajectories (t_k, L_k) to leave the space-time domain Q. The experimentally observed convergence rate for Algorithm 5.1 is in agreement with the theoretical first order convergence in h (though we note that the convergence theorem in [15; 16] is proved under restrictive regularity conditions and the payoff of the trigger swap and the boundary of the space domain G do not satisfy these conditions).

Table 1. *Performance of Algorithm 5.1 for the trigger swap.*

h	error	mean exit time
0.25	$2.22 \times 10^{-2} \pm 6.39 \times 10^{-4}$	12.51
0.2	$1.85 \times 10^{-2} \pm 6.26 \times 10^{-4}$	12.61
0.125	$1.17 \times 10^{-2} \pm 6.01 \times 10^{-4}$	12.78
0.1	$9.56 \times 10^{-3} \pm 5.92 \times 10^{-4}$	12.83
0.0625	$6.03 \times 10^{-3} \pm 5.78 \times 10^{-4}$	12.92
0.05	$4.67 \times 10^{-3} \pm 5.72 \times 10^{-4}$	12.95

6. Barrier swaption

In this section we consider Monte Carlo evaluation of a knock-out swaption under the LMM. We use the knock-out swaption as a guide in our exposition, its treatment is rather general and it can be used to value different barrier options, where the underlying and barrier can be expressed as functionals of some diffusion process.

A European payer (receiver) swaption is an option that gives its holder a right, but not an obligation, to enter a payer (receiver) swap at a future date at a given fixed rate K. The swaption maturity usually coincides with the first reset date T_0 of the underlying swap. The underlying swap length $T_N - T_0$ is called the tenor of the swaption.

Without loss of generality, we concentrate on a knock-out receiver swaption with the first reset date T_0. A knock-out swaption has the structure as a standard swaption except that if the underlying swap rate is above a barrier level R_{up} at any time before T_0 then the swaption expires worthless. The price of the knock-out swaption at time $t = 0$ under the forward measure Q^{T_0} is given by:

$$V_{swaption}(0) = P(0, T_0) E^{\mathrm{Q}^{T_0}} \left[\delta \left(R_{swap}(T_0) - K \right)_+ \sum_{j=1}^{N} P(T_0, T_j) \chi \left(\theta > T_0 \right) \right], \quad (36)$$

where θ is the first exit time of the process $R_{swap}(s)$, $s \geq 0$, from the interval

$(0, R_{up})$. The swap rate $R_{swap}(s)$ can be expressed in terms of the spanning LIBOR rates as

$$R_{swap}(s) = \frac{1 - 1/\prod_{j=0}^{N-1}\left(1 + \delta L^j(s)\right)}{\delta \sum_{i=0}^{N-1} 1/\prod_{j=0}^{i}\left(1 + \delta L^j(s)\right)}. \tag{37}$$

The bond prices $P(T_0, T_j)$ can be expressed via LIBOR rates (see (11)) as well.

Also, let τ be the first exit time of the space-time process $(s, R_{swap}(s))$ from the domain $D = [0, T_0) \times (0, R_{up})$ (obviously, $\tau = \theta \wedge T_0$).

The LMM dynamics of LIBOR rates under Q^{T_0} are given by (cf. (13)):

$$\frac{dL^i(t)}{L^i(t)} = \sigma_i(t) \sum_{j=0}^{i} \frac{\delta L^j(t)}{1 + \delta L^j(t)} \rho_{i,j} \sigma_j(t) dt + \sigma_i(t) dW_i^{T_0}(t), \; i = 0, \ldots, N-1. \tag{38}$$

In this example we deal with pricing the barrier swaption (36) expressed in terms of the spanning LIBOR rates with dynamics in the form of (38). This means that we consider this problem in the coordinate system of the LIBOR rates and the barrier is given as an implicit surface in the LIBOR coordinates. We also introduce the space domain G in the phase space of the SDEs (38) corresponding to the interval $(0, R_{up})$ on the swap-rate semi-line. As usual, the corresponding space-time domain $Q := [0, T_0) \times G$.

For test purposes, let us introduce an analytical approximation for the barrier swaption. To this end, we note that under the Swap Market Model (SMM, see details in [4; 18; 21]) the barrier swaption pricing problem admits the closed-form solution (cf. (19))

$$\begin{aligned} V_{swaption}(0) = \delta \sum_{j=1}^{N} P(0, T_j) \{ & R_{swap}(0) \left[\Phi(\delta_+(R_{swap}(0)/K, v_{R_{swap}})) \right. \\ & -\Phi(\delta_+(R_{swap}(0)/R_{up}, v_{R_{swap}}))] \\ & -K \left[\Phi(\delta_-(R_{swap}(0)/K, v_{R_{swap}})) - \Phi(\delta_-(R_{swap}(0)/R_{up}, v_{R_{swap}})) \right] \\ & -H \left[\Phi(\delta_+(R_{up}^2/(KR_{swap}(0)), v_{R_{swap}})) - \Phi(\delta_+(R_{up}/R_{swap}(0), v_{R_{swap}})) \right] \\ & +KR_{swap}(0)\Phi(\delta_-({R_{up}}^2/(KR_{swap}(0)), v_{R_{swap}})/R_{up}) \\ & \left. -\Phi(\delta_-(R_{up}/R_{swap}(0), v_{R_{swap}})) \right] \}, \end{aligned} \tag{39}$$

where $\delta_\pm$ are from (20),

$$v_{R_{swap}}^2 = \int_0^{T_i} \left(\sigma_{R_{swap}}(s)\right)^2 ds,$$

and $\sigma_{R_{swap}}(s)$ is the instantaneous volatility of the log-normal dynamics of the swap rate.

Using Rebonato's formula [4] , we can compute the "approximate" volatility $v_{R_{swap}}^{LMM}$ for the LMM analogous to the volatility $v_{R_{swap}}$ in the SMM entering (39) as

$$v_{R_{swap}}^{LMM} = \sum_{i,j=0}^{N-1} \frac{\omega_i(0)\omega_j(0)L^i(0)L^j(0)\rho_{ij}}{(R_{swap}(0))^2} \int_0^{T_0} \sigma_i(s)\sigma_j(s)ds, \tag{40}$$

where

$$\omega_i(0) = \frac{1 - 1/\prod_{j=0}^{i-1}\left(1+\delta L^j(0)\right)}{\delta \sum_{k=0}^{N-1} 1/\prod_{j=0}^{k}\left(1+\delta L^j(0)\right)}.$$

The quantity $v_{R_{swap}}^{LMM}$ can be used as a proxy for $v_{R_{swap}}$ in (39) to compute approximate barrier swaption prices under LMM. We will check in our numerical experiments whether an approximation obtained by our algorithm is consistent with this analytical approximation.

6.1. *Algorithm*

Here we exploit Algorithm 2.1. We choose a time step $h > 0$ so that $M = T_0/h$ is an integer. Again inside the domain G we use the weak Euler scheme to simulate trajectories of the log LIBOR rates (38):

$$\begin{aligned} \ln L_{k+1}^i = \ln L_k^i &+ \sigma_i(t_k) \sum_{j=0}^{i} \frac{\delta L_k^j}{1+\delta L_k^j} \rho_{i,j}\sigma_j(t_k)h \\ &- \frac{1}{2}\left(\sigma_i(t_k)\right)^2 h + \sigma_i(t_k)\sqrt{h}\sum_{j=i}^{N-1} U_{i,j}\xi_{j,k+1}, \\ i = 0, \ldots, N-1,& \end{aligned} \tag{41}$$

where $\xi_{j,k}$ are mutually independent random variables distributed by the law $P(\xi = \pm 1) = 1/2$.

For a fixed t_k, we denote by $\ln L_k$ the point with coordinates $\ln L_k^0, \ln L_k^1, \ldots, \ln L_k^{N-1}$, i.e. $\ln L_k = (\ln L_k^0, \ln L_k^1, \ldots, \ln L_k^{N-1})^\top$. As before, we follow the random walk constructed by (41) until we reach the boundary zone $S_{t_k,h}$. Algorithmically, it implies that we implement a check at each step whether the current position of the random walk is in the boundary zone $S_{t_k,h}$. More precisely, we evaluate at time t_k whether the current position $\ln L_k$ is such that the maximum increment from point $\ln L_k$ according to all possible realizations of (41) at the next time level t_{k+1} results in the state of the random walk below the barrier, i.e. in the domain G.

Introduce

$$\ln L_{k,Max} = \max_i \ln L_k^i$$

and

$$\ln \hat{L}_{k+1} = \ln L_{k,Max} + \sigma_{Max}^2 hN - \frac{1}{2}\sigma_{Max}^2 h + \sigma_{Max}\sqrt{hN}, \tag{42}$$

where $\sigma_{Max} = \max_{i,k} \sigma_i(t_k)$. Using the fact that

$$R_{swap}(\hat{L}_{k+1}, ..., \hat{L}_{k+1}) = \hat{L}_{k+1},$$

one can see that the current position of the random walk $\ln L_k$ is inside the domain G if the following condition is satisfied

$$\ln \hat{L}_{k+1} < \ln R_{up}. \tag{43}$$

Algorithmically, we do the following. If condition (43) is true, we evaluate the next position of the random walk at t_{k+1} according to (41) and continue further with the algorithm unless $k+1 = M$ (i.e., we have reached the maturity time T_0 of the swaption).

We note the condition (43) is computationally cheap but it is rather rough. Once this condition fails, we check a finer but computationally more expensive condition based on the maximum increments of each of $L^i(t_k)$ towards the boundary:

$$\begin{aligned} &R_{swap}(L_k^0(1+\sigma_0(t_k)\sigma_{Max,k}h + \sigma_0(t_k)\sqrt{Nh}), L_k^1(1+2\sigma_1(t_k)\sigma_{Max,k}h \\ &+\sigma_1(t_k)\sqrt{(N-1)h}), ..., L_k^{N-1}(1+N\sigma_{N-1}(t_k)\sigma_{Max,k}h + \sigma_{N-1}(t_k)\sqrt{h})) \\ &< R_{up}, \end{aligned} \tag{44}$$

where $\sigma_{Max,k} = \max_j \sigma_j(t_k)$. If the condition (44) holds, we again carry on to the next time step t_{k+1} using (41) and continue further with the algorithm unless $k+1 = M$.

If both conditions (43) and (44) fail, the random walk has reached the boundary zone $S_{t_k,h}$, where as before we apply the different procedure which require us to find the projection $\ln L_k^\pi := (L_k^{\pi,0}, L_k^{\pi,1}, \ldots, L_k^{\pi,N-1})^\top$ of the current position $\ln L_k$ on the boundary given as the implicit function of the spanning LIBOR rates:

$$\ln R_{swap}(t_k) = \ln R_{up}. \tag{45}$$

For completeness of the exposition, let us discuss how the projection $\ln L_k^\pi$ can be simulated before we return to the description of the algorithm. The problem of finding the point $\ln L_k^\pi$ is equivalent to finding the minimum value of the function

$$|\ln L_k^\pi - \ln L_k|^2 = \left(\ln L_k^{\pi,0} - \ln L_k^0\right)^2 + \cdots + \left(L_k^{\pi,N-1} - \ln L_k^{N-1}\right)^2 \tag{46}$$

subject to the constraint

$$\ln\left(\frac{\prod\limits_{j=0}^{N-1}\left(1+\delta L_k^{\pi,j}\right) - 1}{\delta\left(1+\sum\limits_{i=0}^{N-2}\prod\limits_{j=i+1}^{N-1}\left(1+\delta L_k^{\pi,j}\right)\right)}\right) = \ln R_{up}. \tag{47}$$

We regard $\ln L_k^{\pi,1}, \ldots, \ln L^{\pi,N-1}$ as independent variables in the constraint equation (47) and write $\ln L_k^{\pi,0}$ as

$$\ln L_k^{\pi,0} = \ln \left(\frac{R_{up} \cdot \left(1 + \sum_{i=0}^{N-2} \prod_{j=i+1}^{N-1} \left(1 + \delta L_k^{\pi,j}\right)\right) + 1}{\prod_{j=1}^{N-1} \left(1 + \delta L_k^{\pi,j}\right)} - \frac{1}{\delta} \right). \tag{48}$$

Hence the minimization problem is reduced to finding the point $\ln L_k^{\pi,1}$, $\ldots, \ln L^{\pi,N-1}$ at which the function $|\ln L_k^{\pi} - \ln L_k|^2$ from (46) with $\ln L_k^{\pi,0}$ from (48) has its minimum value. This optimization problem can be solved using standard procedures, e.g. the MATLAB function "lsqnonlin()".

Let us now continue with the description of the algorithm. When $\ln L_k$ is in the boundary zone, we either stop the chain at $\ln L_k^{\pi}$ with probability p :

$$p = \frac{\lambda\sqrt{h}}{|\ln L_k^{\pi} - \ln L_k| + \lambda\sqrt{h}}, \tag{49}$$

where

$$\lambda\sqrt{h} = \sqrt{N}\left(\sigma_{Max}^2 hN - \frac{1}{2}\sigma_{Max}^2 h + \sigma_{Max}\sqrt{hN}\right); \tag{50}$$

or we jump inside the domain G to the point $\ln L_k + \lambda\sqrt{h}\frac{\overrightarrow{\ln L_k^{\pi}\ \ln L_k}}{|\ln L_k^{\pi} - \ln L_k|}$ with probability $1-p$, apply the Euler step (41) to evaluate $\ln L_{k+1}$ and continue further with the algorithm unless $k+1 = M$.

The outcome of simulating each trajectory is the point $(t_{\varkappa}, \ln L_{\varkappa})$. In Algorithm 6.1 we present the pseudocode for simulating a single trajectory based on the algorithm we described above.

6.2. *Numerical results*

We give some results for pricing a barrier swaption by Algorithm 6.1. We consider the barrier swaption with the initial LIBOR curve flat at 0.05, constant volatility $\sigma_i(t)$ at 0.1 and the following parameters: $T_0 = 10$, $N = 10$, $\delta = 1$ (i.e., $T_N = T^* = 20$), $K = 0.01$, $R_{up} = 0.075$, $\beta = 0.1$. The simulations use 10^6 Monte Carlo runs.

The pricing problem for the barrier swaption (36) does not admit a closed-form solution. We used the barrier swaption price found by Algorithm 6.1 with $h = 0.01$ and 10^7 of Monte Carlo runs as the reference solution. This reference price is 0.15506 with the Monte Carlo error 1.42×10^{-4}, which gives half of the size of the confidence interval for the corresponding estimator with probability 0.95. The analytical approximation based on (39) and (40) yields the price of the barrier swaption 0.15556, which turns out to be very accurate.

We present results of the experiments in Table 2. As in the previous section, the error column values before "±" are estimates of the bias computed using the

Algorithm 6.1 Barrier swaption: Pseudocode for simulating a single trajectory.

```
FOR k = 1 to M
    calculate ln L̂_{k+1} by (42)
    IF (43) is true:   calculate ln L_{k+1} by (41)
    ELSE
        IF (44) is true: calculate ln L_{k+1} by (41)
        ELSE
            solve the minimisation problem (46) with ln L_k^{π,0} from (48)
            generate u ~ Unif[0, 1]
            calculate probability p by (49)
            IF u < p
                break
            ELSE
                SET ln L_k to ln L_k + λ√h (vec(ln L_k^π ln L_k)) / |ln L_k^π − ln L_k|
                calculate ln L_{k+1} by (41)
            ENDIF
        ENDIF
    ENDIF
ENDFOR
```

reference price value and the values after "±" reflect the Monte Carlo error with probability 0.95. The "mean exit time" is the average time for approximate trajectories to exit the space-time domain Q. It is clear that the results demonstrate the expected first order of convergence.

Table 2. *Performance of Algorithm 6.1 for the barrier swaption.*

h	error	mean exit time
0.25	$1.01 \times 10^{-2} \pm 4.33 \times 10^{-4}$	9.36
0.2	$8.08 \times 10^{-3} \pm 4.37 \times 10^{-4}$	9.40
0.125	$5.15 \times 10^{-3} \pm 4.42 \times 10^{-4}$	9.46
0.1	$4.15 \times 10^{-3} \pm 4.44 \times 10^{-4}$	9.48
0.0625	$2.58 \times 10^{-3} \pm 4.47 \times 10^{-4}$	9.51
0.03125	$1.03 \times 10^{-3} \pm 4.49 \times 10^{-4}$	9.54

Acknowledgment

MVT was partially supported by the Leverhulme Trust.

Bibliography

[1] Andersen, L., Brotherton-Racliffe, R.: Exact exotics. *Risk* **9** (1996), 85–89.
[2] Baldi, P.: Exact asymptotics for the probability of exit from a domain and applications to simulation. *Ann. Prob.* **23** (1995), 1644–1670.

[3] Beaglehole, D. R., Dybvig, P.H., Zhou G.: Going to extremes: Correcting simulation bias in exotic option valuation. *Financial Analyst Journal,* January/February (1997), 62–68.

[4] Brigo, D., Mercurio, F.: *Interest Rate Models: Theory and Practice.* Springer, 2006.

[5] Derman, E., Kani, I., Ergener, D., Bardhan, I.: Enhanced numerical methods for options with barriers. *Quantitative Strategies Research Notes*, Goldman Sachs. May 1995.

[6] Glasserman, P.: *Monte Carlo Methods in Financial Engineering.* Springer, 2003.

[7] Gobet, E.: Weak approximation of killed diffusion using Euler schemes. *Stoch. Process. Appl.* **87** (2000), 167–197.

[8] Gobet, E.: Advanced Monte Carlo methods for barrier and related exotic options. In *Mathematical Modeling and Numerical Methods in Finance*, Eds. A. Bensoussan, Q. Zhang, P. Ciarlet. Elsevier, 2009, 497–528.

[9] Kloeden, P. E., Platen, E.: *Numerical Solution of Stochastic Differential Equations.* Springer, 1992.

[10] Kunitomo, N., Ikeda, M.: Pricing options with curved boundaries. *Math. Fin.* **4** (1992), 275–298.

[11] Milstein, G.N.: *Numerical Integration of Stochastic Differential Equations.* Kluwer Academic Publishers, 1995.

[12] Milstein, G.N.: Solution of the first boundary value problem for equations of parabolic type by means of the integration of stochastic differential equations. *Theor. Probab. Appl.*, **40** (1995), 556–563.

[13] Milstein, G.N., Rybkina, N.F.: An algorithm for random walks over small ellipsoids for solving the general Dirichlet problem. *J. Comp. Math. Math. Phys.*, **33** (1993), 631–647.

[14] Milstein, G.N. , Schoenmakers, J.G.M.: Monte Carlo construction of hedging strategies against multi-asset European claims. *Stoch. Stoch. Rep.* **73** (2002), 125–157.

[15] Milstein, G.N., Tretyakov, M.V.: The simplest random walks for the Dirichlet problem. *Theory Prob. Appl.* **47** (2002), 53–68.

[16] Milstein, G.N., Tretyakov, M.V.: *Stochastic Numerics for Mathematical Physics.* Springer, 2004.

[17] Milstein, G.N., Tretyakov, M.V.: Practical variance reduction via regression for simulating diffusions. *SIAM J. Numer. Anal.* **47** (2009), 887–910.

[18] Rebonato, R.: *Interest-rate Option Models.* Wiley, 1998.

[19] Rubinstein, M., Reiner, E.: Breaking down the barriers. *Risk* **4** (1991), 28–35.

[20] Shevchenko, P.: Addressing the bias in Monte Carlo pricing of multiasset options with multiple barriers through discrete sampling. *J. Comp. Fin.* **6** (2003), 1–20.

[21] Schoenmakers, J.: *Robust Libor Modelling and Pricing of Derivative Products.* Chapman and Hall/CRC, 2005.

Chapter 12

Coupling local currency Libor models to FX Libor models

John Schoenmakers

Weierstrass Institute for Applied Analysis and Stochastics, Mohrenstrasse 39, D-10117 Berlin

Abstract We focus on the coupling of two existing and calibrated single currency Libor models into a joint Libor model that allows for pricing of multiple currency based structured interest rate products. Our main contribution is twofold: On the one hand we provide a method for synthesizing two local currency based correlation structures into a correctly defined joint correlation structure that describes the cross Libor correlations between the two currencies in a realistic way. On the other hand we introduce an (necessary) FX related factor X in order to describe the unified model with respect to one particular numéraire measure. In addition we propose to calibrate this factor to FX instruments in the case where X is modeled via Heston type dynamics.

1. Introduction

Libor interest rate modeling, initially developed by [16], [6], and [12] almost two decades ago, is still considered to be the universal tool for evaluation of structured interest rate products. One of the main reasons is the great flexibility in the choice of the Libor volatilities in the Libor framework. Starting from deterministic volatility structures leading to the Libor market model, many enhancements have been proposed in order to match implied volatility patterns of liquid products such as caps and swaptions. In this respect we mention (among other approaches) the Lévy Libor model by [8] (see for example [3] and [17] for a numerical treatments and practical implementation), displaced diffusion, CEV Libor models, log-normal mixture models, and even random parameter Libor models (e.g. [5] and the references therein for an overview). Another important line of research is the development of (one factor) stochastic volatility models based on CIR type scalar volatilities by [1], [19], and their multi factor extensions by [2] and more recently [14]. Further we mention SABR related Libor models (e.g. [10]) that are based on a different types of scalar volatilities. SABR based Libor models gained popularity because they allow for pricing of European liquids by relatively simple approximation formulas based on heat kernel expansion techniques.

For pricing of structured interest rate products that involve different currencies (quanto style products), Libor models that are jointly defined with respect to these currencies are called for. Although there already appeared some approaches in the literature more recently (e.g. [4]), by now we have not seen a generic approach to connect two existing generally specified and calibrated single currency Libor models.

Here we present a generic approach to melt two given Libor rate models with respect to two different currencies (domestic and foreign) into a unified Libor model. As a key issue we propose a tractable approach to synthesize the Libor correlation structures given in their respective currencies into a joint correlation structure from which the initial domestic and foreign correlations may be retrieved, and moreover the cross correlations between domestic and foreign Libors are modeled as a (re-scaled) suitably defined average of the domestic and foreign correlations. This averaging procedure is based on coupling of a particular square root of the domestic structure with another particular square root of the foreign structure. The coupling is carried out in such a way that joint matrix is a real correlation matrix in the sense that it is positive and has diagonal entries that are all equal to one. In order to describe the unified model with respect to a unified measure, for instance the terminal domestic bond or the terminal foreign bond measure, an additional FX related factor X has to be incorporated. Finally we outline an FFT based procedure by [7] for pricing liquidly traded FX options, in the case where X is driven by a Heston type stochastic volatility process. This procedure may then be used in order to calibrate the dynamics of X. The method is fairly general in the sense that it can be applied to virtually all Libor models driven by a Wiener environment.

2. Resume of Wiener based Libor modeling

For a fixed sequence of tenor dates $0 =: T_0 < T_1 < \ldots < T_n$, called a tenor structure, we consider zero bond processes B_i, $i = 1, \ldots, n$, where each B_i is defined on the interval $[0, T_i]$ and ends up with terminal face value $B_i(T_i) = 1$. We now define a set of forward Libors on the tenor structure by

$$L_i(t) := \frac{1}{\delta_i}\left(\frac{B_i(t)}{B_{i+1}(t)} - 1\right), \quad 0 \le t \le T_i, \ 1 \le i < n, \tag{1}$$

where the $\delta_i := T_{i+1} - T_i$, $i = 1, \ldots, n-1$, denoting the periods between two subsequent tenor dates, are called day-count fractions. L_i is in fact the annualized effective rate corresponding to a forward rate agreement (FRA) contracted at time t, for the period $[T_i, T_{i+1}]$. Here we assume that according to this agreement, the interest rate $\delta_i L_i(T_i)$ par notional 1 has to be payed at T_{i+1}.

In this chapter we consider a framework where the zero-bonds $(B_i)_{i=1,\ldots,n}$ that define the Libors are adapted processes which are defined on a filtered probability space $(\Omega, (\mathcal{F}_t)_{0 \le t \le T_\infty}, P)$ with $T_\infty \ge T_n$ being some finite time horizon. Throughout it is assumed that the filtration $(\mathcal{F}_t)$ is generated by some d-dimensional standard

Brownian motion $\mathcal{W}$ (thus excluding jump type models). Furthermore, we consider predictable (column) processes σ_i with state $\mathbb{R}^d$, that denote the volatility of the bonds B_i respectively, predictable (scalar) drift processes μ_i denoting the drifts of the B_i, and a (scalar) market price of risk process λ, all adapted to the driving Brownian motion $\mathcal{W}$. That is, in the objective measure the zero bond dynamics are of the form

$$\frac{dB_i}{B_i} = \mu_i dt + \sigma_i^\top d\mathcal{W} \quad \text{with} \quad \mu_i = \sigma_i^\top \lambda. \tag{2}$$

Under some further mild technical conditions (see [12] and [13] for details) there now exists for each i, $1 \leq i < n$, an $\mathbb{R}^d$-valued predictable volatility process Γ_i such that the Libor dynamics are given by

$$\frac{dL_i}{L_i} = -\sum_{j=i+1}^{n-1} \frac{\delta_j L_j}{1+\delta_j L_j} \Gamma_i^\top \Gamma_j dt + \Gamma_i^\top d\mathcal{W}^{(n)}, \quad 0 \leq t \leq T_i,\ 1 \leq i < n, \tag{3}$$

where $\mathcal{W}^{(n)}$ is an equivalent standard Brownian motion under the terminal numéraire measure P_n induced by the terminal zero coupon bond B_n, that is, for all j, B_j/B_n are P_n-martingales. (We do not dwell on issues of local versus true martingales here.) In particular

$$\Gamma_i := \delta_i^{-1} L_i^{-1} (1 + \delta_i L_i) \left(\sigma_i - \sigma_{i+1}\right)_i, \ 1 \leq i < n. \tag{4}$$

For some general fixed i, $1 \leq i < n$ we may consider instead the numéraire measure P_{i+1} induced by the bond B_{j+1}, and then for $1 \leq j \leq i$ we obtain from (3) the dynamics

$$\begin{aligned}
\frac{dL_j}{L_j} &= \Gamma_j^\top \left(-\sum_{k=j+1}^{n-1} \frac{\delta_k L_k}{1+\delta_k L_k} \Gamma_k dt + d\mathcal{W}^{(n)} \right) \\
&= -\sum_{k=j+1}^{i} \frac{\delta_k L_k}{1+\delta_k L_k} \Gamma_j^\top \Gamma_k dt + \Gamma_j^\top \left(-\sum_{k=i+1}^{n-1} \frac{\delta_k L_k}{1+\delta_k L_k} \Gamma_k dt + d\mathcal{W}^{(n)} \right) \\
&=: -\sum_{k=j+1}^{i} \frac{\delta_k L_k}{1+\delta_k L_k} \Gamma_j^\top \Gamma_k dt + \Gamma_j^\top d\mathcal{W}^{(i+1)}, \quad 1 \leq j \leq i.
\end{aligned} \tag{5}$$

Since by (1) L_i is a martingale under P_{i+1}, it automatically follows that $\mathcal{W}^{(i+1)}$ in (5) is a standard Brownian motion under the equivalent measure P_{i+1}. Finally we note that in the case where the Γ_j are *deterministic* we have the well documented Libor Market Model (LMM) (see for example [5] and [18] and the references therein).

3. Multi currency extension of the Libor model

In this section we will melt two markets, the domestic and the foreign interest rate market, into just one. That is, we are going to consider zero bonds and more general traded assets in this extended market and determine their *unified* dynamics described by an SDE. Let $(B_{1,...,}B_n, B_1^*, ..., B_{n^*}^*)$ be an arbitrage free joint system of domestic zero bonds B_i and foreign zero bonds B_i^* expressed *in domestic currency,* corresponding to a domestic and foreign tenor structure $0 =: T_0 < T_1 < \ldots < T_n$, and $0 =: T_0^* < T_1^* < \ldots < T_{n^*}^*$, respectively. Since it only makes sense to consider the domestic and foreign bond system on the same joint time interval, we make the following structural assumption,

$$T_n = T_{n^*}^* = T_\infty,$$

that is, we allow both tenor structures to be different, but they both span the same time period. In view of (2) we consider the coupled dynamics

$$\begin{aligned} \frac{dB_i}{B_i} &= \mu_i dt + \sigma_i^\top d\mathbf{W}, \quad 1 \le i \le n, \\ \frac{dB_i^*}{B_i^*} &= \mu_i^* dt + \sigma_i^{*\top} d\mathbf{W}\,, \quad 1 \le i \le n^*, \end{aligned} \tag{6}$$

where now $\mathbf{W}$ is a D-dimensional standard Brownian motion with D being sufficiently large. Connected with (6) we so introduce a general FX-Libor system $(L_1, ..., L_{n-1}, L_1^*, ..., L_{n^*-1}^*, X)$ defined by

$$L_i = \frac{1}{\delta_i}(\frac{B_i}{B_{i+1}} - 1), \quad L_i^* = \frac{1}{\delta_i^*}(\frac{B_i^*}{B_{i+1}^*} - 1), \quad X = \frac{B_{n^*}^*}{B_n} \tag{7}$$

with $\delta_i^* := T_{i+1}^* - T_i^*$. Then with respect to B_n as numéraire we obtain under P_n the joint dynamics

$$\begin{aligned} \frac{dL_i}{L_i} &= -\sum_{j=i+1}^{n-1} \frac{\delta_j L_j}{1+\delta_j L_j} \Gamma_i^\top \Gamma_j dt + \Gamma_i^\top d\mathbf{W}^{(n)}, \quad 1 \le i < n, \\ \frac{dL_i^*}{L_i^*} &= -\Gamma_i^{*\top} \Gamma_X dt - \sum_{j=i+1}^{n^*-1} \frac{\delta_j^* L_j^*}{1+\delta_j^* L_j^*} \Gamma_i^{*\top} \Gamma_j^* dt + \Gamma_i^{*\top} d\mathbf{W}^{(n)}, \quad 1 \le i < n^*, \\ \frac{dX}{X} &= \Gamma_X^\top d\mathbf{W}^{(n)}, \end{aligned} \tag{8}$$

where $\Gamma_X := \sigma_{n^*}^* - \sigma_n$, $\mathbf{W}^{(n)}$ is standard Brownian motion under P_n, and due to (4)

$$\begin{aligned} \Gamma_i &= (\delta_i L_i)^{-1} (1 + \delta_i L_i) (\sigma_i - \sigma_{i+1})_i\,, \quad 1 \le i < n, \\ \Gamma_i^* &= (\delta_i^* L_i^*)^{-1} (1 + \delta_i^* L_i^*) \left(\sigma_i^* - \sigma_{i+1}^*\right)\,, \quad 1 \le i < n^*. \end{aligned}$$

Similarly, with respect to $B^*_{n^*}$ as numéraire we get

$$\frac{dL_i}{L_i} = \Gamma_i^\top \Gamma_{X_n} dt - \sum_{j=i+1}^{n-1} \frac{\delta_j L_j}{1+\delta_j L_j} \Gamma_i^\top \Gamma_j dt + \Gamma_i^\top d\mathbf{W}^{(n^*)}, \quad 1 \le i < n,$$

$$\frac{dL_i^*}{L_i^*} = -\sum_{j=i+1}^{n^*-1} \frac{\delta_j^* L_j^*}{1+\delta_j^* L_j^*} \Gamma_i^{*\top} \Gamma_j^* dt + \Gamma_i^{*\top} d\mathbf{W}^{(n^*)}, \quad 1 \le i < n^*,$$

$$\frac{dX}{X} = \|\Gamma_X\|^2 dt + \Gamma_X^\top d\mathbf{W}^{(n^*)}, \tag{9}$$

where $\mathbf{W}^{(n^*)}$ is a D-dimensional standard Brownian motion under the measure P_{n^*}, corresponding to $B^*_{n^*}$ that satisfies

$$d\mathbf{W}^{(n^*)} = d\mathbf{W}^{(n)} - \Gamma_X dt. \tag{10}$$

The connecting relation (10) is easily verified in the following way. Since X^{-1} is a martingale under P_{n^*}, we derive by Ito's formula and using (8),

$$\frac{dX^{-1}}{X^{-1}} = X\left(-\frac{1}{X^2}dX + \frac{1}{X^3}d\langle X, X\rangle\right) = -\Gamma_X^\top d\mathbf{W}^{(n)} + \|\Gamma_X\|^2 dt$$
$$= -\Gamma_X^\top \left(d\mathbf{W}^{(n)} - \Gamma_X dt\right) = -\Gamma_X^\top d\mathbf{W}^{(n^*)},$$

from which (10) follows.

Due to the above approach, the domestic zero bonds B_i are in general correlated with the foreign zero bonds B_i^* (in domestic currency). We think that this is a natural way of modeling and, indeed, leaving this possibility out of consideration would give rise to a very controversial discussion among practitioners. However, as a consequence, the volatility structures of both the domestic and foreign Libors in (8) (respectively (9)) need to be determined, and moreover also the volatility process Γ_X.

4. Connecting a local general model with a foreign extended market model

Let us assume that we are given a local Libor model (3) where the local volatility processes are of the form

$$\Gamma_i(t) = \|\Gamma_i\|(t) e_i(t), \quad e_i \in \mathbb{R}^d, \quad 1 \le i < n, \tag{11}$$

where the e_i are unit vectors that at most deterministically depend on t, and the $\|\Gamma_i\|(t)$ are given scalar volatility processes adapted to a d-dimensional Brownian motion $\mathcal{W}$. W.l.o.g. we assume that the correlation structure introduced by

$$R_{ij} := \left[(e_i)^\top e_j\right](t), \quad 1 \le i, j < n,$$

has constant rank d. In addition we assume that we are also given a foreign Libor model where the foreign volatility processes are of the more special form

$$\Gamma_i^*(t) = g_i(t, L^*) e_i^*(t), \quad e_i^* \in \mathbb{R}^{d^*}, \quad 1 \le i < n^*, \tag{12}$$

where e_i^* are unit vectors deterministically depending on t, and $g_i(t,\cdot)$ are nonnegative deterministic (scalar) volatility functions. Hence the foreign Libors follow a so called extended market model in the foreign terminal bond measure P_{*n}. For suitably chosen g_i, (12) might represent for example a CEV model, some displaced diffusion model, or a standard market model. It is well known that the distribution of such a model, that is the distribution of $\left(L_j^*\right)_{1\le j<n^*}$ in the P_{n^*} measure, is completely determined by these volatility functions g_i and the correlation structure R^* determined by

$$R_{ij}^* = \left[\left(e_i^*\right)^\top e_j^*\right](t) \quad 1 \le i,j < n^*. \tag{13}$$

W.l.o.g. we assume that R^* has constant rank d^*. Our goal is now to construct a new joint model (8) such that the distribution of $(L_j)_{1\le j<n^*}$ and $\left(L_j^*\right)_{1\le j<n^*}$ coincide with the respective initial ones in their respective measures.

Let us define $C \in \mathbb{R}^{(n-1)\times d}$ by $C_{ik} = e_{i,k}$ $1 \le i < n$, $1 \le k \le d$, hence $CC^\top = R$. Next, let $F \in \mathbb{R}^{(n^*-1)\times d}$, and for some suitable $p \ge 0$ to be determined, $G \in \mathbb{R}^{(n^*-1)\times p}$ with $p = rank(G)$ (with $\mathbb{R}^{(n^*-1)\times 0} := \varnothing$, i.e. an empty matrix), such that F and G solve the following matrix equation,

$$FF^\top + GG^\top = R^*. \tag{14}$$

We then have

$$\begin{aligned}\Sigma &:= \begin{pmatrix} C & \emptyset \\ F & G \end{pmatrix}\begin{pmatrix} C & \emptyset \\ F & G \end{pmatrix}^\top = \begin{pmatrix} C & \emptyset \\ F & G \end{pmatrix}\begin{pmatrix} C^\top & F^\top \\ \emptyset & G^\top \end{pmatrix} \\ &= \begin{pmatrix} R & CF^\top \\ FC^\top & R^* \end{pmatrix}\end{aligned}$$

and it holds

$$\frac{dL_i}{L_i} = -\sum_{j=i+1}^{n-1} \frac{\delta_j L_j}{1+\delta_j L_j}\Gamma_i^\top \Gamma_j dt + \Gamma_i^\top d\mathbf{W}^{(n)}, \quad 1 \le i < n, \tag{15}$$

$$\frac{dL_i^*}{L_i^*} = -\sum_{j=i+1}^{n^*-1} \frac{\delta_j^* L_j^*}{1+\delta_j^* L_j^*}\Gamma_i^{*\top} \Gamma_j^* dt + \Gamma_i^{*\top} d\mathbf{W}^{(n^*)}, \quad 1 \le i < n^*,$$

with respect to an extended Brownian motion $\mathbf{W}^{(n)} := (\mathcal{W}^{(n)}, \widetilde{\mathcal{W}}^{(n)}) \in \mathbb{R}^D$ and $\mathbf{W}^{(*n)} := (\mathcal{W}^{(n^*)}, \widetilde{\mathcal{W}}^{(n^*)}) \in \mathbb{R}^D$, under the measures P_n and P_{n^*}, respectively, with $D = d + p$ and

$$\Gamma_i = \|\Gamma_i\|\,\mathbf{e}_i, \quad 1 \le i < n, \quad \text{and} \quad \Gamma_i^* = g_i \mathbf{e}_i^*, \quad 1 \le i < n^*, \tag{16}$$

where

$$\begin{aligned}&\mathbf{e}_{i,k} := e_{i,k}, \quad 1 \le k \le d, \quad \mathbf{e}_{i,d+k} = 0, \quad 1 \le k \le p, \quad 1 \le i < n, \\ &\mathbf{e}_{i,k}^* := F_{ik}, \quad 1 \le k \le d, \quad \mathbf{e}_{i,d+k}^* = G_{ik}, \quad 1 \le k \le p, \quad 1 \le i < n^*. \end{aligned} \tag{17}$$

There are two edge solutions: i) $F = 0$ implying that $GG^\top = R^*$ (so $p = d^*$ hence $D = d + d^*$) and

$$\Sigma = \begin{pmatrix} R & \emptyset \\ \emptyset & R^* \end{pmatrix},$$

i.e. L and L^* are independent, ii) $G = \varnothing$ ($p = 0$) implying that $FF^\top = R^*$ (so $D = d \geq d^*$),

$$\Sigma = \begin{pmatrix} R & CF^\top \\ FC^\top & R^* \end{pmatrix}, \tag{18}$$

and both the local and the foreign model are driven by $\mathcal{W}^{(n)} \in \mathbb{R}^d$ in fact.

A simple pragmatic solution

By taking $p = d^*$ in the case $d \geq d^*$ and letting F_1 and G_1 be matrices as specified above with $F_1F_1^\top = R^*$ and $G_1G_1^\top = R^*$, we have that for any $|\varrho| \leq 1$, $F_\varrho = \varrho F_1$ and $G_\varrho = \sqrt{1-\varrho^2}G_1$ solve the matrix equation (14). Let us specialize to the case $d = d^* = p$: A matrix G with $GG^\top = R^*$ is determined up to a orthogonal transformation. Indeed, let $G_1 \in \mathbb{R}^{(n^*-1)\times p}$ be the unique lower triangular (Cholesky) root of R^* with $G_{1,ii} > 0$, and $G_1G_1^\top = R^*$, then any G with $G = G_1Q$ for orthogonal $Q \in \mathbb{R}^{p\times p}$, satisfies $GG^\top = R^*$. Now let further C_1 be the unique lower triangular (Cholesky) root of R, hence $C_1C_1^\top = R$, then determine $Q_C \in \mathbb{R}^{p\times p}$ such that $C = C_1Q_C$, and take $F_1 = G_1Q_C$. For the joint FX Libor model we then take for F and G in (17),

$$F_\varrho = \varrho G_1 Q_C \quad \text{and} \quad G_\varrho = \sqrt{1-\varrho^2}G_1, \tag{19}$$

respectively, and the volatilities, say $\Gamma_{\varrho,i}$ in (16) accordingly. For the cross currency Libor correlation matrix we then obtain in (18)

$$CF_\varrho^\top = \varrho C Q_C^\top G_1^\top = \varrho C_1 G_1^\top, \quad |\varrho| \leq 1,$$

hence

$$\Sigma_\varrho := \begin{pmatrix} R & \varrho C_1 G_1^\top \\ \varrho G_1 C_1^\top & R^* \end{pmatrix} \tag{20}$$

being a valid correlation matrix for any $|\varrho| \leq 1$. In the particular case where $R = R^*$ (hence $n = n^*$) we thus obtain by this construction $CF_\varrho^\top = \varrho R$. In the general case where $R \neq R^*$ and possibly $n \neq n^*$ (but $d = d^* = p$) we may consider the matrix $C_1G_1^\top$ as some kind of average between R and R^*. In the next section we will outline the calibration of ϱ to FX rate vanilla options.

5. Calibration to FX market

Let us consider $X := B^*_{n^*}/B_n$. Note that (cf. [9]),

$$B_n(t) = B_{\eta(t)}(t) \prod_{j=\eta(t)}^{n-1} \frac{1}{1+\delta_j L_j(t)}$$

for $0 \le t \le T_n = T_\infty$. and $\eta(t) := \min\{m : T_m \ge t\}$. Thus, by (15),

$$\begin{aligned}\frac{dB_n}{B_n} &= (...)dt + d\ln B_n \\ &= (...)dt + \frac{dB_{\eta(t)}}{B_{\eta(t)}} - \sum_{j=\eta(t)}^{n-1} \frac{\delta_j L_j}{1+\delta_j L_j} \Gamma_j^\top d\mathbf{W}^{(\cdot)} \\ &=: (...)dt + \sigma_{\eta(t)}^\top d\mathbf{W}^{(\cdot)} - \sum_{j=\eta(t)}^{n-1} \frac{\delta_j L_j}{1+\delta_j L_j} \Gamma_j^\top d\mathbf{W}^{(\cdot)}.\end{aligned}$$

In the same way, for $t \le T^*_{n^*} = T_\infty$,

$$B^*_{n^*}(t) = B^*_{\eta^*(t)}(t) \prod_{j=\eta^*(t)}^{n^*-1} \frac{1}{1+\delta^*_j L^*_j(t)}$$

with $\eta^*(t) := \min\{m : T^*_m \ge t\}$ and so

$$\begin{aligned}\frac{dB^*_{n^*}(t)}{B^*_{n^*}(t)} &= (...)dt + d\ln B^*_{n^*} \\ &= (...)dt + \sigma^{*\top}_{\eta^*(t)} d\mathbf{W}^{(\cdot)} - \sum_{j=\eta^*(t)}^{n^*-1} \frac{\delta^*_j L^*_j}{1+\delta^*_j L^*_j} \Gamma_j^{*\top} d\mathbf{W}^{(\cdot)}.\end{aligned}$$

Note that $B^*_{n^*}(t)$ is the foreign terminal bond expressed in domestic currency. We so may set for $t \le T_\infty$, $B^*_{n^*}(t) =: \zeta(t)\widetilde{B}_{n^*}(t)$, where $\widetilde{B}_{n^*}(t)$ is a foreign bond expressed in the foreign currency and $\zeta(t)$ is the FX spot rate. In particular we have $B^*_{n^*}(T_{n^*}) = \zeta(T_{n^*}) = \zeta(T_n) = \zeta(T_\infty)$, and

$$\zeta(t) = \frac{B^*_{n^*}(t)}{\widetilde{B}_{n^*}(t)} = \frac{B^*_{\eta^*(t)}(t)}{\widetilde{B}_{\eta^*(t)}(t)}.$$

We thus have

$$\begin{aligned}\frac{dX}{X} &= (...)dt + \left(\sigma^{*\top}_{\eta^*(t)} - \sigma^\top_{\eta(t)}\right) d\mathbf{W}^{(\cdot)} \\ &\quad + \left(\sum_{j=\eta(t)}^{n-1} \frac{\delta_j L_j}{1+\delta_j L_j} \Gamma_j^\top - \sum_{j=\eta^*(t)}^{n^*-1} \frac{\delta^*_j L^*_j}{1+\delta^*_j L^*_j} \Gamma_j^{*\top} \right) d\mathbf{W}^{(\cdot)} \\ &=: (...)dt + \Gamma_X^\top d\mathbf{W}^{(\cdot)} = \Gamma_X^\top d\mathbf{W}^{(n)}\end{aligned}$$

with

$$\Gamma_X = \sigma^*_{\eta^*(t)} - \sigma_{\eta(t)} + \sum_{j=\eta(t)}^{n-1} \frac{\delta_j L_j}{1+\delta_j L_j}\Gamma_j - \sum_{j=\eta^*(t)}^{n^*-1} \frac{\delta^*_j L^*_j}{1+\delta^*_j L^*_j}\Gamma^*_j. \tag{21}$$

Let us further assume that

$$\frac{d\zeta}{\zeta} = (\cdots)dt + \left(\sigma^{fx}\right)^\top d\mathbf{W}^{(\cdot)}.$$

Since $B^*_{\eta^*(t)}(t) = \zeta(t)\widetilde{B}_{\eta^*(t)}(t)$ we then have $\sigma^*_{\eta^*(t)} = \sigma^{fx} + \widetilde{\sigma}_{\eta^*(t)}$ with $\widetilde{\sigma}_j$ being the volatility of the foreign zero bond maturing at T^*_j. From this we observe that Γ_X is completely determined by specification of σ^{fx} and the difference $\widetilde{\sigma}_{\eta^*(t)} - \sigma_{\eta(t)}$. Conversely, specifying Γ_X implicitly determines $\sigma^{fx} + \widetilde{\sigma}_{\eta^*(t)} - \sigma_{\eta(t)}$ via (21).

Remark 12.1. Moreover, in practice one may neglect the volatility of $B_{\eta(t)}(t)$ and $\widetilde{B}_{\eta^*(t)}(t)$, respectively, being the volatilities of zero bonds less than one period before maturity. We then have in approximation

$$\sigma^*_{\eta^*(t)} - \sigma_{\eta(t)} \approx \sigma^{fx}(t) \tag{22}$$

in (21).

More generally, for $i \le n^*$ and $j \le n$ we may consider the process $X_{i,j} := B^*_i(t)/B_j(t)$, $0 \le t \le T^*_i \wedge T_j$ (hence $X \equiv X_{n^*,n}$), and for its volatility $X_{i,j}$ we derive in a similar way,

$$\begin{aligned}\Gamma_{X_{i,j}} &= \sigma^*_{\eta^*(t)} - \sigma_{\eta(t)} + \sum_{k=\eta(t)}^{j-1} \frac{\delta_k L_k}{1+\delta_k L_k}\Gamma_k - \sum_{k=\eta^*(t)}^{i-1} \frac{\delta^*_k L^*_k}{1+\delta^*_k L^*_k}\Gamma^*_k \\ &= \Gamma_X + \sum_{k=i}^{n^*-1} \frac{\delta^*_k L^*_k}{1+\delta^*_k L^*_k}\Gamma^*_k - \sum_{k=j}^{n-1} \frac{\delta_k L_k}{1+\delta_k L_k}\Gamma_k, \quad t \le T^*_{i-1} \wedge T_{j-1}. \end{aligned} \tag{23}$$

So any $\Gamma_{X_{i,j}}$ is determined by Γ_X via (23).

Let us now consider an option to buy one unit of foreign currency for K units of domestic currency at time T^*_i, $i \le n^*$, and assume that $T^*_i = T_{i'}$ for a certain i'. Clearly, the net payoff of this option is

$$(\zeta(T^*_i) - K)^+ = (B^*_i(T^*_i) - K)^+ = \left(\frac{B^*_i(T_{i'})}{B_{i'}(T_{i'})} - K\right)^+ = (X_{i,i'}(T_{i'}) - K)^+,$$

and the option value in domestic currency at time $t = 0$ is given by

$$\begin{aligned} C_i(K) &:= B_n(0)E_n \frac{(\zeta(T^*_i) - K)^+}{B_n(T^*_i)} \\ &= B_{i'}(0)E_{i'}\left(X_{i,i'}(T_{i'}) - K\right)^+. \end{aligned}$$

For $i = n^*$, $i' = n$, we thus obtain by $T^*_{n^*} = T_n = T_\infty$,

$$C_{n^*}(K) := B_n(0) E_n \frac{(\zeta(T^*_{n^*}) - K)^+}{B_n(T_n)}$$
$$= B_n(0) E_n \left(X(T_\infty) - K\right)^+ . \tag{24}$$

We thus conclude that any standard FX option maturing on a joint tenor date $T^*_i = T_{i'}$ as described above may be priced, once the volatility process Γ_X is specified, via the formula

$$C_i(K) = B_{i'}(0) E_{i'} \exp\left[-\frac{1}{2}\int_0^{T_{i'}} \left\|\Gamma_{X_{i,j}}\right\|^2 dt + \int_0^{T_{i'}} \Gamma^\top_{X_{i,j}} d\mathbf{W}^{(i')}\right], \tag{25}$$

where $\Gamma_{X_{i,j}}$ follows from (23). Needless to say that a particular evaluation procedure for (25) largely depends on the specific structure of the respective volatility specifications for Γ, Γ^*, and Γ_X.

Example 1. In the special case of a (multi-factor) domestic and foreign *Libor Market Model,* that is Γ and Γ^* are deterministic vector functions, the $\Gamma_{X_{i,j}}$ may be obtained from Γ_X by standardly freezing the Libors in (23). If moreover Γ_X is taken to be deterministic as well, we may then compute all prices (25) by the Black 76 formula.

Remark 12.2. We further observe, for instance, that when Γ_X has a Heston type structure, like $\Gamma_X =: \beta_X \sqrt{V} \mathbf{e}_X$ for some deterministic β_X, unit vector $\mathbf{e}_X$, and square-root volatility process V, then due to (23) $\Gamma_{X_{i,j}}$ is essentially **not** of Heston type for $i < n^*$. In this respect we should note that the approach in [4], where simultaneously all the $\mathrm{FX}_i = \Gamma_{X_{i,i}}$ have a Heston type volatility structure seems to be inconsistent with this observation.

Remark 12.3. In the case where both the domestic and foreign model is a *one-factor Libor Market Model,* i.e. both Γ and Γ^* are deterministic scalar volatilities connected to a one dimensional Brownian motion, we are in a setting related to the one in [4] in a sense.

We continue with a further mild structural assumption on the process Γ_X, namely that it is of the form

$$\Gamma_X = \|\Gamma_X\| \left(\sum_{j=1}^{n-1} \rho_X \mathbf{e}_j + \sum_{j=1}^{n^*-1} \rho^*_X \mathbf{e}^*_j\right) =: \|\Gamma_X\| \, \mathbf{e}_X, \tag{26}$$

under the normalization condition

$$\|\mathbf{e}_X\|^2 = \left\|\sum_{j=1}^{n-1} \rho_X \mathbf{e}_j + \sum_{j=1}^{n^*-1} \rho^*_X \mathbf{e}^*_j\right\|^2 =$$

$$\rho_X^2 \sum_{j,j'=1}^{n-1} R_{jj'} + 2\varrho \rho_X \rho^*_X \sum_{j=1}^{n-1} \sum_{j'=1}^{n^*-1} \left[C_1 G_1^\top\right]_{jj'} + (\rho^*_X)^2 \sum_{j=1}^{n^*-1} \sum_{j'=1}^{n^*-1} R^*_{jj'} = 1, \tag{27}$$

where ρ_X and ρ_X^* are considered to be some kind of uniform partial correlation of the FX market with the domestic and foreign Libor rates, respectively. Further, in (26) $\|\Gamma_X\|$ is in general a scalar stochastic process that is still to be specified. In (8) we now have by (26) and (19),

$$\Gamma_i^{*\top}\Gamma_X = \|\Gamma_X\| \left(\sum_{j=1}^{n-1} \rho_X \Gamma_i^{*\top} \mathbf{e}_j + \sum_{j=1}^{n^*-1} \rho_X^* \Gamma_i^{*\top} \mathbf{e}_j^* \right)$$

$$\|\Gamma_X\| \left(\varrho\rho_X \|\Gamma_i^*\| \sum_{j=1}^{n-1} \left[C_1 G_1^\top\right]_{ji} + \rho_X^* \|\Gamma_i^*\| \sum_{j=1}^{n^*-1} R_{ij}^* \right). \tag{28}$$

In particular the correlations of X with the domestic and foreign Libors are given by

$$\begin{aligned} \operatorname{Corr}_{X,L_i} &= \sum_{j=1}^{n-1} \rho_X \mathbf{e}_j \cdot \mathbf{e}_i + \sum_{j=1}^{n^*-1} \rho_X^* \mathbf{e}_j^* \cdot \mathbf{e}_i \qquad (29)\\ &= \rho_X \sum_{j=1}^{n-1} R_{ij} + \varrho\rho_X^* \sum_{j=1}^{n^*-1} \left[C_1 G_1^\top\right]_{ij} \\ &=: \rho_X P_i + \varrho\rho_X^* Q_i, \quad 1 \le i < n, \quad \text{and} \\ \operatorname{Corr}_{X,L_i^*} &= \sum_{j=1}^{n-1} \rho_X \mathbf{e}_j \cdot \mathbf{e}_i^* + \sum_{j=1}^{n^*-1} \rho_X^* \mathbf{e}_j^* \cdot \mathbf{e}_i^* \qquad (30)\\ &= \varrho\rho_X \sum_{j=1}^{n-1} \left[C_1 G_1^\top\right]_{ji} + \rho_X^* \sum_{j=1}^{n^*-1} R_{ij}^* \\ &=: \varrho\rho_X Q_i^* + \rho_X^* P_i^*, \quad 1 \le i < n^*, \end{aligned}$$

respectively. Since all processes X, L_i, and L_i^* are observable at the market and the constants P_i, Q_i, and P_i^*, Q_i^*, in (29) and (30) are in principle known from the respective calibrations of the domestic and foreign Libor system, it seems natural to estimate the correlations in (29) and (30) from historical data. This may be done by minimizing the total square distance

$$\sum_{i=1}^{n} \left(\varrho\rho_X Q_i^* + \rho_X^* P_i^* - \widehat{\operatorname{Corr}}_{X,L_i} \right)^2$$

$$+ \sum_{i=1}^{n^*} \left(\varrho\rho_X Q_i^* + \rho_X^* P_i^* - \widehat{\operatorname{Corr}}_{X,L_i^*} \right)^2 \to \min_{\varrho,\rho_X,\rho_X^*}$$

with $\widehat{\operatorname{Corr}}_{X,L_i}$ and $\widehat{\operatorname{Corr}}_{X,L_i^*}$ being the respectively estimated correlations, under the normalization restriction

$$\rho_X^2 \sum_{i=1}^{n-1} P_i + 2\varrho\rho_X\rho_X^* \sum_{i=1}^{n-1} Q_i + (\rho_X^*)^2 \sum_{i=1}^{n-1} P_i^* = 1$$

(note that $\sum_{i=1}^{n-1} Q_i = \sum_{i=1}^{n^*-1} Q_i^*$ and cf. (27)). Note that after determination of Γ_X, the dynamics of the FX rate ζ are implicitly determined by (21) and (22), and are in particular driven by the total Brownian motion $\mathbf{W}$.

After identifying $\mathbf{e}_X$ in (26) in the above way, the norm process $\|\Gamma_X\|$ has to be modeled appropriately, such that calibration to a suitably large set of plain vanilla FX options is feasible. The most simple way is to assume that $\|\Gamma_X\|$ is deterministic (cf. Example 1). However the typically observed skew patterns in implied volatilities of vanilla FX options may not be captured in this way. Therefore more sophisticated choices are called for. Below we will sketch the procedure in the context of a Heston type model for X and a fairly generally structured domestic and foreign Libor model as described in Section 4.

Let us assume that

$$\frac{dX}{X} = \beta_X \sqrt{V} \mathbf{e}_X^\top d\mathbf{W}^{(n)}, \tag{31}$$

where V follows the square-root dynamics

$$dV = \kappa_X(\theta_X - V)dt + \sigma_X \sqrt{V} \left(\rho_X \mathbf{e}_X^\top d\mathbf{W}^{(n)} + \sqrt{1 - \rho_X^2} dW_X \right). \tag{32}$$

In (31) and (32) the parameters β_X, κ_X, θ_X, σ_X, and ρ_X are assumed to be constants, and W_X is an additional independent standard Brownian motion to inforce decorrelation between X and V. (Formally one might extend the vector $\mathbf{W}$ with an extra Brownian component and extend correspondingly all the unit vectors $\mathbf{e}_i$, $\mathbf{e}_i^*$, $\mathbf{e}_X$, with an extra zero component.) Subsequently we may calibrate the system (31)-(32) to a family of vanilla FX options (24) with different strikes and common maturity $T_n = T_{n*}^* = T_\infty$. This may be done in a standard way by using a relatively fast Fourier based pricing procedure. Although this pricing procedure is more or less standard, we still present it here for the convenience of the reader (cf. also [14]).

Let us write (24) as

$$C(K) := C_{n^*}(K) = B_n(0) E_n \left(X(0) e^{\ln \frac{X(T_\infty)}{X(0)}} - K \right)^+. \tag{33}$$

We may then apply the Fourier pricing method of Carr-Madan (spelled out later on) to the triple

$$\varphi(z\,; v), \quad X(0), \quad K,$$

where the characteristic function

$$\varphi(z\,; v) := E_n \left[e^{\mathrm{i} z \ln \frac{X(T_\infty)}{X(0)}} \,\middle|\, V(0) = v \right] \tag{34}$$

may be obtained as follows. Consider the logarithm of (31),

$$d \ln X = -\frac{1}{2} \beta_X^2 V dt + \beta_X \sqrt{V} \mathbf{e}_X^\top d\mathbf{W}^{(n)}, \tag{35}$$

along with the square-root dynamics (32). Let us then abbreviate $Y^{0,y,v}(t) := \ln X(t)$ with $Y^{0,y,v}(0) = \ln X(0) =: y$, and $V^{0,y,v}(t) := V(t)$ with $V^{0,y,v}(0) = V(0) =: v$. Then by (35), the generator of the vector process (Y, V) is given by

$$A := A_{y,v} := -\frac{1}{2}\beta_X^2 v dt \frac{\partial}{\partial y} + \kappa_X (\theta_X - v) \frac{\partial}{\partial v} + \frac{1}{2} v \beta_X^2 \frac{\partial^2}{\partial y^2} + v\beta_X \sigma_X \rho_X \frac{\partial^2}{\partial y \partial v} + \frac{1}{2}\sigma_X^2 v \frac{\partial^2}{\partial v^2}.$$

Let $\widehat{p}(z, z'; t, y, v)$ satisfy the Cauchy problem

$$\frac{\partial \widehat{p}}{\partial t} = A\widehat{p}, \qquad \widehat{p}(z, z'; 0, y, v) = e^{\mathrm{i}(zy + z'v)}. \tag{36}$$

Then

$$\widehat{p}(z, z'; t, y, v) = E e^{\mathrm{i}\left(zY^{0,y,v}(t) + z'V^{0,x,v}(t)\right)}.$$

We are only interested in the solution for $z' = 0$. Let us therefore consider the ansatz

$$\widehat{p}(z; t, y, v) = \exp\left(A(z;t) + B_0(z;t)y + B(z;t)v\right)$$

with

$$A(z;0) = 0, \qquad B_0(z;0) = \mathrm{i}z, \quad B(z;0) = 0. \tag{37}$$

Substitute this ansatz into (36) yields,

$$\left(\frac{\partial A}{\partial t} + \frac{\partial B_0}{\partial t} x + \frac{\partial B}{\partial t} v\right) = -\frac{1}{2} v\beta_X^2 B_0 + \kappa_X (\theta_X - v) B + \frac{1}{2} v\beta_X^2 B_0^2 + v\beta_X\sigma_X\rho_X B_0 B + \frac{1}{2}\sigma_X^2 v B^2,$$

and we so obtain the Riccati system

$$\begin{aligned}
\frac{\partial A}{\partial t} &= \kappa_X \theta_X B \\
\frac{\partial B_0}{\partial t} &= 0 \\
\frac{\partial B}{\partial t} &= -\frac{1}{2}\beta_X^2 B_0 - \kappa_X B + \frac{1}{2}\beta_X^2 B_0^2 + \beta_X\sigma_X\rho_X B_0 B + \frac{1}{2}\sigma_X^2 B^2.
\end{aligned}$$

In view of (37) we then get

$$\begin{aligned}
\frac{\partial A}{\partial t} &= \kappa_X \theta_X B \\
\frac{\partial B}{\partial t} &= -\frac{1}{2}\beta_X^2 \left(\mathrm{i}z + z^2\right) - \left(\kappa_X - \mathrm{i}z\beta_X\sigma_X\rho_X\right) B + \frac{1}{2}\sigma_X^2 B^2.
\end{aligned}$$

As a well known fact (see [11]) this system can be explicitly solved, but there are different representations for its solution depending on the chosen branch of the complex logarithm. We here use Lord and Kahl's representation due to the principal branch, see [15]*, to get

$$B(z;t) = \frac{a+d}{\sigma_X^2}\frac{1-e^{dt}}{1-ge^{dt}}$$

and

$$A(z;t) = \frac{\kappa_X\theta_X}{\sigma_X^2}\left\{(a-d)\,t - 2\ln\left[\frac{e^{-dt}-g}{1-g}\right]\right\},$$

where

$$\begin{aligned} a &= \kappa_X - \mathrm{i}z\beta_X\sigma_X\rho_X \\ d &= \sqrt{a^2 + (\mathrm{i}z + z^2)\,\beta_X^2\sigma_X^2} \\ g &= \frac{a+d}{a-d}. \end{aligned}$$

Taking all together we have with $t = T_\infty$ for (34),

$$\varphi(z\,;v) = e^{-\mathrm{i}z\ln X(0)}\widehat{p}(z\,;T_\infty, \ln X(0), v) = \exp\left(\widetilde{A}(z;T_\infty) + B(z;T_\infty)v\right) \tag{38}$$

with

$$\begin{aligned} B(z;T_\infty) &= \frac{a+d}{\sigma_X^2}\frac{1-e^{dT_\infty}}{1-ge^{dT_\infty}}, \quad \text{and} \\ \widetilde{A}(z;t) &:= \frac{\kappa_X\theta_X}{\sigma_X^2}\left\{(a-d)\,T_\infty - 2\ln\left[\frac{e^{-dT_\infty}-g}{1-g}\right]\right\}. \end{aligned}$$

Carr & Madan inversion formula

Due to Carr and Madan [7], the FX vanilla option price may be obtained by the following inversion formula,

$$\begin{aligned} C(K) &= B_n(0)(X(0)-K)^+ + \\ &\frac{B_n(0)X(0)}{2\pi}\int_{-\infty}^{\infty}\frac{1-\varphi(z-\mathrm{i};V(0))}{z(z-\mathrm{i})}e^{-\mathrm{i}z\ln\frac{K}{X(0)}}dz, \end{aligned} \tag{39}$$

where φ is given by (38). The integrand in (39) decays with rate z^{-2} if $|z| \to \infty$, which is relatively slow from a numerical point of view. Therefore it is better to modify the inversion formula in the following way. Let $\varphi^{\mathcal{B}}$ be the characteristic function (34) due to some Black model,

$$X(T_\infty) = X(0)e^{-\frac{1}{2}(\sigma^B)^2T_\infty + \sigma^B\sqrt{T_\infty}\varsigma}, \quad \varsigma \in N(0,1)$$

in the measure P_n, for a particular suitably chosen volatility σ^B. We then have (cf. Black's 76 formula)

$$E_n\left(X(T_\infty)-K\right)^+ = \mathcal{B}(X(0), T_\infty, \sigma^B, K),$$

*Roger Lord confirmed a typo in the published version in a personal communication and therefore referred to the preprint version.

where

$$\mathcal{B}(X,T,\sigma,K) := X\mathcal{N}(d_+) - K\mathcal{N}(d_-)\,, \quad \text{with}$$
$$d_\pm := \frac{\ln\frac{X}{K} \pm \frac{1}{2}\sigma^2 T}{\sigma\sqrt{T}}, \quad \text{and}$$
$$\varphi^{\mathcal{B}}(z\,;v) = \varphi^{\mathcal{B}}(z) = E_n e^{\mathrm{i}z\left(-\frac{1}{2}(\sigma^B)^2 T_\infty + \sigma^B\sqrt{T_\infty}\varsigma\right)}$$
$$= e^{-\frac{1}{2}(\sigma^B)^2 T_\infty (z^2+\mathrm{i}z)}.$$

By application of Carr and Madan's formula to the Black model we get,

$$C^{\mathcal{B}}(K) := B_n(0)\mathcal{B}(X(0),T_\infty,\sigma^B,K) = B_n(0)(X(0)-K)^+ \qquad (40)$$
$$+ \frac{B_n(0)X(0)}{2\pi}\int_{-\infty}^{\infty}\frac{1-\varphi^{\mathcal{B}}(z-\mathrm{i})}{z(z-\mathrm{i})}e^{-\mathrm{i}z\ln\frac{K}{X(0)}}dz,$$

and then by subtracting (40) from (39) we obtain,

$$C(K) = C^{\mathcal{B}}(K) + \qquad (41)$$
$$\frac{B_n(0)X(0)}{2\pi}\int_{-\infty}^{\infty}\frac{\varphi^{\mathcal{B}}(z-\mathrm{i};\cdot)-\varphi(z-\mathrm{i};V(0))}{z(z-\mathrm{i})}e^{-\mathrm{i}z\ln\frac{K}{X(0)}}dz.$$

Inversion formula (41) is usually much more efficient due to the typically much faster decaying integrand in comparison with (39).

Acknowledgements

The author thanks Dr. Suso Kraut (HSH Nordbank), Dr. Marcus Steinkamp (HSH Nordbank), and Dr. Stanley Mathew for fruitful discussions on this topic. Further, partial support of DFG Research Center MATHEON 'Mathematics for Key Technologies' in Berlin is gratefully acknowledged.

Bibliography

[1] Andersen, L. and Brotherton-Ratcliffe, R. (2001). Extended Libor Market Models with Stochastic Volatility. Working paper, Gen Re Securities.

[2] Belomestny, D., Mathew, S. and Schoenmakers, J. (2011). Multiple stochastic volatility extension of the Libor market model and its implementation. *Monte Carlo Methods Appl.* (2009), 15, no. 4, 285–310.

[3] Belomestny, D. and Schoenmakers, J. (2006). *A Jump-Diffusion Libor Model and its Robust Calibration, Quant. Finance,* **11**, pp. 529–546.

[4] Benner, W. and Zyapkov, L. (2008). A multifactor Cross-Currency LIBOR Market Model. *Banks and Bank Systems,* **3** (4), 73–84.

[5] Brigo, D. and Mercurio, F. (2001). *Interest rate models—theory and practice.* Springer Finance. Springer-Verlag, Berlin.

[6] Brace, A., Gatarek, D. and Musiela, M. (1997). The Market Model of Interest Rate Dynamics. *Mathematical Finance,* **7** (2), 127–155.

[7] Carr, P. and Madan, D. (1999). Option Valuation Using the Fast Fourier Transform, *Journal of Computational Finance*, **2**, 61–74.

[8] Eberlein, E. and Özkan, F. (2005). The Lévy Libor model, *Finance Stoch.* **7**, no. 1, 1–27.

[9] Glasserman, P. (2004). *Monte Carlo methods in financial engineering.* Applications of Mathematics (New York), **53**. Stochastic Modelling and Applied Probability. Springer-Verlag, New York.

[10] Hagan, P. and Lesniewski, A. (2008). LIBOR market model with SABR style stochastic volatility. Working paper.

[11] Heston, S. (1993). A closed-form solution for options with stochastic volatility with applications to bond and currency options. *The Review of Financial Studies*, **6**, No. 2, 327-343.

[12] Jamshidan, F. (1997). LIBOR and swap market models and measures. *Finance and Stochastics*, **1**, 293–330.

[13] Jamshidian, F. (2001). LIBOR Market Model with Semimartingales, in "Option Pricing, Interest Rates and Risk Management", Cambridge Univ.

[14] Ladkau, M., Schoenmakers, J. and Zhang, J. (2012). Libor market model with expiry-wise stochastic volatility and displacement. WIAS Preprint 1702.

[15] Lord, R. and Kahl, C. (2010). Complex logarithms in Heston-like models. *Math. Fin.*, **20**, Issue 4, 671–694.

[16] Miltersen, K., Sandmann, K. and Sondermann, D. (1997). Closed-form solutions for term structure derivatives with lognormal interest rates. *Journal of Finance*, 409-430.

[17] Papapantoleon, A., Schoenmakers, J. and Skovmand, D. (2011). Efficient and accurate log-Levy approximations to Levy driven LIBOR models. *J. of Computational Finance* (to appear).

[18] Schoenmakers, J. (2005). *Robust Libor Modelling and Pricing of Derivative Products.* BocaRaton London NewYork Singapore: Chapman & Hall – CRC Press.

[19] Wu, L. and Zhang, F. (2006). Libor Market Model with Stochastic Volatility. *Journal of Industrial and Management Optimization*, **2**, 199–207.

Chapter 13

Dimension-wise decompositions and their efficient parallelization

Philipp Schröder, Peter Mlynczak and Gabriel Wittum
Goethe-Center for Scientific Computing, Goethe-University,
Kettenhofweg 139, D-60325 Frankfurt am Main, Germany

1. Introduction

Many problem classes in finance lead to high dimensional partial differential equations which need to be solved efficiently. To circumnavigate the *curse of dimension* several methods exist, e.g. dimension-wise decomposition techniques. In this chapter we will present an overview over the different methods available to cope with high dimensional problems. The class of dimension-wise decomposition methods, which we will discuss in detail, decomposes a high dimensional problem into a set of low dimensional problems. Dependent on the dimension d of the total problem, and the order of the decomposition method the number of low-dimensional problems can be quite large. This makes efficient parallelization techniques necessary.

2. High-dimensional problems in computational finance

In the following section we will introduce two example problems from computational finance which are of high dimension. The first will be the standard Black-Scholes equation for the pricing of an option contract. The other will be an implementation of the Libor Market Model (LMM) for the pricing of interest rate derivatives.

2.1. *Equity options: The Black-Scholes Model*

In one dimension, the development of an asset over time is described in financial theory using a geometric Brownian motion follows the stochastic differential equation:

$$dS(t) = S(t)\mu dt + S(t)\sigma dW(t) \quad t \geq 0, S(t = 0) = a. \tag{1}$$

In higher dimensions the developement of a set of (correlated) assets over time is equally described by a n-dimensional geometric Brownian motion. Using the Feynman-Kac formula, the partial differential equation, which is the solution to this SDE can be easily calculated, and we arrive at the multidimensional expansion to the famous Black-Scholes equation.

$$\frac{\partial u}{\partial t} - r\sum_{i=1}^{d} S_i \frac{\partial u}{\partial S_i} - \frac{1}{2}\sum_{i,j=1}^{d} \sigma_i \sigma_j \rho_{ij} S_j S_j \frac{\partial^2 u}{\partial S_i \partial S_j} + ru = 0 \tag{2}$$

with initial condition

$$u(\mathbf{S},0) = g(\mathbf{S}) := \begin{cases} (\sum_i \mu_i S_i - K)_+ & \forall \mathbf{S} \in \mathbb{R}_+^d \\ (K - \sum_i \mu_i S_i)_+ & \forall \mathbf{S} \in \mathbb{R}_+^d \end{cases} . \tag{3}$$

The initial condition is specific to the pricing problem which is to be solved. In this case the initial condition presented here corresponds to a European style put option with arithmetic average.

By Girsanov's theorem this PDE is closely related to the expected value

$$u(\mathbf{S},T) = e^{-rt} E_Q[g(\mathbf{S}(T))] \tag{4}$$

under the risk-neutral measure Q.

2.1.1. *Transformations of the Black-Scholes equation*

Several coordinate transformations of the Black-Scholes equation are known which offer different advantages compared to the original approach. We will present the most common transformations here.

The most common transformation is the so called *Log Price* transformation. It is achieved by defining $y_i := log(S_i)$. This leads to transformations

$$\partial S_i \to S_i \partial y_i$$

and

$$\sum_{i,j} \partial S_i \partial S_j \to \sum_{i,j} \left(-\delta_{ij} \frac{1}{S_i^2} + \frac{1}{S_i S_j} \partial y_i \partial y_j \right)$$

by which we arrive at the Log-Price representation:

$$\frac{\partial u}{\partial t} - \sum_{i=1}^{d} (r - \frac{1}{2}\sigma_i^2) \frac{\partial u}{\partial y_i} - \frac{1}{2}\sum_{i,j=1}^{d} \sigma_i \sigma_j \rho_{ij} \frac{\partial^2 u}{\partial y_i \partial y_j} + ru = 0 \quad \forall (\mathbf{y},t) \in \mathbb{R}^d \times (0,T) \tag{5}$$

with initial conditions

$$u(\mathbf{y},0) = \left(K - \sum_{i=1}^{d} \mu_i e^{y_i} \right) \quad \forall \mathbf{y} \in \mathbb{R}^d. \tag{6}$$

This representation of (2) has the advantage, that the diffusion as well as the convection term have constant coefficients. A disadvantage of the transformation, however, is that the orginally only one-sided unbounded domain is now unbounded on both sides.

Another possible transformation is closely related to a principal component analysis of the covariance matrix. The starting point is the representation with constant coefficients. Transforming the equation into the eigensystem of the covariance matrix

$$\mathbf{\Sigma}_{ij} := \rho_{ij}\sigma_i\sigma_j \tag{7}$$

by an orthogonal transformation with $\mathbf{Q}$ bringing $\mathbf{\Sigma}$ to diagonal form

$$\mathbf{Q}\mathbf{\Sigma}\mathbf{Q}^T = diag(\lambda_i). \tag{8}$$

To receive this we transform using

$$\mathbf{z} = \mathbf{Q}\mathbf{y} \tag{9}$$

and receive the equation

$$\frac{\partial u}{\partial t} - \sum_{i,j=1}^{d} q_{ij}(r - \frac{1}{2}\sigma_i^2)\frac{\partial u}{\partial z_i} - \frac{1}{2}\sum_{i=1}^{d}\lambda_i\frac{\partial^2 u}{\partial z_i^2} + ru = 0 \quad \forall(\mathbf{z}, t) \in \mathbb{R}^d \times (0, T) \tag{10}$$

with initial condition

$$u(\mathbf{z}, 0) = \left(K - \sum_{i=1}^{d}\mu_i e^{\sum_{j=1}^{d} q_j i z_j}\right)_+ . \tag{11}$$

Performing an additional translation

$$\mathbf{x} := \mathbf{z} + t\mathbf{b} \tag{12}$$

with

$$b_i := \sum_{j=1}^{d} q_{ij}(r - \frac{1}{2}\sigma_j^2) \tag{13}$$

and after the substitution $v = e^{rt}$ we arrive at the heat-equation

$$\frac{\partial u}{\partial t} - \frac{1}{2}\sum_{i=1}^{d}\lambda_i\frac{\partial^2 u}{\partial x_i^2} = 0 \quad \forall(\mathbf{x}, t) \in \mathbb{R}^d \times (0, T) \tag{14}$$

with intial condition

$$u(\mathbf{x}, 0) = \left(K - \sum_{i=1}^{d}\mu_i e^{\sum_{j=1}^{d} q_j i x_j}\right)_+ . \tag{15}$$

At this point another relation to the representation in integral form, respectively via an expectation value can be seen. As it is known, the heat equation has a solution by integration with Green's function:

$$u(\mathbf{x_0}, t) = \frac{e^{-rt}}{\prod_{i=1}^{d}(2\pi t\lambda_i)^{1/2}} \times \int_{\mathbb{R}^d}\left(K - \sum_{i=1}^{d}\mu_i e^{\sum_{j=1}^{d} q_{ji} x'_j}\right)_+ e^{-\frac{1}{2t}\sum_{i=1}^{d}(x_i^0 - x'_i)^2/\lambda_i} d\mathbf{x}'. \tag{16}$$

Here the relation to the original variables $\mathbf{S_0}$ is given by

$$\begin{aligned} x_i &= z_i + tb_i \\ &= \sum_{j=1}^{d} q_{ij} y_j + t \sum_{j=1}^{d} q_{ij}(r - \frac{1}{2}\sigma_j^2) \\ &= \sum_{j=1}^{d} q_{ij} \left(y_j + t(r - \frac{1}{2}\sigma_j^2) \right) \\ &= \sum_{j=1}^{d} q_{ij} \left(lnS_j^0 + t(r - \frac{1}{2}\sigma_j^2) \right). \end{aligned}$$

2.2. *Interest-rate derivatives: The Libor-Market-Model*

The Libor-Market-Model, also LMM or BGM (Brace, Gatarek, Musiela) Model has first been described in 1997 by the authors who gave the model its name in [4]. Today the model is the market standard for the pricing of interest rate derivatives depending on several interest rates (e.g. the 3-month EURIBOR rate and the 6-month EURIBOR rate). The model assumes the developement of the individual forward-rates dL_i by

$$dL_i = \mu_i dt + \sigma_i dW^i. \tag{17}$$

Note, that in this case μ_i and σ_i can also be time dependent. The modeled forward-rates can be directly observed in the market. This is also the main advantage compared to the approach of the Heath-Jarrow-Morton Model, which models instantaneous forward rates, which can not be directly observed.

For pricing within the LMM-framework, the different forward-rates have to be seen under a common probability measure, which again will be the martingale measure Q for risk-neutral pricing. One possible choice for this measure would be the so called terminal measure Q_{T_N} which is the martingale measure for the forward-rate with the longest maturity. All other forward-rates have to be transformed into this measure by use of the Radon-Nikodym theorem. We arrive at the formulation of the model under the terminal measure:

$$dL_i = L_i\mu_i + L_i\sigma_i dW^N \tag{18}$$

with

$$\mu_i(t) := \begin{cases} -\sigma_i(t) \sum_{j=i+1}^{N-1} \frac{\delta_j L_j}{1+\delta_j L_j} \rho_{ij}\sigma_j & j < N-1 \\ 0 & j = N-1 \end{cases}. \tag{19}$$

Note that only the terminal forward-rate is a martingale under this measure.

Having derived the dynamics of the forward-rates under a common measure, we can use Itō's lemma to arrive at following set of partial differential equations:

$$\frac{\partial u}{\partial t} + \frac{1}{2} \sum_{i,j=1}^{N-1} \rho_{ij}\sigma_i\sigma_j L_i L_j \frac{\partial^2 u}{\partial L_i \partial L_j} + \sum_{i=1}^{N-1} \mu_i L_i \frac{\partial u}{\partial L_i} = 0 \tag{20}$$

with initial conditions

$$\mu_i := \begin{cases} -\sigma_i \sum_{j=i+1}^{N-1} \frac{\delta_j L_j}{1+\delta_j L_j} \rho_{ij}\sigma_j & j < N-1 \\ 0 & j = N-1 \end{cases}. \tag{21}$$

2.2.1. *Pricing example: Bermuda swaption*

We conclude this section with an example for the pricing of an interest-rate derivative with the Libor-Market-model. Additionally to the partial differential equation and the model parameters, we also need to specify the option-type specific initial conditions.

A Bermuda swaption gives the bearer the right but not the obligation to enter into an interest-rate swap contract at a predetermined set of dates T_i. The value of the Bermuda swaption therefore is the maximum of an interest-rate swap entered at date T_i and a Bermudan swaption with the next (and all of the remaining exercise dates) T_{i+1}:

$$u_{BSw}(T_i, \ldots, T_N) = \max\left(u(T_i, \ldots, T_N); u_{BSw}(T_{i+1}, \ldots, T_N)\right). \tag{22}$$

Depending on the number of periods, the problem results in a $N - i$ dimensional problem. Since normal swaps have maturities of ten and more years as well as quarterly payment dates, we can easily reach dimensions of 40 and more. First attempts to solve such high dimensional problems were discussed in [3] and [16]. Using sparse grids the authors were able to solve problems with up to five spatial dimensions.

3. Methods for high dimensional partial differential equations

In this section we will discuss some of the up-to-date methods for high dimensional partial differential equations. We will focus on methods with applications to finance but will also present methods with possible applications to finance which to our knowledge have not yet been realized

Methods for high dimensional equations – except for analytical methods – can be put into three categories.

The first category is a reduction of the problem dimension itself. By use of the Karhunen-Loeve transformation the stochastical differential equation which governs the dynamics of the model can be reduced to a lower dimensional equation. This results naturally in a lower dimensional partial differential equation which describes the solution to the SDE.

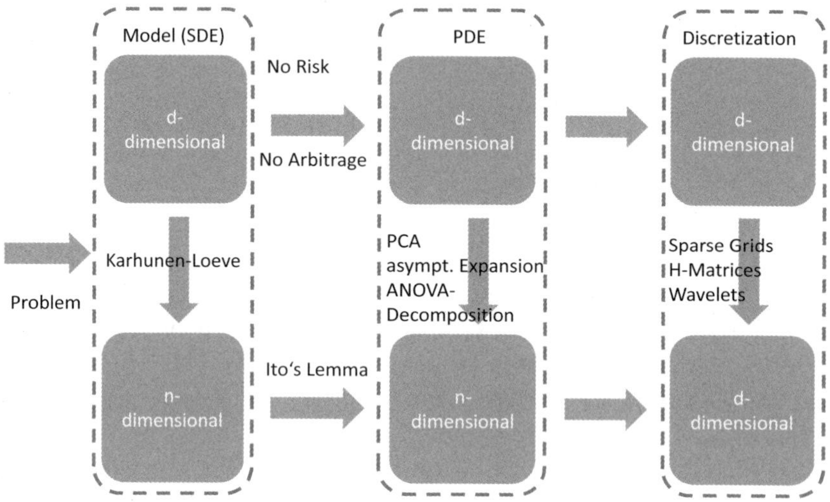

Figure 1. Overview over methods for high dimensional PDEs.

As a second possibility, the dimension of the partial differential equation can be reduced. By transformation it is possible to derive a lower dimensional approximation to the original equation. Several different approaches exist, with different properties and approximation errors. The method set of dimension-wise decompositions, which we will discuss in length later in this chapter belongs to this class of methods.

The last possibility to reduce the complexity, respectively the high dimensionality of the problem is the discretization layer. This is also the area where the broadest range of different methods exits. The method of sparse grids, which we will discuss and parallelize later, belongs to this class of methods. An overview of the different method classes can be seen in Figure 1.

3.1. *Principal Component Analysis*

Principal Component Analysis (short PCA) is a well known technique from statistics. In it an optimized representation of the variables is found (with respect to the orthogonal distance). This corresponds to an eigenvalue decomposition of the variables covariance matrix.

PCA furthermore has a close connection to linear regression. Whereas with linear regression one tries to minimize the distance in the y-direction (or in direction of the dependent variable) the principal component analysis minimizes the orthogonal distance, as can be seen in figure 2. This also ensures that the resulting variables (in the eigensystem of the covariance matrix) are uncorrelated.

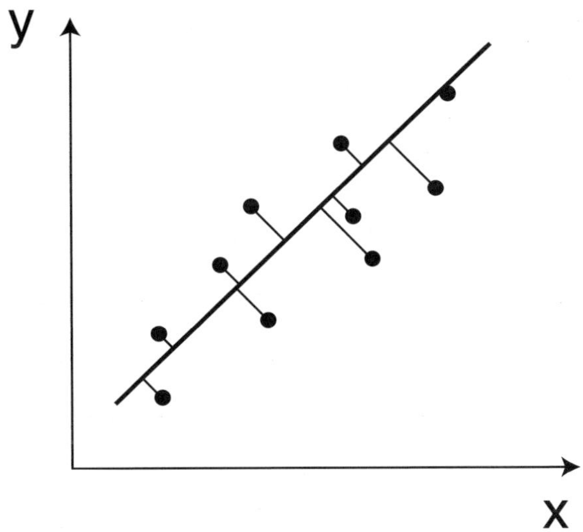

Figure 2. Orthogonal distance of sample datapoints.

Mathematically we are looking for a transformation, which finds a new coordinate system by

$$\mathbf{x} = \mathbf{Q}\mathbf{y}. \tag{23}$$

while transforming in a way such that the variance of each individual variables variance

$$Var(x_i) = \mathbf{q_i^T \Sigma q_i} \tag{24}$$

is maximized under the constraint

$$(\mathbf{q_i}, \mathbf{q_j}) = 0 \qquad \forall i \neq j \tag{25}$$

$$(\mathbf{q_i}, \mathbf{q_i}) = 1. \tag{26}$$

Since this corresponds to the eigenvalue decomposition of the covariance matrix, $\mathbf{Q}$ consists of the eigenvectors and the λ_i are the corresponding eigenvalues of the covariance matrix. Furthermore, the magnitude of the λ_i defines the ratio of total variance that is captured within the variable x_i.

A reduction of the total dimension can the be reached by omitting variables in the transform (by omitting the respective eigenvectors). To minimize the projection error, one omits the eigenvectors corresponding to the smalles eigenvalues.

3.2. *Tensor-product methods*

In the next two subsections we will discuss two methods which do not try to reduce the model equation but try to minimize the complexity of the numerical approximation of the solution function in the domain by a clever choice of basis functions. The first approach of this kind is the tensor-product approximation.

The tensor product approximation was first introduced by Beylkin and Mohlenkamp in [1] and [2], where the authors introduce the so called *separation representation.* The idea is to broaden the *separation of variables* approach; instead of using the simple representation for the function $f(\mathbf{x})$

$$f(x_1, \dots, x_d) \approx \phi_1(x_1) \cdot \ldots \cdot \phi_d(x_d) \tag{27}$$

we consider the linear combination

$$f(x_1, \dots, f_d) = \sum_{i=1}^{r} s_i \phi_1^i(x_1) \cdot \ldots \cdot \phi_d^i(x_d) + \mathcal{O}(\varepsilon). \tag{28}$$

Here r denotes the so called separation rank and s_i the separation coefficient.

Let $\mathbf{F}$ be the representation of a discrete function on a d-dimensional rectangular grid

$$\mathbf{F} = F(j_1, \dots, j_d) \tag{29}$$

with

$$j_i = 1, \dots, M_i, \tag{30}$$

then one can find a representation for $\mathbf{F}$ by

$$\mathbf{F} \approx \sum_{i=1}^{r} s_i F_1^i(j_1) \cdot \ldots \cdot F_d^i(j_d) \tag{31}$$

or using tensor product notation

$$\sum_{i=1}^{r} s_i \mathbf{F}_1^i \otimes \ldots \otimes \mathbf{F}_d^i. \tag{32}$$

Linear operators in d dimensions represented in their discrete form by matrices

$$\mathcal{A} = A(j_1, j_1'; \dots; j_d, j_d'), \tag{33}$$

can be written in separation representation

$$\mathcal{A} = \sum_{i=1}^{r} s_i A_1^i(j_1, j_1') \cdot \ldots \cdot A_d^i(j_d, j_d'), \tag{34}$$

where the A_l^i are one-dimensional matrices, i.e. linear operators on one-dimensional functions in the sense as defined above.

As it is shown in [2] the following operations can be performed on operators in separation representation

Remark 13.1 (basic linear algebra). *Let d be the dimension, r the separation rank and M as defined above. The following estimations hold for linear operators in the separation representation:*

- ***Memory requirements:*** *The memory requirement for $\mathbf{F}$ in separated representation is of order $\mathcal{O}(d \cdot r \cdot M)$.*
- ***Multiplication:*** *The product of two grid functions in separation representation is calculated with a number of operations of order $\mathcal{O}(d \cdot r \cdot \tilde{r} \cdot M)$ by*

$$(\mathbf{F}, \tilde{\mathbf{F}}) = \sum_{i=1}^{r} \sum_{\tilde{i}=1}^{\tilde{r}} s_i \tilde{s}_{\tilde{i}} (\mathbf{F}_1^i, \tilde{\mathbf{F}}_1^{\tilde{i}}) \cdot \ldots \cdot (\mathbf{F}_d^i, \tilde{\mathbf{F}}_d^{\tilde{i}}). \tag{35}$$

- ***Matrix-Vector multiplication:*** *The complexity of a matrix-vector multiplication in separation representation is of order $\mathcal{O}(d \cdot r_{\mathcal{A}} \cdot r_{\mathbf{F}} \cdot M^2)$ and the product can be calculated through*

$$\mathbf{G} := \mathcal{A}\mathbf{F} = \sum_{i=1}^{r_{\mathcal{A}}} \sum_{\tilde{i}=1}^{r_{\mathbf{F}}} s_i^{\mathcal{A}} s_{\tilde{i}}^{\mathbf{F}} (\mathcal{A}_1^i \mathbf{F}_1^{\tilde{i}}) \otimes \ldots \otimes (\mathcal{A}_d^i \mathbf{F}_d^{\tilde{i}}). \tag{36}$$

One can easily see that the complexity of all operations is only linear in the total dimension d. Furthermore, the respective separation ranks are relevant for the complexity estimates. In general the resulting object, while still in representation will have a higher separation rank than the original objects. In the case of a matrix-vector multiplication the separation rank will be

$$r_{\mathbf{G}} = r_{\mathcal{A}} r_{\mathbf{F}}. \tag{37}$$

This is, of course, only a worst case estimation, however the need for a method to minimize the separation rank becomes obvious. The authors in [2] propose an alternating least-squares approach to minimize the separation rank. We note here, that a nearly optimal representation may be preferable to a full optimization of the separation rank due to a much higher computational effort of the full optimization.

3.2.1. *Kronecker-tensor-product representation*

In [10] the authors define a tensor product representation on basis of the kronecker product

$$\mathcal{A} \times \mathcal{B} = [a_{ij}]\mathcal{B}, \tag{38}$$

whiche means that each element of the matrix $\mathcal{A}$ is multiplied with matrix $\mathcal{B}$. If $\mathcal{A} \in \mathbb{R}^{n \times m}$ and $\mathcal{B} \in \mathbb{R}^{r \times p}$, then the resulting matrix is of dimensionality $\mathbb{R}^{nr \times mp}$.

The representation of a matrix in *hierarical Kronecker-tensor-product* format (HKT) is being formulated by the sum Kronecker-tensor-products

$$\mathcal{A} = \sum_{i=1}^{r} U_i \times V_i, \tag{39}$$

where the individual Kronecker factors U_i and V_i are being represented by $\mathcal{H}$-matrices. By this representation we are able to compute a matrix vector multiplication

$$\mathcal{A}\mathbf{F} = \sum_{i=1}^{r} \sum_{j=1}^{r_\mathbf{F}} (U_i \mathbf{F}_j^1) \times (V_i \mathbf{F}_j^2) \tag{40}$$

with a complexity of $\mathcal{O}(r_\mathbf{F} rs\sqrt{N}\, log\, N)$ if $\mathbf{F}$ has a representation like

$$\mathbf{F} = \sum_{i=1}^{r} \mathbf{F}_i^1 \times \mathbf{F}_i^2. \tag{41}$$

As before, r denotes the representation rank of the tensor product representation and s the maximal rank of the $\mathcal{H}$-matrices. Further literature on $\mathcal{H}$-matrices can be found in [9].

As we can see, by using $\mathcal{H}$-matrices the complexity estimate for a matrix-vector multiplication in separation representation can be reduced from $\mathcal{O}(dr_\mathcal{A} r_\mathbf{F} N^2)$ to $\mathcal{O}(r_\mathbf{F} rs\sqrt{N}\, log\, N)$ and is no longer quadratically dependent on the number of nodes in each dimension. The HKT representation therefore enables the efficient solution of problems with several hundred dimensions (see [10]).

3.3. *Sparse Grids*

3.3.1. *The foundations of sparse grids*

Sparse grids are commonly used throughout numerical analysis for the discretization of high dimensional problems. They were first described in [17] for the numerical integration of high dimensional problems. Since then several extensions to the sparse grid quadrature have been developed, e.g. in [5]. The idea behind sparse grids is to use a hierachical tensor product representation to reduce the complexity of the numerical discretization and thereby the complexity of the numerical solution of the problem. We define a hierachical difference space $W_\mathbf{l}$ by

$$W_\mathbf{l} := V_\mathbf{l} \setminus \bigoplus_{j=1}^{d} V_{\mathbf{l}-\mathbf{e_j}} \tag{42}$$

with $\mathbf{l} \in \mathbb{N}^d$ as multi-index and $\mathbf{e_j}$ as *jth* unit vector. With n being the number of refinements, we can describe a cartesian grid V_n as

$$V_n := \bigoplus_{|\mathbf{l}|_\infty \leq n} W_\mathbf{l} \tag{43}$$

where

$$|\mathbf{l}|_\infty := \max_{1\leq j\leq n} \mathbf{l_j}. \tag{44}$$

Since our aim is to reduce the number of nodes needed for calculation of the solution we define the sparse grid function space $V_n^s \subset V_n$ via

$$V_n^s := \bigoplus_{|\mathbf{l}|_1 \leq n} W_{\mathbf{l}}. \tag{45}$$

The motivation behind this choice is to neglect those basis function with a small support in the function representation.

The advantage of sparse grids compared to cartesian grids is that the number of nodes is being reduced, as can be seen in figure 3. We note, however, that only the number of nodes inside the domain is being reduced, whereas the number of nodes on the edge is equal to the case of cartesian grid representation.

A more detailed investigation shows, that in d dimensions and on a grid level n

$$\binom{n+d-1}{d-1} \tag{46}$$

grids each with $\mathcal{O}(2^n)$ nodes are needed. The respective grids are highlighted in figure 4 on the right side. The complexity of a problem discretized using sparse grids can therefore be estimated by $\mathcal{O}(2^n n^{d-1})$ instead of $\mathcal{O}(2^{nd})$ as in the case of cartesian grids.

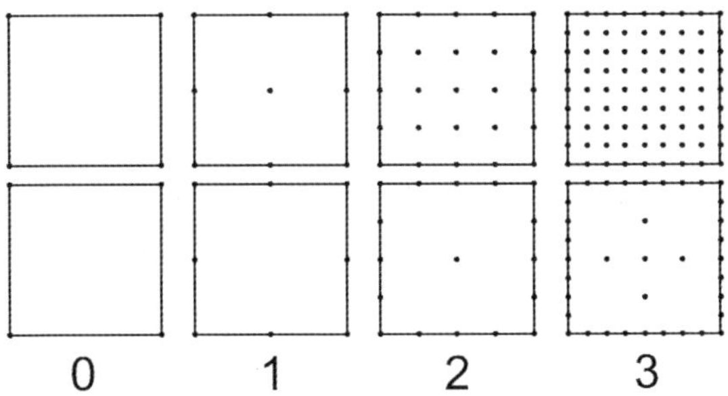

Figure 3. Grid points in a cartesian and in a sparse grid.

The numerical error of sparse grids can be estimated via

$$||u - u_n||_\infty \leq \frac{1}{6^d} \left\| \frac{\partial^{2d} u}{\partial x_1^2 \cdots \partial x_d^2} \right\|_\infty \cdot \left(1 + \sum_{i=1}^{d-1} \left(\frac{3}{4}\right)^i \cdot \binom{n+i-1}{i} \right) h_n^2. \tag{47}$$

Here it is interesting to note the requirement for bounded mixed derivatives. In particular for option pricing problems with their non-smooth initial conditions this can pose a limit to the convergence of sparse grids (see [14] for more details).

A visualization of the subspaces in a cartesian and a sparse grid are represented in figure 4. Here $\mathbf{i} = (i_1, .., i_n)$ denotes the position in the "grid table" and $\leq_i$ is to be interpreted component wise. We show the cartesian grid on the left side and the corresponding sparse grid on the right side. A position $\mathbf{i} = (i_1, .., i_n)$ corresponds to a grid which has been refined i_j-times in the x_j direction.

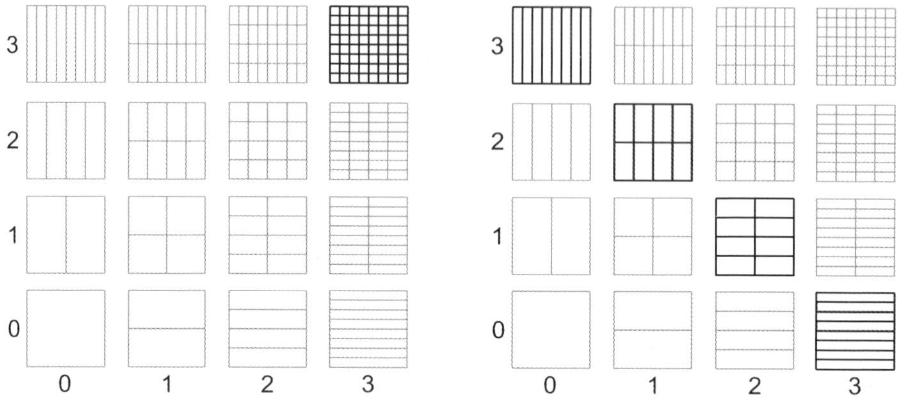

(a) cartesian grid (b) sparse grid

Figure 4. Grid table for a two-dimensional grid.

3.3.2. *Combination technique for sparse grids*

By using the combination technique for sparse grids introduced in [8] one reaches a given complexity of $\mathcal{O}(2^n n^{d-1})$ for the calculations and, as it is shown in [15] an asymptotic error equal to the one from equation (47). This is possible by using only the subgrids on the diagonal and the subdiagonal in figure 4. The solutions on the highlighted grids can be computed independently. The total solution can be calculated by summation over the subgrids on the diagonal in figure 4. By the substraction of the grids on the subdiagonal, the correct weighting of the nodes can be assured. For the sake of transparency, we only show the 2D case with grid level 2 here. Higher grid levels and dimensions can be calculated in the same way.

The numerical error of the sparse grid combination technique has been derived in [15]:

$$|u - u_n| \leq \frac{2K}{(d-1)!} \left(\frac{5}{2}\right)^{d-1} (n + 2(d-1))^{d-1} 4^{-n}. \tag{48}$$

The combination technique, compared to the usual sparse grid approach, has two main advantages: First, the complexity of the problem is further reduced since it is decomposed onto several subgrids. Secondly, the solutions on the individual subgrids are independent, which enables an easy parallelization approach to be discussed later on.

4. Dimension-wise decomposition methods

4.1. *ANOVA Decompositions*

In general, the ANOVA decomposition expands the multivariate function $f(\mathbf{x})$ into the sum

$$f(\mathbf{x}) = \sum_{u \subseteq \mathcal{D}} f_u(\mathbf{x_u}). \tag{49}$$

The individual terms here can be obtained via the projection

$$P_u f(\mathbf{x_u}) := \int_{\Omega^{d-|u|}} f(\mathbf{z}) d\mu_{\mathcal{D}\setminus u}(\mathbf{x}) \tag{50}$$

and the recursive definition

$$f_{\mathbf{u}}(\mathbf{x_u}) := P_{\mathbf{u}} f(\mathbf{x_u}) - \sum_{\mathbf{v} \subset \mathbf{u}} f_{\mathbf{v}}(\mathbf{x_v}). \tag{51}$$

Here $\mathbf{x_u}$ denotes the $|\mathbf{u}|$-dimensional vector that consists solely of the components of the index $\mathbf{u}$, where $|\mathbf{u}|$ is the length of $\mathbf{u}$ which is also called order of $\mathbf{u}$. In addition, μ_j denotes probability measures on the borel sets of $\Omega_j \subseteq \mathbb{R}$ and $\Omega^{\mathbf{u}}$ the $|\mathbf{u}|$-dimensional product set

$$\Omega^{\mathbf{u}} := \bigotimes_{j \in \mathbf{u} \subseteq \mathcal{D}} \Omega_j \tag{52}$$

with $\mathcal{D} = \{1, \ldots, d\}$. Furthermore, we define a d-dimensional product measure by

$$d\mu(\mathbf{x}) = \prod_{j=1}^{d} d\mu_j(x_j). \tag{53}$$

Let $V^d := \mathcal{L}^2(\Omega^d, \mu)$ be the Hilbert space of square integrable functions with the scalar product

$$(f, g)_\mu := \int_{\Omega^d} f(\mathbf{z}) g(\mathbf{x}) d\mu(\mathbf{x}). \tag{54}$$

4.1.1. *Classical ANOVA decomposition*

By choosing the Lebesgue measure $\mu(x_i) = dx_i$ as probability measure and $\Omega_i = [0, 1]$, one can construct the so called classical ANOVA decomposition with the following projections

$$P_{\mathbf{u}} f(\mathbf{x_u}) = \int_{[0,1]^{d-|\mathbf{u}|}} f(\mathbf{x}) d\mathbf{x}_{\mathcal{D}\setminus\mathbf{u}}. \tag{55}$$

While the classical ANOVA provides many desirable features, e.g. easy access to error estimators and smoothing properties of the projection, it has the clear disadvantage, that even for the constant term $f_\emptyset$ one has to solve an integration problem of dimension d. Since this is neither desirable and mostly simply not possible, we introduce the Anchor ANOVA decomposition in the following section, whose projections require much less effort.

4.1.2. *Anchor ANOVA decomposition*

Due to its complexity advantages, the Anchor ANOVA decomposition is commonly used when handling high dimensional problems. It can be constructed by choosing the Dirac measure $\mu(x_i) = \delta(x_i - a_i)dx_i$ as probability measure as well as $\Omega_i = [0,1]$, where $\mathbf{a} \in \Omega^d$ designates the so called anchor point which gives the decomposition its name. The Anchor ANOVA decomposition can be constructed through the following projections:

$$P_{\mathbf{u}} f(\mathbf{x_u}) = \int_{[0,1]^{d-|\mathbf{u}|}} f(\mathbf{x})\delta(\mathbf{x} - \mathbf{a})d\mathbf{x}_{\mathcal{D}\setminus\mathbf{u}} \tag{56}$$

$$= f(\mathbf{x})|_{\mathbf{x}=\mathbf{a}\setminus\mathbf{x_u}}. \tag{57}$$

Here $f(\mathbf{x})|_{\mathbf{x}=\mathbf{a}\setminus\mathbf{x_u}}$ means that

$$x_i = a_i \qquad \forall i \notin \mathbf{u}, \tag{58}$$

where $f_{\mathbf{u}}$ can be understood as behavior of $f(\mathbf{x})$ in the hyperplane through $\mathbf{a}$. In this context the decomposition is also named CUT-HDMR in [13]. [7] provides a good summary of the methodology and some generalizations, as well as a link between the Anchor ANOVA decomposition and the multivariate Taylor-expansion.

For a better understanding of the effects of the Anchor ANOVA's projections, we show the one- and two-dimensional projections of a simple forward contract as well as a geometric basket option. The one-dimensional projections are shown in figure 5, while the two-dimensional projections $f_{12}(\mathbf{x})$ are shown in figure 6.

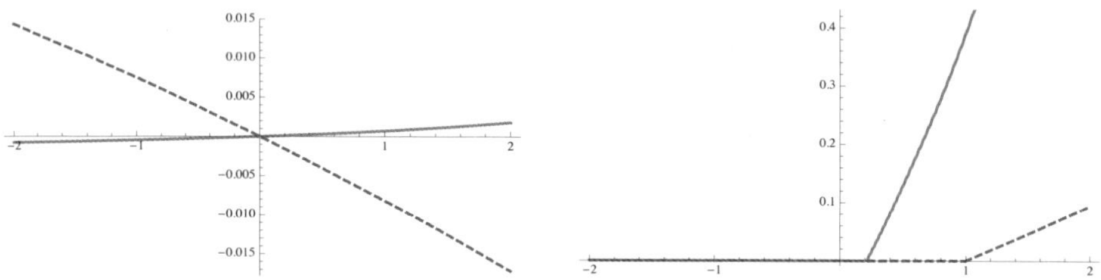

(a) Forward (b) Basket option

Figure 5. One-dimensional projections of the Anchor ANOVA decomposition.

For the basket option, in both the one-dimensional and the two-dimensional case, one can clearly observe the non-smoothing effect of the Anchor ANOVA.

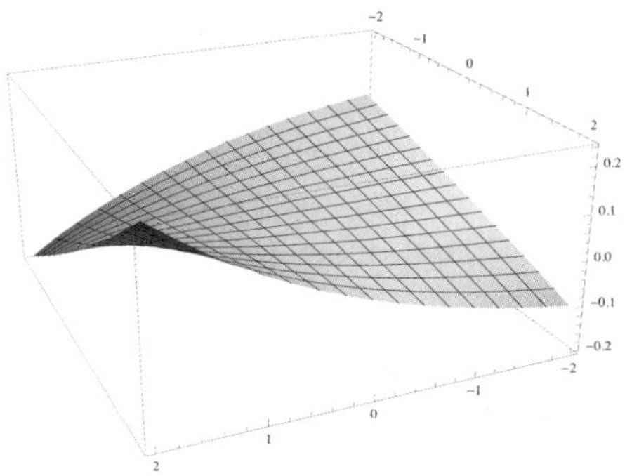

(a) Forward

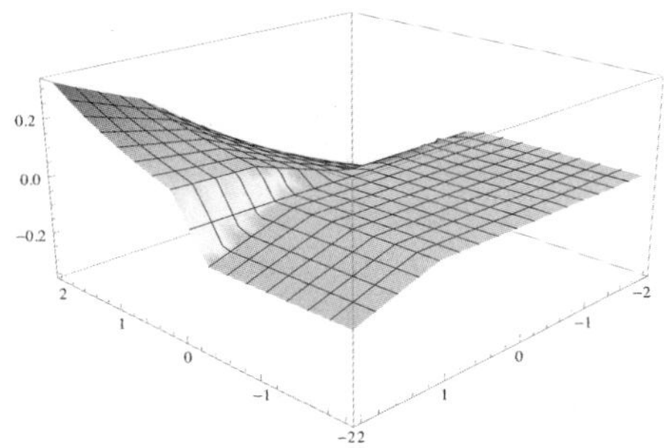

(b) Basket option

Figure 6. Two-dimensional projections of the Anchor ANOVA decomposition.

Remark 13.2 (Anchor ANOVA with normal distribution).
By choosing $\Omega = \mathbb{R}^d$ *and* $d\mu(\mathbf{x}) = \delta(\mathbf{x}-\mathbf{a})\varphi_{d,\boldsymbol{\lambda}}(\mathbf{x})d\mathbf{x}$ *with* $\varphi_{d,\boldsymbol{\lambda}}(\mathbf{x})$ *as the multi-dimensional normal distribution*

$$\varphi_{d,\boldsymbol{\lambda}}(\mathbf{x}) := \frac{1}{\prod_i \sqrt{2\pi\lambda_i}} e^{-\sum_{i=1}^d x_i^2/2\lambda_i}, \tag{59}$$

we define the projections

$$P_{\mathbf{u}} f(\mathbf{x}_{\mathbf{u}}) = \left(f(\mathbf{x})\varphi_{d-|\mathbf{u}|,\boldsymbol{\lambda}}(\mathbf{x}_{\mathbf{u}})\right)\Big|_{\mathbf{x}=\mathbf{a}\setminus\mathbf{x}_{\mathbf{u}}}, \tag{60}$$

which result in the Anchor ANOVA with the normal distribution.

4.2. *Taylor-like ANOVA decomposition*

Taking a closer look at the variance distribution of the individual anchor ANOVA projections, it becomes evident that not all the two-dimensional terms are of equal importance. Therefore we propose an approximation to the Standard ANOVA decomposition. Taking only some terms of the Standard ANOVA decomposition into account we define the following approximation, which we will call Taylor-like ANOVA expansion:

Proposition 13.1 (Taylor-like ANOVA expansion of first order). *Let*

$$v_{\mathbf{u}} := I f_{\mathbf{u}}, \tag{61}$$

where $f_{\mathbf{u}}$ *denotes the ANOVA term resulting from an ANOVA decomposition of the payoff-function* $g(\mathbf{x})$ *and*

$$If := \int_\Omega f(\mathbf{x})d\mathbf{x} \tag{62}$$

denotes the integral operator. Then

$$u(\mathbf{x},t) \approx v_0 + \sum_{i=1}^d v_i + \sum_{i=2}^d v_{1,i} \tag{63}$$

represents a Taylor-like approximation to the Standard ANOVA decomposition.

As we will see later this is, indeed a very good approximation to the full solution. Compared to the ordinary ANOVA expansion, the Taylor-like ANOVA expansion has a clear complexity advantage since it depends only linearly on the dimension of the full problem. Furthermore, as we will see later, it also has a precision advantage compared to the ordinary ANOVA expansion of similar dimension. The complexity advantage is even greater if we look at equation (63) in terms of the solutions $u^{(n,j)}(\mathbf{x},t)$ and not the individual ANOVA terms $v_{\mathbf{u}}$.

4.3. *Taylor-like ANOVA expansion of higher order*

Two obvious possibilities exist to generalize the Taylor-like ANOVA expansion to expansions of higher order:

i) A Taylor approximation of higher order, e.g. 2nd order;
ii) a Taylor approximation of first order around a higher dimensional (e.g. 2D) problem.

Proposition 13.2. ***(Taylor approximation of first order around a higher dimensional problem)***
Beginning from

$$u(\mathbf{x},t) \approx u^{(2)}(\mathbf{x},t) + \sum_{j=3}^{d} \left(u^{(2,j)}(\mathbf{x},t) - u^{(2)}(\mathbf{x},t) \right), \tag{64}$$

we again substitute the corresponding ANOVA terms and obtain

$$u(\mathbf{x},t) \approx v_0 + \sum_{i=1}^{d} v_i + \sum_{i=2}^{d} v_{1,i} + \sum_{i=3}^{d} (v_{2,i} + v_{1,2,i}). \tag{65}$$

This formulation is still linear in the dimension d of the full problem. The complexity advantage of a formulation in terms of the solutions $u^{(n)}(\mathbf{x},t)$ is even clearer here. While in the formulation using the numerical solutions $u^{(n)}(\mathbf{x},t)$ still one n-dimensional and $d-n$ (n+1) dimensional problems need to be solved. In terms of a formulation in ANOVA terms, this corresponds to d one-dimensional, $(2d-3)$ two-dimensional and $(d-2)$ three-dimensional problems.

Proposition 13.3 (Taylor approximation of second order).
If we take higher order Taylor approximations into consideration, in the case of a second order approximation we get

$$u(\mathbf{x},t) = u^{(n)}(\mathbf{x},t) + \sum_{j=n+1}^{d} \lambda_j \frac{\partial u}{\partial \lambda_j}(\mathbf{x},t)\Big|_{\lambda=\lambda^{(\mathbf{n})}} \tag{66}$$

$$+ \frac{1}{2} \sum_{i,j=n+1}^{d} \lambda_i \lambda_j \frac{\partial^2 u}{\partial \lambda_i \partial \lambda_j}(\mathbf{x},t)\Big|_{\lambda=\lambda^{(\mathbf{n})}} + \mathcal{O}\left(||\lambda - \lambda^{(\mathbf{n})}||^4 \right). \tag{67}$$

We will approximate the second derivative by difference quotients using central differences:

$$\frac{\partial^2 u}{\partial\lambda_i\partial\lambda_j}(\mathbf{x},t)\Big|_{\lambda=\lambda^{(\mathbf{n})}} \approx \frac{u^{(n,i,j)}(\mathbf{x},t) - u^{(n,i)}(\mathbf{x},t) - u^{(n,j)}(\mathbf{x},t) + u^{(n)}(\mathbf{x},t)}{\lambda_i\lambda_j}, \tag{68}$$

and end up with the following approximation for n=1:

$$\begin{aligned} u(\mathbf{x},t) &\approx u^{(1)}(\mathbf{x},t) + \sum_{i=2}^{d}(u^{(1,i)}(\mathbf{x},t) - u^{(1)}(\mathbf{x},t)) \\ &\quad + \frac{1}{2}\sum_{i,j=2}^{d}(u^{(1,i,j)}(\mathbf{x},t) - u^{(1,i)}(\mathbf{x},t) - u^{(1,j)}(\mathbf{x},t) + u^{(1)}(\mathbf{x},t)) \\ &\underset{63}{=} v_0 + \sum_{i=1}^{d} v_i + \sum_{i=2}^{d} v_{1,i} \\ &\quad + \frac{1}{2}\sum_{i,j=2}^{d}(u^{(1,i,j)}(\mathbf{x},t) - u^{(1,i)}(\mathbf{x},t) - u^{(1,j)}(\mathbf{x},t) + u^{(1)}(\mathbf{x},t)), \end{aligned}$$

which can be written as

$$u(\mathbf{x},t) \approx v_0 + \sum_{i=1}^{d} v_i + \sum_{i=2}^{d} v_{1,i} + \sum_{\substack{i=2\\ i<j}}^{d}(v_{1ij} + v_{ij}), \tag{69}$$

since $v_{ij} = v_{ji}$.

Since a representation in terms of numerical solutions is preferable, we rewrite the original formulation using the Gaussian formula

$$\sum_{k=1}^{n} k = \frac{n(n+1)}{2}, \tag{70}$$

and $u^{(n,i,j)}(\mathbf{x},t) = u^{(n,j,i)}(\mathbf{x},t)$ *as*

$$\begin{aligned} u(\mathbf{x},t) &\approx \left(1 - (d-n) + \frac{(d-n-1)(d-n)}{2}\right) u^{(n)}(\mathbf{x},t) \\ &\quad - \sum_{i=n+1}^{d} (d-n-2)u^{(n,i)}(\mathbf{x},t) \\ &\quad + \frac{1}{2}\sum_{i,j=n+1}^{d} u^{(n,i,j)}(\mathbf{x},t). \end{aligned}$$

It is easy to see that this approximation contains $\frac{(d-n-1)(d-n)}{2}$ $(n+2)$-dimensional problems, hence is no longer of linear order in d.

5. Efficient parallelization of dimension-wise decompositions

In the previous sections we introduced examples from the class of high dimensional problems in finance and presented various methods on how to cope with the *curse of dimension*. In this rather technical section we will elaborate on an efficient implementation scheme that significantly reduces computational time. The reduction is achieved by combining the sparse grids and dimension-wise decomposition methods (chapter 3.3 and 4) using parallelization.

We will shortly introduce parallel processes and inter-process communication in section 5.1. On this basis, we will explain an object-oriented concept for parallelizing solvers for high dimensional problems using sparse grids (section 5.2) and present an enhancement that allows for parallel solving of multiple problems in section 5.3. The concepts introduced are implemented in the simulation software $\mathcal{SG}2$ [6] at the Goethe Center for Scientific Computing. The pseudo code here is in C++ syntax, a widely used programming language in numerical software development.

5.1. *Parallel Processes Using the Message Passing Interface*

The purpose of this section is to depict the basic principles of parallelization. We use the "Message Passing Interface" (MPI), that manages communication between different processes running in parallel, for explanation. The interface has implementations in various programming languages (e.g. C++, Java, Python, ...). For clarity, we will describe a few interface methods but will not present a comprehensive documentation of the interface. For this the reader is referred to [12].

The fundamental approach when parallelizing program code is the following: when the program is executed MPI will start several parallel processes executing the same program code. Every process has an unique identifier and processes communicate using interface methods provided by MPI which must be integrated into the program code by the programmer. Taking advantage of the parallelility of the processes is thus a duty of the user of the interface, i.e. the programmer. For example, she can utilize the process identifier to enforce the execution of specific parts of code only in specific processes. Figure 7 gives an indication of the concept.* The programmer distributes different parts of the calculation to different processes and gathers the solution of every process on a so-called *Master* process. The Master then continues processing the calculation. Note, that all other processes may still be executing the same code the Master is executing, but might hold different data.

Basic operations supported by the interface are, for example, waiting on other processes to arrive at the same line of code or sending and receiving of messages and variables from/to other processes. There exists 1:1-communication as well as 1:n-communication routines, whereas the latter covers the distribution from (*Broadcast*) and the gathering on a single process (*Gather* or *Collect*) of Integer, Double or Char

*We assume that only one MPI process is started on one Central Processing Unit (CPU). This is – by default – invoked by MPI upon startup, but may be varied by the programmer.

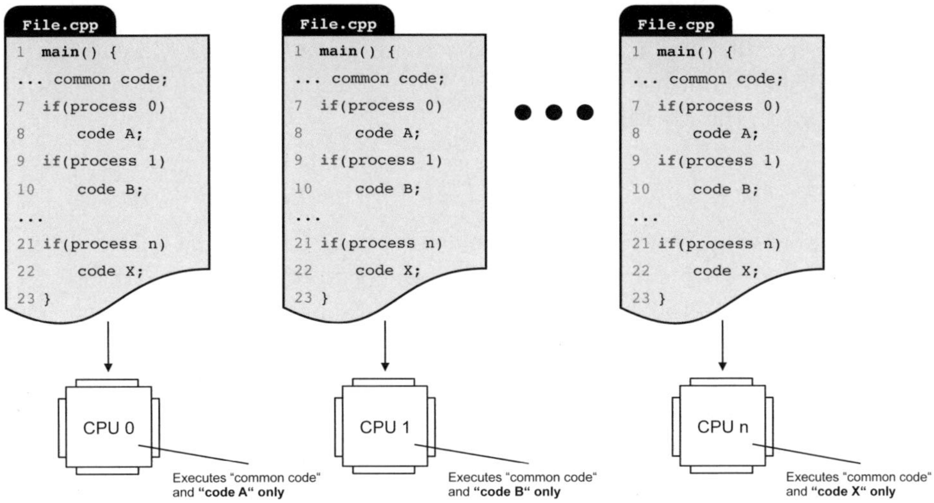

Figure 7. Parallel processes in the Message Passing Interface.

data types. Which process receives which data is managed using *Communicators*. In principle, only processes sharing a common Communicator are able to communicate with each other.

In the following sections we assume that extended interface methods for broadcasting and gathering of Arrays and Vectors from/on a single process exist. Both methods can be implemented using only the basic MPI operations just described.

5.2. *Sparse Grid Parallelization*

As already seen in section 3.3, partial differential equations on sparse grids can be solved by discretizing and solving the equation on every sparse grid independently and separately. The solution on the full grid is then represented as a linear combination of all partial solutions and is obtained via extrapolation.

The parallelization of these decoupled problems is *embarrassingly simple*: Every MPI process calculates the solution on a single sparse grid. When all processes are finished a Master process gathers the individual solutions on the grids of the sparse grid table and combines them to the full solution using a combination technique. The number of calculating processes is bounded from above by the number of sparse grids for a given dimension d of the full problem and the refinements n; as given by equation (46).*

For the dimension-wise decomposition method from section 4 it would be favorable to solve multiple partial differential equations in parallel using the described sparse grid parallelization for the solution of every single lower dimensional prob-

*Of course, one can start additional processes but their calculations will be redundant.

lem. In section 5.3 we present our technical solution: the *MPI Scheduler* class, that organizes the parallelization of sparse grid solver objects. First we will elaborate on an object-oriented implementation scheme of the Message Passing Interface which is fundamental to the MPI Scheduler.

5.2.1. *Object-Oriented Implementation of the Message Passing Interface*

An object-oriented approach for MPI implementation could be the following: when starting MPI, i.e., when starting the parallel processes, we create an MPI object in every process. It is initialized with some basic parameters of the process.* These may be the total number of processes started or the processes ID – its *Rank*. The programmer can then access the parameters of the process objects, e.g. by Getter and Setter methods, to execute conditional instructions in different processes; recall figure 7. Extended interface methods however can be realized as static methods as they must not differ in different processes. Figure 8 depicts a diagrammatic process object.

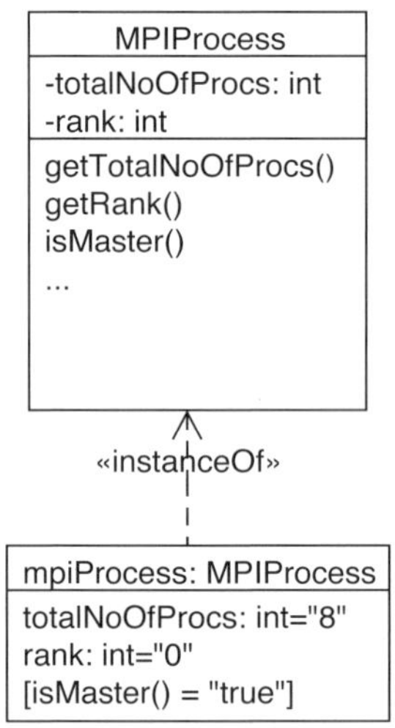

Figure 8. Class diagram of the MPI object that exists on every process. A common approach is to assign the Master status to the process whose Rank is zero.

5.2.2. *Parallel Solutions to Sparse Grids*

Having introduced the basic functionality of the MPI process object and extended communication routines mentioned in section 5.1, we present the sparse grid par-

*This may be realized as a static instantiation of the object in *this* process.

allelization approach from the software package [6]. We are aware that there are many ways to implement sparse grid parallelization techniques, however, the following approach demonstrates well the simplicity of parallelizing decoupled problems.

For a given refinement and the dimension of the problem we set up the sparse grid table (compare figure 4) in every process, but not a grid vector (i.e. the discretized solution) for every grid of the sparse grid table. When initializing the discretization of the respective sparse grid, every process checks whether there exists a grid vector for the grid. In this case, the process executes code that creates the discretization and solves the partial differential equation on this specific grid. When all processes have solved the equations on their grid, a Master process gathers all solutions from the processes and extrapolates them to the solution of the full problem.

If enough processes exist, only one grid vector per process will be created and thus every process solves the partial differential equation on a different grid. We present again equation (46) from section 3.3 as it formulates the number of processes needed at best as a function $f(n, d)$ of refinement n and dimension d:

$$f(n, d) = \binom{n + d - 1}{d - 1}.$$

In case of fewer processes, one process covers multiple grid vectors and calculates the solutions for multiple grids.

It must be emphasized that the assignment of sparse grid solutions to processes should not be arbitrary. For example, the isotropy of a (sparse) grid has major influence on the time needed to solve the equation on that grid ([11]). In addition, several factors, e.g., vastly varying diffusion coefficients, influence the runtime of the solvers. At the end, the total runtime of solving the full problem is bounded from below by the process(es) with the longest calculation time for the partial solution(s).

5.3. *Parallelization of Sparse Grid Solvers*

Here we present a new implementation concept to solve multiple high dimensional problems in parallel. The fundamental idea of the the MPI Scheduler class is to group processes and to solve, again in parallel, one or more problems of lower dimension using the aforementioned sparse grid parallelization.

Firstly, we introduce extensions to the MPI process object which allow for grouping of processes. Then we illustrate how to fit solution methods for partial differential equations – like sparse grids – into an object-oriented framework. Putting these pieces together, we conclude by demonstrating the functionality of the MPI Scheduler. It groups MPI processes and assigns solver objects for partial differential equations to these process groups for solving.

5.3.1. *Extensions to the MPI Objects*

To encapsulate the parallelization within each process group we add a group-wide process ID to the MPI process object. The communication within a group is organized using group specific Communicators. In addition, we add an attribute to the object storing the number of processes in the same group that *this* process is in. This is realized by a group ID so every process knows its membership. Figure 9 shows the extended process object. Note that one might also have to modify

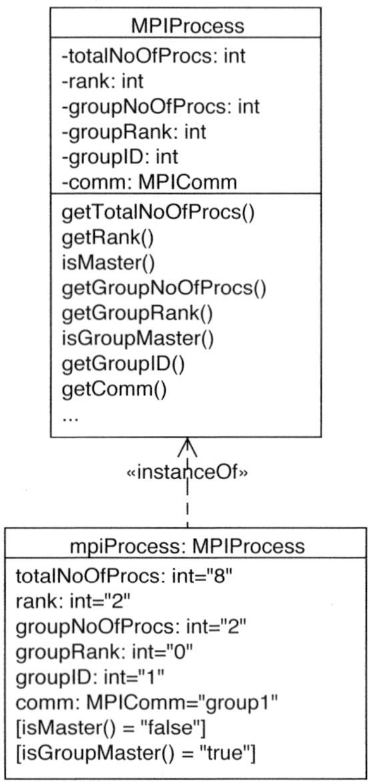

Figure 9. Class diagram of the extended MPI object, which allows for grouping of MPI processes.

the extended interface methods for broadcasting and gathering of Arrays and Vectors. This task is not too difficult as inter-process communication is organized using Communicators and every group has its own. The following function call gives an example:

```
template <typename T> void broadcastArray(Array<T>, Communicator);
```

5.3.2. *Object-Oriented Solution of Partial Differential Equations*

In order to keep the parallelization flexible and re-usable for various types of solvers, the MPI Scheduler assigns solver objects to different process groups. It starts the solution process centrally and manages the gathering and combination of all solutions on the Master when finished. To be manageable, the solver objects essentially need a group ID to enable their assignment to groups and two further methods:

- a method `solve()`
- and a method `getSolution()`.

Which problem type the solvers implement and how they are solved is of no importance to the MPI Scheduler.

Even though the actual design of the solver objects is negligible, we suggest the following fundamental components:

i) the discretization (in space),
ii) the time stepping scheme,
iii) and a linear (or non-linear) solver.

A solver object for sparse grids additionally needs a grid table. Figure 10 depicts two possible solver objects: one ordinary and one using sparse grids.

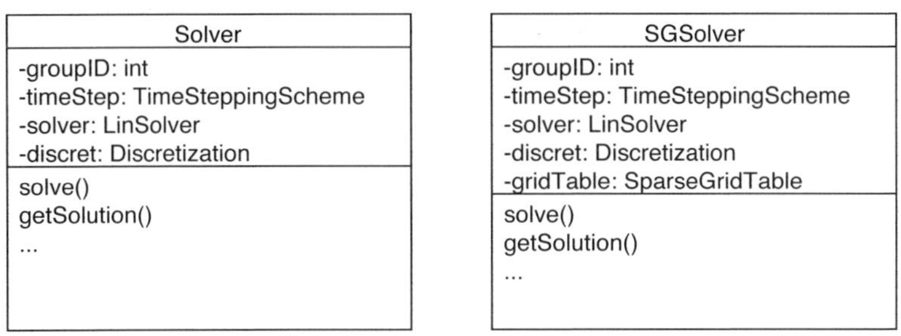

(a) An ordinary solver object. (b) A solver object for sparse grids.

Figure 10. Class diagrams for possible solver objects for partial differential equations.

5.3.3. *The MPI Scheduler*

An MPI Scheduler object stores a reference to an Array of solver objects, which contains the problems to be solved. Figure 11 shows the schematic design of the MPI Scheduler. We would like to discuss the following methods in more detail:

> `createGroups(Mode)`: Groups processes given different modes. In principle applicable to other tasks than the parallel solving of (sparse grid) problems.
> `schedule(Method)`: Assigns solver objects to groups by setting group IDs. If there are more solver objects than groups, multiple solver objects may be assigned to one group. It implements further scheduling methods that adjusts to the grouping modes.

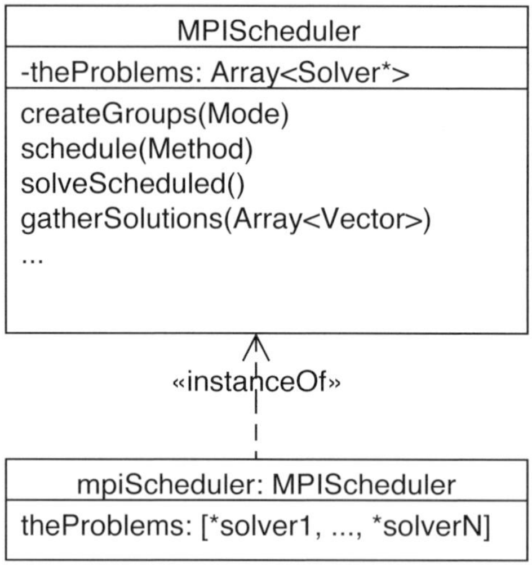

Figure 11. Class diagram of the MPI Scheduler. The MPI Scheduler object stores references to the solver objects in an Array. It groups the processes, manages and initializes the parallel solving and gathers the solutions at the end.

> `solveScheduled()`: Calls the `solve()` method of the solver objects. Thus it triggers the solving of the problems.
>
> `gatherSolutions(Array<Vector>)`: Gathers the solutions of all groups in an Array on the Master. Recall that the solution on every sparse grid is a grid vector.*

As illustrated in section 5.1 the program code is executed in all processes but some instructions are conditional to specific processes. We can use this design principle of MPI to minimize the communication overhead between the processes when creating the groups. [Grouping of processes] When executing the `createGroups(Mode)` method, all processes run through a loop and decide locally, using their own Ranks, which group they join. When entering a group, the group ID and the group Communicator are set for each process. Therefore the programmer may define an Array of intervals after which the processes arrange in order:

```
Array<Array<int> > range;

...
```

*Of course, the solution may be evaluated at a given point in the domain, i.e. a grid point, or at any given point in time during execution of the time stepping scheme. If needed, the MPI Scheduler implementation should also provide respective gathering methods.

```
for(int i = 0; i < range.length(); i++){
  if(   mpiProcess.getRank() <  range[i].max()
     && mpiProcess.getRank() >= range[i].min()) {

    mpiProcess.setGroupID(i);
    mpiProcess.setComm(i);

  }
}
```

This way, e.g., eight processes may be split up into two groups using a two-dimensional Array containing the intervals `[0;3]` and `[4;7]`. Solving the scheduled problems also can be realized locally by looping over all solver objects. Every process checks if its group ID matches the group ID of the current solver object. In this case it triggers the calculation of the solution by calling the `solve()` method of the solver object. Thus every process only solves problems assigned to its group ID.

For the rest of this section, we will focus on the operation principle of the MPI Scheduler. Figure 12 illustrates the functioning of the MPI Scheduler. In the initial situation all solver objects as well as the MPI Scheduler object are created in every process. The MPI Scheduler is instantiated with references on the solver objects (figure 12a).* As explained in example 5.3.3, the MPI Scheduler object firstly groups the processes by local decision and determines Master processes within each group (figure 12b). Calling `schedule(Method)` results in a mapping of groups to solver objects. For this the MPI Scheduler sets the process group ID to that solver's group ID attribute that will calculate the solution to the problem associated with the solver object (figure 12c). Subsequently the MPI Scheduler initializes the calculations of the solutions in every group by executing `solveScheduled()` (figure 12d). If sparse grid parallelization is used within the groups, the sparse grids will now be assigned to the processes in the respective group and the partial differential equation will be solved (figure 12e). Finally, figure 12f depicts how the MPI Scheduler gathers the sparse grid solutions from the group Masters on the overall Master using `gatherSolutions(Array<Vector>)`.†

When combining MPI Scheduler and sparse grid parallelization it might happen that there are fewer CPUs than "number of sparse grids $\times$ number of groups". In this case one can either *overcommit* the cores, i.e., assign more than one MPI process to one CPU, or reduce the number of processes per group. As described in section 5.2 the sparse grid parallelization then assigns multiple grids to one process. Generally, the latter would be faster in terms of runtime as it saves communication overhead between the processes.

---

*For simplicity, we set one process equal to one CPU.

†The solution Array is created on every process and the MPI Scheduler must ensure that it copies the correct Vectors to the overall Master's Array.

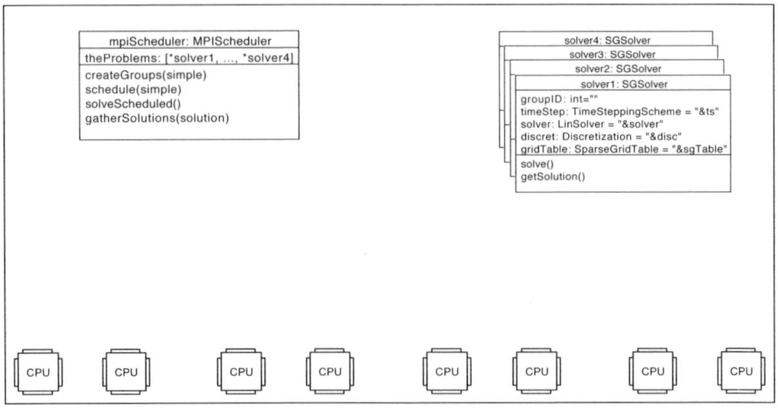

(a) Initial situation.

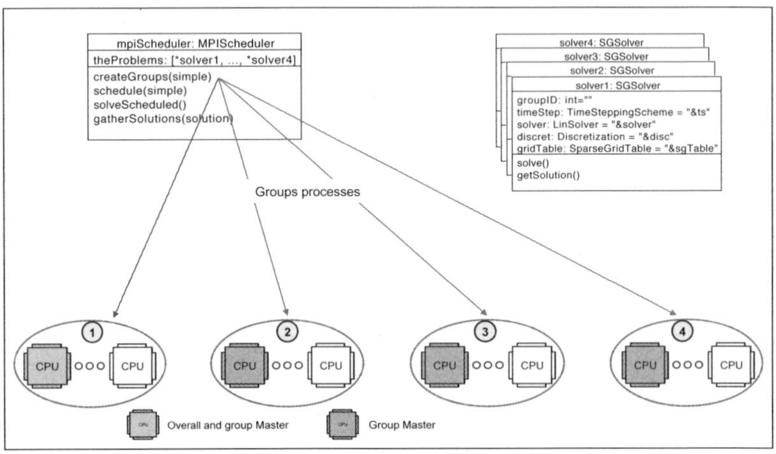

(b) Grouping.

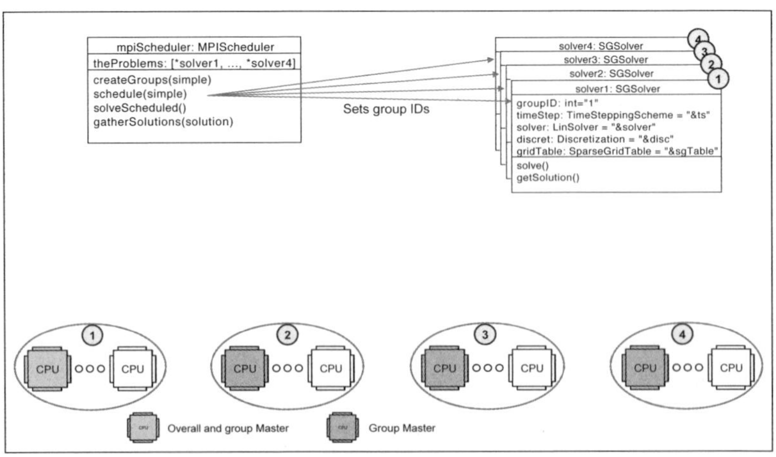

(c) Scheduling.

Figure 12. Interaction of the MPI Scheduler and the solver objects.

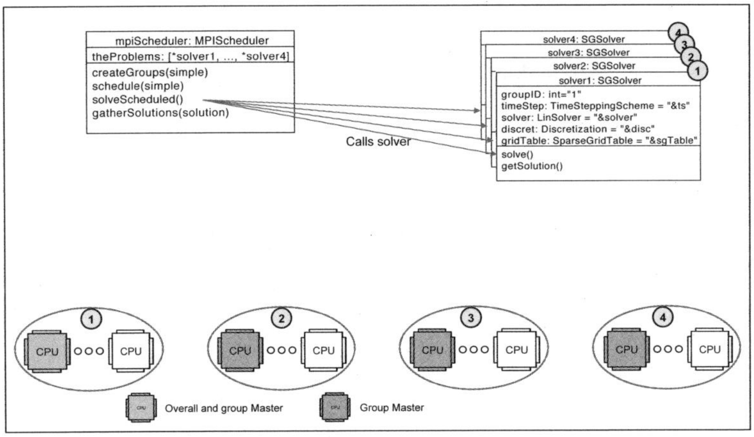

(d) Initializing.

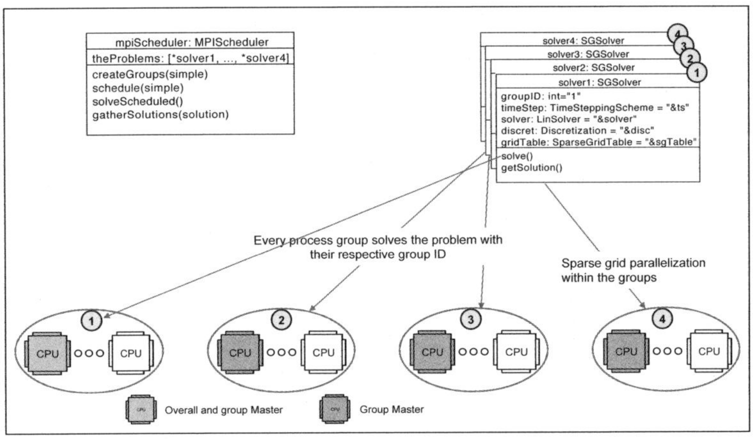

(e) Solving.

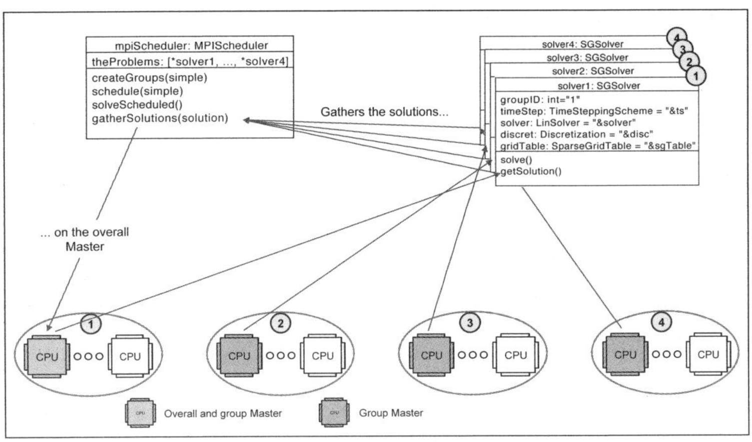

(f) Gathering.

Figure 12. Interaction of the MPI Scheduler and the solver objects.

Bibliography

[1] Beylkin, G. and Mohlenkamp, M.J: Numerical operator calculus in higher dimensions. *Proceedings of the National Academy of Sciences*, 99(16):10246, 2002.

[2] Beylkin, G. and Mohlenkamp, M.J.: Algorithms for numerical analysis in high dimensions. *SIAM Journal on Scientific Computing*, 26(6):2133–2159, 2005.

[3] Blackham, J.: Sparse grid solutions to the libor market model. Master's thesis, University of Oxford, 2004.

[4] Brace, A., Gatarek, D. and Musiela, M.: The market model of interest rate dynamics. *Mathematical finance*, 7(2):127–155, 1997.

[5] Gerstner, T. and Griebel, M.: Dimension–adaptive tensor–product quadrature. *Computing*, 71(1):65–87, 2003.

[6] Goethe Center for Scientific Computing (G-CSC). Numerical Simulation Software Library SG 2. http://sg2.sourceforge.net/.

[7] Griebel, M.: *Sparse grids and related approximation schemes for higher dimensional problems*. Foundations of Computational Mathematics (FoCM05), 2005.

[8] Griebel, M., Schneider, M. and Zenger, C.: *A combination technique for the solution of sparse grid problems*. Elsevier, 1992.

[9] Hackbusch, W.: A sparse matrix arithmetic based on h-matrices. part i: Introduction to-matrices. *Computing*, 62(2):89–108, 1999.

[10] Hackbusch, W., Khoromskij, B.N. and Tyrtyšnikov, E.E.: Hierarchical kronecker tensor product approximations. *J. Numer. Math*, 13:119–156, 2005.

[11] Hackbusch, W.: *Iterative Lösung großer schwachbesetzter Gleichungssysteme*. Teubner Verlag, 1993.

[12] Message Passing Interface Forum. MPI: A Message-Passing Interface Standard. Version 2.2, Message Passing Interface Forum, September 2009.

[13] Rabitz, H. and Aliş, O.F.: General foundations of high-dimensional model representations. *Journal of Mathematical Chemistry*, 25(2):197–233, 1999.

[14] Reisinger, C.: Analysis of linear difference schemes in the sparse grid combination technique. *Arxiv preprint arXiv:0710.0491*, 2007.

[15] Reisinger, C.: *Numerische Methoden für hochdimensionale parabolische Gleichungen am Beispiel von Optionspreisaufgaben*. PhD thesis, Universität Heidelberg, 2004.

[16] Sergeev, V.: Sparse grid method in the libor market model. Master's thesis, Universiteit van Amsterdam, 2006.

[17] SA Smolyak: Quadrature and interpolation formulas for tensor products of certain classes of functions. In *Dokl. Akad. Nauk SSSR*, volume 4, page 123, 1963.